Lecture Notes in Computer Science   6225

Commenced Publication in 1973
Founding and Former Series Editors:
Gerhard Goos, Juris Hartmanis, and Jan v

Stefan Mangard
François-Xavier Standaert (Eds.)

Cryptographic Hardware and Embedded Systems – CHES 2010

12th International Workshop
Santa Barbara, USA, August 17-20, 2010
Proceedings

 Springer

Volume Editors

Stefan Mangard
Chip Card & Security, Infineon Technologies
Am Campeon 1-12, 85579 Neubiberg, Germany
E-mail: stefan.mangard@infineon.com

François-Xavier Standaert
UCL Crypto Group, Université catholique de Louvain
Place du Levant 3, 1348 Louvain-la-Neuve, Belgium
E-mail: fstandae@uclouvain.be

Library of Congress Control Number: Applied for

CR Subject Classification (1998): E.3, D.4.6, K.6.5, E.4, C.2, H.2.7, G.2.1

LNCS Sublibrary: SL 4 – Security and Cryptology

ISSN	0302-9743
ISBN-10	3-642-15030-6 Springer Berlin Heidelberg New York
ISBN-13	978-3-642-15030-2 Springer Berlin Heidelberg New York

springer.com

© International Association for Cryptologic Research 2010
Printed in Germany

Typesetting: Camera-ready by author, data conversion by Scientific Publishing Services, Chennai, India
Printed on acid-free paper   06/3180

Preface

Since 1999, the workshop on Cryptographic Hardware and Embedded Systems (CHES) is the foremost international scientific event dedicated to all aspects of cryptographic hardware and security in embedded systems. Its 12th edition was held in Santa Barbara, California, USA, August 17–20, 2010. Exceptionally this year, it was co-located with the 30th International Cryptology Conference (CRYPTO). This co-location provided unique interaction opportunities for the communities of both events. As in previous years, CHES was sponsored by the International Association for Cryptologic Research (IACR).

The workshop received 108 submissions, from 28 different countries, of which the Program Committee selected 30 for presentation. Each submission was reviewed by at least 4 committee members, for a total of 468 reviews. Two invited talks completed the technical program. The first one, given by Ivan Damgård and Markus Kuhn, was entitled "Is Theoretical Cryptography Any Good in Practice?", and presented jointly to the CRYPTO and CHES audiences, on Wednesday, August 18, 2010. The second one, given by Hovav Shacham, was entitled "Cars and Voting Machines: Embedded Systems in the Field."

The Program Committee agreed on giving a best paper award to Alexandre Berzati, Cécile Canovas-Dumas and Louis Goubin, for their work "Public Key Perturbation of Randomized RSA Implementations." These authors will also be invited to submit an extended version of their paper to the *Journal of Cryptology*, together with the authors of two other contributions. First, Jean-Philippe Aumasson, Luca Henzen, Willi Meier and María Naya-Plasencia, authors of "Quark: a Lightweight Hash." Second, Luca Henzen, Pietro Gendotti, Patrice Guillet, Enrico Pargaetzi, Martin Zoller and Frank K. Gürkaynak, for their paper entitled "Developing a Hardware Evaluation Method for SHA-3 Candidates." These papers illustrate three distinct areas of cryptographic engineering research, namely: physical (aka implementation) security, the design of lightweight primitives and the efficient hardware implementation of cryptographic algorithms.

We would like to express our deepest gratitude to the various people who helped in the organization of the conference and made it a successful event. In the first place, we thank the authors who submitted their works. The quality of the submissions and the variety of the topics that they cover are reflective of an evolving and growing research area, trying to bridge the gap between theoretical advances and their practical application in commercial products. The selection of 30 papers out of these strong submissions was a challenging task and we sincerely thank the 41 Program Committee members, as well as the 158 external reviewers, who volunteered to read and discuss the papers over several months. They all contributed to the review process with a high level of professionalism, expertise and fairness. We also acknowledge the great contribution of our invited speakers. We highly appreciated the assistance of Çetin Kaya Koç and

Jean-Jacques Quisquater, the General Co-chairs of CHES 2010, and the help of the local staff at the University of California Santa Barbara. A big thank-you to Tal Rabin, the Program Chair of CRYPTO 2010, for the good collaboration and discussions which allowed a nice interaction between CRYPTO and CHES. We owe our gratitude to Shai Halevi, for maintaining the review website, to Jens-Peter Kaps, for maintaining the CHES website, and to the staff at Springer, for making the finalization of these proceedings an easy task. We also express our gratitude to our generous sponsors, namely: Cryptography Research, Riscure, Technicolor, Oberthur Technologies, the Research Center for Information Security and Telecom ParisTech. And finally, we would like to thank the CHES Steering Committee for allowing us to serve at such a prestigious workshop.

August 2010   Stefan Mangard
                                                    François-Xavier Standaert

CHES 2010

Workshop on Cryptographic Hardware and Embedded Systems
Santa Barbara, California, USA, August 17–20, 2010

Sponsored by *International Association for Cryptologic Research*

General Co-chairs

Çetin Kaya Koç   University of California Santa Barbara, USA
Jean-Jacques Quisquater Université catholique de Louvain, Belgium

Program Co-chairs

Stefan Mangard   Infineon Technologies, Germany
François-Xavier Standaert   Université catholique de Louvain, Belgium

Program Committee

Lejla Batina   Radboud University Nijmegen,
                           The Netherlands and
                        KU Leuven, Belgium
Daniel J. Bernstein   University of Illinois at Chicago, USA
Guido Bertoni   STMicroelectronics, Italy
Jean-Luc Beuchat   University of Tsukuba, Japan
Christophe Clavier   Université de Limoges, France and Institut
                           d'Ingénierie Informatique de Limoges, France
Jean-Sébastien Coron   University of Luxembourg, Luxembourg
Josep Domingo-Ferrer   Universiat Rovira i Virgili, Catalonia
Hermann Drexler   Giesecke & Devrient, Germany
Viktor Fischer   Université de Saint-Étienne, France
Wieland Fischer   Infineon Technologies, Germany
Pierre-Alain Fouque   ENS, France
Kris Gaj   George Mason University, USA
Louis Goubin   Université de Versailles, France
Aline Gouget   Gemalto, France
Johann Großschädl   University of Luxembourg, Luxembourg
Jorge Guajardo   Philips Research, The Netherlands
Kouichi Itoh   Fujitsu Laboratories, Japan
Marc Joye   Technicolor, France

Çetin Kaya Koç	University of California Santa Barbara, USA
François Koeune	Université catholique de Louvain, Belgium
Soonhak Kwon	Sungkyunkwan University, South Korea
Kerstin Lemke-Rust	University of Applied Sciences Bonn-Rhein-Sieg, Germany
Marco Macchetti	Nagravision SA, Switzerland
Mitsuru Matsui	Mitsubishi Electric, Japan
Michael Neve	Intel, USA
Elisabeth Oswald	University of Bristol, UK
Máire O'Neill	Queens University Belfast,UK
Christof Paar	Ruhr-Universität Bochum, Germany
Eric Peeters	Texas Instruments, Germany
Axel Poschmann	Nanyang Technological University, Singapore
Emmanuel Prouff	Oberthur Technologies, France
Pankaj Rohatgi	Cryptography Research, USA
Akashi Satoh	Research Center for Information Security, Japan
Erkay Savas	Sabanci University, Turkey
Patrick Schaumont	Virginia Tech, USA
Werner Schindler	Bundesamt für Sicherheit in der Informationstechnik (BSI), Germany
Sergei Skorobogatov	University of Cambridge, UK
Tsuyoshi Takagi	Kyushu University, Japan
Stefan Tillich	Graz University of Technology, Austria
Mathias Wagner	NXP Semiconductors, Germany
Colin Walter	Royal Holloway, UK

External Reviewers

Manfred Aigner
Abdulkadir Akin
Toru Akishita
Jean-Philippe Aumasson
Aydin Aysu
Jean-Claude Bajard
Selçuk Baktir
Brian Baldwin
Alessandro Barenghi
Timo Bartkewitz
Adolf Baumann
Florent Bernard
Alexandre Berzati
Peter Birkner
Markus Bockes
Andrey Bogdanov
Lilian Bossuet

David A. Brown
Cécile Canovas-Dumas
Jiun-Ming Chen
Zhimin Chen
Chen-Mou Cheng
Jung Hee Cheon
Sylvain Collange
Guillaume Dabosville
Joan Daemen
Jean-Luc Danger
Blandine Debraize
Jérémie Detrey
Sandra Dominikus
Emmanuelle Dottax
Benedikt Driessen
Miloš Drutarovský
Nicolas Estibals

Junfeng Fan
Benoît Feix
Martin Feldhofer
Georges Gagnerot
Berndt Gammel
Max Gebhardt
Laurie Genelle
Benedikt Gierlichs
Christophe Giraud
Tim Güneysu
Guy Gogniat
Gilbert Goodwill
Sylvain Guilley
Jian Guo
Xu Guo
Dong-Guk Han
Takuya Hayashi

Stefan Heyse
Naofumi Homma
Yohei Hori
Michael Hutter
Arni Ingimundarson
Josh Jaffe
Pascal Junod
Marcelo Kaihara
Dina Kamel
Markus Kasper
Michael Kasper
Timo Kasper
Toshihiro Katashita
Tino Kaufmann
Yuto Kawahara
Chang Hoon Kim
Inyoung Kim
Mario Kirschbaum
Ilya Kizhvatov
Miroslav Knezevic
Kazuyuki Kobayashi
Noboru Kunihiro
Taekyoung Kwon
Yun-Ki Kwon
Cédric Lauradoux
Mun-Kyu Lee
Manfred Lochter
Patrick Longa
Liang Lu
Yingxi Lu
Raimondo Luzzi
Abhranil Maiti
Marcel Medwed
Nicolas Meloni
Filippo Melzani
Giacomo de Meulenaer

Marine Minier
Amir Moradi
Ernst Mülner
Elke De Mulder
Takao Ochiai
Rune Odegard
Siddika Berna Örs
David Oswald
Pascal Paillier
Young-Ho Park
Hervé Pelletier
Ludovic Perret
Carlo Peschke
Christophe Petit
Thomas Peyrin
Gilles Piret
Thomas Plos
Thomas Popp
Jürgen Pulkus
Bo Qin
Michael Quisquater
Denis Réal
Francesco Regazzoni
Christof Rempel
Mathieu Renauld
Matthieu Rivain
Thomas Roche
Francisco Rodríguez-H.
Mylène Roussellet
Vladimir Rožić
Heuisu Ryu
Minoru Saeki
Kazuo Sakiyama
Gokay Saldamli
Jörn-Marc Schmidt
Peter Schwabe

Yannick Seurin
Martin Seysen
Saloomeh Shariati
Hideo Shimizu
Takeshi Shimoyama
Masaaki Shirase
Abdulhadi Shoufan
Chang Shu
Hervé Sibert
Yannick Sierra
Michal Sramka
Oliver Stein
Marc Stöttinger
Takeshi Sugawara
Daisuke Suzuki
Alexander Szekely
Robert Szerwinski
Masahiko Takenaka
Yannick Teglia
Arnaud Tisserand
Lionel Torres
Leif Uhsadel
Gilles Van Assche
Jérôme Vasseur
Vincent Verneuil
David Vigilant
Yi Wang
Lei Wei
Ralf-Philipp Weinmann
Jiang Wu
Qianhong Wu
Jun Yajima
Dai Yamamoto
Lei Zhang
Ralf Zimmermann

Table of Contents

Tamper Resistance and Hardware Trojans

Efficient Implementations II

SHA-3

Fault Attacks and Countermeasures

PUFs and RNGs

New Designs

Side-Channel Attacks and Countermeasures II

QUARK: A Lightweight Hash[*]

Jean-Philippe Aumasson[1], Luca Henzen[2],
Willi Meier[3,**], and María Naya-Plasencia[3,***]

[1] Nagravision SA, Cheseaux, Switzerland
[2] ETH Zurich, Switzerland
[3] FHNW, Windisch, Switzerland

Abstract. The need for lightweight cryptographic hash functions has been repeatedly expressed by application designers, notably for implementing RFID protocols. However not many designs are available, and the ongoing SHA-3 Competition probably won't help, as it concerns general-purpose designs and focuses on software performance. In this paper, we thus propose a novel design philosophy for lightweight hash functions, based on a single security level and on the sponge construction, to minimize memory requirements. Inspired by the lightweight ciphers Grain and KATAN, we present the hash function family QUARK, composed of the three instances U-QUARK, D-QUARK, and T-QUARK. Hardware benchmarks show that QUARK compares well to previous lightweight hashes. For example, our lightest instance U-QUARK conjecturally provides at least 64-bit security against all attacks (collisions, multicollisions, distinguishers, etc.), fits in 1379 gate-equivalents, and consumes in average 2.44 µW at 100 kHz in 0.18 µm ASIC. For 112-bit security, we propose T-QUARK, which we implemented with 2296 gate-equivalents.

1 Introduction

In 2006, Feldhofer and Rechberger [1] pointed out the lack of lightweight hash functions for use in RFID protocols, and gave recommendations to encourage the design of such primitives. But as recently observed in [2] the situation has not much evolved in four years[1], despite a growing demand; besides RFID, lightweight hashes are indeed relevant wherever the cost of hardware matters, be it in embedded systems or in smartcards.

---

[*] This work was partially supported by European Commission through the ICT programme under contract ICT-2007-216676 ECRYPT II.

[**] Supported by the Hasler foundation www.haslerfoundation.ch under project number 08065.

[***] This work was carried out during the tenure of an ERCIM "Alain Bensoussan" Fellowship Programme.

[1] Note the absence of "Hash functions" category in the ECRYPT Lightweight Cryptography Lounge (http://www.ecrypt.eu.org/lightweight/).

S. Mangard and F.-X. Standaert (Eds.): CHES 2010, LNCS 6225, pp. 1–15, 2010.

The ongoing NIST "SHA-3" Hash Competition [3] aims to develop a general-purpose hash function, and received as many as 64 original and diverse submissions. Most of them, however, cannot reasonably be called "lightweight", as most need more than (say) 10 000 gate equivalents (GE)^2. An exception is CubeHash [5], which can be implemented with 7630 GE in 0.13 μm ASIC [6] to return digests of up to 512 bits. For comparison, Feldhofer and Wolkerstorfer [7] reported a 8001 GE implementation of MD5 (128-bit digests, 0.35 μm ASIC), O'Neill [8] implemented SHA-1 (160-bit, 0.18 μm) with 6122 GE, and the compression function MAME by Yoshida et al [9] (256-bit, 0.18 μm) fits in 8100 GE. These designs, however, are still too demanding for many low-end environments.

A major step towards lightweight hashing is the work by Bogdanov et al. [10], which presented constructions based on the lightweight block cipher PRESENT [11]; they for example proposed to instantiate the Davies-Meyer ($E_m(h) \oplus h$) construction with PRESENT-80, giving a hash function with 64-bit digests implemented with as few as 1600 GE, in 0.18 μm ASIC.

Another interesting approach was taken with Shamir's SQUASH [12], which processes short strings only, offers 64-bit preimage resistance, and is expected to need fewer than 1000 GE. However, SQUASH is not collision resistant—as it targets RFID authentication protocols where collision resistance is unnecessary—and so is inappropriate for applications requiring a collision-resistant hash function.

In this paper, we present a novel approach to design lightweight hashes, illustrated with the proposal of a new family of functions, called QUARK.

2 Description of the QUARK Hash Family

2.1 Sponge Construction

QUARK uses the sponge construction, as depicted in Fig. 1. Following the notations introduced in [13], it is parametrized by a *rate* (or block length) r, a *capacity* c, and an *output length* n. The *width* of a sponge construction is the size of its internal state $b = r + c \geq n$.

Given an initial state, the sponge construction processes a message m as follows:

1. **Initialization**: The message is padded by appending a '1' bit and sufficiently many zeroes to reach length a multiple of r.
2. **Absorbing phase**: The r-bit message blocks are xored into the last r bits of the state (i.e., $Y_{b/2-r}, \ldots, Y_{b/2-1}$), interleaved with applications of the permutation P.
3. **Squeezing phase**: The last r bits of the state are returned as output, interleaved with applications of the permutation P, until n bits are returned.

---

^{2} For example, the "5000 GE" implementation of Keccak reported in [4, §7.4.3] is only that of the coprocessor, without the memory storing the 1600-bit internal state. Hence the total gate count of the complete design is well above 10 000.

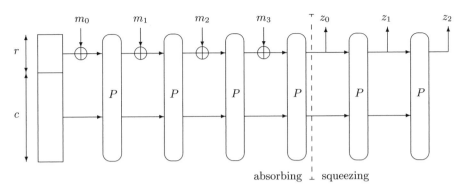

Fig. 1. The sponge construction as used by QUARK, for the example of a 4-block (padded) message

2.2   Permutation

QUARK uses a permutation P inspired by the stream cipher Grain and by the block cipher KATAN (see §3.3 for details), as depicted in Fig. 2.

The permutation P relies on three non-linear Boolean functions f, g, and h, on a linear Boolean function p, and on an internal state composed, at epoch $t \geq 0$, of

- an NFSR X of $b/2$ bits set to $X^t = (X_0^t, \ldots, X_{b/2-1}^t)$;
- an NFSR Y of $b/2$ bits set to $Y^t = (Y_0^t, \ldots, Y_{b/2-1}^t)$;
- an LFSR L of $\lceil \log 4b \rceil$ bits set to $L^t = (L_0^t, \ldots, L_{\lceil \log 4b \rceil -1}^t)$.

P processes a b-bit input in three stages, as described below:

Initialization. Upon input $s = (s_0, \ldots, s_{b-1})$, P initializes its internal state as follows:

- X is initialized with the first $b/2$ input bits: $(X_0^0, \ldots, X_{b/2-1}^0) := (s_0, \ldots, s_{b/2-1})$;
- Y is initialized with the last $b/2$ input bits: $(Y_0^0, \ldots, Y_{b/2-1}^0) := (s_{b/2}, \ldots, s_{b-1})$;
- L is initialized to the all-one string: $(L_0^0, \ldots, L_{\lceil \log 4b \rceil -1}^0) := (1, \ldots, 1)$.

State update. From an internal state (X^t, Y^t, L^t), the next state $(X^{t+1}, Y^{t+1}, L^{t+1})$ is determined by *clocking* the internal mechanism as follows:

1. The function h is evaluated upon input bits from X^t, Y^t, and L^t, and the result is written h^t: $h^t := h(X^t, Y^t, L^t)$;
2. X is clocked using Y_0^t, the function f, and h^t:

$$(X_0^{t+1}, \ldots, X_{b/2-1}^{t+1}) := (X_1^t, \ldots, X_{b/2-1}^t, Y_0^t + f(X^t) + h^t) ;$$

3. Y is clocked using the function g and h^t:

$$(Y_0^{t+1}, \ldots, Y_{b/2-1}^{t+1}) := (Y_1^t, \ldots, Y_{b/2-1}^t, g(Y^t) + h^t) ;$$

4. L is clocked using the function p:

$$(L_0^{t+1}, \ldots, L_{\lceil \log 4b \rceil}^{t+1}) := (L_1^t, \ldots, L_{\lceil \log 4b \rceil -1}^t, p(L^t)) .$$

Computation of the output. Once initialized, the state of QUARK is updated $4b$ times, and the output is the final value of the NFSR's X and Y, using the same bit ordering as for the initialization.

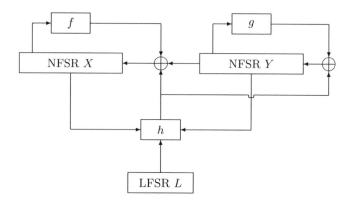

Fig. 2. Diagram of the permutation of QUARK (for clarity, the feedback of the LFSR with the function p is omitted)

2.3 Proposed Instances

There are three different flavors[3] of QUARK: U-QUARK, D-QUARK, and T-QUARK. For each, we give its rate r, capacity c, width b, digest length n, and its functions f, g, and h. The function p, used by the data-independent LFSR, is the same for all three instances: given a 10-bit register L, p returns $L_0 + L_3$.

U-QUARK is the lightest flavor of QUARK. It was designed to provide *64-bit security*, and to admit a parallelization degree of 8. It has sponge numbers $r = 8, c = 128, b = 2 \times 68, n = 128$.

Given a 68-bit register X, f returns

$$X_0 + X_9 + X_{14} + X_{21} + X_{28} + X_{33} + X_{37} + X_{45} + X_{50} + X_{52} + X_{55}$$
$$+ X_{55}X_{59} + X_{33}X_{37} + X_9X_{15} + X_{45}X_{52}X_{55} + X_{21}X_{28}X_{33}$$
$$+ X_9X_{28}X_{45}X_{59} + X_{33}X_{37}X_{52}X_{55} + X_{15}X_{21}X_{55}X_{59}$$
$$+ X_{37}X_{45}X_{52}X_{55}X_{59} + X_9X_{15}X_{21}X_{28}X_{33} + X_{21}X_{28}X_{33}X_{37}X_{45}X_{52} \ .$$

Given a 68-bit register Y, g returns

$$Y_0 + Y_7 + Y_{16} + Y_{20} + Y_{30} + Y_{35} + Y_{37} + Y_{42} + Y_{49} + Y_{51} + Y_{54}$$
$$+ Y_{54}Y_{58} + Y_{35}Y_{37} + Y_7Y_{15} + Y_{42}Y_{51}Y_{54} + Y_{20}Y_{30}Y_{35}$$
$$+ Y_7Y_{30}Y_{42}Y_{58} + Y_{35}Y_{37}Y_{51}Y_{54} + Y_{15}Y_{20}Y_{54}Y_{58}$$
$$+ Y_{37}Y_{42}Y_{51}Y_{54}Y_{58} + Y_7Y_{15}Y_{20}Y_{30}Y_{35} + Y_{20}Y_{30}Y_{35}Y_{37}Y_{42}Y_{51} \ .$$

---

[3] In particle physics, the u-quark is lighter than the d-quark, which itself is lighter than the t-quark; our eponym hash functions compare similarly.

Given 68-bit registers X and Y, and a 10-bit register L, h returns

$$L_0 + X_1 + Y_2 + X_4 + Y_{10} + X_{25} + X_{31} + Y_{43} + X_{56} + Y_{59}$$
$$+ Y_3 X_{55} + X_{46} X_{55} + X_{55} Y_{59} + Y_3 X_{25} X_{46} + Y_3 X_{46} X_{55}$$
$$+ Y_3 X_{46} Y_{59} + L_0 X_{25} X_{46} Y_{59} + L_0 X_{25} .$$

D-QUARK is the second-lightest flavor of QUARK. It was designed to provide *80-bit security*, and to admit a parallelization degree of 8. It has sponge numbers $r = 16, c = 160, b = 2 \times 88, n = 160$.

D-QUARK uses the same function f as U-QUARK, but with taps 0, 11, 18, 19, 27, 36, 42, 47, 58, 64, 67, 71, 79 instead of 0, 9, 14, 15, 21, 28, 33, 37, 45, 50, 52, 55, 59, respectively.

D-QUARK uses the same function g as U-QUARK, but with taps 0, 9, 19, 20, 25, 38, 44, 47, 54, 63, 67, 69, 78 instead of 0, 7, 15, 16, 20, 30, 35, 37, 42, 49, 51, 54, 58, respectively.

Given 88-bit registers X and Y, and a 10-bit register L, h returns

$$L_0 + X_1 + Y_2 + X_5 + Y_{12} + Y_{24} + X_{35} + X_{40} + X_{48} + Y_{55} +$$
$$Y_{61} + X_{72} + Y_{79} + Y_4 X_{68} + X_{57} X_{68} + X_{68} Y_{79} + Y_4 X_{35} X_{57} +$$
$$Y_4 X_{57} X_{68} + Y_4 X_{57} Y_{79} + L_0 X_{35} X_{57} Y_{79} + L_0 X_{35} .$$

T-QUARK is the less light flavor of QUARK. It was designed to provide *112-bit security*, and to admit a parallelization degree of 16. It has sponge numbers $r = 32, c = 224, b = 2 \times 128, n = 224$.

T-QUARK uses the same function f as U-QUARK, but with taps 0, 16, 26, 28, 39, 52, 61, 69, 84, 94, 97,103,111 instead of 0, 9, 14, 15, 21, 28, 33, 37, 45, 50, 52, 55, 59, respectively.

T-QUARK uses the same function f as U-QUARK, but with taps 0, 13, 28, 30, 37, 56, 65, 69, 79, 92, 96,101,109 instead of 0, 7, 15, 16, 20, 30, 35, 37, 42, 49, 51, 54, 58, respectively.

Given 128-bit registers X and Y, and a 10-bit register L, h returns

$$L_0 + X_1 + Y_3 + X_7 + Y_{18} + Y_{34} + X_{47} + X_{58} + Y_{71} + Y_{80} + X_{90} + Y_{91} +$$
$$X_{105} + Y_{111} + Y_8 X_{100} + X_{72} X_{100} + X_{100} Y_{111} + Y_8 X_{47} X_{72} + Y_8 X_{72} X_{100} +$$
$$Y_8 X_{72} Y_{111} + L_0 X_{47} X_{72} Y_{111} + L_0 X_{47} .$$

3 Design Rationale

3.1 Single Security Level

An originality of QUARK is that its expected security level against second preimages differs from its digest length (see §4.2 for a description of the generic attack). In particular it offers a similar security against generic collision attacks and generic second preimage attacks, that is, approximately $2^{c/2}$. Note that the sponge construction ensures a resistance of approximately $\max(c, n)$ bits against preimages, as recently shown in [14, §4.2][4]. Hence, QUARK provides increased resistance to preimage attacks.

A disadvantage of this approach is that one "wastes" half the digest bits, as far as second preimage resistance is concerned. However, this little penalty brings dramatic performance gains, for it reduces memory requirements of about 50 % compared to classical designs with a same security level. For instance, U-QUARK provides 64-bit security against collisions and second preimages using 146 memory bits (i.e., the two NFSR's plus the LFSR), while DM-PRESENT provides 64-bit security against preimages but only 32-bit security against collisions with 128 bits of required memory.

3.2   Sponge Construction

The sponge construction [13] seems the only real alternative to the classical Merkle-Damgård construction based on a compression function (although several "patched" versions were proposed, adding counters, finalization functions, etc.). It rather relies on a *single permutation*, and message blocks are integrated with a simple XOR with the internal state. Sponge functions do not require storage of message blocks nor of "feedforward" intermediate values, as in Davies-Meyer constructions, however they need a larger state to achieve traditional security levels, which compensates those memory savings.

The sponge construction was proven indifferentiable from a random oracle (up to some bound) when instantiated with a random permutation or transformation, which is the highest security level a hash construction can achieve. But its most interesting feature is its flexibility: given a fixed permutation P, varying the parameters r, c, and n offers a wide range of trade-offs efficiency/security[5].

3.3   Permutation Algorithm

We now briefly justify our design choices regarding the permutation P. To avoid "reinventing the wheel", we borrowed most design ideas from the stream cipher Grain and from the block cipher KATAN, as detailed below.

Grain. The stream cipher Grain-v1 was chosen in 2008 as one of the four "promising new stream ciphers" by the ECRYPT eSTREAM Project[6]. It consists of two 80-bit shift registers combined with three Boolean functions, which makes it one of the lightest designs ever. Grain's main advantages are its simplicity and its performance flexibility (due to the possibility of parallelized implementations). However, a direct reuse of Grain fails to give a secure permutation for a hash function, because of "slide distinguishers" (see §4.5), of the existence of differential characteristics [16], and of (conjecturally) statistical distinguishers for Grain-128 [17].

---

[4] The expected workload to find a preimage was previously estimated to $2^n + 2^{c-1}$ in [15, §5.3], although that was not proven optimal.

[5] See the interactive page "Tune KECCAK to your requirements" at http://keccak.noekeon.org/tune.html

[6] See http://www.ecrypt.eu.org/stream

KATAN. The block cipher family KATAN [18] (CHES 2009) is inspired by the stream cipher Trivium [19] and builds a keyed permutation with two NFSR's combined with two light quadratic Boolean functions. Its small block sizes (32, 48, and 64 bits) plus the possibility of "burnt-in key" (with the KTANTAN family) lead to very small hardware footprints. KATAN's use of two NFSR's with short feedback delay contributes to a rapid growth of the density and degree of implicit algebraic equations, which complicates differential and algebraic attacks. Another interesting design idea is its use of a LFSR acting both as a counter of the number of rounds, and as an auxiliary input to the inner logic (to simulate two distinct types of rounds). Like Grain, however, KATAN is inappropriate for a direct reuse in a hash function, because of its small block size.

Taking the best of both. Based on the above observations, QUARK borrows the following design decisions from Grain:

- A mechanism in which each register's update depends on both registers.
- Boolean functions of high degree (up to six, rather than two in KATAN) and high density.

And KATAN inspired us in choosing

- Two NFSR's instead of a NFSR and a LFSR; Grain's use of a LFSR was motivated by the need to ensure a long period during the keystream generation (where the LFSR is autonomous), but this seems unnecessary for hashing. Moreover, the dissimetry in such a design is a potential threat for a secure permutation.
- An auxiliary LFSR to act as a counter and to avoid self-similarity of the round function.

Choice of the Boolean functions. The quality of the Boolean functions in P determines its security. We thus first chose the functions in QUARK according to their individual properties, according to known metrics (see, e.g., [20]). The final choice was made by observing the empirical resistance of the *combination* of the three functions to known attacks (see §4.3-4.4).

In QUARK, we chose f and g functions similar to the non-linear function of Grain-v1. These functions achieve good, though suboptimal, non-linearity and resilience (see Table 1). They have degree six and include monomials of each degree below six. Note that having a relatively high degree (e.g., six rather than two) is cheap, since logical AND's need fewer gates than XOR's (respectively, approximately one and 2.5). The distinct taps for each register break the symmetry of the design (note that KATAN also employs similar functions for each register's feedback).

As h function, distinct for each flavor of QUARK, we use a function of lower degree than f and g, but with more linear terms to increase the cross-diffusion between the two registers.

Table 1. Properties of the Boolean functions of each QUARK instance (for h, we consider that the parameter L_0 is zero)

Hash function	Boolean function	Var.	Deg.	Non-lin. (max)	Resil.
QUARK (all)	f	13	6	3440 (4056)	3
QUARK (all)	g	13	6	3440 (4056)	3
U-QUARK	h	12	3	1280 (2016)	6
D-QUARK	h	15	3	10240 (16320)	9
T-QUARK	h	16	3	20480 (32640)	10

4   Preliminary Security Analysis

4.1   The Hermetic Sponge Strategy

Like for the hash function KECCAK [4], we follow the *hermetic sponge strategy*, which consists in adopting the sponge construction with a permutation that should not have exploitable properties. The indifferentiability proof of the sponge construction [13] implies that any non-generic attack on a QUARK hash function implies a distinguisher for its permutation P (but a distinguisher for P does not necessarily leads to an attack on QUARK). This reduces the security of P to that of the hash function that uses it.

More precisely, the indifferentiability proof of the sponge construction ensures an expected complexity at least $\sqrt{\pi}2^{c/2}$ against any *differentiating attack*, independently of the digest length (of course, practical attacks of complexity below that bound exist when short digests are used, but these apply as well to a random oracle). This proof covers collision and (second) preimage attacks, as well as more specific attacks as multicollision or herding attacks [21]. As QUARK follows the hermetic sponge strategy, the proof of [13] is directly applicable.

4.2   Generic Second Preimage Attack

The generic second preimage attack against QUARK is similar to the generic preimage attack against the hash function CubeHash [5], which was described in [22] and discussed in [23]. It has complexity equivalent to more than $2^{c/2+1}$ evaluations of P and so to more than $b2^{c/2+3}$ clocks of P's mechanism, that is, 2^{74}, 2^{90}, and 2^{123} clocks for U-, D-, and T-QUARK respectively.

4.3   Resistance to Cube Attacks and Cube Testers

The recently proposed "cube attacks" [24] and "cube testers" [25] are higher-order differential cryptanalysis techniques that exploit weaknesses in the algebraic structure of a cryptographic algorithm. These techniques are mostly relevant for algorithms based on non-linear components whose ANF has low degree and low density (e.g., the feedback function of an NFSR). Cube testers were for example applied [17] to the stream cipher Grain-128 [26]. Cube attacks/testers are thus tools of choice to attack (reduced version of) QUARK's permutation, since it resembles Grain-128.

Recall that QUARK targets security against any nontrivial structural distinguisher for its permutation P. We thus applied cube testers—i.e., distinguishers—rather than cube attacks—i.e., key recovery attacks—to the permutation of each QUARK flavor. We followed a methodology inspired by [17], using bitsliced C implementations of P and an evolutionary algorithm to optimize the parameters of the attack.

In our simplified attack model, the initial state is chosen uniformly at random to try our distinguishers. Table 2 reports our results.

Table 2. Highest number of rounds t such that the state (X^t, Y^t) could be distinguished from random using a cube tester with the given complexity. Percentage of the total number of rounds is given in parentheses.

Instance	Total rounds	Rounds attacked		
		in 2^8	in 2^{16}	in 2^{24}
U-QUARK	544	109	111	114 (21.0 %)
D-QUARK	704	144	144	148 (21.0 %)
T-QUARK	1024	213	220	222 (21.7 %)

One observes from Table 2 that all QUARK flavors showed a similar resistance to cube testers, with a fraction $\approx 21\%$ of the total number of rounds attacked with an effort 2^{24}. How many rounds would we break with a higher complexity? It cannot be determined analytically (to our present knowledge), however heuristical arguments can be given, based on previous results [24, 25, 17]: the number of rounds attackable seems indeed to evolve logarithmically rather than linearly, as a function of the number of variables used. A worst-case assumption (for the designers) is thus that of a linear evolution. Under this assumption, one could attack 126 rounds of U-QUARK in 2^{64} (23.2 % of the total), 162 rounds of D-QUARK in 2^{80} (23.0 %), and 271 rounds of T-QUARK in 2^{112} (26.5 %). Therefore, QUARK is unlikely to be broken by cube attacks or cube testers.

Note that 220 of Grain-128's 256 rounds could be attacked in [17] in 2^{24}; this result, however, should not be compared to the value 222 reported in Table 2, since the latter attack concerns *any bit of the internal state*, while the former concerns *the first keystream bit* extracted from the internal state after 220 rounds. One could thus attack at least $220 + 127 = 347$ rounds of Grain-128 by observing any bit of the internal state. Therefore, although T-QUARK uses registers of same length as Grain-128, it is significantly more resistant to cube testers.

4.4 Resistance to Differential Attacks

Differential attacks covers all attacks that exploit nonideal propagation of differences in a cryptographic algorithm (or components thereof). A large majority of attacks on hash functions are at least partially differential, starting with the breakthrough results on MD5 and SHA-1. It is thus crucial to analyze the resistance of new designs to differential attacks, which means in our case to analyze the permutation of QUARK against differential distinguishers.

We consider a standard attack model, where the initial state is assumed chosen uniformly at random and where we seek differences in the initial state that give biased differences in the state obtained after the (reduced-round) permutation. We focus on "truncated" differentials in which the output difference concerns a small subset of bits (e.g., a single bit), because these are sufficient to distinguish the (reduced-round) permutation from a random one, and are easier to find for an adversary than differentials on all the b bits of the state.

First, we observe that it is easy to track differences during the first few rounds, and in particular to find probability-1 (truncated) differential characteristics for reduced-round versions. For example, in U-QUARK, a difference in the bit Y_{29}^0 in the initial state never leads to a difference in the output of f or of h at the 30th round; hence after $(67 + 30) = 97$ rounds, X_0^{97} will be unchanged. Similar examples can be given for 117 rounds of D-QUARK and 188 rounds of T-QUARK. For higher number of rounds, however, it becomes difficult to manually track differences, and so an automated search is necessary. As a heuristical indicator of the resistance to differential attacks, we programmed an automated search for high-probability truncated differentials, given an input difference in a single bit. Table 3 presents our results, showing that we could attack approximately as many rounds with truncated differentials as with cube testers (see Table 2).

Table 3. Highest number of rounds t such that the state (X^t, Y^t) could be distinguished from random using a simple differential distinguisher with the given complexity. Percentage of the total number of rounds is given in parentheses.

Instance	Total rounds	Rounds attacked in 2^8	in 2^{16}	in 2^{24}
U-QUARK	544	109	116	119 (21.9 %)
D-QUARK	704	135	145	148 (21.0 %)
T-QUARK	1024	206	211	216 (21.1 %)

We expect advanced search techniques to give differential distinguishers for more rounds (e.g., where the sparse difference occurs slightly later in the internal state, as in [16]). However, such methods seem unlikely to apply to the $4b$ rounds of QUARK's permutation. For example, observe that [16] presented a characteristic of probability 2^{-96} for the full 256-round Grain-128; for comparison, T-QUARK makes 1024 rounds, uses more complex feedback functions, and targets a security level of 112 bits; characteristics of probability greater than 2^{-112} are thus highly improbable, even assuming that the adversary can control differences during (say) the first 256 rounds.

4.5   Resistance to Slide Distinguishers

Suppose that the initial state of the LFSR of QUARK is not the all-one string, but instead is determined by the input of P—that is, P is redefined to accept $(b + 10)$ rather than b input bits. It is then straightforward to distinguish P from a random transform: pick a first initial state (X^0, Y^0, L^0), and consider the second initial state $(X'^0, Y'^0, L'^0) = (X^1, Y^1, L^1)$, i.e., the state obtained after clocking the first state once. Since all rounds are identical, the shift will be preserved between the two states, leading to final states (X^{4b}, Y^{4b}, L^{4b}) and $(X'^{4b}, Y'^{4b}, L'^{4b}) = (X^{4b+1}, Y^{4b+1}, L^{4b+1})$. One thus obtains two input/output pairs satisfying a nontrivial relation, which is a distinguisher for the modified P considered. The principle of the attack is that of slide attacks on block ciphers [27]; we thus call the above a *slide distinguisher*.

The above idea is at the basis of "slide resynchronization" attacks on the stream cipher Grain [16, 28], which are related-key attacks using as relation a rotation of the key, to simulate a persistent shift between two internal states.

To avoid the slide distinguisher, we use a trick previously used in KATAN: making each round dependent on a bit coming from a LFSR initialized to a fixed value, in order to simulate two distinct types of rounds. It is thus impossible to have two valid initial states shifted of one or more clocks, and such that the shift persists through the $4b$ rounds.

5   Hardware Implementation

This section reports our hardware implementation of the QUARK instances. Note that QUARK is not optimized for software (be it 64- or 8-bit processors), and other types of designs are preferable for such platforms. We thus focus on hardware efficiency. Our results arise from pure simulations, and are thus not supported by real measurements on a fabricated chip. However, we believe that this evaluation gives a fair and reliable overview of the overall VLSI performance of QUARK.

5.1   Architectures

Three characteristics make QUARK particularly attractive for lightweight hashing: first, the absence in its sponge construction of "feedforward" values, which normally would require additional dedicated memory components; second, its use of shift registers, which are extremely easy to implement in hardware; and third, the possibility of several space/time implementation trade-offs. Based on the two extremal trade-off choice, we designed two architecture variants of U-QUARK, D-QUARK, and T-QUARK:

- **Serial:** Only one permutation module, hosting the circuit for the functions f, g, and h, is implemented. Each clock cycle, the bits of the registers X, Y, and L are shifted by one. These architectures corresponds to the most compact designs. They contain the minimal circuitry needed to handle incoming messages and to generate the correct output digests.

– **Parallel:** The number of the implemented permutation modules corresponds to the parallelization degree given in §2.3. The bits in the registers are accordingly shifted. These architectures increase the number of rounds computed per cycle—and therefore the throughput—at extra area costs.

In addition to the three feedback shift registers, each design has a dedicated controller module that handles the sponge process. This module is made up of a finite-state machine and of two counters for the round and the output digest computation. After processing all message blocks during the absorbing phase, the controller switches automatically to the squeezing phase (computation of the digest), if no further r-bit message blocks are given. This implies that the message has been externally padded.

5.2   Methodology

We described the serial and parallel architectures of each QUARK instance in functional VHDL, and synthesized the code with Synopsys Design Vision-2009.06 targeting the UMC 0.18 µm 1P6M CMOS technology with the FSA0A_C cell library from Faraday Technology Corporation. We used the generic process (at typical conditions), instead of the low-leakage for two reasons: first the leakage dissipation is not a big issue in 0.18 µm CMOS, and second, for such small circuits the leakage power is about two orders of magnitude smaller than the total power. To provide a thorough and more reliable analysis, we extended the implementation up to the back-end design. Place and route have been carried out with the help of Cadance Design Systems Velocity-9.1. In a square floorplan, we set a 98 % row density, i.e., the utilization of the core area. Two external power rings of 1.2 µm were sufficient for power and ground distribution. In this technology six metal layers are available for routing. However, during the routing phase, the fifth and the sixth layers were barely used. The design flow has been placement, clock tree synthesis, and routing with intermediate timing optimizations.

Each architecture was implemented at the target frequency of 100 kHz. As noted in [10, 18], this is a typical operating frequency of cryptographic modules in RFID systems. Power simulation was measured for the complete design under real stimuli simulations (two consecutive 512-bit messages) at 100 kHz. The switching activity of the circuit's internal nodes was computed generating Value Change Dump (VCD) files. These were then used to perform statistical power analysis in the velocity tool. Besides the mean value, we also report the peak power consumption, which is a limiting parameter in RFID systems ([7] suggests a maximum of 27 µW). Table 4 reports the performance metrics obtained from our simulations at 100 kHz.

5.3   Discussion and Comparison with PRESENT-Based Designs

As reported in Table 4, the three serial designs need fewer than 2300 GE, thus making 112-bit security affordable for restricted-area environments. Particularly appealing for ultra-compact applications is the U-QUARK function, which offers 64-bit security but requires only 1379 GE and dissipates less than 2.5 μW. To the best of our knowledge, U-QUARK is lighter than all previous designs with comparable security claims. We expect an instance of QUARK with 256-bit security (e.g., with $r = 64, c = 512$) to fit in 4500 GE.

Note that in the power results of the QUARK circuits, the single contributions of the mean power consumption are 68 % of internal, 30 % of switching, and 2 % of leakage power. Also important is that the peak value exceeds maximally 27 % of the mean value.

As reported in Table 4, DM-PRESENT-80/128 and H-PRESENT-128 also offer implementation trade-offs. For a same (second) preimage resistance of at least 64 bits, U-QUARK fits in a smaller area, and even the 80-bit-secure D-QUARK does not need more GE than DM-PRESENT-128. In terms of throughput, however, QUARK underperforms PRESENT-based designs. This may be due to its high security margin (note that 26 of the 31 rounds of PRESENT, as a block cipher, were attacked [29], suggesting a thin security margin against distinguishers in the "open key" model of hash functions).

Table 4. Compared hardware performance of PRESENT-based and QUARK lightweight hash functions. Security is expressed in bits (e.g., "128" in the "Pre." column means that preimages can be found within approximately 2^{128} calls to the function). Throughput and power consumption are given for a frequency of 100 kHz.

Hash function	Security Pre.	Security Coll.	Block [bits]	Area[a] [GE]	Lat. [cycles]	Thr. [kbps]	Power [μW] Mean	Power [μW] Peak
DM-PRESENT-80	64	32	80	1600	547	14.63	1.83	-
DM-PRESENT-80	64	32	80	2213	33	242.42	6.28	-
DM-PRESENT-128	64	32	128	1886	559	22.90	2.94	-
DM-PRESENT-128	64	32	128	2530	33	387.88	7.49	-
H-PRESENT-128	128	64	64	2330	559	11.45	6.44	-
H-PRESENT-128	128	64	64	4256	32	200.00	8.09	-
U-QUARK	128	64	8	1379	544	1.47	2.44	2.96
U-QUARK×8	128	64	8	2392	68	11.76	4.07	4.84
D-QUARK	160	80	16	1702	704	2.27	3.10	3.95
D-QUARK×8	160	80	16	2819	88	18.18	4.76	5.80
T-QUARK	224	112	32	2296	1024	3.13	4.35	5.53
T-QUARK×16	224	112	32	4640	64	50.00	8.39	9.79

[a] One GE is the area of a 2-input drive-one NAND gate, i.e., in the target 0.18 μm technology, 9.3744 μm^2.

Acknowledgments

We would like to thank Gilles Van Assche for helpful comments.

References

1. Feldhofer, M., Rechberger, C.: A case against currently used hash functions in RFID protocols. In: Meersman, R., Tari, Z., Herrero, P. (eds.) OTM 2006 Workshops. LNCS, vol. 4277, pp. 372–381. Springer, Heidelberg (2006)
2. Preneel, B.: Status and challenges of lightweight crypto. Talk at the Early Symmetric Crypto (ESC) seminar (January 2010)
3. NIST: Cryptographic hash algorithm competition,
   http://www.nist.gov/hash-competition
4. Bertoni, G., Daemen, J., Peeters, M., Assche, G.V.: Keccak sponge function family main document. Submission to NIST, Round 2 (2009),
   http://keccak.noekeon.org/Keccak-main-2.0.pdf
5. Bernstein, D.J.: CubeHash specification (2.B.1). Submission to NIST, Round 2 (2009), http://cubehash.cr.yp.to/submission2/spec.pdf
6. Bernet, M., Henzen, L., Kaeslin, H., Felber, N., Fichtner, W.: Hardware implementations of the SHA-3 candidates Shabal and CubeHash. In: CT-MWSCAS. IEEE, Los Alamitos (2009)
7. Feldhofer, M., Wolkerstorfer, J.: Strong crypto for RFID tags - a comparison of low-power hardware implementations. In: ISCAS, pp. 1839–1842. IEEE, Los Alamitos (2007)
8. O'Neill, M.: Low-cost SHA-1 hash function architecture for RFID tags. In: Workshop on RFID Security RFIDsec. (2008)
9. Yoshida, H., Watanabe, D., Okeya, K., Kitahara, J., Wu, H., Kucuk, O., Preneel, B.: MAME: A compression function with reduced hardware requirements. In: ECRYPT Hash Workshop (2007)
10. Bogdanov, A., Leander, G., Paar, C., Poschmann, A., Robshaw, M.J.B., Seurin, Y.: Hash functions and RFID tags: Mind the gap. In: Oswald, E., Rohatgi, P. (eds.) CHES 2008. LNCS, vol. 5154, pp. 283–299. Springer, Heidelberg (2008)
11. Bogdanov, A., Knudsen, L.R., Leander, G., Paar, C., Poschmann, A., Robshaw, M.J.B., Seurin, Y., Vikkelsoe, C.: PRESENT: An ultra-lightweight block cipher. In: Paillier, P., Verbauwhede, I. (eds.) CHES 2007. LNCS, vol. 4727, pp. 450–466. Springer, Heidelberg (2007)
12. Shamir, A.: SQUASH - a new MAC with provable security properties for highly constrained devices such as RFID tags. In: Nyberg, K. (ed.) FSE 2008. LNCS, vol. 5086, pp. 144–157. Springer, Heidelberg (2008)
13. Bertoni, G., Daemen, J., Peeters, M., Assche, G.V.: On the indifferentiability of the sponge construction. In: Smart, N.P. (ed.) EUROCRYPT 2008. LNCS, vol. 4965, pp. 181–197. Springer, Heidelberg (2008)
14. Bertoni, G., Daemen, J., Peeters, M., Assche, G.V.: Sponge-based pseudo-random number generators. In: CHES (to appear, 2009)
15. Bertoni, G., Daemen, J., Peeters, M., Assche, G.V.: Sponge functions,
    http://sponge.noekeon.org/SpongeFunctions.pdf
16. Cannière, C.D., Kücük, O., Preneel, B.: Analysis of Grain's initialization algorithm. In: SASC 2008 (2008)
17. Aumasson, J.P., Dinur, I., Henzen, L., Meier, W., Shamir, A.: Efficient FPGA implementations of highly-dimensional cube testers on the stream cipher Grain-128. In: SHARCS (2009)
18. Cannière, C.D., Dunkelman, O., Knezevic, M.: KATAN and KTANTAN - a family of small and efficient hardware-oriented block ciphers. In: Clavier, C., Gaj, K. (eds.) CHES 2009. LNCS, vol. 5747, pp. 272–288. Springer, Heidelberg (2009)

19. Cannière, C.D., Preneel, B.: Trivium. In: Robshaw, M.J.B., Billet, O. (eds.) New Stream Cipher Designs. LNCS, vol. 4986, pp. 84–97. Springer, Heidelberg (2008)
20. Sarkar, P., Maitra, S.: Construction of nonlinear boolean functions with important cryptographic properties. In: Preneel, B. (ed.) EUROCRYPT 2000. LNCS, vol. 1807, pp. 485–506. Springer, Heidelberg (2000)
21. Kelsey, J., Kohno, T.: Herding hash functions and the Nostradamus attack. In: Vaudenay, S. (ed.) EUROCRYPT 2006. LNCS, vol. 4004, pp. 183–200. Springer, Heidelberg (2006)
22. Bernstein, D.J.: CubeHash appendix: complexity of generic attacks. Submission to NIST (2008), http://cubehash.cr.yp.to/submission/generic.pdf
23. Aumasson, J.-P., Brier, E., Meier, W., Naya-Plasencia, M., Peyrin, T.: Inside the hypercube. In: Boyd, C., Nieto, J.M.G. (eds.) ACISP 2009. LNCS, vol. 5594, pp. 202–213. Springer, Heidelberg (2009)
24. Dinur, I., Shamir, A.: ube attacks on tweakable black box polynomials. In: Joux, A. (ed.) EUROCRYPT 2009. LNCS, vol. 5479, pp. 278–299. Springer, Heidelberg (2010)
25. Aumasson, J.P., Dinur, I., Meier, W., Shamir, A.: Cube testers and key recovery attacks on reduced-round MD6 and Trivium. In: Dunkelman, O. (ed.) Fast Software Encryption. LNCS, vol. 5665, pp. 1–22. Springer, Heidelberg (2009)
26. Hell, M., Johansson, T., Maximov, A., Meier, W.: A stream cipher proposal: Grain-128. In: IEEE International Symposium on Information Theory, ISIT 2006 (2006)
27. Biryukov, A., Wagner, D.: Slide attacks. In: Knudsen, L. (ed.) FSE 1999. LNCS, vol. 1636, pp. 245–259. Springer, Heidelberg (1999)
28. Lee, Y., Jeong, K., Sung, J., Hong, S.: Related-key chosen IV attacks on Grain-v1 and Grain-128. In: Mu, Y., Susilo, W., Seberry, J. (eds.) ACISP 2008. LNCS, vol. 5107, pp. 321–335. Springer, Heidelberg (2008)
29. Cho, J.Y.: Linear cryptanalysis of reduced-round PRESENT. In: Pieprzyk, J. (ed.) CT-RSA 2010. LNCS, vol. 5985, pp. 302–317. Springer, Heidelberg (2010)

PRINTcipher: A Block Cipher for IC-Printing

Lars Knudsen[1], Gregor Leander[1],
Axel Poschmann[2,*], and Matthew J.B. Robshaw[3]

[1] Technical University Denmark, DK-2800 Kgs. Lyngby, Denmark
[2] School of Physical and Mathematical Sciences,
Nanyang Technological University, Singapore
[3] Orange Labs, Issy les Moulineaux, France
{Lars.R.Knudsen,G.Leander}@mat.dtu.dk,
aposchmann@ntu.edu.sg, matt.robshaw@orange-ftgroup.com

Abstract. In this paper we consider some cryptographic implications of *integrated circuit (IC) printing*. While still in its infancy, IC-printing allows the production and personalisation of circuits at very low cost. In this paper we present two block ciphers PRINTcipher-48 and PRINTcipher-96 that are designed to exploit the properties of IC-printing technology and we further extend recent advances in lightweight block cipher design.

Keywords: symmetric cryptography, block cipher, IC-printing, hardware implementation.

1 Introduction

New technologies open new applications and often bring challenging new problems at the same time. Most recently, advances in device manufacture have opened the possibility for extremely low-cost RFID tags. However, at the same time, their exceptional physical and economic constraints mean that we must leave behind much of our conventional cryptography. This has spurred the development of the field of lightweight cryptography.

This paper considers another technological advance, that of *integrated circuit printing* or *IC-printing*. Using silicon inks, circuits can quite literally be printed onto a range of materials using high-definition printing processes. The technology remains in its infancy and its true potential is yet to be fully understood. But the claimed advantages include the ability to print on to thin and flexible materials and, since the conventional fabrication process is by-passed, to be much cheaper than silicon-based deployments [19]. Since the main driver for IC-printing is economic, the typically-cited areas of application overlap closely with the typical domains for lightweight cryptography. Indeed, one of the oft-stated applications of IC-printing is in the fabrication of cheap RFID tags [13]. Therefore there

---

* The research was supported in part by the Singapore National Research Foundation under Research Grant NRF-CRP2-2007-03.

S. Mangard and F.-X. Standaert (Eds.): CHES 2010, LNCS 6225, pp. 16–32, 2010.

is much in common between some of the techniques proposed for conventional RFID tags and those that will be used on printed tags.

However IC-printing has some interesting properties and these allow us to take a fresh look at our cryptography and to see how it might be adapted to this new field. In this paper, therefore, we consider the task of adding some simple security functionality to a printed tag, and following what has now become a reasonably well-trodden path, we start out with the design of a block cipher.

Block ciphers make a natural starting point for several reasons. Not only can they be used in many different ways, but as a community we feel somewhat more at ease with their design and analysis. That said, for such extreme environments as IC-printing, we are working right at the edge of established practice and we are forced to consider and highlight some interesting problems. This is the purpose behind the block cipher PRINTCIPHER.

2 Design Approach to PRINTCIPHER

Just as for other constrained environments, the size of implementation will be a dominant issue in IC-printing. Our work will therefore have close links with other block cipher work in the field of lightweight cryptanalysis. In fact our starting point for the work in this paper will be the block cipher PRESENT [1] which appears to offer a range of design/implementation trade-offs. However we will re-examine the structure of PRESENT in the particular context of IC-printing.

Conceptually we can imagine that within a block cipher we need an "encryption computation" and a "subkey computation". For the first, there are limits to the short-cuts we can make since we are constrained by the attentions of the cryptanalyst. This means, for the most part, that proposals for a given security level and a given set of block cipher parameters would occupy pretty much the same space. If we wanted to reduce the space occupied by an implementation then we would most likely reduce the block size, something that has been proposed independently elsewhere [2]. However, for the "subkey computation" things are a little different and exactly how a key should be used is not always clear. This highlights two separate issues.

The first issue is whether a key is likely to be changed in an application. In fact there is probably not too much debate about this issue and many commentators over the years [1,21] have made the point that for RFID applications it is very unlikely that one would want to change the key. Indeed some other RFID implementation work [20] has demonstrated that the overhead in supporting a change of key can be significant.

The second issue is the exact form of the key schedule. Some block ciphers, *e.g.* IDEA [14], have a very simple key schedule in which subkeys are created by sampling bits of the user-supplied key. This is, in effect, the approach used in the KTANTAN family of ciphers. The advantage of this approach is that no working memory is needed for the subkey computations. Other lightweight block ciphers have some key schedule computation, *e.g.* PRESENT, while another proposal CGEN [21] proposes to use no key schedule; the user-supplied key is used without any sampling or additional computation.

Returning to the situation at hand, conventional silicon manufacturing uses lithographic techniques to massively duplicate an implementation across a silicon wafer. This gives the economy of scale to offset the fabrication costs but at the same time requires that all implementations are identical. In this paper, we take advantage of the properties of IC-printing to propose another approach. Regular IC manufacture requires all versions of the cipher to be identical and so while a specific tag can be personalised with a unique key, this is a post-fabrication step. With a printer, however, there is essentially no cost in changing the circuit that is printed at each run. This means that part—or all—of the secret key can be embedded into the algorithm description. The algorithms that appear on different printed labels will be subtly different from one another.

The PRINTCIPHER family was designed with this approach in mind. PRINTCIPHER-48 is a 48-bit block cipher which uses a fixed 48-bit secret key and derives an additional 32 bits of security via the secret algorithm variability. Different trade-offs can be established either reducing the effective security, say to 64 bits, and/or independently increasing the block size to 96 bits. In fact this is a particularly useful block size since it matches the length of an *electronic product code* (EPC) [5]. However we will tend to concentrate our attentions in this paper on two proposals PRINTCIPHER-48 and PRINTCIPHER-96. Given the amount of work in the area of block ciphers, some points of similarity with other proposals in the literature are inevitable. For instance, 3-bit S-boxes have been used in 3-way [4] and the Scaleable Encryption Algorithm (SEA) [26] while key-dependent algorithm features have appeared in a variety of block ciphers including Blowfish [24], Twofish [25], and GOST [10].

3   PRINTCIPHER-48 and PRINTCIPHER-96

PRINTCIPHER is a block cipher with b-bit blocks, $b \in \{48, 96\}$, and an effective key length of $\frac{5}{3} \times b$ bits. The essential structure of PRINTCIPHER is that of an

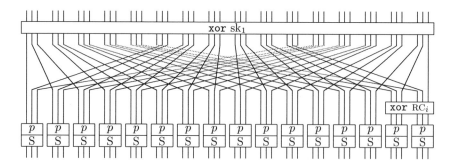

Fig. 1. One round of PRINTCIPHER-48 illustrating the bit-mapping between the 16 3-bit S-boxes from one round to the next. The first subkey is used in the first **xor**, the round counter is denoted RC_i, while key-dependent permutations are used at the input to each S-box.

SP-network with $r = b$ rounds. It follows that PRINTCIPHER-48 operates on 48-bit blocks, uses an 80-bit key and consists of 48 rounds while PRINTCIPHER-96 operates on 96-bit blocks, uses a 160-bit key and consists of 96 rounds. Each round of encryption consists of the following steps:

1. The cipher state is combined with a round key using bitwise exclusive-or (xor).
2. The cipher state is shuffled using a fixed linear diffusion layer.
3. The cipher state is combined with a round constant using bitwise xor.
4. The three-bit entry to each S-box is permuted in a key-dependent permutation layer.
5. The cipher state is mixed using a layer of $\frac{b}{3}$ non-linear S-box substitutions.

Key xor. The current state of the cipher is combined using bitwise xor with an b-bit subkey sk_1. This subkey is identical in all rounds.

Linear diffusion. The pLayer is a simple bit permutation that is specified in the following way. Bit i of the current state is moved to bit position $P(i)$ where

$$P(i) = \begin{cases} 3 \times i \bmod b - 1 & \text{for } 0 \leq i \leq b - 2, \\ b - 1 & \text{for } i = b - 1. \end{cases}$$

Round counter RC_i. The round counter RC_i for $1 \leq i \leq r$ is combined using xor to the least significant bits of the current state. The values of the round counter are generated using an n-bit shift register ($n = \lceil \log_2 r \rceil$) in the following way. Denote the state of the register as $x_{n-1} || \ldots || x_1 || x_0$ and compute the update as follows:

$$\begin{aligned} t &= 1 + x_{n-1} + x_{n-2} \\ x_i &= x_{i-1} \qquad \text{for } n - 1 \geq i \geq 1 \\ x_0 &= t \end{aligned}$$

The shift register is initialised to all zeros, *i.e.* 000000 or 0000000, and is then incremented at the start of every round. The round counter RC_i takes the current value of the register $x_{n-1} || \ldots || x_1 || x_0$.

Keyed permutation. Each set of three bits, namely the input bits to each of the S-boxes, are permuted among themselves. For each of the $\frac{b}{3}$ S-boxes the permutation can be the same or different and it is chosen in a key-dependent manner from a set of four. However for each S-box the same permutation— once chosen—is used in the same position in every round. In other words, $\frac{b}{3}$ permutations (of three bits) are picked from a set of four in a key-dependent manner. This gives $4^{b/3}$ possible mini-permutation layers which is equivalent to $\frac{2}{3} \times b$ key bits.

sBoxLayer. A single 3- to 3-bit S-box is applied $\frac{b}{3}$ times in parallel. For the sBoxLayer the current state is considered as $\frac{b}{3}$ 3-bit words, each word is processed using the same S-box, and the next state is the concatenation of the outputs. The action of the S-box is given by the following table.

x	0	1	2	3	4	5	6	7
$S[x]$	0	1	3	6	7	4	5	2

3.1 Deriving the Permutations from the User Key

The $\frac{5}{3} \times b$-bit user-supplied key k is considered as consisting of two subkey components $k = sk_1 || sk_2$ where sk_1 is b bits long and sk_2 is $\frac{2}{3} \times b$ bits long. The first subkey is used, unchanged, within the xor layer of each and every round.

The second subkey sk_2 is used to generate the key-dependent permutations in the following way. The $\frac{2}{3} \times b$-bits are divided into $\frac{b}{3}$ sets of two bits and each two-bit quantity $a_1 || a_0$ is used to pick one of four of the six available permutations of the three input bits. Specifically, the three input bits $c_2 || c_1 || c_0$ are permuted to give the following output bits according to the value of $a_1 || a_0$. Of course one can combine the bitwise permutation with the fixed S-box to give, conceptually, four virtual S-boxes. These are given below and testvectors for both PRINTCIPHER variants can be found in the appendix.

$a_1 || a_0$	
00	$c_2 || c_1 || c_0$
01	$c_1 || c_2 || c_0$
10	$c_2 || c_0 || c_1$
11	$c_0 || c_1 || c_2$

x	0	1	2	3	4	5	6	7
$V_0[x]$	0	1	3	6	7	4	5	2
$V_1[x]$	0	1	7	4	3	6	5	2
$V_2[x]$	0	3	1	6	7	5	4	2
$V_3[x]$	0	7	3	5	1	4	6	2

3.2 Security Goals

Our security goals behind PRINTCIPHER are the usual security claims for a block cipher with the operational parameters of PRINTCIPHER. Note that in the case of PRINTCIPHER-48 even though we have a regular sized 80-bit key, we only have a 48-bit block cipher and this greatly limits the opportunities for an attacker.

We follow much of the established literature on lightweight cryptography and do not consider side-channel attacks in this paper. While this is certainly a factor for consideration, typical applications are very low-cost and the potential gains for an attacker are minor. Even in a relatively well-developed field such as RFID tags for the supply chain it is not clear what level of protection is really appropriate for most deployments of lightweight cryptography. For IC-printing this is even more unclear, and there are some concerns that are particular to IC-printing for which appropriate precautions will likely be needed, such as the use of opaque masks to shield the circuit from simple inspection. Note that, shielding protection is not exclusively an issue for PRINTCIPHER where the key is part of the circuit, but also for more standard ciphers where the key is stored in memory, as it is in principle possible to inspect memory in similar ways (see for example [23]) .

Where we differ from some other work in the field, however, is that for PRINTCIPHER we are not particularly concerned by related-key attacks. This is not because we believe that PRINTCIPHER is in some way particularly vulnerable to them (see Section 4.3 for details). Instead it is because we believe that related-key attacks are so alien to the intended use of PRINTCIPHER that there is no point in considering them. Recall that a (printed) device will be initialised with a key in a random way. To mount a related-key attack one has to somehow find a pair of deployed devices that, by chance, satisfy a stated condition. We consider this to be an entirely unrealistic threat.

3.3   Some Features of the Design

During the design of PRINTCIPHER there were some interesting choices to make. Certainly, to improve the implementation efficiency we required that each round was identical, even as far as having an identical subkey in each round. However having the same round key in every round meant that we were restricted to 48-bit keys. So to increase the effective key length we used some additional permutation steps that could be key-dependent. Permutations cost nothing in hardware and, for our application of IC-printing, they incur effectively no additional cost during the printing of the cipher. It can be shown that there are no equivalent keys in the sense that there are no two pairs of subkey components (sk_1, sk_2) that will yield the same round function. Note that since every round is identical—to the point of having the same round key—we needed to introduce a round-dependent value to prevent slide attacks and this was done using a shift register-based counter as outlined above.

The S-box. The 3-bit S-box that we chose is optimal with respect to linear and differential properties. However we cannot avoid the existence of single-bit to single-bit differences or masks and so our specific choice of S-box minimizes there occurrence. That is, for a given single-bit input difference (resp. mask) exactly one single-bit output difference (resp. mask) occurs with non-zero probability (resp. non-zero bias). We generated all 3-bit S-boxes with this property and it turns out that there are exactly 384 such S-boxes in total.

Clearly, permuting the input bits and (xor) adding constants before or after the S-box preserves the desired properties. Up to these changes, there is only one possible choice of S-box, *i.e.* all 384 S-boxes fulfilling the desired criteria can be constructed from any one of them by permuting the input bits and adding constants before and after the S-box (indeed $384 = 6 \cdot 2^3 \cdot 2^3$).

Thus in the design of PRINTCIPHER there is, in effect, only one suitable choice of S-box. Choosing any other of the 384 possible S-boxes would result in the same cipher for a different key, up to an additional xor with a constant to the plaintext and the ciphertext. More formally, given two S-boxes S_0, S_1 out of the 384 possible choices and any key (sk_1, sk_2) there exist a key (sk_1', sk_2') and constants c_1, c_2 such that

$$\text{PRINTCIPHER}_{S_0, sk_1, sk_2}(p) = \text{PRINTCIPHER}_{S_1, sk_1', sk_2'}(p \oplus c_1) \oplus c_2$$

for any plaintext p.

Those observations imply another interesting property of the S-box of PRINTCIPHER. Namely, instead of permuting the input bits of the S-box one could permute the output bits of the S-box and xor suitable constants before and after the S-box. More precisely, denoting the PRINTCIPHER S-box by S, for any bit permutation P, there exist constants c and d such that

$$S(P(x)) = P(S(x \oplus c)) \oplus d \quad \forall x.$$

Note that, while this might give some freedom in implementing the cipher we did not see any security implications of this.

The bit permutation. We choose the permutation so as to give the potential for full dependency after a minimal number of rounds, *i.e.* after $4 = \lceil \log_3 48 \rceil$ rounds. Note that in general, given an SP-network with block size b and s bit Sboxes, where s divides b, it can be shown that the bit permutation

$$P(i) = \begin{cases} s \times i \bmod b - 1 & \text{for } 0 \leq i \leq b - 2, \\ b - 1 & \text{for } i = b - 1. \end{cases}$$

provides optimal diffusion in the sense that full dependency is reached after $\lceil \log_s b \rceil$ rounds. The bit permutation – or rather its inverse – used for the block cipher PRESENT is a special case of this general result.

4   Security Analysis

In this section we analyze the security of our proposal with respect to the main cryptanalytic methods known. Though we focus on PRINTCIPHER-48, the security analysis can be easily extended to PRINTCIPHER-96.

4.1   Differential and Linear Characteristics

Let p be the probability of a linear characteristic, then define the correlation of the linear characteristic as $q = (2p - 1)^2$ [18]. As mentioned above, the S-box in PRINTCIPHER was chosen with good differential and linear properties. These properties are inherited by the other three virtual S-boxes, and so if we combine the key-dependent permutation with the S-box operation any differential characteristic over any S-box has a probability of at most $1/4$, and any linear characteristic over any S-box has a correlation of at most $1/4$.

Any characteristic over s rounds of PRINTCIPHER would have at least one active S-box per round. Consequently, an s-round differential characteristic will have a probability of at most 2^{-2s} and any s-round linear characteristic will have a correlation of at most 2^{-2s}. Thus, conventional differential and linear characteristics are unlikely to play a role in the cryptanalysis of PRINTCIPHER with the specified 48 respectively 96 rounds.

We furthermore experimentally checked for differential effects, i.e., the probability of differentials compared to the probability of characteristics. Consider the following one-round iterative characteristic (octal representation):

$$(0\ 0\ 0\ 0\ 0\ 0\ 0\ 0\ 0\ 0\ 0\ 0\ 0\ 0\ 0\ 1\ ) \rightarrow (0\ 0\ 0\ 0\ 0\ 0\ 0\ 0\ 0\ 0\ 0\ 0\ 0\ 0\ 0\ 1\ ).$$

Only the S-box in the least significant bits is active. This characteristic has probability $1/4$ when the active S-box is V_0 or V_1. The iterative characteristic above has an expected probability of 2^{-24} for 12 rounds.

We implemented experiments with 20 keys, each randomly chosen but such that the S-box in the least significant bits is either V_0 or V_1. For each key we generated 2^{28} pairs of texts of the above difference. The number of pairs of the expected difference after 12 rounds of encryption was 16.6 on the average, where 16 is expected for the characteristic. In similar tests over 14 rounds using 2^{30} pairs, the average number of pairs obtained was 4.5 on average, where 4 was expected. Here the expected probability of the iterative characteristic is 2^{-28}. These tests suggest that there is no significant differential effect for the characteristic. Computing the exact differential effect for a characteristic over many more rounds of PRINTCIPHER is a very complex task. However since the probability of the iterative characteristic is very low, e.g. 2^{-80} for 40 rounds, we expect that good probability differentials are unlikely to exist for PRINTCIPHER.

4.2   High Order Differentials and Algebraic Attacks

The algebraic degree of the S-box is 2 and due to the large number of 48 rounds we expect the total degree of the cipher to be close to the maximum. This assumption is supported by the following experiments. It is well-known that for a function of algebraic degree d, a dth-order differential will be a constant, and the value of a $(d+1)^{st}$-order differential will be zero. Consequently, if a dth-order differential over s rounds for one key is not zero, then the algebraic degree of this encryption function is at least $d-1$. For seven rounds of PRINTCIPHER and for ten randomly chosen keys we computed the values of two different 25th-order differentials. In all cases the values were nonzero. The experiments suggest that the algebraic degree of PRINTCIPHER reaches its maximum after much less than the specified 48 rounds. Due to this observations, we expect PRINTCIPHER to be secure against higher order differential attacks.

Regarding the so-called algebraic attacks, first observe that there exist quadractic equations over all 3-bit S-boxes, also those of PRINTCIPHER. Therefore, the secret key of one particular encryption can be described as the solution to a number of quadratic equations. However such a system of equation for PRINTCIPHER will be huge because of the large number of rounds, and with the techniques known today, there is not much hope that such systems can be solved in time faster than simply trying all values of the key. Moreover, the key dependent permutations potentially make the resulting systems of equation even more complex and harder to solve.

4.3   Related-Key Attacks

As stated above, we consider related-key attacks to be an entirely unrealistic threat. However, in the spirit of academic completeness, we consider the issue here.

The four S-boxes in PRINTCIPHER are closely related. As an example, S-box 0 and S-box 1 produce the same output for each of four inputs and similarly for S-boxes 2 and 3 and for S-boxes 4 and 5. Consider two keys different only in the selection of one S-box, say, the leftmost one. Assume further that one key selects S-box V_0 and the other key selects S-box V_1. It follows that for one round of encryption, the encryption function induced by the two keys will be equal for half the inputs. Consequently, the encryption functions over s rounds can be expected to produce identical ciphertexts for one in 2^s texts.

There are other related keys. Consider two keys different only in xor halves and only in the input to one S-box. For such two keys it may be possible to specify a keyed differential characteristic where the differences in the texts are canceled by the differences in the xor key in every second round. If in all other rounds it is assumed that there is only one active S-box and that the difference in the inputs equal the difference in the outputs, then one gets an s-round differential characteristic of probability 2^{-s} (for even s).

The observations in this section can potentially be used to devise related-key attacks which could recover a key for PRINTCIPHER using a little less than 2^b texts. It is clear, however, that if the keys of PRINTCIPHER are chosen uniformly at random it is very unlikely that one would find keys related as described above.

4.4   Statistical Saturation Attacks

Statistical saturation attacks have been presented in [3] and successfully applied to round-reduced versions of PRESENT. The key idea for statistical saturation attacks is to make use of low diffusion trails in the linear layer of PRESENT. As PRINTCIPHER uses a very similar linear layer, it seems natural that the attack applies to reduced round versions of PRINTCIPHER as well. We identified low diffusion trails for any number of S-boxes involved, see Table 1 for examples of the most promising ones using up to eight S-boxes in a trail. One example of such a low diffusion trail is given below.

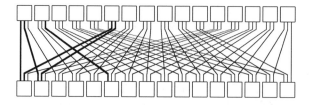

As explained in [3] increasing the number of S-boxes in the trail makes estimating the complexity of the attack very complicated. Thus, in our experiments we focused only on the case of three active S-boxes in the trail. All four possible

Table 1. Promising trails of different sizes

S-boxes in the trail	# of bits in the trail	Ratio
$\{0,1,5\}$	5	5/9
$\{0,1,5,15\}$	7	7/12
$\{4,10,12,14,15\}$	9	9/15
$\{0,1,2,5,6,7\}$	12	12/18
$\{3,8,9,10,11,13,15\}$	14	14/21
$\{0,1,4,5,10,12,14,15\}$	18	18/24

Table 2. Estimated squared distance ($\log_2$) for low diffusion trails with ratio 5/9

S-boxes in the trail	$\{0,1,5\}$	$\{2,6,7\}$	$\{10,14,15\}$	$\{8,9,13\}$
Round 1	0	0	0	0
Round 2	-3.72	-3.72	-3.74	-3.73
Round 3	-6.67	-6.74	-6.89	-6.65
Round 4	-9.14	-8.98	-9.19	-9.05
Round 6	-13.08	-13.17	-13.17	-13.10
Round 8	-16.92	-17.25	-16.96	-17.10
Round 10	-21.02	-20.88	-20.87	-21.03
Round 12	-25.38	-25.33	-24.82	-24.93
Round 14	-28.72	-28.94	-29.19	-28.94
Round 16	-32.83	-33.05	-33.27	-33.00

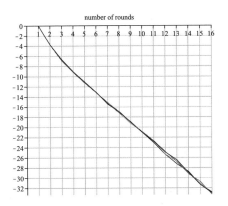

Fig. 2. The estimated squared bias with the number of rounds on the x-axis and the $\log_2$ of the squared bias given on the y-axis

trails gave very similar results. We estimated the bias for 50 randomly chosen keys for up to 10 rounds and for 20 randomly chosen keys for up to 15 rounds. Table 2 and Figure 2 show the squared euclidian distance between the distribution in the trail and the uniform distribution. The data complexity for attacking $r + 3$, *resp.* $r + 4$ rounds, depending on how many key bits are guessed, is approximately the reciprocal of the squared euclidian distance for r rounds. While our experiments are certainly limited, the results strongly suggest that no more than 30 rounds of PRINTCIPHER can be broken using this attack.

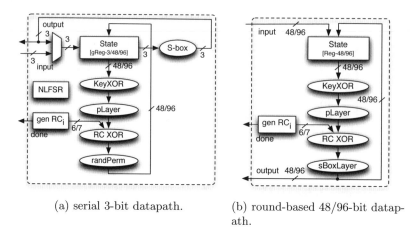

(a) serial 3-bit datapath.   (b) round-based 48/96-bit datapath.

Fig. 3. Two architectures for PRINTCIPHER

5   Implementation Results

To demonstrate the efficiency of our proposal we have implemented both PRINTCIPHER variants in VHDL and used *Synopsys DesignVision 2007.12* [27] to synthesize them using the *Virtual Silicon* (VST) standard cell library *UMCL18G212T3*, which is based on the *UMC L180 0.18μm 1P6M* logic process and has a typical voltage of 1.8 Volt [29].

Before presenting the results we stress the unique deployment environment offered by IC-printing. While our implementation efforts allow us to obtain a reasonable estimate of the space required, in terms of gate equivalents (GE), for an IC-printing implementation of PRINTCIPHER, any attempts to compare the likely power consumption with other implementations of lightweight cryptography are not just difficult (as is usually the case), but they are essentially meaningless. For this reason our performance results and comparisons will concentrate on the space occupied by an implementation.

Figure 3 depicts two architectures that were implemented: a serialized one with a datapath of 3-bits and a round-based one with a datapath of 48 or 96 bits. Components that contain mainly sequential logic are presented in rectangles while purely combinational components are presented in ovals.

The first serialized implementation of PRINTCIPHER-48 used a *finite state machine* (FSM) that required 120 GE out of which 95 GE were occupied by two arithmetic counters: 59 GE were occupied by the 6-bit round counter and additional 36 GE were required for a 4-bit counter to keep track of the 3-bit chunks of the serialized state. Similar to KATAN [2] we replaced the arithmetic round counter by a shift register-based counter, which saved 28 GE (or 47%) while having better distribution properties. The second counter was also replaced by a register-based counter which decreased the gate count by another 12 GE (35%). Finally we completely omitted the FSM and replaced it with some combinatorial gates to generate the control signals required (*e.g.* for the MUX). In total, by omitting the FSM and optimizing the control logic, we were able to save 54 GE (45%).

As part of our quest for a minimal S-box, we used the Boolean minimization tool BOOM II [7,8] to obtain the boolean functions of all 48 S-box variants that can be generated from a 3-bit S-box, by permuting the output bits and XORing a hardwired constant. Our synthesis results show that the results vary between 10.67 and 12 GE, and we chose a minimal S-box.

In order to be able to present a detailed break down for each component of PRINTCIPHER (see the accompanying table), we advised the compiler to *compile simple, i.e.* to keep the hierarchy of the components. The smallest area footprint is achieved, however, if the compiler uses the *compile ultra* command, which allows the merging and optimization of different components simultaneously. Since the key xor is hardwired, the area requirements for the KeyXOR component are dependent on the Hamming weight of the key. The implementation figures of Figure 4 used a key with Hamming weight 24, thus the area footprint of a serialized implementation of PRINTCIPHER-48 is bounded by 386 GE and 418 GE for keys with Hamming weight 0 and 48, respectively (694 GE and 758 GE for PRINTCIPHER-96). The results show that both PRINTCIPHER

PRINTcipher-n		$n = 48$		$n = 96$	
		serial	round	serial	round
cycles		768	48	3072	96
throughput @100 KHz (Kbps.)		6.250	100	3.125	100
compile ultra	sum	402	503	726	967
compile simple	sum	411	528	733	1011
sequential:	State	288	288	576	576
	genRC$_i$	31	31	36	36
	NLFSR	23	0	30	0
combinational:	MUX	7	0	7	0
	KeyXOR	16	16	32	32
	pLayer	0	0	0	0
	RC XOR	16	16	19	19
	sBoxLayer	11	171	11	342
	control	12	4	15	4
	other	7	2	7	2

Fig. 4. Implementation figures for PRINTCIPHER

variants scale nicely; by spending more area for additional S-boxes, the throughput can be scaled (nearly) linearly. At this point it is noteworthy to highlight the significant overhead (43 GE or 10.5%) that is required for additional control logic in a serialized PRINTCIPHER-48 implementation. This shows that it is hard to gain further area reductions. Furthermore, note that a 6-bit xor with the round constant RC$_i$ requires the same area as the 48-bit hardwired xor with a key with a typical Hamming weight of 24.

While observing our earlier caveats about the use of power estimates in the context of IC-printing, we did make some measurements for the likely power consumption of more conventional silicon-based implementations. We used *Synopsys PowerCompiler* version *A-2007.12-SP1* [28] to estimate the performance of our implementations. Measurements using the smallest wire-load model (10K GE) at a supply voltage of 1.8 Volt and a frequency of 100 KHz suggested a power consumption below 2.6 μW; a good indication that all PRINTCIPHER variants are well-suited to demanding applications including printed passive RFID tags. It is a well-known fact that at low frequencies, as typical for low-cost applications, the power consumption is dominated by its static part, which is proportional to the amount of transistors involved. Furthermore, the power consumption strongly depends on the used technology and greatly varies with the simulation method. Thus we refer to the area figures (in GE) as the most important measure and to have a fair comparison we do not include the power values in Table 3.

Table 3 compares a selection of lightweight block and stream cipher implementations that have been optimized for a minimal area footprint. It can be seen that the serialized implementation of PRINTCIPHER requires the least amount of area for its block and key sizes (402 GE). Moreover, spending additional 100 GE (or 25%) the throughput can be increased 16 fold to 100 Kpbs at a frequency

Table 3. Hardware implementation results of some symmetric encryption algorithms

Algorithm		key size	block size	cycles/ block	Throughput (@100 KHz)	Tech. [μm]	Area [GE]
Stream Ciphers							
Trivium	[9]	80	1	1	100	0.13	2,599
Grain	[9]	80	1	1	100	0.13	1,294
Block Ciphers							
PRESENT	[22]	80	64	547	11.7	0.18	1,075
SEA	[17]	96	96	93	103	0.13	3,758
mCrypton	[16]	96	64	13	492.3	0.13	2,681
HIGHT	[12]	128	64	34	188	0.25	3,048
AES	[6]	128	128	1,032	12.4	0.35	3,400
AES	[11]	128	128	160	80	0.13	3,100
DESXL	[15]	184	64	144	44.4	0.18	2,168
KATAN32	[2]	80	32	255	12.5	0.13	802
KATAN48	[2]	80	48	255	18.8	0.13	927
KATAN64	[2]	80	64	255	25.1	0.13	1054
KTANTAN32	[2]	80	32	255	12.5	0.13	462
KTANTAN48	[2]	80	48	255	18.8	0.13	588
KTANTAN64	[2]	80	64	255	25.1	0.13	688
PRINTCIPHER-48		80	48	768	6.25	0.18	402
PRINTCIPHER-48		80	48	48	100	0.18	503
PRINTCIPHER-96		160	96	3072	3.13	0.18	726
PRINTCIPHER-96		160	96	96	100	0.18	967

of 100 KHz, while still having a remarkably small area footprint. The resulting throughput per area ratio of 198.8 Kpbs per GE is even suited for high-speed applications though our main focus is on a low area footprint.

It is noteworthy to stress that we designed PRINTCIPHER to be secure even in the absence of a key schedule. This allows for significant area savings, because no flipflops to store the key state are required. Of course one could hardwire all the roundkeys for any cipher with a key schedule and, theoretically, this would allow for similar savings. In practice, however, this is not the case. Since all low-area implementations are serialized or round-based designs, one needs complex additional logic to select the right roundkey or even the right part of the roundkey. For a serialized AES for example, one would need a 128-bit wide 11-to-1 MUX to select the correct roundkey plus an 8-bit wide 16-to-1 MUX to select the right chunk of the roundkey. Our experiments reveal that a 128-bit wide 8-to-1 MUX already consumes 1276 GE, which makes it more efficient to store the key state in flipflops (768 GE) than to hardwire the roundkeys.

Though they have not been the focus of our design, for those interested in software implementations we estimate the performance of PRINTCIPHER on a 64-bit platform to be around 5-10 times slower than an optimized AES implementation: merging the permutation and using 6-bit S-boxes could give an implementation with 9-12 cycles per round. With 48 rounds this amounts to 72-95 cycles per byte while AES runs in 10-20 cycles per byte.

6   Conclusions

In this paper we have considered the technology of IC-printing and we have seen how it might influence the cryptography that we use. In particular we have proposed the lightweight block cipher PRINTCIPHER that explicitly takes advantage of this new manufacturing approach. Naturally it must be emphasized that PRINTCIPHER-48 is intended to be an object of research rather than being suitable for deployment. It is also intended to be a spur to others who might be interested in considering this new technology. Certainly we believe that the properties of IC-printing could be an interesting line of work and we feel that it helps to highlight several intriguing problems in cryptographic design, most notably how best to use a cipher key.

References

1. Bogdanov, A., Knudsen, L.R., Leander, G., Paar, C., Poschmann, A., Robshaw, M.J.B., Seurin, Y., Vikkelsoe, C.: PRESENT - An Ultra-Lightweight Block Cipher. In: Paillier, P., Verbauwhede, I. (eds.) CHES 2007. LNCS, vol. 4727, pp. 450–466. Springer, Heidelberg (2007)
2. de Cannière, C., Dunkelman, O., Knezević, M.: KATAN and KTANTAN–A Family of Small and Efficient Hardware-Oriented Block Ciphers. In: Clavier, C., Gaj, K. (eds.) CHES 2009. LNCS, vol. 5747, pp. 272–288. Springer, Heidelberg (2009)
3. Collard, B., Standaert, F.-X.: A Statistical Saturation Attack against the Block Cipher PRESENT. In: Fischlin, M. (ed.) CT-RSA 2009. LNCS, vol. 5473, pp. 195–211. Springer, Heidelberg (2009)

4. Daemen, J., Govaerts, R., Vandewalle, J.: A new approach to block cipher design. In: Anderson, R. (ed.) FSE 1993. LNCS, vol. 809, pp. 18–32. Springer, Heidelberg (1994)
5. EPCglobal. Organisation information, http://www.epcglobal.com
6. Feldhofer, M., Wolkerstorfer, J., Rijmen, V.: AES Implementation on a Grain of Sand. IEE Proceedings of Information Security 152(1), 13–20 (2005)
7. Fišer, P., Hlavička, J.: BOOM - A Heuristic Boolean Minimizer. Computers and Informatics 22(1), 19–51 (2003)
8. Fišer, P., Hlavička, J.: Two-Level Boolean Minimizer BOOM-II. In: Proceedings of 6th Int. Workshop on Boolean Problems – IWSBP'04, pp. 221–228 (2004)
9. Good, T., Benaissa, M.: Hardware Results for Selected Stream Cipher Candidates. In: State of the Art of Stream Ciphers (SASC 2007), Workshop Record (February 2007), www.ecrypt.eu.org/stream
10. GOST. Gosudarstvennyi standard 28147-89, cryptographic protection for data processing systems. Government Committee of the USSR for Standards (1989) (in Russian)
11. Hämäläinen, P., Alho, T., Hännikäinen, M., Hämäläinen, T.D.: Design and Implementation of Low-Area and Low-Power AES Encryption Hardware Core. In: DSD, pp. 577–583 (2006)
12. Hong, D., Sung, J., Hong, S., Lim, J., Lee, S., Koo, B.S., Lee, C., Chang, D., Lee, J., Jeong, K., Kim, H., Kim, J., Chee, S.: HIGHT: A New Block Cipher Suitable for Low-Resource Device. In: Goubin, L., Matsui, M. (eds.) CHES 2006. LNCS, vol. 4249, pp. 46–59. Springer, Heidelberg (2006)
13. Kovio. Company information available via, http://www.kovio.com
14. Lai, X., Massey, J., Murphy, S.: Markov ciphers and differential cryptanalysis. In: Davies, D. (ed.) EUROCRYPT 1991. LNCS, vol. 547, pp. 17–38. Springer, Heidelberg (1991)
15. Leander, G., Paar, C., Poschmann, A., Schramm, K.: New Lightweight DES Variants. In: Biryukov, A. (ed.) FSE 2007. LNCS, vol. 4593, pp. 196–210. Springer, Heidelberg (2007)
16. Lim, C., Korkishko, T.: mCrypton - A Lightweight Block Cipher for Security of Low-cost RFID Tags and Sensors. In: Song, J., Kwon, T., Yung, M. (eds.) WISA 2005. LNCS, vol. 3786, pp. 243–258. Springer, Heidelberg (2006)
17. Mace, F., Standaert, F.-X., Quisquater, J.-J.: ASIC Implementations of the Block Cipher SEA for Constrained Applications. In: RFID Security — RFIDsec 2007, Workshop Record, Malaga, Spain, pp. 103–114 (2007)
18. Matsui, M.: New Structure of Block Ciphers with Provable Security against Differential and Linear Cryptanalysis. In: Gollmann, D. (ed.) FSE 1996. LNCS, vol. 1039, pp. 205–218. Springer, Heidelberg (1996)
19. PolyIC. Information available via, http://www.polyIC.com
20. Poschmann, A., Robshaw, M.J.B., Vater, F., Paar, C.: Lightweight Cryptography and RFID: Tackling the Hidden Overheads. In: Lee, D., Hong, S. (eds.) Proceedings of ICISC '09. Springer, Heidelberg (to appear, 2009)
21. Robshaw, M.J.B.: Searching for Compact Algorithms: CGEN. In: Nguyên, P.Q. (ed.) VIETCRYPT 2006. LNCS, vol. 4341, pp. 37–49. Springer, Heidelberg (2006)
22. Rolfes, C., Poschmann, A., Leander, G., Paar, C.: Ultra-Lightweight Implementations for Smart Devices - Security for 1000 Gate Equivalents. In: Grimaud, G., Standaert, F.-X. (eds.) CARDIS 2008. LNCS, vol. 5189, pp. 89–103. Springer, Heidelberg (2008)

23. Samyde, D., Skorobogatov, S., Anderson, R., Quisquater, J.: On a New Way to Read Data from Memory. In: SISW '02: Proceedings of the First International IEEE Security in Storage Workshop, pp. 65–69. IEEE Computer Society, Los Alamitos (2002)

24. Schneier, B.: Description of a new variable-length key, 64-bit block cipher (Blowfish). In: Anderson, R. (ed.) FSE 1993. LNCS, vol. 809, pp. 191–204. Springer, Heidelberg (1994)

25. Schneier, B., Kelsey, J., Whiting, D., Wagner, D., Hall, Ferguson., N.: Twofish: A 128-bit block cipher. Submitted as candidate for AES, www.nist.gov/aes

26. Standaert, F.-X., Piret, G., Gershenfeld, N., Quisquater, J.-J.: SEA: A Scalable Encryption Algorithm for Small Embedded Applications. In: Domingo-Ferrer, J., Posegga, J., Schreckling, D. (eds.) CARDIS 2006. LNCS, vol. 3928, pp. 222–236. Springer, Heidelberg (2006)

27. Synopsys. Design Compiler User Guide - Version A-2007.12 (December 2007), http://tinyurl.com/pon88o

28. Synopsys. Power Compiler User Guide - Version A-2007.12 (March 2007), http://tinyurl.com/lfqhy5

29. Virtual Silicon Inc. 0.18 μm VIP Standard Cell Library Tape Out Ready, Part Number: UMCL18G212T3, Process: UMC Logic 0.18 μm Generic II Technology: 0.18μm (July 2004)

Appendix A: Testvectors

Table 4. Testvectors for PRINTCIPHER-96 in hexadecimal notation

	Testvector 1	Testvector 2
plaintext	5A97E895A9837A50CDC2D1E1	A83BB396B49DAA6286CD7834
key	953DDBBFA9BF648FF6940846	D83F1CEF1084E8131AA14510
permkey	70F22AF090356768	62C67A890D558DD0
ciphertext	45496A1283EF56AFBDDC8881	EE5A079934D98684DE165AC0
	Testvector 3	Testvector 4
plaintext	5CED2A5816F3C3AC351B0B4B	61D7274374499842690CA3CC
key	EC5ECFEF020442CF3EF50B8A	2F3F647A9EE6B4B5BAF0B173
permkey	68EA816CEBA0EFE5	A07CF36902B48D24
ciphertext	7F49205AF958DD440ED35D9E	3EB4830D385EA369C1C82129

Table 5. Sequence of RC$_i$ for PRINTCIPHER-96 in hexadecimal notation

i	1	2	3	4	5	6	7	8	9	10	11	12	13	14	15	16	17	18	19	20	21	22	23	24
RC$_i$	01	03	07	0F	1F	3F	7E	7D	7B	77	6F	5F	3E	7C	79	73	67	4F	1E	3D	7A	75	6B	57
i	25	26	27	28	29	30	31	32	33	34	35	36	37	38	39	40	41	42	43	44	45	46	47	48
RC$_i$	2E	5C	38	70	61	43	06	0D	1B	37	6E	5D	3A	74	69	53	26	4C	18	31	62	45	0A	15
i	49	50	51	52	53	54	55	56	57	58	59	60	61	62	63	64	65	66	67	68	69	70	71	72
RC$_i$	2B	56	2C	58	30	60	41	02	05	0B	17	2F	5E	3C	78	71	63	47	0E	1D	3B	76	6D	5B
i	73	74	75	76	77	78	79	80	81	82	83	84	85	86	87	88	89	90	91	92	93	94	95	96
RC$_i$	36	6C	59	32	64	49	12	25	4A	14	29	52	24	48	10	21	42	04	09	13	27	4E	1C	39

Table 6. Cipher example for PRINTCIPHER-48 in hexadecimal notation

plaintext	key	permkey	ciphertext
4C847555C35B	C28895BA327B	69D2CDB6	EB4AF95E7D37

Rd. RC	keyAddition	pLayer	RC XOR	S-box perm.	S-box
1 01	8E0CE0EFF120	ED9921498D92	ED9921498D93	ED92A24B0AE3	5B12FB6E89BE
2 03	999A6ED4BBC5	A9DE9DEC68E1	A9DE9DEC68E2	65BF1EEC6991	C765F5585F59
3 07	05ED60E26D22	0D8345DB891C	0D8345DB891B	0D88C67B886B	1B0F85D50E66
4 0F	D987106F3C1D	90FA448917F7	90FA448917F8	517A442917F8	7DA8472D9C90
5 1F	BF20D297AEEB	EAEB7C66A29B	EAEB7C66A284	E76AFCC6C484	46D997A676C7
6 3E	8451021C44BC	84015030F6C4	84015030F6FA	4800D09277B9	6C0198FFAD51
7 3D	AE890D459F2A	A21FEB888A8E	A21FEB888AB3	621FEB2A0CB3	C394A62F8AEE
8 3B	011C3395B895	21B2166079C3	21B2166079F8	21D815C079F8	234E1CA02E90
9 37	E1C6891A1CEB	D21646B4A739	D21646B4A70E	D21C45B6464D	BF9A4493FC4C
10 2F	7D12D129CE37	7E6B4E01D46B	7E6B4E01D444	BE6ACD033304	8BD98C02E3C7
11 1E	495119B8D1BC	343C0771F644	343C0771F65A	385607D37729	344C02BEADE1
12 3C	F6C497049F9A	F273FBB01388	F273FBB013B4	F279FBB014F4	5FAE969C17AF
13 39	9D26032625D4	80C9572711F0	80C9572711C9	4122D78591CA	61B39AE7108B
14 33	A33B0F5D22F0	8284BE2EFCA6	8284BE2EFC95	43053D8EFAA6	6287D4FA28FD
15 27	A00F41401A86	8A120A28096C	8A120A28094B	46180988094B	679E09EC0F0E
16 0E	A5169C563D75	C2B7C50CF1F1	C2B7C50CF1FF	C35DC60E71FF	A2CA851BA092
17 1D	604210A192E9	323008758223	32300875823E	325008D7043D	3FC008B28614
18 3A	FD489D08B46F	F2FDC6148E49	F2FDC6148E73	F377C5160F33	5EAC841385EE
19 35	9C2411A9B795	A0F94B631543	A0F94B631576	6172CBC19375	C1A38EA3132C
20 2B	032B1B192157	00A437063C6F	00A437063C44	01443704DB04	01F62A04C9C7
21 16	C37EBFBEFBBC	F5B6BF73FFF0	F5B6BF73FFE6	F9DD3FD3FFD5	574BD2BD249C
22 2C	95C3470716E7	8851DEB480FF	8851DEB480D3	4431DDB601A3	6460B493817E
23 18	A6E82129B305	A3912B930C43	A3912B930C5B	6390AB318B2B	C110E63F09E6
24 30	039873853B9D	09B23FE05AC3	09B23FE05AF3	05D83FE03DB3	074E12406B6E
25 21	C5C687FA5915	D41397D93571	D41397D93550	D81997795360	B41F5AD5C338
26 02	7697CF6FF143	7ED5B38D45BF	7ED5B38D45BD	BF35B32D22FE	8AE76E39B395
27 05	486FFB8381EE	792C13768B7E	792C13768B7B	B4C613D68D7B	90FC1EB20B16
28 0B	52748B08396D	50D63316C741	50D63316C74A	913C3316A749	FDEA2E123D09
29 17	3F62BBA80F72	436F7F579428	436F7F57943F	82EEFF57923F	E25592711212
30 2E	20DD07CB2069	028092DCCF17	028092DCCF39	0301117E2E7A	0281D9CBB453
31 1C	C0094C718628	B804C809AA06	B804C809AA1A	7405480B4C29	D007080EFA21
32 38	128F9DB4C85A	6466A2C53BAC	6466A2C53B94	A86D21655CE4	8C5BF9C5CBBF
33 31	4ED36C7FF9C4	3D9FA1BD64F6	3D9FA1BD64C7	3D9FA2BD6387	2B157B89D342
34 23	E99DEE33E139	FF8C9581FB17	FF8C9581FB34	FF8716237C74	490DDD22AA6F
35 06	8B8548989814	A81E24C03544	A81E24C03542	641E24605341	C4143FC04301
36 0D	069CAA7A717A	4595318DFF18	4595318DFF15	8994B22F7E66	EF16EB3AA47D
37 1B	2D9E7E809606	2B3DDCC04968	2B3DDCC04973	26D7DC602973	264CB7C03F2E
38 36	E4C4227A0D55	9303519D3551	9303519D3567	5288D23D5357	7E0F9B29C31A
39 2D	BC870E93F161	A6DD91C4A137	A6DD91C4A11A	6B3792664069	CEED5BC7F061
40 1A	0C65CE7DC21A	6C0D981B378E	6C0D981B3794	AC079819D6E4	980D70174DBF
41 34	5A85E5AD7FC4	5DDAEBE505C6	5DDAEBE505F2	9DBB6BE503F1	EB69264582A9
42 29	29E1B3FFB0D2	63B816FF349E	63B816FF34B7	A3D215FDD2B7	81421C4B42EA
43 12	43CA89F17091	549426F93823	549426F93831	991425F95832	F5963C55CE2B
44 24	371EA9EFFC50	67D7664D5DB2	67D7664D5D96	ABBCE54D3AE5	8D6BBC79E9BC
45 08	4FE329C3DBC7	351F2FFE007F	351F2FFE0077	389EAFFC8137	3494E24881EA
46 11	F61C77F2B391	BBF1BB697911	BBF1BB697900	77F1BBC97840	D12156ADAE40
47 22	13A9C3179C3B	68527682BA9F	68527682BABD	A4387522DCBE	846E6C224AD5
48 04	46E6F99878AE	5DB722F2A768	5DB722F2A76C	9DDCA1F2C75C	EB4AF95E7D37

Sponge-Based Pseudo-Random Number Generators

Guido Bertoni[1], Joan Daemen[1], Michaël Peeters[2], and Gilles Van Assche[1]

[1] STMicroelectronics
[2] NXP Semiconductors

Abstract. This paper proposes a new construction for the generation of pseudo-random numbers. The construction is based on sponge functions and is suitable for embedded security devices as it requires few resources. We propose a model for such generators and explain how to define one on top of a sponge function. The construction is a novel way to use a sponge function, and inputs and outputs blocks in a continuous fashion, allowing to interleave the feed of seeding material with the fetch of pseudo-random numbers without latency. We describe the consequences of the sponge indifferentiability results to this construction and study the resistance of the construction against generic state recovery attacks. Finally, we propose a concrete example based on a member of the KECCAK family with small width.

Keywords: pseudo-random numbers, hash function, stream cipher, sponge function, indifferentiability, embedded security device, Keccak.

1 Introduction

In various cryptographic applications and protocols, random numbers are used to generate keys or unpredictable challenges. While randomness can be extracted from a physical source, it is often necessary to provide many more bits than the entropy of the physical source. A pseudo-random number generator (PRNG) provides a way to do so. It is initialized with a seed, generated in a secret or truly random way, and it then expands the seed into a sequence of bits.

For cryptographic purposes, it is required that the generated bits cannot be predicted, even if subsets of the sequence are revealed. In this context, a PRNG is pretty similar to a stream cipher. If the key is unknown, it must be infeasible to infer anything on the key stream, even if it is partially known.

The state of the PRNG must have sufficient entropy, from the point of view of the adversary, so that the prediction of the output bits cannot rely on simply guessing the state. Hence, the seeding material must provide sufficient entropy. Physical sources of randomness usually provide seeding material with relatively low entropy rate due to imbalance of or correlations between bits. To increase entropy, one may use the seeding material from several randomness sources. However, this entropy must be transferred to the finite state of the PRNG. Hence, we need a way to gather and combine seeding material coming from

S. Mangard and F.-X. Standaert (Eds.): CHES 2010, LNCS 6225, pp. 33–47, 2010.

several sources into the state of the PRNG. Loading different seeds into the PRNG shall result in different output sequences. The latter implies that different seeds result in different state values. In this respect, a PRNG is similar to a cryptographic hash function that should be collision-resistant.

It is convenient for a pseudo-random number generator to be reseedable, i.e., one can bring an additional source of entropy after pseudo-random bits have been generated. Instead of throwing away the current state of the PRNG, reseeding combines the current state of the generator with the new seeding material. From a user's point of view, a reseedable PRNG can be seen as a black box with an interface to request pseudo-random bits and an interface to provide fresh seeds.

The remainder of this paper is organized as follows. We continue our introduction with the advantages and limitations of our construction and an illustrative example of a pseudo-random number generator mode of a hash function. We then define the reference model of a reseedable PRNG in Section 2 and specify and motivate our sponge-based construction in Section 3. We discuss the security aspects of our proposal in Section 4 and provide a concrete example in Section 5.

1.1   Advantages and Limitations of Our Construction

With their variable-length input and variable-length output, sponge functions combine in a unified way the functionality of hash functions and stream ciphers. They make therefore a natural candidate for building PRNGs, taking the seeding material as input and producing a sequence of pseudo-random bits as output.

In this paper, we provide a clean and efficient way to construct a reseedable PRNG with a sponge function. The main idea is to integrate in the same construction the combination of the various sources of seeding material and the generation of pseudo-random output bits. The only requirement for seeding material is to be available as bit sequences, which can be presented as such without any additional preprocessing. So both seeding and random generation can work in a continuous fashion, making the implementation simple and avoiding extra iterations when providing additional seeding material.

In the context of an embedded security device, the efficiency and the simplicity of the implementation is important. In our construction we can keep the state size small thanks to two reasons. First, the use of a permutation preserves the entropy of the state (see Section 1.2). Second, we have strong bounds on the expected complexity of generic state recovery attacks (see Section 4.2).

Making sure that the seeding material provides enough entropy is out of scope of this paper. This aspect has been studied in the literature, e.g., [10,16] and is fairly orthogonal to the problem of combining various sources and generating pseudo-random bits.

In our construction, forward security must be explicitly activated. Forward security (also called forward secrecy) requires that the compromise of the current state does not enable the attacker to determine the previously generated pseudo-random bits [2,9]. As our construction is based on a permutation, revealing the state immediately allows the attacker to backtrack the generation up to the previous combination of that state and seeding material. Nevertheless, reseeding regularly with sufficient entropy already prevents the attacker from going backwards. Also, an embedded security device such as a smartcard in which such a PRNG would be used is designed to protect the secrecy of keys and therefore reading out the state is expected to be difficult. Yet, we propose in Section 4.3 a simple solution to get forward secrecy at a small extra cost. Hence, if forward security is required, one can apply this mechanism at regular intervals.

1.2   Using a Hash Function for Pseudo-Random Number Generation

Sponge functions are a generalization of hash functions and using the latter for generating pseudo-random bits is not new, e.g., [12,14]. For instance, NIST published a recommendation for random number generation using deterministic random bits generators [14]. They specify how to implement a PRNG using a hash function, a keyed hash function, a block cipher or an elliptic curve. When using a hash function H, the state of the PRNG is essentially determined by two values, V and C, each of the size of the input block of H.

- At initialization, both V and C are obtained by hashing the seeding material, a nonce and an optional personalization string. If V and C are larger than the output size of H, a specific derivation function is used to produce a longer digest.
- The pseudo-random bits are produced by hashing V. If more than one output block is requested, further blocks are produced by hashing $V + i$, where i is the index of the produced output block. The value of V is then updated by combining it with, amongst others, $H(V)$ and C. The value C is not modified in this process.
- When reseeding, the new value of V is obtained by hashing the old value of V together with the new seeding material. The value C is derived from the new value of V by hashing.

For a PRNG based on a hash function, there are two aspects we wish to draw attention to.

First, due to the requirements they must satisfy, cryptographic hash function are not injective. Iterating the function, i.e., computing $H(H(\ldots H(x))\ldots)$ reduces the size of the range resulting in entropy loss. To prevent this, one can for instance keep the original seed along with the evolving state. In the hash function based PRNG specified in [14], the value V evolves by iterated hashing every time output bits are produced, but the value C does not and therefore keeps the full entropy of the seed. This comes at the cost of keeping a state twice the block size of the hash function.

Second, when reseeding, the current state or the original seed must be hashed together with the seeding material. However, the current state V and the seed C are already the result of a hashing process.

The sponge-based construction we propose below addresses these two aspects more efficiently. First, by using a P-sponge, i.e., a sponge function based on a permutation, no entropy is lost when iterating the permutation and this allows one to have a smaller state for the same security level. Second, the current state of our construction is precisely the state of the sponge function. Hence, reseeding is more efficient than in the example above, as the current state can be reused immediately instead of being hashed again.

Finally, the use of a sponge function for PRNG is conceptually simpler than existing constructions.

2   Modeling a Reseedable Pseudo-Random Number Generator

We define a reseedable PRNG as a stateful entity that supports two types of requests, in any order:

- *feed* request, feed(σ), injects a seed consisting of a non-empty string $\sigma \in \mathbb{Z}_2^+$ into the state of the PRNG;
- *fetch* request, fetch(l), instructs the PRNG to return l bits.

The *seeding material* is the concatenation of the σ's received in all feed requests.

Informally, the requirements for a reseedable PRNG can be stated as follows. First, its output (i.e., responses to fetch requests) must depend on all seeding material fed (i.e., payload of feed requests). Second, for an adversary not knowing the seeding material and that has observed part of the output, it must be infeasible to infer anything on the remaining part of the output.

To have more formal security requirements, one often defines a reference system that behaves ideally. For sponge functions, hash functions and stream ciphers the appropriate reference system is the random oracle [1]. For reseedable PRNG we cannot just use a random oracle as it has a different interface. However, we define an ideal PRNG as a particular *mode of use* of a random oracle.

The mode we define is the following. It keeps as state the sequence of all feed and fetch requests received, the *history* h. Upon receipt of a feed request feed(σ), it updates the history by incorporating it. Upon receipt of a fetch request fetch(l), it queries the random oracle with a string that encodes the history and returns the bits z to $z + l - 1$ of its response to the requester, with z the number of bits requested in the fetch requests since the last feed request. Hence, concatenating the responses of a run of fetch requests is just the response of the random oracle to a single query. This is illustrated in Figure 1. We call this mode the *history-keeping* mode with encoding function $e(h)$. The definition of a history-keeping mode hence reduces to the definition of this encoding function.

As the output of the PRNG must depend on the whole seeding material received, the encoding function $e(h)$ must be injective in the seeding material. In other words, for any two sequences of requests with different seeding materials, the two images through $e(h)$ must be different. We call this property *seed-completeness*. With a seed-complete encoding function, the response of the mode to a fetch request corresponds with non-overlapping parts of the response of the random oracle to different input strings. It follows that the PRNG returns independent and a priori uniformly distributed bits.

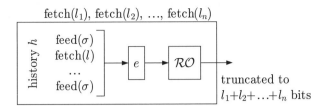

Fig. 1. Response of an ideal reseedable PRNG to fetch requests

We thus propose the following definition of an ideal PRNG. In the sequel, we will use PRNG to indicate a reseedable pseudo-random number generator.

Definition 1. *An* ideal PRNG *is a history-keeping mode calling a random oracle with an encoding function $e(h)$ that is seed-complete.*

3   Constructing a PRNG Using a Sponge Function

In general, the history-keeping mode is not practical as it needs to store all past queries and hence requires ever growing amounts of memory. In this section we will show that if we use a sponge function instead of a random oracle we can define an encoding function that can work with a limited amount of memory.

3.1   The Sponge Construction

The sponge construction [3] is a simple iterated construction for building a function $\mathcal{S}[f]$ with variable-length input and arbitrary output length based on a fixed-length transformation (or permutation) f operating on a fixed number b of bits. Here b is called the width. A sponge function, i.e., a function implementing the sponge construction provides a particular way to generalize hash functions and has the same interface as a random oracle.

For given values of r and c, the sponge construction operates on a state of $b = r + c$ bits. The value r is called the *bitrate* and the value c the *capacity*. First, all the bits of the state are initialized to zero. The input message is padded and cut into blocks of r bits. The sponge construction then proceeds in two phases: the *absorbing phase* followed by the *squeezing phase*.

- In the absorbing phase, the r-bit input message blocks are XORed into the first r bits of the state, interleaved with applications of the function f. When all message blocks are processed, the sponge construction switches to the squeezing phase.
- In the squeezing phase, the first r bits of the state are returned as output blocks, interleaved with applications of the function f. The number of output blocks is chosen at will by the user.

The last c bits of the state are never directly affected by the input blocks and are never output during the squeezing phase. The capacity c actually determines the attainable security level of the construction [4].

3.2   Reusing the State for Multiple Feed and Fetch Phases

It seems natural to translate the feed of seeding material into the absorbing
phase and the fetch of pseudo-random numbers into the squeezing phase of a
sponge function, as illustrated in Figure 2. However, as such, a sponge function
execution has only one absorbing phase (i.e., one input), followed by a single
squeezing phase (i.e., one output, of arbitrary length), and thus cannot be used
to provide multiple "absorbing" phases and multiple "squeezing" phases.

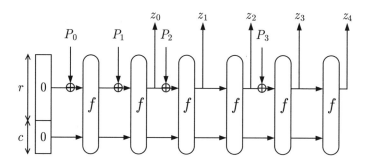

Fig. 2. The sponge construction with multiple feed and fetch phases

This apparent difficulty is easy to circumvent. Conceptually, it suffices to
consider that each time pseudo-random bits are fetched, a different execution of
the sponge function is queried with a different input, as illustrated in Figure 3.
When entering the squeezing phase of each of these queries (so before pseudo-
random bits are requested), one must thus guarantee that the data absorbed so
far compose a valid sponge input, i.e., the input is properly padded [3]. This can
be achieved by defining an encoding function adapted to the particular sponge.

In the sponge construction, an input message $m \in \mathbb{Z}_2^*$ must be cut into blocks
of r bits and padded. Let us denote as $p(m)$ the function that does this, and we
assume that this function only appends bits after m (as in the padding of most,
if not all, practical hash functions). Let us assume that we wish to reuse the
state of the sponge whose input was the string m_1 and from which $l > 0$ output
bits have been squeezed. The state of the sponge function at this point is as if
the partial message $m_1' = p(m_1)||0^{r(\lceil l/r \rceil - 1)}$ was absorbed. Note that the zero
blocks account for the extra iterations due to the squeezing phase. Restarting
the sponge from this point means that the input is going to be a message m_2 of
which m_1' is a prefix.

3.3   Constructing a Reseedable Pseudo-Random Number Generator

To define a PRNG formally, we need to specify a seed-complete encoding function
$e(h)$ that maps the sequence h of feed and fetch requests onto a string of bits,
as in Section 2. The output of $e(h)$ is then used as input to the sponge function.
In practice, the idea is not to call the sponge function with the whole $e(h)$ every
time a fetch is requested. Instead, the construction uses the sponge function in

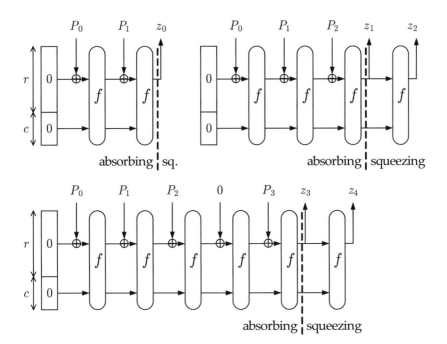

Fig. 3. The multiple feed and fetch phases of Figure 2 can be viewed as a sponge function queried multiple times, each having only one absorbing and one squeezing phase. In this example, $P_0||P_1$, $P_0||P_1||P_2$ and $P_0||P_1||P_2||0^r||P_3$ must all be valid sponge inputs.

a cascaded way, reusing the state as explained in Section 3.2. To allow the state of the sponge function to be reused as described above, $e(h)$ must be such that if $h' = h||\texttt{fetch}(l)||\texttt{feed}(\sigma)$, then $p(e(h))||0^{r(\lceil l/r \rceil - 1)}$ is a prefix of $e(h')$.

We now explain how to link a mode to a practical implementation. To make the description easier, we describe a mode with two restrictions. We later discuss how to implement a more elaborate mode without these restrictions. The first restriction is on the length of the seed requests. For a fixed integer k, we require that the length of the seeding material σ in any feed request $\texttt{feed}(\sigma)$ is such that $|p(\sigma)| = kr$. In other words, after padding, the seeding material covers exactly k blocks of r bits. The second restriction is that the first request must be $\texttt{feed}$.

The mode is stateful, and its state is composed of $m \in \mathbb{N}$, the number of bits fetched since the last feed. We start with a new execution of a sponge function, and we set $m = 0$. Depending on the type of requests, the following operations are done on the sponge function on the one hand and on the encoding function $e(h)$ on the other. We denote by e a string that reflects $e(h)$ as the requests are appended to the history h.

- If the request is $\texttt{fetch}(l)$, the following is done.
  - The implementation produces l output bits by squeezing them from the sponge function. Formally, e will be adapted during the next feed request.
  - The value of m is adapted: $m \leftarrow m + l$.

– If the request is feed(σ), the following is done.
  • Formally, this feed request triggers a query to the sponge function with e as input. If it is not the first request, e is up-to-date only up to the last feed request. So, the effect of the fetch requests since the last feed request must be incorporated into e, as if e was previously absorbed. First, e becomes $p(e)$ to simulate the padding when switching to the squeezing phase after the previous feed request. Then $\lceil m/r \rceil - 1$ blocks of r zeroes are appended to e to account for the extra calls to the f function during the subsequent fetch requests. Now m is reset: $m \leftarrow 0$. (This part affects only e formally; nothing needs to be done in the implementation.)
  • Then, the implementation absorbs σ. Formally, this is reflected by appending σ to e.
  • Finally, the implementation switches the sponge function to the squeezing phase. This means that the absorbed data must be padded and the permutation f is applied to the state. (Formally, this does not change e, as the padding is by definition performed when switching to the squeezing phase.)

To show that the encoding function is seed-complete, let us demonstrate how to find the seeding material from it. If $e(h)$ is empty, no feed request has been done and the seeding material is the empty string. If $e(h)$ is not empty, it necessarily ends with the fixed amount of seeding material from the last feed request, which we extract. Before that, there can be one or more blocks of r bits equal to zero. This can only come from blocks that simulate fetch requests, as the padding function p would necessarily create a non-zero block. So, we can skip backwards consecutive blocks of zeroes, until the beginning of $e(h)$ is reached or a non-zero block is encountered. In this last case, we can extract the seeding material from the k blocks of r bits and move backwards by the same amount. Finally, we repeat this process until the beginning of $e(h)$ is reached.

The construction, described directly on top of the permutation f, is given in Algorithm 1. For completeness, we also give in Algorithm 2 an implementation of the squeezing phase of a sponge function, although it follows in a straight-forward way from the definition [3]. The cost of a feed request is always k calls to the permutation f. Consecutive fetch requests totalling m bits of output cost $\lceil m/r \rceil - 1$ calls to f. So a fetch(l) with $l \leq r$ just after a feed request is free.

The restriction of fixed-size feed requests is not essential and can be removed. The description of the mode would be only a bit more complex, but would distract the reader from the aspects of this construction that tightly integrate to a sponge function and its underlying function f. In fact, the restriction of fixed-size feed requests makes it easy to ensure and to show that the encoding function is seed-complete. To allow for variable length seeding materials and retain seed-completeness, some form of padding within the encoding function must be introduced to make sure that the boundaries of the seeding material can be identified. Furthermore, one may have to add a way to distinguish blocks of zero-valued seeding material from zero blocks due to fetch requests. This can be done, e.g., by putting a bit 1 in every block that contains seeding material.

Algorithm 1. Direct implementation of the PRNG using the permutation f

---

$s = 0^{r+c}$
$m = 0$
while requests are received **do**
    if the request is `fetch`(σ) with $|p(\sigma)| = kr$ **then**
        $P_1||\ldots||P_k = p(\sigma)$
        for $i = 1$ to k **do**
            $s = s \oplus (P_i||0^c)$
            $s = f(s)$
        end for
        $m = 0$
    end if
    if the request is `fetch`(l) **then**
        Squeeze l bits from the sponge function (see Algorithm 2)
    end if
end while

Algorithm 2. Implementation of the squeezing of l bits from the sponge function

---

Let a be the number of available bits, i.e., $a = r$ if $m = 0$ or $a = (-m \bmod r)$
otherwise
while $l > 0$ **do**
    if $a = 0$ (we need to squeeze the sponge further) **then**
        $s = f(s)$
        $a = r$
    end if
    Output $l' = \min(a,l)$ bits by taking bits $r - a$ to $r - a + l' - 1$ of the state
    Subtract l' from a and from l, and add l' to m
end while

The restriction of the first request being a feed request can be removed, even though it makes little sense generating pseudo-random bits without first feeding seeding material. If the first request is a fetch, the implementation immediately pads the (empty string) input, switches the sponge function to the squeezing phase and produces output bits by squeezing. Formally, in the next feed request, this must be accounted for in e by setting e to $p(\text{empty string})||0^{r(\lceil m/r \rceil - 1)}$.

4   Security

Hash functions are often designed in two steps. In the first step, one chooses a mode of operation that relies on a cryptographic primitive with fixed input size (e.g., a compression function or a permutation) and builds a function that can process a message of arbitrary size. If the security of the mode of operation can be proven, it then guarantees that any potential flaw can only come from the underlying cryptographic primitive, and thereby reduces the scope of cryptanalysis.

We proceed similarly to assess the security of the PRNG, in two steps. First, we look at the security of the construction against generic attacks, i.e., against attacks that do not use the specific properties of f. We do this in the following subsections. Then, the security of the PRNG depends on the actual function f and we give an example in Section 5.

4.1   Indifferentiability

Indifferentiability is a concept developed by Maurer, Renner and Holenstein and allows one to compare the security of a system to that of an ideal object, such as the random oracle [11]. The system can use an underlying cryptographic primitive (e.g., a compression function or a permutation) as a public subsystem. For instance, many hash function constructions have been proven to be indifferentiable from a random oracle when using an ideal compression function or a random permutation as public subsystem (e.g., [8]).

By using indifferentiability, one can build a construction that does not have any generic flaw, i.e., any undesired property or attack that does not rely on the specific properties of the underlying primitive.

Theorem 1. *The pseudo-random number generator $\mathcal{P}[\mathcal{F}]$ that uses a permutation $\mathcal{F}$ is (t_D, t_S, N, ϵ)-indifferentiable from an ideal PRNG, for any t_D, $t_S = O(N^2)$, $N < 2^c$ and any ϵ with $\epsilon > N^2/2^{c+1}$ when $1 \ll N$.*

Proof. The proof follows immediately from [4, Theorem 2], where the (t_D, t_S, N, ϵ)-indifferentiability is proven between the sponge construction and a random oracle. In [4, Theorem 2], the adversary has access to two interfaces: one to the permutation $\mathcal{F}$ or its simulator, and one to input a message $m \in \mathbf{Z}_2^*$. In the context of this theorem, the same settings apply, except that the adversary does not have a direct access to the latter interface but only through the encoding function $e(h)$. The same restriction applies both on the side of the sponge construction and on the side of the random oracle. Since the adversary has no better access than in [4, Theorem 2], her probability of success cannot be higher.   □

Distinguishing the sponge-based PRNG calling a random permutation from an ideal PRNG defined in Section 2 takes about $2^{c/2}$ operations. In other words, the former is as secure as the latter if c is large enough.

4.2   Resistance against State Recovery

Indifferentiability provides a proof of resistance against all possible generic attacks on the construction. However, in practice, we can also look at the resistance of the construction against generic attacks with a specific goal. In this case, the resistance cannot be lower than $2^{c/2}$ but may be higher.

The main purpose of the PRNG is to avoid that an adversary, who has seen some of the generated bits, can predict other values. A way to predict other output bits is to recover the state of the PRNG by observing the generated pseudo-random bits. In fact, since we use a permutation, the adversary can equivalently recover the state at any time during a fetch request. She can also determine the state before or after a feed request if she can guess the seeding material input during that request.

Let the state of a sponge function be denoted as (a, x), where a is the outer part (i.e., the r-bit part output during the squeezing phase) and x represents inner part (i.e., the remaining c bits). Let $(a_0, a_1, \ldots, a_\ell)$ be a sequence of known output blocks. The goal of the adversary is to find a value x_0 such that $f(a_{i-1}, x_{i-1}) = (a_i, x_i)$ for $1 \leq i \leq \ell$ and some values x_i. Notice that once x_0 is fixed, the values x_i, $1 \leq i \leq \ell$ follow immediately. Furthermore, since f is a permutation, the adversary can choose to first determine x_i for some index i and then compute all the other $x_{j \neq i}$ from $(a_i, x_i$ by applying f and f^{-1}.

An instance of the *passive state recovery problem* is given by a vector $(a_0, a_1, \ldots, a_\ell)$ of r-bit values. We focus on the case where such a sequence of values was actually observed, so that we are sure there is at least one solution. Also, we assume that there is only one solution, i.e., one value x_0. This is likely if $\ell r > c$, and the probability that more than one solution exists decreases exponentially with $\ell r - c$. The adversary wants to determine unseen output blocks, so she wants to have only one solution anyway and will ask for more output blocks to remove any ambiguity.

The adversary can query the permutation f with values (a, x) and get $f(a, x)$ or its inverse to get $f^{-1}(a, x)$. If f is a random permutation, we wish to compute an upper bound on the success probability after N queries.

Theorem 2. *Given an instance of the passive state recovery problem $A = (a_0, a_1, \ldots, a_\ell)$ and knowing that there is one and only one solution x_0, the success probability after N queries is at most $N2^{-c}m(A)$, with $m(A)$ the multiplicity defined as*

$$m(A) = \max\{m_f(A), m_b(A)\}, \text{ with}$$
$$m_f(A) = \max_{a \in \mathbf{Z}_2^r} |\{i \ : \ 0 \leq i < \ell \wedge a_i = a\}|, \text{ and}$$
$$m_b(A) = \max_{a \in \mathbf{Z}_2^r} |\{i \ : \ 1 \leq i \leq \ell \wedge a_i = a\}|.$$

Proof. Let $F_1(A)$ be the set of permutations f such that there is only one solution to the state recovery problem with instance A. For a given value (a, x), within $F_1(A)$, the inner part of $f(a, x)$ (or $f^{-1}(a, x)$) can be symmetrically chosen among the 2^c possible values as the problem instance does not express any constraints on the inner parts. In other words, if x is such that the outer part of $f(a, x)$ is b, then for any $x' \neq x$ there exists another permutation $f' \in F_1(A)$ where x' is such that the outer part of $f'(a, x')$ is b too. Such symmetries exist also for multiple inner values, independently of each other, as long as the corresponding outer values are different. E.g., if $a_1 \neq a_2$ and (x_1, x_2) is such that the outer parts of $f(a_i, x_i)$ are b_i for $i = 1, 2$, then for any $(x_1', x_2') \neq (x_1, x_2)$ there exists another permutation $f' \in F_1(A)$ where (x_1', x_2') verifies the same equality.

Let us first consider that $\ell = 1$. In this case, $m(A) = 1$.

Let $F_1(A, x_0, x_1)$ be the subset of $F_1(A)$ where the value x_0 is the solution and $f(a_0, x_0) = (a_1, x_1)$. The sets $F_1(A, x_0, x_1)$ partition the set $F_1(A)$ into 2^{2c}

subsets of equal size identified by x_0 and x_1, or in other words, x_0 and x_1 cut the set in an orthogonal way.

The goal of the adversary is to determine in which subset $F_1(A, x_0, x_1)$ the permutation f is. To do so, she is going to make queries of the form (a_0, x_0) and check if the outer part of $f(a_0, x_0)$ is a_1 (called *forward queries*), or she can make queries to the inverse permutation and check if $f^{-1}(a_1, x_1)$ gives a_0 as outer part (called *backward queries*). As the subsets $F_1(A, x_0, x_1)$ cut $F_1(A)$ orthogonally in x_0 and x_1, forward queries help determine whether x_0 is the solution but without reducing the set of possible values for x_1, and vice-versa for backward queries. So, after N_f forward queries and N_b backward queries, the probability that one of them gives the solution is $1 - (1 - N_f/2^c)(1 - N_b/2^c) \leq N/2^c$, where the probability is taken over all permutations f drawn uniformly from $F_1(A)$.

Let us now consider the general case where $\ell > 1$. The reasoning can be generalized in a straightforward way if all the a_i are different, but some adaptations have to be made to take into account the values appearing multiple times. Given a set of indexes $\{i_1, \ldots, i_m\}$ such that $a_{i_1} = a_{i_2} = \cdots = a_{i_m}$, there may or may not be constraints on the possible values that the corresponding inner values $x_{i_1}, x_{i_2}, \ldots, x_{i_m}$ can take. For instance, if $a_{i_1-1} \neq a_{i_2-1}$ or if $a_{i_1+1} \neq a_{i_2+1}$, then necessarily $x_{i_1} \neq x_{i_2}$. In another example, A can be periodic, allowing the x_i values to be equal.

Let $i(j, k)$ be a partition of the indexes 0 to ℓ such that $a_{i(j,k)} = a_{i(j',k')}$ iff $j = j'$, i.e., the j index identifies the subsets and the k index the indices within that subset. Let $F_1(A, x_0, x_1, \ldots, x_\ell)$ be the subset of $F_1(A)$ such that $(x_0, x_1, \ldots, x_\ell)$ is the solution. Here, the set $F_1(A)$ is again cut into subsets of equal size if we use the n vectors $(x_{i(j,1)}, \ldots, x_{i(j,m_j)})$ as identifiers, and each of these vectors cut $F_1(A)$ in an orthogonal way. (In general, however, the values x corresponding to identical values a do not cut $F_1(A)$ in an orthogonal way.)

The adversary can make a forward query to check whether $f(a_{i(j,k)}, x_{i(j,k)})$ gives $a_{i(j,k)+1}$ as outer value. Using the same query, she can also check whether $f(a_{i(j,k')}, x_{i(j,k)})$ yields $a_{i(j,k')+1}$ for any other k' (as long as $i(j, k') < \ell$). The same reasoning goes for backward queries: does $f^{-1}(a_{i(j,k')}, x_{i(j,k)})$ yield $a_{i(j,k')-1}$ for any k' (as long as $i(j, k') > 0$). So, a forward (resp. backward) query can count as up to $m_f(A)$ (resp. $m_b(A)$) chances to hit the correct outer value. After N queries, the probability that one of them gives the solution is at most $m(A)N/2^c$, where the probability is taken over all permutations f drawn uniformly from $F_1(A)$.   □

The previous theorem also imposes an upper bound on the success probability of preimage attacks, generically against a sponge function. This follows from the fact that finding a preimage implies that the state can be recovered.

This theorem covers the case of a passive adversary who observes output blocks. Now, the PRNG implementation could allow seeding material to be provided from outside, hence allowing an active adversary to absorb blocks of his choice. This case is covered in the next theorem. We assume that the adversary controls the blocks b_i that are injected at each iteration, i.e., the PRNG computes

$f(a_i \oplus b_i, x_i) = (a_{i+1}, x_{i+1})$ and the adversary observes a_{i+1}. Now an instance of the problem is also determined by the injected blocks $B = (b_0, b_1, \ldots, b_\ell)$.

Theorem 3. *Given an instance of the active state recovery problem $A = (a_0, a_1, \ldots, a_\ell)$, $B = (b_0, b_1, \ldots, b_\ell)$ and knowing that there is one and only one solution x_0, the success probability after N queries is at most $N2^{-c}\ell$.*

Proof. The reasoning is the same as in Theorem 2, except that the queries are slightly different. In a forward query, the adversary checks if the outer part of $f(a_i \oplus b_i, x_i)$ is a_{i+1}. In a backward query, she checks if the outer part of $f^{-1}(a_i, x_i)$ is $a_{i-1} \oplus b_{i-1}$. Another difference is that now the forward multiplicity to be considered is

$$m_{\mathrm{f}}(A, B) = \max_{a \oplus b \in \mathbf{Z}_2^r} |\{i \ : \ 0 \leq i < \ell \wedge a_i \oplus b_i = a \oplus b\}|,$$

as one forward query can be used to check inner values at up to $m_{\mathrm{f}}(A, B)$ indexes at once. Furthermore, the adversary can influence the multiplicity, e.g., by making sure $a_i \oplus b_i$ is always the same value. So $m(A) \leq \ell$ and the success probability after N queries is at most $N2^{-c}\ell$. $\square$

An active attacker can use $\ell = 2^{c/2}$ output blocks and the complexity of her attack is going to be $N = 2^{c/2}$, a result in agreement with the indifferentiability result of Theorem 1. However, here we can distinguish between the *data complexity*, i.e., the available number of output data of the PRNG and the *time complexity*, the number of queries to f, of the attack. If the implementation of a PRNG limits the number of output blocks to some value $\ell_{\max} < 2^{c/2}$, the time complexity of a generic attack is bounded by $N = 2^c/\ell_{\max} > 2^{c/2}$.

4.3   Forward Security

Our construction does not inherently provide forward security, but it can be explicitly triggered by using the following technique. One can fetch $r' \leq r$ bits out of the current PRNG and feed them immediately afterwards. This way, the r' bits of the outer part of the state will be set to zero, making this process an irreversible step. By repeating this process $\lceil c/r' \rceil$ times, the adversary has to guess at least c bits when evaluating the state backwards. This process can be activated, for instance, at regular intervals.

5   A Concrete Example with KECCAK

KECCAK is a family of sponge functions submitted to the SHA-3 contest organized by NIST [13,6,7]. The family uses seven permutations ranging from a width of 25 bits to a width of 1600 bits. While the SHA-3 proposal uses KECCAK-f[1600] only, other members of the family with a smaller width can be interesting in the context of a PRNG in an embedded device. For instance, KECCAK[$r=96, c=104$] and KECCAK[$r=64, c=136$] both use KECCAK-f[200]

as underlying permutation. This permutation is suitable for devices with scarse resources as the state can be stored in only 25 bytes. In hardware it can be built in a very compact core and in software it can be implemented with bitwise Boolean instructions and rotations within bytes only. These sponge functions can produce 96 and 64 pseudo-random bits, resp., per call to KECCAK-f[200].

In terms of security, KECCAK follows what is called the hermetic sponge strategy [7,5]. This means that the KECCAK-f permutations are designed with the target that they cannot be distinguished from a randomly-chosen permutation. Biased output bits on one of the KECCAK members, for instance, would imply a distinguisher on the underlying permutation KECCAK-f and would therefore contradict the design strategy.

Against passive state recovery attacks in the generic case, Theorem 2 proves a resistance of $2^c/m(A)$. If a sequence of ℓr output bits is known, the expected value of $m(A)$ is close to 1 unless $\ell > 2^{r/2}$. One can limit to $r2^{r/2}$ the number of output bits between times where the state has gained at least c bits of fresh seeding material. This way, KECCAK[$r=96, c=104$] and KECCAK[$r=64, c=136$] provides a resistance of about 2^{104} and 2^{136}, resp., against state recovery, at least as long as no distinguisher on KECCAK-f[200] is found.

If the PRNG allows the user to provide seeding material, active state recovery attacks must also be considered. Here, the implementation can limit, e.g., to $\ell_{\max} = 2^{24}$ or 2^{32} output blocks before the state has again been fed with c bits of fresh seeding material. In this case, KECCAK[$r=64, c=136$] provides a resistance of about 2^{112} and 2^{104}, respectively.

We have implemented our PRNG based on KECCAK[$r=96, c=104$] and KECCAK[$r=64, c=136$] and passed the statistical tests proposed by NIST [15]. The tests were performed on 200 sequences of 10^6 bits each. The sequences were generated by squeezing 2×10^8 bits after providing the empty string as input, namely $\lfloor$KECCAK[$r=96, c=104$]$()\rfloor_{2 \times 10^8}$ and $\lfloor$KECCAK[$r=64, c=136$]$()\rfloor_{2 \times 10^8}$.

6 Conclusions

We have presented a construction for building a reseedable pseudo-random number generator using a sponge function. This construction is efficient in terms of memory use and processing, and inherits the provable security properties of the sponge construction. We have provided bounds on generic state recovery attacks allowing the use of a small state. We have given a concrete example of such a PRNG based on KECCAK with a state of only 25 bytes that is particularly suitable for embedded devices.

References

1. Bellare, M., Rogaway, P.: Random oracles are practical: A paradigm for designing efficient protocols. In: ACM Conference on Computer and Communications Security 1993, pp. 62–73 (1993)
2. Bellare, M., Yee, B.: Forward-security in private-key cryptography, Cryptology ePrint Archive, Report 2001/035 (2001), http://eprint.iacr.org/

3. Bertoni, G., Daemen, J., Peeters, M., Van Assche, G.: Sponge functions. In: Ecrypt Hash Workshop 2007 (May 2007), also available as public comment to NIST,
   http://www.csrc.nist.gov/pki/HashWorkshop/Public_Comments/2007_May.html
4. Bertoni, G., Daemen, J., Peeters, M., Van Assche, G.: On the indifferentiability of the sponge construction. In: Smart, N.P. (ed.) EUROCRYPT 2008. LNCS, vol. 4965, pp. 181–197. Springer, Heidelberg (2008), http://sponge.noekeon.org/
5. Bertoni, G., Daemen, J., Peeters, M., Van Assche, G.: Cryptographic sponges (2009), http://sponge.noekeon.org/
6. Bertoni, G., Daemen, J., Peeters, M., Van Assche, G.: KECCAK specifications, version 2, NIST SHA-3 Submission (September 2009), http://keccak.noekeon.org/
7. Bertoni, G., Daemen, J., Peeters, M., Van Assche, G.: KECCAK sponge function family main document, NIST SHA-3 Submission (updated) (September 2009), http://keccak.noekeon.org/
8. Coron, J., Dodis, Y., Malinaud, C., Puniya, P.: Merkle-Damgård revisited: How to construct a hash function. In: Shoup, V. (ed.) CRYPTO 2005. LNCS, vol. 3621, pp. 430–448. Springer, Heidelberg (2005)
9. Desai, A., Hevia, A., Yin, Y.L.: A practice-oriented treatment of pseudorandom number generators. In: Knudsen, L.R. (ed.) EUROCRYPT 2002. LNCS, vol. 2332, pp. 368–383. Springer, Heidelberg (2002)
10. Ferguson, N., Schneier, B.: Practical cryptography. John Wiley & Sons, Chichester (2003)
11. Maurer, U., Renner, R., Holenstein, C.: Indifferentiability, impossibility results on reductions, and applications to the random oracle methodology. In: Naor, M. (ed.) TCC 2004. LNCS, vol. 2951, pp. 21–39. Springer, Heidelberg (2004)
12. NIST: Federal information processing standard 186-2, digital signature standard (DSS) (May 1994)
13. NIST: Announcing request for candidate algorithm nominations for a new cryptographic hash algorithm (SHA-3) family. Federal Register Notices 72(212), 62212–62220 (2007),
    http://csrc.nist.gov/groups/ST/hash/index.html
14. NIST: NIST special publication 800-90, recommendation for random number generation using deterministic random bit generators (revised) (March 2007)
15. NIST: NIST special publication 800-22, a statistical test suite for random and pseudorandom number generators for cryptographic applications (revision 1) (August 2008)
16. Viega, J.: Practical random number generation in software. In: ACSAC '03: Proceedings of the 19th Annual Computer Security Applications Conference, Washington, DC, USA, p. 129. IEEE Computer Society, Los Alamitos (2003)

A High Speed Coprocessor for Elliptic Curve Scalar Multiplications over $\mathbb{F}_p$

Nicolas Guillermin[1,2]

[1] DGA Information Superiority, Bruz, France
[2] IRMAR, Université Rennes 1, France

Abstract. We present a new hardware architecture to compute scalar multiplications in the group of rational points of elliptic curves defined over a prime field. We have made an implementation on Altera FPGA family for some elliptic curves defined over randomly chosen ground fields offering classic cryptographic security level. Our implementations show that our architecture is the fastest among the public designs to compute scalar multiplication for elliptic curves defined over a general prime ground field. Our design is based upon the Residue Number System, guaranteeing carry-free arithmetic and easy parallelism. It is SPA resistant and DPA capable.

Keywords: elliptic curve, high speed, RNS, prime field, FPGA.

1 Introduction

Twenty five years after their introduction for cryptographic applications [14], elliptic curves are well established in the field of public key cryptography. A standard of the National Institute of technology (NIST) recommends their use for digital signature [17]. The most time consuming operation in elliptic curve based protocols is the scalar multiplication. As a consequence, scalar multiplication has attracted a lot of attention in public literature. Available designs may differ greatly depending on the target implementation (GPGPUs, CPUs, ASICs, FPGAs) and the aim they try to achieve which may be related to speed, size, power consumption or security issues. We refer the reader to [3,25,10,21] for example of known implementations.

In this paper, we describe the fastest available architecture for computing $[k]G$ over curves defined over $\mathbb{F}_p$ for general prime p in FPGA. Our architecture is based on Residue Number Systems (RNS) and is resistant against side channel attacks. We have made a FPGA implementation of our design. Actually, FPGA implementations are particularly interesting for at least two reasons : they are well suited to provide a good local protection level, and they constitute generally the first step towards faster ASIC implementations.

Target application of such special purpose designs are all the fields where both high speed, low latency and high level resistance against attacks are required (example : IPSEC set-top box).

S. Mangard and F.-X. Standaert (Eds.): CHES 2010, LNCS 6225, pp. 48–64, 2010.

Related work : A great overview of high speed hardware accelerator for ECC is given by [13]. Designs can be split in two categories : those which support elliptic curves over $\mathbb{F}_{2^n}$, and those over $\mathbb{F}_p$. Architectures of the first group give the best speed to security ratio. It is mostly due to the field structure (No carry is propagated). State of the art implementations show a latency under 20 μs for a 2^{80} security [8]. Nevertheless large characteristic remain interesting, mostly because $\mathbb{F}_p$ offers less structure than $\mathbb{F}_{2^n}$, and may be safer. Some architecture can also support both field types [22].

Some implementations are specific to a pseudo-Mersenne prime [7]. These implementations may be faster than the one which do not depend on the relying field. Nevertheless the ability of changing the curve is also an asset for security (finding weak curves is still an active research area). Our architecture is of this kind. To our knowledge the best architecture is the one of Mentens [11], which computes a 1 ms 160 bit scalar multiplication on a Xilinx Virtex 2 pro. Most of the implementations are based on a multi-precision Montgomery representations of numbers, allowing reduction without expensive divisions.

Another way to represent big numbers is the Residue Number System. It provides fast and carry free arithmetic. A modified version of the Montgomery algorithm [1] makes it suitable for arithmetic in $\mathbb{F}_p$. Szerwinski et al [25] used it to produce the fastest software implementation of scalar multiplication. Kawamura et al [10,16] proposed a very efficient architecture for RSA signature on an ASIC. Their contribution is analysed in section 3, since it is the starting point of our work.

The higher p is, the more efficient RNS is (because its high parallelization ability). Then we could think that applying RNS to ECC will be less interesting than RSA. We show in this paper that this drawback is compensated by 2 advantages. First, the RNS ability to execute patterns like $AB + CD$ in only one reduction, while both products are almost free, reduces the time of point operations in ECC while it is useless for RSA (2.2). Second, ECC operations are parallelizable, therefore we can deepen the pipelines while keeping a high rate occupation (3.2).

Our contribution : We present a complete redefinition of main module of Kawamura. Thanks to it we can use elliptic curve and we reach high speed on a FPGA. We design the first architecture to break 1 ms for 160 bit elliptic curve scalar multiplication over prime field of any characteristic, even on a 130 nm node FPGA (the Altera Stratix family in our case). Our scalable architecture keeps its advantage even for larger groups (up to 512 bits). We also propose an algorithm for RNS-Radix transformation that does not cost a single gate, and base choice considerations for RNS use with elliptic curves.

Structure of the paper : The section 2 deals with mathematical backgrounds of RNS and elliptic curves. The section 3 describes and analyses the choices that are made to improve Kawamura's architecture, and and the section 4 gives the results of implementations, and compares it to other existing design. Design schemes are at the end of the paper.

2   Mathematical Background

Notations : In all the paper, for $a, b \in \mathbb{N}$, we denote by $|a|_b$ the result of a modulo b.

2.1   RNS

Overview : Let $B = \{m_1, \cdots, m_n\}$ be a set of co-prime natural integers, and $M = \prod_{i=1}^{n} m_i$. The residue number system (RNS) representation $\{X\}_B$ of $X \in \mathbb{N}$ such that $0 \leq X < M$ is the unique set of positive integers $\{x_1, \cdots, x_n\}$ with $x_i = |X|_{m_i}$. This representation allows fast arithmetic in $\mathbb{Z}/M\mathbb{Z}$ since

$$\{X \odot Y\}_B = \{|x_1 \odot y_1|_{m_1}, \cdots, |x_n \odot y_n|_{m_n}\} \tag{1}$$

for $\odot \in (+, -, \times, /)$, $/$ being only available for Y coprime with M. The integer X is recovered thanks to the Chinese remainder theorem :

$$X = \left| \sum_{i=1}^{n} |x_i \times M_i^{-1}|_{m_i} \times M_i \right|_M \quad \text{where } M_i = \frac{M}{m_i}. \tag{2}$$

Note that M_i^{-1} is then well defined in $\mathbb{Z}/m_i\mathbb{Z}$. In the rest of the paper, B is called a RNS base and the $\{m_i\}_{i=1,\dots,n}$ are called channels of B, since every calculation are done independently modulo these channels.

RNS Montgomery reduction algorithm : The Montgomery reduction application was first introduced in [15] for the purpose of multiprecision arithmetic. The paper [1] presents an adaptation in the context of RNS representation.

In the following we recall the main results of this last paper. Let p be a prime, $\alpha > 2$ an integer, B and $\tilde{B}$ be two RNS bases with their channel products M and $\tilde{M}$ such that $M > \alpha p$ and $\tilde{M} > 2p$. For all input $a < \alpha p^2$ given in B and $\tilde{B}$ the Montgomery algorithm computes S in B and $\tilde{B}$ such that $S < 2p$ and $|S|_p = |a \times M^{-1}|_p$. The factor M^{-1} is not a concern if a lot of computation in modular arithmetic are to be done in a row, which is the case in most applications related to cryptography. Actually, thanks to the use of Montgomery representative $\phi(X) = |XM|_p$ (see [15]), one only needs a transformation at the beginning of any calculation $\phi(X) = Red_{Montg}(X \times |M^2|_p, p, B, \tilde{B})$ and the corresponding invert transformation $\phi^{-1}(Y) = Red_{Montg}(Y, p, B, \tilde{B})$ at the end.

The base change $B_{ext}(X, B, \tilde{B})$ is due to the fact that one can not divide by M in B. The second base extension computes $\{S\}_B$ from $\{S\}_{\tilde{B}}$. Both are then available for another computation. Unlike the multiprecision Montgomery algorithm, we can not execute the final reduction, since it is not easy to know if S is more or less than p (comparison is a greedy operation in RNS representation). The result S is then kept between 0 and $2p$. The main consequence is that $\tilde{M}$ has to be up to $2p$. The choice of B depends on the maximal number we wish to reduce. Proposition 1 shows that $\tilde{M}$ does not need to be more than M. Even if $\tilde{Q}$ is not equal to Q but to $|Q|_{\tilde{M}}$, the algorithm still gives the correct output.

Algorithm 1. $Red_{Montg}(X, p, B, \tilde{B})$

Require: B and $\tilde{B}$ RNS bases with $M > \alpha p$ and $\tilde{M} > 2p$
Require: p co-prime with M and $\tilde{M}$
Require: $\{X\}_B$ and $\{X\}_{\tilde{B}}$ RNS representation of $X < \alpha p^2$ in B and $\tilde{B}$
Require: precalculations : $\{|-p^{-1}|_M\}_B, \{|M^{-1}|_{\tilde{M}}\}_{\tilde{B}}$ and $\{p\}_{\tilde{B}}$
Require: algorithm $B_{ext}(A, B_1, B_2)$ computing $\{|A|_{M_2}\}_{B_2}$ from $\{A\}_{B_1}$
Ensure: $\{S\}_B$ and $\{S\}_{\tilde{B}}$ such that $|S|_p = |XM^{-1}|_p$ and $S < 2p$
 1: $Q \leftarrow X \times |-p^{-1}|_M$ in B
 2: $\tilde{Q} \leftarrow B_{ext}(Q, B, \tilde{B})$
 3: $\tilde{R} \leftarrow X + \tilde{Q} \times p$ in $\tilde{B}$
 4: $\tilde{S} \leftarrow \tilde{R} \times M^{-1}$ in $\tilde{B}$
 5: $S \leftarrow B_{ext}(S, \tilde{B}, B)$
 6: **return** S and $\tilde{S}$

Proposition 1. *Given $\alpha > 2$, if $M > \alpha p$ and $\tilde{M} > 2p$ then $Red_{Montg}(X, p, B, \tilde{B})$ gives the correct output for every X between 0 and αp^2.*

Proof. Be $X < \alpha p^2$, $M > \alpha p$ and $\tilde{M} > 2p$. $Q = |XP^{-1}|_M$, therefore $Q < M$. By B_{ext}, $\tilde{Q} = |Q|_{\tilde{M}}$. As $X_{\tilde{B}} = |X|_{\tilde{M}}$ and $p < \tilde{M}$, $\tilde{S}$ is equal to $|(X + Qp)/M|_{\tilde{M}}$. As $Q < M$, $T = (X + Qp)/M < 2p$. Therefore $\tilde{S} = T$, and as $\alpha > 2$ $S = T$ too. As $T \equiv |XM^{-1}|_p$, we can conclude that S and $\tilde{S}$ are the expected results.

RNS base extension : The greediest steps of this algorithm are the two base extensions $B_{ext}(X, B, \tilde{B})$ and $B_{ext}(X, \tilde{B}, B)$. They are a classical $O(n^2)$ algorithm, where n is the RNS base size, and the elementary operation is a modular multiplication/addition on a channel. The main concern of every algorithm implementing B_{ext} is to provide a way to compute the final reduction by M, to calculate γ such that

$$X = \sum_{i=1}^{n} |x_i \times M_i^{-1}|_{m_i} \times M_i - \gamma M. \tag{3}$$

Once γ is calculated, X is easily recovered on every channel $\tilde{m}_i$ by multiplying and accumulating the result. Three different algorithm are proposed by literature :

- a Mixed Radix System (MRS) approach [2] which natively avoids final reduction, but is hard to implement in hardware because of the structure of the algorithm, but remains a good alternative in software,
- an extra modulus approach proposed by Shenoy and Kumaresan [24]. The idea of this algorithm is to have a m_e coprime with M and $\tilde{M}$, and to use $X[m_e]$ to compute γ. The main drawback of this approach is that we need to keep $|X|_{m_e}$ all along during the calculation, while we just need it during the base extension,
- a floating point approach proposed by Posch and Posch [19], and improved by Kawamura et al. [10]. The main idea is to transform the equation 3 as follow:

$$X = \sum_{i=1}^{n} \frac{M\xi_i}{m_i} - M\lfloor \sum_{i=1}^{n} \frac{\xi_i}{m_i} \rfloor \text{with } \xi_i = |x_i M_i^{-1}|_{m_i} \tag{4}$$

Algorithm 2. $Montg - ladder(k, G, \mathbf{C}_{p,a_4,a_6})$

Require: $k \in \mathbb{N} = \sum k_i.2^i, G$ a point of $\mathbf{C}_{p,a_4,a_6}$
Ensure: $R = (k)G$
 1: $R \leftarrow \mathcal{O}$; $S \leftarrow G$
 2: **for** i from $\log_2(k)$ to 0 **do**
 3:   **if** $(k_i = 0)$ **then**
 4:     $R \leftarrow 2R$; $S \leftarrow R + S$
 5:   **else**
 6:     $S \leftarrow 2S$; $R \leftarrow R + S$
 7:   **end if**
 8: **end for**
 9: **return** R

The main drawback of the floating point approach is a potential emergence of an offset due to the approximation while computing $\lfloor \sum_{i=1}^{n} \xi_i/m_i \rfloor$. In [10] ξ_i/m_i is approximated by $\xi_i/2^r$ where r is chosen as word depth in the proposed architecture. As m_i is chosen as a pseudo-Mersenne prime near 2^r, the offset of the calculation may be easily limited to $1/2$. In [10] is explained how this possible error can be without consequences for the result, as soon as $\tilde{M} < 6p$. The output of Red_{Montg} will be less than $3p$, whatever the input is (proposition 1 is then easily adapted, with $\tilde{M} > 6p$ and $\alpha > 3$ the output of the algorithm 1 will be less than $3p$).

2.2   Elliptic Curves

Overview : In this paper, considered elliptic curves $\mathbf{C}_{p,a_4,a_6}$ are seen as sets of couples $(x, y) \in \mathbb{F}_p^2$ verifying the following equation, with p prime and extra conditions on a_4 and a_6 which are not discussed here.

$$y^2 = x^3 + a_4 x + a_6 \tag{5}$$

Together with the point at infinity $\mathcal{O}$, $\mathbf{C}_{p,a_4,a_6}$ is an abelian group. The composition law has a geometric meaning described by the vertical and tangent. Some specific curve shapes (forms of the equation) spare multiplications and reduction while computing $P + Q$ and $2P$ over $\mathbf{C}_{p,a_4,a_6}$ [9,5].Nevertheless the Weierstrass form represents all elliptic curves over prime fields (through isomorphism over $\mathbb{F}_p^2$). Other representations can only represent curves with subgroups (e.g order 4 for Edwards curves and Montgomery form, order 3 for Hessian curves...). Here we consider general curves in Weierstrass form, given by 5.

Addition and doubling formulæ : The Montgomery ladder [9] algorithm is a SPA-resistant square and multiply algorithm, computing $[k]G$ over $\mathbf{C}_{p,a_4,a_6}$ using one double and one add per bit of k.

Moreover, one can use projective coordinates X_P, Y_P, Z_P of point $P = x_P, y_P$ where $x_P = X_P/Z_P$ and $y_P = Y_P/Z_P$ when $Z_P \neq 0$. With projective coordinates every point of the curve has $p-1$ different representation. This can be used to execute leak-resistant computation (by changing the point representation before realizing the scalar multiplication) [4]. Point additions and doubling are then computed without inversion in $\mathbb{F}_p$.

Combined with algorithm 2, we can spare the computation of $Y_{[k]G}$ ($y_{[k]G}$ is recomputed at the end of the algorithm if necessary). Formulæ for adding and doubling points optimized for RNS are given in [20]. We briefly recall them in the following table.

$P+Q$	$2P$
$A \leftarrow Z_P X_Q + X_P Z_Q$	$E \leftarrow Z_P^2$
$B \leftarrow 2X_P X_Q$	$F \leftarrow 2X_P Z_P$
$C \leftarrow 2Z_P Z_Q$	$G \leftarrow X_P^2$
$D \leftarrow a_4 A + a_6 C$	$H \leftarrow -4a_6 E$
$Z_{P+Q} \leftarrow A^2 - BC$	$I \leftarrow a_4 E$
$X_{P+Q} \leftarrow BA + CD + 2x_{P-Q}Z_{P+Q}$	$X_{2P} \leftarrow FH + (G-I)^2$
	$Z_{2P} \leftarrow 2F(G+I) - EH$

At the end of the scalar multiplication, an inversion is needed to recompute $x_{[k]G} = X_{[k]G}/Z_{k[G]}$. In our results this final inversion is taken in account, considering that we use little Fermat's theorem to compute the inversion (which is possible with our design, and does not cost any gate except in the sequencer).

The main feature of RNS compared to classical representation is that multiplication is almost free while all the computation complexity is on reduction. Therefore, it is interesting to find $AB+CD$ pattern in the addition and doubling law of the curve. This is done by the table given above in 13 reductions for 1 Montgomery ladder step (1 point-addition and 1 point doubling).

2.3   Base Choice

Our purpose is to use Kawamura's base extension in a massively parallel architecture. A value r is set as the word depth, and B and $\tilde{B}$ are chosen to be pseudo-Mersenne values $m_i = 2^r - \epsilon_i$, with $\epsilon_i < 2^q$ and $q < r/2$. $\tilde{B}$ is chosen exactly the same manner. Regarding addition and doubling formulæ, we can set α of algorithm 1 to 45. Indeed, the maximum value we have to reduce is $Z_{2P} \leftarrow 2F(G+I) - EH$. As F,G,H and I are less than $3p$, $2F(G+I)$ is at most $36p^2$. Since we can not afford to set negative input in Red_{Montg}, and we are unable to verify that $2F(G+I) > EH$, we have to calculate $(3p - E)H$ which is positive, and less than $9p^2$. Therefore $M > 45p$. As it is shown in Radix-RNS transformation subsection, 2^r is set as the m_0 value. In order to spare gates in our design, q has to be as small as possible. If the targeted technology is a Stratix family FPGA, we will use 18×18 or 36×36 multipliers. Here are the main features of chosen bases for the use of 18×18 and 36×36 multipliers (r is the word size in bits, n is the number of parallel rower modules and q is the max size of the ϵ_i in bits) :

curve	160	192	256	384	512	160	192	256	384	512
r	17	18	18	18	18	34	33	33	36	35
n	10	11	15	22	29	5	6	8	12	15
q	7	7	8	8	9	5	6	6	7	8

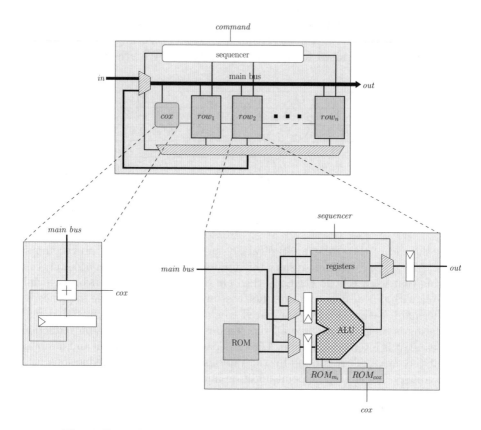

Fig. 1. General architecture and focus on the Cox and Rower design

3 Hardware Architecture

3.1 Architecture Overview

Already published architecture using RNS : Kawamura [10] proposed an
architecture suitable for RSA calculation using his base extension algorithm.
His general architecture is the same as ours and is given by the upper outline
of the figure 1. He divided his design in multiple "Rower" modules, which were
in charge of calculating $|\sum_{i=1}^{n}(M\xi_i)/m_i|_{m_j}$. and a "Cox" module in charge of
calculating $\lfloor\sum_{i=1}^{n}(\xi_i/2^r)\rfloor$. "Cox" design is very simple (a small adder). In [16]
an improvement took the advantage of setting one cox per Rower. He then
spared one cycle per reduction, computing γ in the same cycle. The results were
interesting, but the Rower pipeline was not deep enough to reach high clock
frequency (3 stages).

Improvements in our design : Our architecture is an improvement of Kawamura's [10] [16] which makes it

- suitable for elliptic curves,
- able to provide protections against side-channel attacks,
- designed to reach high clock frequencies.

The first limitation of Kawamura's architecture is the usage of only one RAM per channel. It only can execute a squaring, or multiply by a ROM data. This is not a limitation for RSA, the exponentiation algorithm only executes a square and a multiply by a constant, and so does the base reduction. This limitation is no more acceptable for elliptic curves. A general purpose register file (GPRs) must be added in order to multiply 2 local variables. Also the needed precalulation must be redefined. This point is focused in the subsection 3.3.

The second limitation of Kawamura's architecture is the design of the pipeline core, which executes the operation $acc = |x \times y + acc|_{m_i}$. Kawamura's goal was to keep busy every pipeline stage 100% of the cycles. To do so, he designed a 3 stage pipeline, and needed to use 3 times less Rower than the number of channels he had in B and $\tilde{B}$. Our architecture increases the pipeline depth to reach higher clock frequencies. We show that a 100% pipeline occupation is not really necessary for elliptic curves : it is easy to keep a good pipeline occupation even with deeper pipelines, with as many Rower as channels in B and $\tilde{B}$, and with less channels for elliptic curves than for RSA (considering 160 bits curves versus 1024 bits bases for RSA). This point is focused in the subsection 3.2.

Our architecture is showed on figure 1. It is the same as in [10] for the upper scheme part. Thus the Rower design (lower part of the scheme) is completely different. It is mainly composed with n parallel channels which execute $acc = |x \times y + acc|_{m_i}$ at each cycle and get operands and put the result from/in one of the 16 General Purpose Registers (GPR). Therefore, our architecture is able to compute a multiplication in $\mathbb{Z}/M\mathbb{Z}$ at each clock cycle. Our Rower architecture is described in 3.2.

We propose a RNS-Radix transformation in subsection 3.4, and eventually discuss about consequences of our choice for resistance against side channel attacks in 3.5.

3.2 Pipeline Architecture

In order to get high speed, the Arithmetical and Logical Unit(ALU) must execute $|r_1 \times r_2|_m$ and accumulate the result at each clock cycle.

Considering this constraint, algorithm 3 computes the modular multiplication for any pseudo-Mersenne number. The accumulation may be executed at every step of the algorithm. For every value P, P_{lsb} are the r less significant bits of P, while $P_{msb} = \lfloor P/2^r \rfloor$.

As it is shown in algorithm 3 , there are in the proposed pipeline structure only 5 generic operands:

- a $r \times r$ multiplier,
- a $r \times q$ multiplier,
- a $q \times q$ multiplier,

Algorithm 3. $MM(r_1, r_2, m_i)$

Require: r_1 and $r_2 < 2^r$
Require: $m_i = 2^r - \epsilon_i$ a pseudo-Mersenne number
Ensure: $|r_1 \times r_2|_{m_i}$
  1: $P = r_1 \times r_2$
  2: $Q = P_{msb} \times \epsilon_i$ $R = |P_{lsb}|_{m_i}$
  3: $R = Q_{msb} \times \epsilon_i$ $S = |Q_{lsb} + P_{lsb}|_{m_i}$
  4: $T = |R + S|_{m_i}$
  5: **return** T

Algorithm 4. $3_{add}(P, Q, R)$

Require: P,Q and R : vector(r)
Ensure: $P + Q + R$
  1: X : vector(r) $= P$ xor Q xor R
  2: C : vector(r) $= $ MAJ(P,Q,R)
  3: **return** $X + 2C$

- 2 r modulo-adder taking two entries less than m_i, (4 if we consider accumulation, see the pipeline subsection)
- a r modulo-adder taking one over two entries less than m_i, the other being less than 2^r.

The multipliers are not an issue for FPGA, since both Altera Stratix and Xilinx Virtex families have got multiplier blocks. The following assumption are taken :

- A $a \times b$ multiplier will be implemented in a single FPGA DSP block, for every a and $b < 36$, even if b is 9 bit wide.
- A $a \times b$ multiplier will be implemented in a single 9×9 blocks if both a and b are smaller than 9.

These facts have been verified for every synthesis during this study.

Modular addition takes advantage on the fact that an addition by three operands of size r only costs one LUT pass through and one addition of $r + 1$ operands, by using algorithm 4. Then, if a and b are less than m_i, $|a + b|_{m_i}$ can be computed by computing in parallel $r_1 = a + b$ and $r_2 = a + b + \epsilon_i$ and by considering the carry of r_2 to choose the correct result. Figure 2 describes the adder design. If $m_i < a \leq 2^r$, an extra addition $a + b + 2\epsilon_i$ is required.

Two pipeline architectures are proposed in this paper (figure 2). Both try to balance the pipeline stages with one another, but make different assumptions :

- The first one makes the assumption that the a modular addition is twice faster than a multiplication.
- The second one makes the assumption that a modular addition runs as fast as multiplication.

To increase the pipeline occupation, we overlap independent operations : for example, the computation of B and C start before the one of A is finished. Then,

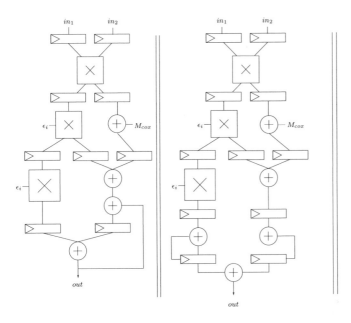
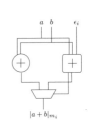

Fig. 2. 5 and 6 stage pipeline, and an adder modulo m_i

the wait states of the calculation of A, are taken up by B and C. This technique increases the pipeline occupation, but may increase the number of needed general purpose registers. This is a trade-off. It is analysed in the subsection 3.3.

On an Altera Stratix II chip, the maximum clock frequency of the 5 stage pipeline is 110 MHz, while the 6 stage reaches 158 MHz. At is is shown in subsection 3.3, the percentage of idle states is respectively 95% and 90% for a pipeline of depth 5 and 6, and a design with 5 channels (for a 160 bit curve with a channel length of 33 bits). This is the worst case of this study for pipeline occupation. We can then conclude that the 6 stage pipeline is the best choice for Stratix II technology. This study has to be done again for each target technology (including ASICs).

3.3   Memory

Precalculations and ROM content : The Rower main ROM is filled with the precalculated values described in this subsection.

Not considering reduction, we need during the calculation the following 3 variables : a_4, a_6 and $3p$: a_4 and a_6 to compute D and I, $3p$ to compute any subtraction (for example Z_{P+Q} and Z_{2P}). To compute H it is possible to precompute $-4a_6$. Eventually we need $|M^2|_p$ to compute Montgomery representatives. For computation we also need 0, 1 and -1 for every channel. For radix-RNS transformation we need 2^r and for RNS-radix we need 2^{-r} and -2^{-r} only on $\tilde{B}$ (see subsection 3.4).

Algorithm 5. $Reduction(GPR_1, GPR_2)$ on a single Rower

Require: X a value we wish to reduce $GPR_1 = |X|_{m_i}$ and $GPR_2 = |X|_{\tilde{m}_i}$
Ensure: $Red_{Montg}(X, p, B, \tilde{B})$ in GPR_1 and GPR_2
1: cycle 1 : $GPR_1 \leftarrow |GPR_1 \times p^{-1} M_i^{-1}|_{m_i}$
2: wait 1 : wait for GPR_1
3: cycle 1' : $out \leftarrow GPR_1$
4: cycle $2 + j$ ($j \in [0, \cdots, n-1]$) : GPR_1 accumulates $|in \times M_j p(M\tilde{M}_i)^{-1} + M_{cox}|_{\tilde{m}_i}$
5: cycle $2 + n$: GPR_1 accumulates $|GPR_2 \times (M\tilde{M}_i)^{-1}|_{\tilde{m}_i}$
6: wait 2 : wait for GPR_1
7: cycle $3 + n$: $out \leftarrow GPR_1$ and $GPR_2 \leftarrow |GPR_1 \times \tilde{M}_i|_{\tilde{m}_i}$
8: cycle $4 + n + j$ ($j \in [0, \cdots, n-1]$) : GPR_1 accumulates $|in \times \tilde{M}_j + M_{cox}|_{m_i}$
9: wait 3 : wait for GPR_1

Algorithm 5 gives a fast version of the algorithm in our architecture. Cycle 1' can be executed with another instruction. During cycles $2 + j$ and $4 + n + j$, the main bus is set to $out[j]$ the output of the jth channel. The needed precomputed values are

- for a channel m_i of B (modulo m_i) :
  $p^{-1} M_i^{-1}$, $\tilde{M}_j$ for $j \in [0; \cdots; n-1]$.
- for a channel $\tilde{m}_i$ of $\tilde{B}$ (modulo $\tilde{m}_i$) :
  $M_j p(M\tilde{M}_i)^{-1}$ for $j \in [0; \cdots; n-1]$, $(M\tilde{M}_i)^{-1}$.

For a FPGA implementation, all these values may be set in ROMs, a curve change can be done by loading a different bitstream. Our results are given for fixed ROM. If it is necessary to change the curve during runtime, or if an ASIC implementation is needed, user may choose between 2 options :

- use RAMs instead of ROMs. This allows to change bases too, but all the values above have to be computed.
- reduce the number of curve-dependant values. Only 5 precomputed values per channel are needed : p, a_4, a_6, $|M^2|_p$ and $|-p^{-1}|_M$. It costs 2 extra cycles and 2 extra wait per reduction.

For each Rower design, two extra small ROMs are needed, each one containing two values. The ROM_{m_i} holds m_i and $\tilde{m}_i$. The ROM_{cox} holds the value $|-Mp\tilde{M}_i^{-1}|_{\tilde{m}_i}$ and $|-\tilde{M}|_{m_i}$ when the cox module set up the signal cox, these values are injected in the pipeline.

General purpose register file : Elliptic curves formulæ use local variables which have to be multiplied with one another, contrary to RSA which only has to square or multiply by a constant. That is why our architecture uses for every channel a general purpose register file (GPRs). Since every local variable has to be evaluated in M and $\tilde{M}$, one need twice as GPR as the maximum of local variables.

As it is explained in the pipeline subsection, operations are overlapped : intermediate result computation may start before the previous is finished. This may implicate an increase of the number of needed registers.

The following table shows that 16 GPR are enough : local variables are limited to 7 and reductions are at least executed by 2, most of time by 3. An 8th local variable is taken by x_G the exponentiated point abscissa. This leads to a pipeline fill rate of 90% for the 6 stage pipeline and 16 GPR per channels. This is a good trade-off.

step	calculation	living variables
1	$A\ B\ C$	$X_P\ Z_P\ A\ B\ C$
2	$D\ Z_{P+Q}$	$X_P\ Z_P\ A\ B\ C\ D\ Z_{P+Q}$
3	$X_{P+Q}\ E\ F$	$X_P\ E\ F\ Z_{P+Q}\ X_{P+Q}$
4	$G\ H\ I$	$E\ F\ G\ H\ I\ Z_{P+Q}\ X_{P+Q}$
5	$X_{2P}\ Z_{2P}$	$X_{2P}\ Z_{2P}\ Z_{P+Q}\ X_{P+Q}$

3.4   Radix-RNS Transformation

The RNS representation is not practical for using outside the design. Moreover, there is no need for extra material either to transform a number X from its classical multiprecision representation $(X_{n-1}, \cdots, X_0)$ where $X = \sum_{i=0}^{n-1} X_i 2^i$) to its RNS representation $(x_1, ... x_n)$, nor the contrary. Using our architecture, the Radix-RNS transformation is trivial, if the sequencer can set up the main bus to the X_i values, and $|2^r|_{m_i}$ is in ROM.

For RNS-Radix transformation, the main idea is to set the channel m_0 of B to 2^r. If it is so, $X_0 = x_0$. To find X_1, all we need to do is to compute $X \leftarrow (X - x_0)/2^r$. As 2^r is not co-prime with M, this computation is done over $\tilde{B}$, and the use of $B_{ext}(X, \tilde{B}, B)$ gives $X_1 \leftarrow x_0$. By repeating this operation we compute $X_2, \cdots, X_n$.

This algorithm is not the most efficient but does not cost a single gate in our architecture. It never costs more than 0.3% of the total scalar multiplication time.

3.5   Side Channel Attacks

Our architecture supports an inherent capability to treat simple power analysis (SPA), or differential power analysis (DPA) and fault threats. Indeed, the Montgomery ladder is particularly efficient to counter both side channel attacks and fault attacks (no operation is dummy). Our finite state sequencer does not have any branch capability, bits of k are only read at the beginning of each Montgomery-ladder step to invert registers. Therefore no information leaks from the computation time.

Moreover, randomness can be introduced at the very beginning of the algorithm by changing G representation, replacing $(x_G, 1)$ by $(x_G \times a_1, a_1)$ and $\mathcal{O} = (a_2, 0)$, where a_1 and a_2 are random values. This countermeasure avoids Fouque's attacks [6] on collisions. The only SPA vulnerability is address-bit SPA attack [12], which is difficult to realize on real design. Moreover, to be DPA resistant k may be added with $a_3 \times \#C_{p,a_4,a_6}$ where a_3 is a random value. The impact on speed is $log(a_3)/log(p)$ on the speed. There is no impact on the design size. More robust randomizations of k are also possible.

4   Result and Comparison

In this section we give the overall results, and compare it to different architecture given in the open literature.

4.1   Results

Our target technology is the Altera Stratix family. This choice has very few impact on results compared to Xilinx Virtex family [18]. We chose Altera because of the availability of the Quartus toolchain during the study. Among all the Altera products, we focus on 2 generations. First Stratix are the Altera FPGA at the 130 nm process node. The Xilinx equivalent is the Virtex II-pro. We also have fit our design in the Stratix II generation FPGA (the 90 nm process node), much more efficient. The equivalent is the Virtex IV by Xilinx.

We randomly chose elliptic curves of the following size : 160, 192, 256, 384 and 512 bits. No restriction were given for p, a_4 and a_6 but to be a valid elliptic curve. The result given below does not depend on the effective choice of p but only on $log_2(p)$. The number of Rower n as well as the word depth r are also given. These values are the most efficient considering the curve size.

For every fit we give the considered FPGA family and the exact reference of the chip. The maximum reachable frequency as well as the computation time for a whole scalar multiplication $[k]G$ are given. The size of the exponent k is the same as the size of p. The final inversion, the $y_{[k]G}$ recalculation, the Radix-RNS and RNS-Radix transformations are included in the result.

Eventually the FPGA occupation is given. The Stratix FPGA are composed by logic elements (LE). LE are equivalent to the Look-Up Table (LUT) of the Virtex II-pro. Therefore a Virtex II-pro slice counts for two LE. No equivalent to the slice exists in the Stratix family. On the Stratix II family Altera introduced the Altera Logic Module (ALM), the Virtex IV slice equivalent. The number of used DSP blocks (multipliers) is also given. Altera gives the DSP occupation in numbers of 9×9 multipliers used. Stratix and Stratix II DSP blocks can indeed be configured as one 36×36 multiplier, two 18×18 or eight 9×9 multipliers.

Family	curve	model	n	r	size	DSP	frequency	speed
Stratix	160	EP1S20F484C5	5	34	11431 LE	74	92.6	0.57 ms
	192	EP1S30F780C5	6	33	12480 LE	80	89.6	0.72 ms
	256	EP1S60F780C5	8	33	16200 LE	125	90.7	1.17 ms
	384	EP1S80F1020C5	11	36	25279 LE	176	90.0	2.25 ms
	512 [1]	EP1S80F1020C5	15	35	48305 LE	176	79.6	4.03 ms
Stratix II	160	EP2S30F484C3	5	34	5896 ALM	74	165.5	0.32 ms
	192	EP2S30F484C3	6	33	6203 ALM	92	160.5	0.44 ms
	256	EP2S30F484C3	8	33	9177 ALM	96	157.2	0.68 ms
	384	EP2S60F484C3	11	36	12958 ALM	177	150.9	1.35 ms
	512	EP2S60F484C3	15	35	17017 ALM	244	144.97	2.23 ms

[1] The EP1S80 does not have enough DSP blocks, multipliers are fitted in the LE blocks, and frequency falls.

The results given above show some properties of the chosen design. First the maximum frequency hardly falls with the size of the design. Indeed, no carry is propagated on the whole size of the operands due to RNS. The critical path is then not related to the design structure. For some fit it is in the sequencer, and for some other in a Rower. No further instruction has been given to the fitter except the pursuit of the maximal frequency.

As speed was our main concern, no RAM blocks were used. If size is a matter, ROMs and GPR may be fitted in RAM blocks, to spare logic. The Stratix family lacks some DSP for 512 bit curves, while Stratix II have far enough resources to fit for any cryptographic size (the EP2S60F484C3 is a middle size matrix in the Stratix 2 family).

4.2   Comparison

In this subsection we compare our design with other papers in the open literature. Our architecture supports any elliptic curve over $\mathbb{F}_p$ and is resistant to side channel attacks. We compare it with 3 papers we consider significant:

- the first one is another design based on RNS. In this paper, modular multiplication is realized through an Horner scheme. To our knowledge it is the only implementation using RNS for curves. It is realized by Schinianakis et al [23]. It is implemented on an ASIC and on a FPGA.
- the second one is described in [11], and is based on a multi-precision algorithm. The main idea is an important work on the pipeline architecture for the classical multiprecision algorithm and on long word additions, through carry-look adder. It is realized on a Virtex II-pro. The scalar multiplication is based on a NAF recoding. This is to our knowledge the fastest implementation of elliptic curve scalar multiplication with generic curves. It outperforms every implementation given in [13] and is as fast as [21] using less resources.
- the third one is described in [7]. The design is specific to a particular curve, p being a pseudo Mersenne value. The main idea is a dual clock design, according DSP blocks to run at their maximum speed in Virtex 4 design, 500 MHz. It is the fastest FPGA implementation of elliptic curve scalar multiplication over $\mathbb{F}_p$, but with restrictions on p.

paper	curve	FPGA family	FPGA model	size	freq.(MHz)	speed
This work	160 any	Stratix	EP1S20F484C5	11431 LE	92.6	0.57 ms
	256 any	Stratix	EP1S60F780C5	16200 LE	90.7	1.17 ms
	160 any	Stratix II	EP2S30F484C3	6203 ALM	165.5	0.32 ms
	256 any	Stratix II	EP2S30F484C3	9177 ALM	157.2	0.68 ms
[23]	160 any	Virtex	XCV1000E-8	21000 LUT	58	1.77 ms
	256 any	Virtex	XCV1000E-8	36000 LUT	39.7	3.95 ms
[11]	160 any	Virtex II-pro	XC2VP30	2171 sl.	72	1 ms
	256 any	Virtex II-pro	XC2VP30	3529 sl.	67	2.27 ms
[7]	224 NIST	Virtex 4	XC4VFX12	1580 sl.	487	0.36 ms
	256 NIST	Virtex 4	XC4VFX12	1715 sl.	490	0.49 ms

As a conclusion, our implementation is largely faster and smaller than [23]. Architecture [11] has got a better time surface trade-off (the ratio is about 1.5 if we consider that a Virtex 2 pro slice is equal to 2 Stratix LE, slices having not correspondant in Stratix products). Nevertheless it does not compute $k[G]$ as fast as ours, even on comparable technologies. It is eventually not resistant against side channel attacks. An overhead is needed to be SPA resistant. Ours is natively SPA resistant.

Architecture described in [7] is faster and presents a better size-area tradeoff regarding to ours (assuming that a Stratix II ALM and a Virtex IV slice are equivalent, and only considering slices). Guneysu's work only computes $[k]P$ over NIST curves, using pseudo-mersenne p. We can consider that using pseudo mersenne primes reduces the time complexity by a factor between 2 and 1.68 if lazy reduction is used. Then we can see that our results is competitive in term of latency regarding to his for 224 bit, and becomes better for 256 bits. Moreover, Guneysu's clock speed (500 MHz) is particularly high and may represent an obstacle for industrial integration, and for an ASIC implementation (making the multipliers work twice as fast as the rest of the design would be impossible). Of course it is difficult to realize a fair comparison at this point since the two designs do not target the same curves, but RNS is a competitive alternative for general F_p curves, especially for high security levels.

5 Conclusion

In this paper is presented a hardware architecture realizing elliptic curve scalar multiplication over any curve in $\mathbb{F}_p$, which uses RNS representations to speed up the computation. RNS supports a wide parallelization capability for arithmetic in $\mathbb{F}_p$. The overhead given by the elementary operation ($|a \times b|_m$) is well pipelineable, even with large pipelines (6 stages), and contrary to RSA, the inherent parallelism of elliptic curve operations allows to easily fill the pipeline. Capability to support high clock frequency does not fall with the curve size.

In our future work, we will study other technologies like ASICs (with the usage of RAMs instead of the actual ROMs, other curve shapes, or other operations in elliptic curve cryptography, like pairings.

Acknowledgement : We would like to thank the anonymous referees for their detailed review of this paper and their helpful suggestion. Thanks to David Lubicz, Sylvain Duquesne and Jeremie Detrey for their contribution to this work.

References

1. Bajard, J.-C., Didier, L.-S., Kornerup, P.: An rns montgomery modular multiplication algorithm. IEEE Transactions on Computers 47(7), 766–776 (1998)
2. Bajard, J.-C., Imbert, L., Liardet, P.-Y., Teglia, Y.: Leak resistant arithmetic. In: Joye, M., Quisquater, J.-J. (eds.) CHES 2004. LNCS, vol. 3156, pp. 116–145. Springer, Heidelberg (2004)

3. Chen, L., Yanpu, C., Zhengzhong, B.: An implementation of fast algorithm for elliptic curve cryptosystem over GF(p). Journal of Electronics (China) 21(4), 346–352 (2004)
4. Coron, J.-S.: Resistance against differential power analysis for elliptic curve cryptosystems. In: Koç, Ç.K., Paar, C. (eds.) CHES 1999. LNCS, vol. 1717, pp. 292–302. Springer, Heidelberg (1999)
5. Edwards, H.: A normal form for elliptic curves. Bull. Amer. Math. Soc. 44 (2007)
6. Fouque, P.-A., Valette, F.: The doubling attack – why upwards is better than downwards. In: Walter, C.D., Koç, Ç.K., Paar, C. (eds.) CHES 2003. LNCS, vol. 2779, pp. 269–280. Springer, Heidelberg (2003)
7. Güneysu, T., Paar, C.: Ultra high performance ecc over nist primes on commercial fpgas. In: Oswald, E., Rohatgi, P. (eds.) CHES 2008. LNCS, vol. 5154, pp. 62–78. Springer, Heidelberg (2008)
8. Jarvinen, K.U., Skytta, J.O.: High-speed elliptic curve cryptography accelerator for koblitz curves. In: Annual IEEE Symposium on Field-Programmable Custom Computing Machines, pp. 109–118 (2008)
9. Joye, M., Sung-Min-Yen: The montgomery powering ladder. In: Kaliski Jr., B.S., Koç, Ç.K., Paar, C. (eds.) CHES 2002. LNCS, vol. 2523, pp. 291–302. Springer, Heidelberg (2003)
10. Kawamura, S., Koike, M., Sano, F., Shimbo, A.: Cox-rower architecture for fast parallel montgomery multiplication. In: Preneel, B. (ed.) EUROCRYPT 2000. LNCS, vol. 1807, pp. 523–538. Springer, Heidelberg (2000)
11. Mentens, N.: Secure and Efficient Coprocessor Design for Cryptographic Applications on FPGAs. PhD thesis, Ruhr-University Bochum (2007)
12. Messerges, T.S., Dabbish, E.A., Sloan, R.H.: Power analysis attacks of modular exponentiation in smartcards. In: Koç, Ç.K., Paar, C. (eds.) CHES 1999. LNCS, vol. 1717, pp. 144–157. Springer, Heidelberg (1999)
13. de Dormale, G.M., Quisquater, J.-J.: High-speed hardware implementations of elliptic curve cryptography: A survey. J. Syst. Archit. 53(2-3), 72–84 (2007)
14. Miller, V.S.: Use of elliptic curves in cryptography. In: Williams, H.C. (ed.) CRYPTO 1985. LNCS, vol. 218, pp. 417–426. Springer, Heidelberg (1986)
15. Montgomery, P.L.: Modular multiplication without trial division. Mathematics of Computation 44, 519–521 (1985)
16. Nozaki, H., Motoyama, M., Shimbo, A., Kawamura, S.-i.: Implementation of rsa algorithm based on rns montgomery multiplication. In: Koç, Ç.K., Naccache, D., Paar, C. (eds.) CHES 2001. LNCS, vol. 2162, pp. 364–376. Springer, Heidelberg (2001)
17. National Institute of Science and Technology. The digital signature standard. Technical report,
   http://csrc.nist.gov/publications/fips/archive/fips186-2/fips186-2.pdf
18. White Paper. Stratix vs. virtex-ii pro fpga performance analysis. Technical report,
   http://www.altera.com/literature/wp/wpstxvrtxII.pdf
19. Posch, K.C., Posch, R.: Modulo reduction in residue number systems. IEEE Trans. Parallel Distrib. Syst. 6(5), 449–454 (1995)
20. Ecegovac, M., Duquesne, S., Bajard, J.C.: Combining leak-resistant arithmetic for elliptic curves define over $\mathbb{F}_p$ and rns representation
21. Sakiyama, K., Mentens, N., Batina, L., Preneel, B., Verbauwhede, I.: Reconfigurable modular arithmetic logic unit for high-performance public-key cryptosystems. In: Bertels, K., Cardoso, J.M.P., Vassiliadis, S. (eds.) ARC 2006. LNCS, vol. 3985, pp. 347–357. Springer, Heidelberg (2006)

22. Satoh, A., Takano, K.: A scalable dual-field elliptic curve cryptographic processor. IEEE Transactions on Computers 52, 449–460 (2003)
23. Schinianakis, D.M., Fournaris, A.P., Michail, H.E., Kakarountas, A.P., Stouraitis, T.: An rns implementation of an fpelliptic curve point multiplier. Trans. Cir. Sys. Part I 56(6), 1202–1213 (2009)
24. Shenoy, P.P., Kumaresan, R.: Fast base extension using a redundant modulus in rns. IEEE Trans. Comput. 38(2), 292–297 (1989)
25. Szerwinski, R., Gayneysu, T.: Exploiting the power of GPUs for asymmetric cryptography. In: Oswald, E., Rohatgi, P. (eds.) CHES 2008. LNCS, vol. 5154, pp. 79–99. Springer, Heidelberg (2008)

Co-Z Addition Formulæ and Binary Ladders on Elliptic Curves

(Extended Abstract)

Raveen R. Goundar[1], Marc Joye[2], and Atsuko Miyaji[1]

[1] Japan Advanced Institute of Science and Technology
1-1 Asahidai, Nomi, Ishikawa 923-1292, Japan
raveen.rg@gmail.com, miyaji@jaist.ac.jp
[2] Technicolor, Security & Content Protection Labs
1 avenue de Belle Fontaine, 35576 Cesson-Sévigné Cedex, France
marc.joye@technicolor.com

Abstract. Meloni recently introduced a new type of arithmetic on elliptic curves when adding projective points sharing the same Z-coordinate. This paper presents further co-Z addition formulæ for various point additions on Weierstraß elliptic curves. It explains how the use of conjugate point addition and other implementation tricks allow one to develop efficient scalar multiplication algorithms making use of co-Z arithmetic. Specifically, this paper describes efficient co-Z based versions of Montgomery ladder and Joye's double-add algorithm. Further, the resulting implementations are protected against a large variety of implementation attacks.

Keywords: Elliptic curves, Meloni's technique, Jacobian coordinates, regular binary ladders, implementation attacks, embedded systems.

1 Introduction

Elliptic curve cryptography (ECC), introduced independently by Koblitz [16] and Miller [23] in the mid-eighties, shows an increasing impact in our everyday lives where the use of memory-constrained devices such as smart cards and other embedded systems is ubiquitous. Its main advantage resides in a smaller key size. The efficiency of ECC is dominated by an operation called *scalar multiplication*, denoted as $k\boldsymbol{P}$ where $\boldsymbol{P} \in E(\mathbb{F}_q)$ is a rational point on an elliptic curve $E/\mathbb{F}_q$ and k acts as a secret scalar. This means adding a point $\boldsymbol{P}$ on elliptic curve E, k times. In constrained environments, scalar multiplication is usually implemented through binary methods, which take on input the binary representation of scalar k.

There are many techniques proposed in the literature aiming at improving the efficiency of ECC. They rely on explicit addition formulæ, alternative curve parameterizations, extended point representations, extended coordinate systems, or higher-radix or non-standard scalar representations. See e.g. [1] for a survey of some techniques.

S. Mangard and F.-X. Standaert (Eds.): CHES 2010, LNCS 6225, pp. 65–79, 2010.

In this paper, we target the basic operation, namely the point addition. More specifically, we propose *new co-Z addition formulæ*. Co-Z arithmetic was introduced by Meloni in [22] as a means to efficiently add two projective points sharing the same Z-coordinate. The initial co-Z addition formula proposed by Meloni greatly improves on the general point addition. The drawback is that this fast formula is by construction limited to Euclidean addition chains. The efficiency being dependent on the length of the chain, Meloni suggests to represent scalar k in the computation of kP with the so-called Zeckendorf's representation and proposes a "Fibonacci-and-add" algorithm. The resulting algorithm is efficient but still slower than its binary counterparts. We take a completely different approach in this paper and consider *conjugate point addition* [11,19]. The basic observation is that the addition of two points, $R = P + Q$, yields almost for free the value of their difference, $S = P - Q$. This combined operation is referred to as a conjugate point addition. We propose efficient conjugate point addition formulæ making use of co-Z arithmetic and develop a new strategy for the efficient implementation of scalar multiplications. Specifically, we show that the Montgomery ladder [24] and its dual version [14] can be adapted to accommodate our new co-Z formulæ. As a result, we get efficient co-Z based scalar multiplication algorithms using the regular binary representation.

Last but not least, our scalar multiplication algorithms resist against certain implementation attacks. Because they are built on *highly* regular algorithms, our algorithms inherit of their security features. In particular, they are naturally protected against SPA-type attacks [17] and safe-error attacks [26,27]. Moreover, they can be combined with other known countermeasures to protect against other classes of attacks. Finally, we note that, unlike [5,9,13,21], our version of the Montgomery ladder makes use of the complete point coordinates and so offers a better resistance against (regular) fault attacks [4].

2   Preliminaries

Let $\mathbb{F}_q$ be a finite field with characteristic $\neq 2, 3$. Consider an elliptic curve E over $\mathbb{F}_q$ given by the Weierstraß equation $y^2 = x^3 + ax + b$, with discriminant $\Delta = -16(4a^3 + 27b^2) \neq 0$. This section explains how to get efficient arithmetic on elliptic curves over $\mathbb{F}_q$. The efficiency is measured in terms of field multiplications and squarings. The cost of field additions is neglected. We let M and S denote the cost of a multiplication and of a squaring in $\mathbb{F}_q$, respectively. A typical ratio is S/M = 0.8.

2.1   Jacobian Coordinates

In order to avoid the computation of inverses in $\mathbb{F}_q$, it is advantageous to make use of Jacobian coordinates. A finite point (x, y) is then represented by a triplet $(X : Y : Z)$ such that $x = X/Z^2$ and $y = Y/Z^3$. The curve equation becomes

$$E_{/\mathbb{F}_q} : Y^2 = X^3 + aXZ^4 + bZ^6 \ .$$

The point at infinity, O, is the only point with a Z-coordinate equal to 0. It is represented by $O = (1 : 1 : 0)$. Note that, for any nonzero $\lambda \in \mathbb{F}_q$, the triplets $(\lambda^2 X : \lambda^3 Y : \lambda Z)$ represent the same point.

It is well known that the set of points on an elliptic curve form a group under the chord-and-tangent law. The neutral element is the point at infinity O. Let $P = (X_1 : Y_1 : Z_1)$ and $Q = (X_2 : Y_2 : Z_2)$ be two points on E, with $P, Q \neq O$. The inverse of P is $-P = (X_1 : -Y_1 : Z_1)$. If $P = -Q$ then $P + Q = O$. If $P \neq \pm Q$ then their sum $P + Q$ is given by $(X_3 : Y_3 : Z_3)$ where

$$X_3 = R^2 + G - 2V, \quad Y_3 = R(V - X_3) - 2K_1 G, \quad Z_3 = ((Z_1 + Z_2)^2 - I_1 - I_2)H$$

with $R = 2(K_1 - K_2)$, $G = FH$, $V = U_1 F$, $K_1 = Y_1 J_2$, $K_2 = Y_2 J_1$, $F = (2H)^2$, $H = U_1 - U_2$, $U_1 = X_1 I_2$, $U_2 = X_2 I_1$, $J_1 = I_1 Z_1$, $J_2 = I_2 Z_2$, $I_1 = Z_1^2$ and $I_2 = Z_2^2$ [7].[1] We see that that the addition of two (different) points requires $\underline{11M + 5S}$.

The double of $P = (X_1 : Y_1 : Z_1)$ (i.e., when $P = Q$) is given by $(\mathrm{X}(2P) : \mathrm{Y}(2P) : \mathrm{Z}(2P))$ where

$$\mathrm{X}(2P) = M^2 - 2S, \quad \mathrm{Y}(2P) = M(S - \mathrm{X}(2P)) - 8L, \quad \mathrm{Z}(2P) = (Y_1 + Z_1)^2 - E - N$$

with $M = 3B + aN^2$, $S = 2((X_1 + E)^2 - B - L)$, $L = E^2$, $B = X_1^2$, $E = Y_1^2$ and $N = Z_1^2$ [2]. Hence, the double of a point can be obtained with $\underline{1M + 8S + 1c}$, where c denotes the cost of a multiplication by curve parameter a.

An interesting case is when curve parameter a is $a = -3$, in which case point doubling costs $3M + 5S$ [6]. In the general case, point doubling can be sped up by representing points $(X_i : Y_i : Z_i)$ with an additional coordinate, namely $T_i = aZ_i^4$. This extended representation is referred to as *modified Jacobian coordinates* [7]. The cost of point doubling drops to $3M + 5S$ at the expense of a slower point addition.

2.2 Co-Z Point Addition

In [22], Meloni considers the case of adding two (different) points having the same Z-coordinate. When points P and Q share the same Z-coordinate, say $P = (X_1 : Y_1 : Z)$ and $Q = (X_2 : Y_2 : Z)$, then their sum $P + Q = (X_3 : Y_3 : Z_3)$ can be evaluated faster as

$$X_3 = D - W_1 - W_2, \quad Y_3 = (Y_1 - Y_2)(W_1 - X_3) - A_1, \quad Z_3 = Z(X_1 - X_2)$$

with $A_1 = Y_1(W_1 - W_2)$, $W_1 = X_1 C$, $W_2 = X_2 C$, $C = (X_1 - X_2)^2$ and $D = (Y_1 - Y_2)^2$. This operation is referred to as the ZADD operation. The key observation in Meloni's addition is that the computation of $R = P + Q$ yields for free an equivalent representation for input point P with its Z-coordinate equal to that of output point R, namely

$$(X_1(X_1 - X_2)^2 : Y_1(X_1 - X_2)^3 : Z_3) = (W_1 : A_1 : Z_3) \sim P .$$

---

[1] Actually, Cohen et al. in [7] reports formulæ in $12M + 4S$. The above formulæ in $11M + 5S$ are essentially the same: A multiplication is traded against a squaring in the expression of Z_3 by computing $Z_1 \cdot Z_2$ as $(Z_1 + Z_2)^2 - Z_1^2 - Z_2^2$. See [2,18].

The corresponding operation is denoted ZADDU (i.e., ZADD with update) and is presented in Algorithm 1. It is readily seen that it requires $\underline{5M + 2S}$.

Algorithm 1. Co-Z point addition with update (ZADDU)

Require: $P = (X_1 : Y_1 : Z)$ and $Q = (X_2 : Y_2 : Z)$
Ensure: $(R, P) \leftarrow \text{ZADDU}(P, Q)$ where $R \leftarrow P + Q = (X_3 : Y_3 : Z_3)$ and $P \leftarrow (\lambda^2 X_1 : \lambda^3 Y_1 : Z_3)$ with $Z_3 = \lambda Z_1$ for some $\lambda \neq 0$

> **function** ZADDU(P, Q)
> $\quad C \leftarrow (X_1 - X_2)^2$
> $\quad W_1 \leftarrow X_1 C; W_2 \leftarrow X_2 C$
> $\quad D \leftarrow (Y_1 - Y_2)^2; A_1 \leftarrow Y_1(W_1 - W_2)$
> $\quad X_3 \leftarrow D - W_1 - W_2; Y_3 \leftarrow (Y_1 - Y_2)(W_1 - X_3) - A_1; Z_3 \leftarrow Z(X_1 - X_2)$
> $\quad X_1 \leftarrow W_1; Y_1 \leftarrow A_1; Z_1 \leftarrow Z_3$
> **end function**

---

3 Binary Scalar Multiplication Algorithms

This section discusses known scalar multiplication algorithms. Given a point P in $E(\mathbb{F}_q)$ and a scalar $k \in \mathbb{N}$, the *scalar multiplication* is the operation consisting in calculating $Q = kP$ —that is, $P + \cdots + P$ (k times).

We focus on binary methods, taking on input the binary representation of scalar k, $k = (k_{n-1}, \ldots, k_0)_2$ with $k_i \in \{0, 1\}, 0 \leqslant i \leqslant n - 1$. The corresponding algorithms present the advantage of demanding low memory requirements and are therefore well suited for memory-constrained devices like smart cards.

A classical method for evaluating $Q = kP$ exploits the obvious relation that $kP = 2(\lfloor k/2 \rfloor P)$ if k is even and $kP = 2(\lfloor k/2 \rfloor P) + P$ if k is odd. Iterating the process then yields a scalar multiplication algorithm, left-to-right scanning scalar k. The resulting algorithm, also known as *double-and-add algorithm*, is depicted in Algorithm 2. It requires two (point) registers, R_0 and R_1. Register R_0 acts as an accumulator and register R_1 is used to store the value of input point P.

Algorithm 2. Left-to-right binary method	**Algorithm 3.** Montgomery ladder
Input: $P \in E(\mathbb{F}_q)$ and $k=(k_{n-1}, \ldots, k_0)_2 \in \mathbb{N}$	**Input:** $P \in E(\mathbb{F}_q)$ and $k=(k_{n-1}, \ldots, k_0)_2 \in \mathbb{N}$
Output: $Q = kP$	**Output:** $Q = kP$
1: $R_0 \leftarrow O; R_1 \leftarrow P$	1: $R_0 \leftarrow O; R_1 \leftarrow P$
2: **for** $i = n - 1$ down to 0 **do**	2: **for** $i = n - 1$ down to 0 **do**
3: $\quad R_0 \leftarrow 2R_0$	3: $\quad b \leftarrow k_i; R_{1-b} \leftarrow R_{1-b} + R_b$
4: $\quad$ **if** $(k_i = 1)$ **then** $R_0 \leftarrow R_0 + R_1$	4: $\quad R_b \leftarrow 2R_b$
5: **end for**	5: **end for**
6: **return** R_0	6: **return** R_0

Although efficient (memory- and computation-wise), the left-to-right binary method is subject to SPA-type attacks [17]. From a power trace, an adversary able to distinguish between point doublings and point additions can easily recover the value of scalar k.

A simple countermeasure is to insert a dummy point addition when scalar bit k_i is 0. Using an additional (point) register, say R_{-1}, Line 4 in Algorithm 2 can be replaced with $R_{-k_i} \leftarrow R_{-k_i} + R_1$. The so-obtained algorithm, called *double-and-add-always algorithm* [8], now appears as a regular succession of a point doubling followed by a point addition. However, it also becomes subject to safe-error attacks [26,27]. By timely inducing a fault at iteration i during the point addition $R_{-k_i} \leftarrow R_{-k_i} + R_1$, an adversary can determine whether the operation is dummy or not by checking the correctness of the output, and so deduce the value of scalar bit k_i. If the output is correct then $k_i = 0$ (dummy point addition); if not, $k_i = 1$ (effective point addition).

A scalar multiplication algorithm featuring a regular structure without dummy operation is the so-called *Montgomery ladder* [24] (see also [15]). It is detailed in Algorithm 3. Each iteration is comprised of a point addition followed by a point doubling. Further, compared to the double-and-add-always algorithm, it only requires two (point) registers and all involved operations are effective. Montgomery ladder provides thus a natural protection against SPA-type attacks and safe-error attacks. A useful property of Montgomery ladder is that its main loop keeps invariant the difference between R_1 and R_0. Indeed, if we let $R_b{}^{(\text{new})} = R_b + R_{1-b}$ and $R_{1-b}{}^{(\text{new})} = 2R_{1-b}$ denote the registers after the updating step, we observe that $R_b{}^{(\text{new})} - R_{1-b}{}^{(\text{new})} = (R_b + R_{1-b}) - 2R_{1-b} = R_b - R_{1-b}$. This allows one to compute scalar multiplications on elliptic curves using the x-coordinate only [24] (see also [5,9,13,21]).

Algorithm 4. Right-to-left binary method	**Algorithm 5.** Joye's double-add
Input: $P \in E(\mathbb{F}_q)$ and $k = (k_{n-1}, \dots, k_0)_2 \in \mathbb{N}$ **Output:** $Q = kP$	**Input:** $P \in E(\mathbb{F}_q)$ and $k = (k_{n-1}, \dots, k_0)_2 \in \mathbb{N}$ **Output:** $Q = kP$
1: $R_0 \leftarrow O;\ R_1 \leftarrow P$ 2: **for** $i = 0$ to $n - 1$ **do** 3: **if** $(k_i = 1)$ **then** $R_0 \leftarrow R_0 + R_1$ 4: $R_1 \leftarrow 2R_1$ 5: **end for** 6: **return** R_0	1: $R_0 \leftarrow O;\ R_1 \leftarrow P$ 2: **for** $i = 0$ to $n - 1$ **do** 3: $b \leftarrow k_i$ 4: $R_{1-b} \leftarrow 2R_{1-b} + R_b$ 5: **end for** 6: **return** R_0

There exists a right-to-left variant of Algorithm 2. This is another classical method for evaluating $Q = kP$. It stems from the observation that, letting $k = \sum_{i=0}^{n-1} k_i 2^i$ the binary expansion of k, we can write $kP = \sum_{k_i=1} 2^i P$. A first (point) register R_0 serves as an accumulator and a second (point) register R_1 is used to contain the successive values of $2^i P$, $0 \leqslant i \leqslant n - 1$. When $k_i = 1$, R_1 is added to R_0. Register R_1 is then updated as $R_1 \leftarrow 2R_1$ so that at iteration i it contains $2^i P$. The detailed algorithm is presented in Algorithm 4. It suffers from the same deficiency as the one of the left-to-right variant (Algorithm 2); namely, it is not protected against SPA-type attacks. Again, the insertion of a dummy point addition when $k_i = 0$ can preclude these attacks. Using an additional (point) register, say R_{-1}, Line 3 in Algorithm 4 can be replaced with $R_{k_i-1} \leftarrow R_{k_i-1} + R_1$. But the resulting implementation is then prone to safe-error attacks. The right way to implement it is to effectively make use of *both*

R_0 and R_{-1} [14]. It is easily seen that in Algorithm 4 when using the dummy point addition (i.e., when Line 3 is replaced with $R_{k_i-1} \leftarrow R_{k_i-1} + R_1$), register R_{-1} contains the "complementary" value of R_0. Indeed, before entering iteration i, we have $R_0 = \sum_{k_j=1} 2^j P$ and $R_{-1} = \sum_{k_j=0} 2^j P$, $0 \leq j \leq i-1$. As a result, we have $R_0 + R_{-1} = \sum_{j=0}^{i-1} 2^j P = (2^i - 1)P$. Hence, initializing R_{-1} to P, the successive values of $2^i P$ can be equivalently obtained from $R_0 + R_{-1}$. Summing up, the right-to-left binary method becomes

1: $R_0 \leftarrow O$; $R_{-1} \leftarrow P$; $R_1 \leftarrow P$
2: **for** $i = 0$ to $n-1$ **do**
3:      $b \leftarrow k_i$; $R_{b-1} \leftarrow R_{b-1} + R_1$
4:          $R_1 \leftarrow R_0 + R_{-1}$
5: **end for**
6: **return** R_0

Performing a point addition when $k_i = 0$ in the previous algorithm requires one more (point) register. When memory is scarce, an alternative is to rely on *Joye's double-add algorithm* [14]. As in Montgomery ladder, it always repeats a same pattern of [effective] operations and requires only two (point) registers. The algorithm is given in Algorithm 5. It corresponds to the above algorithm where R_{-1} is renamed as R_1. Observe that the for-loop in the above algorithm can be rewritten into a single step as $R_{b-1} \leftarrow R_{b-1} + R_1 = R_{b-1} + (R_0 + R_{-1}) = 2R_{b-1} + R_{-b}$.

4   New Implementations

In [22], Meloni exploited the ZADD operation to propose scalar multiplications based on Euclidean addition chains and Zeckendorf's representation. In this section, we aim at making use of ZADD-like operations when designing scalar multiplication algorithms based on the classical binary representation. The crucial factor for implementing such an algorithm is to generate two points with the same Z-coordinate at every bit execution of scalar k.

To this end, we introduce a new operation referred to as *conjugate co-Z addition* and denoted ZADDC (for ZADD conjugate), using the efficient caching technique as described in [11,19]. This operation evaluates $(X_3 : Y_3 : Z_3) = P + Q = R$ with $P = (X_1 : Y_1 : Z)$ and $Q = (X_2 : Y_2 : Z)$, together with the value of $P - Q = S$ where S and R share the same Z-coordinate equal to Z_3. We have $-Q = (X_2 : -Y_2 : Z)$. Hence, letting $(\overline{X_3} : \overline{Y_3} : Z_3) = P - Q$, it is easily verified that $\overline{X_3} = (Y_1 + Y_2)^2 - W_1 - W_2$ and $\overline{Y_3} = (Y_1 + Y_2)(W_1 - \overline{X_3}) - A_1$, where W_1, W_2 and A_1 are computed during the course of $P + Q$ (cf. Algorithm 1). The additional cost for getting $P - Q$ from $P + Q$ is thus of only $1M + 1S$. Hence, the total cost for the ZADDC operation is of $\underline{6M + 3S}$. The detailed algorithm is given hereafter.

Algorithm 6. Conjugate co-Z point addition (ZADDC)

Require: $P = (X_1 : Y_1 : Z)$ and $Q = (X_2 : Y_2 : Z)$
Ensure: $(R, S) \leftarrow \text{ZADDC}(P, Q)$ where $R \leftarrow P + Q = (X_3 : Y_3 : Z_3)$ and $S \leftarrow P - Q = (\overline{X_3} : \overline{Y_3} : Z_3)$

---

   function $\text{ZADDC}(P, Q)$
      $C \leftarrow (X_1 - X_2)^2$
      $W_1 \leftarrow X_1 C; W_2 \leftarrow X_2 C$
      $D \leftarrow (Y_1 - Y_2)^2; A_1 \leftarrow Y_1(W_1 - W_2)$
      $X_3 \leftarrow D - W_1 - W_2; Y_3 \leftarrow (Y_1 - Y_2)(W_1 - X_3) - A_1; Z_3 \leftarrow Z(X_1 - X_2)$
      $\overline{D} \leftarrow (Y_1 + Y_2)^2$
      $\overline{X_3} \leftarrow \overline{D} - W_1 - W_2; \overline{Y_3} \leftarrow (Y_1 + Y_2)(W_1 - \overline{X_3}) - A_1$
   end function

4.1 Left-to-Right Scalar Multiplication

The main loop of Montgomery ladder (Algorithm 3) repeatedly evaluates the same two operations, namely

$$R_{1-b} \leftarrow R_{1-b} + R_b; R_b \leftarrow 2R_b \ .$$

We explain hereafter how to efficiently carry out this computation using co-Z arithmetic for elliptic curves.

First note that $2R_b$ can equivalently be rewritten as $(R_b + R_{1-b}) + (R_b - R_{1-b})$. So if T represents a temporary (point) register, the main loop of Montgomery ladder can be replaced with

$$T \leftarrow R_b - R_{1-b}$$
$$R_{1-b} \leftarrow R_b + R_{1-b}; R_b \leftarrow R_{1-b} + T \ .$$

Suppose now that R_b and R_{1-b} share the same Z-coordinate. Using Algorithm 6, we can compute $(R_{1-b}, T) \leftarrow \text{ZADDC}(R_b, R_{1-b})$. This requires $6M + 3S$. At this stage, observe that R_{1-b} and T have the same Z-coordinate. Hence, we can directly apply Algorithm 1 to get $(R_b, R_{1-b}) \leftarrow \text{ZADDU}(R_{1-b}, T)$. This requires $5M + 2S$. Again, observe that R_b and R_{1-b} share the same Z-coordinate at the end of the computation. The process can consequently be iterated. The total cost per bit amounts to $11M + 5S$ but can be reduced to $\underline{9M + 7S}$ (see § 4.4) by trading two (field) multiplications against two (field) squarings.

In the original Montgomery ladder, registers R_0 and R_1 are respectively initialized with point at infinity O and input point P. Since O is the only point with its Z-coordinate equal to 0, assuming that $k_{n-1} = 1$, we start the loop counter at $i = n - 2$ and initialize R_0 to P and R_1 to $2P$. It remains to ensure that the representations of P and $2P$ have the same Z-coordinate. This is achieved thanks to the DBLU operation (see § 4.3).

Putting all together, we so obtain the following implementation of the Montgomery ladder. Remark that register R_b plays the role of temporary register T.

Algorithm 7. Montgomery ladder with co-Z addition formulæ

Input: $P \in E(\mathbb{F}_q)$ and $k = (k_{n-1}, \ldots, k_0)_2 \in \mathbb{N}$ with $k_{n-1} = 1$
Output: $Q = kP$

1: $R_0 \leftarrow P; (R_1, R_0) \leftarrow \mathrm{DBLU}(R_0)$
2: **for** $i = n - 2$ down to 0 **do**
3:     $b \leftarrow k_i$
4:     $(R_{1-b}, R_b) \leftarrow \mathrm{ZADDC}(R_b, R_{1-b})$
5:     $(R_b, R_{1-b}) \leftarrow \mathrm{ZADDU}(R_{1-b}, R_b)$
6: **end for**
7: **return** R_0

4.2   Right-to-Left Scalar Multiplication Algorithm

As noticed in [14], Joye's double-add algorithm (Algorithm 5) is to some extent the dual of the Montgomery ladder. This appears more clearly by performing the double-add operation of the main loop, $R_{1-b} \leftarrow 2R_{1-b} + R_b$, in two steps as

$$T \leftarrow R_{1-b} + R_b; R_{1-b} \leftarrow T + R_{1-b}$$

using some temporary register T. If, at the beginning of the computation, R_b and R_{1-b} have the same Z-coordinate, two consecutive applications of the ZADDU algorithm allows one to evaluate the above expression with $2 \times (5M + 2S)$. Moreover, one has to take care that R_b and R_{1-b} have the same Z-coordinate at the end of the computation in order to make the process iterative. This can be done with an additional 3M.

But there is a more efficient way to get the equivalent representation for R_b. The value of R_b is unchanged during the evaluation of

$$(T, R_{1-b}) \leftarrow \mathrm{ZADDU}(R_{1-b}, R_b); (R_{1-b}, T) \leftarrow \mathrm{ZADDU}(T, R_{1-b})$$

and thus $R_b = T - R_{1-b}$ — where R_{1-b} is the initial input value. The latter ZADDU operation can therefore be replaced with a ZADDC operation; i.e.,

$$(R_{1-b}, R_b) \leftarrow \mathrm{ZADDC}(T, R_{1-b})$$

to get the expected result. The advantage of doing so is that R_b and R_{1-b} have the same Z-coordinate without additional work. This yields a total cost per bit of $11M + 5S$ for the main loop.

It remains to ensure that registers R_0 and R_1 are initialized with points sharing the same Z-coordinate. For the Montgomery ladder, we assumed that k_{n-1} was equal to 1. Here, we will assume that k_0 is equal to 1 to avoid to deal with the point at infinity. This condition can be automatically satisfied using certain DPA-type countermeasures (see § 5.2). Alternative strategies are described in [14]. The value $k_0 = 1$ leads to $R_0 \leftarrow P$ and $R_1 \leftarrow P$. The two registers have obviously the same Z-coordinate but are not different. The trick is to start the loop counter at $i = 2$ and to initialize R_0 and R_1 according the bit value of k_1. If $k_1 = 0$ we end up with $R_0 \leftarrow P$ and

$R_1 \leftarrow 3P$, and conversely if $k_1 = 1$ with $R_0 \leftarrow 3P$ and $R_1 \leftarrow P$. The TPLU operation (see §4.3) ensures that this is done so that the Z-coordinates are the same.

The complete resulting algorithm is depicted below. As for our implementation of the Montgomery ladder (Algorithm 7), remark that temporary register T is played by register R_b.

Algorithm 8. Joye's double-add algorithm with co-Z addition formulæ

Input: $P \in E(\mathbb{F}_q)$ and $k = (k_{n-1}, \ldots, k_0)_2 \in \mathbb{N}$ with $k_0 = 1$
Output: $Q = kP$

1: $b \leftarrow k_1$; $R_b \leftarrow P$; $(R_{1-b}, R_b) \leftarrow \mathrm{TPLU}(R_b)$
2: **for** $i = 2$ to $n - 1$ **do**
3:     $b \leftarrow k_i$
4:     $(R_b, R_{1-b}) \leftarrow \mathrm{ZADDU}(R_{1-b}, R_b)$
5:     $(R_{1-b}, R_b) \leftarrow \mathrm{ZADDC}(R_b, R_{1-b})$
6: **end for**
7: **return** R_0

It is striking to see the resemblance (or duality) between Algorithm 7 and Algorithm 8: they involve the same co-Z operations (but in reverse order) and scan scalar k in reverse directions.

4.3 Point Doubling and Tripling

Algorithms 7 and 8 respectively require a point doubling and a point tripling operation updating the input point. We describe how this can be implemented.

Initial Doubling Point. We have seen in Section 2 that the double of point $P = (X_1 : Y_1 : Z_1)$ can be obtained with $1\mathrm{M} + 8\mathrm{S} + 1\mathrm{c}$. By setting $Z_1 = 1$, the cost drops to $1\mathrm{M} + 5\mathrm{S}$:

$$\mathrm{X}(2P) = M^2 - 2S, \quad \mathrm{Y}(2P) = M(S - \mathrm{X}(2P)) - 8L, \quad \mathrm{Z}(2P) = 2Y_1$$

with $M = 3B + a$, $S = 2((X_1 + E)^2 - B - L)$, $L = E^2$, $B = X_1{}^2$, and $E = Y_1{}^2$. Since $Z(2P) = 2Y_1$, it follows that

$$(S : 8L : Z(2P)) \sim P \quad \text{with } S = 4X_1Y_1{}^2 \text{ and } L = Y_1{}^4$$

is an equivalent representation for point P. Updating point P such that its Z-coordinate is equal to that of $2P$ comes thus for free. We let $(2P, \tilde{P}) \leftarrow \mathrm{DBLU}(P)$ denote the corresponding operation, where $\tilde{P} \sim P$ and $\mathrm{Z}(\tilde{P}) = \mathrm{Z}(2P)$. The cost of DBLU operation (doubling with update) is $\underline{1\mathrm{M} + 5\mathrm{S}}$.

Initial Tripling Point. The triple of $P = (X_1 : Y_1 : 1)$ can be evaluated as $3P = P + 2P$ using co-Z arithmetic [20]. From $(2P, \tilde{P}) \leftarrow \mathrm{DBLU}(P)$, this can be obtained as

$\mathrm{ZADDU}(\tilde{\boldsymbol{P}}, 2\boldsymbol{P})$ with $5M + 2S$ and no additional cost to update $\boldsymbol{P}$ for its Z-coordinate becoming equal to that of $3\boldsymbol{P}$. The corresponding operation, tripling with update, is denoted $\mathrm{TPLU}(\boldsymbol{P})$ and its total cost is of $\underline{6M + 7S}$.

4.4   Combined Double-Add Operation

A point doubling-addition is the evaluation of $\boldsymbol{R} = 2\boldsymbol{P} + \boldsymbol{Q}$. This can be done in two steps as $\boldsymbol{T} \leftarrow \boldsymbol{P} + \boldsymbol{Q}$ followed by $\boldsymbol{R} \leftarrow \boldsymbol{P} + \boldsymbol{T}$. If $\boldsymbol{P}$ and $\boldsymbol{Q}$ have the same Z-coordinate, this requires $10M + 4S$ by two consecutive applications of the ZADDU function (Algorithm 1).

Things are slightly more complex if we wish that $\boldsymbol{R}$ and $\boldsymbol{Q}$ share the same Z-coordinate at the end of the computation. But if we compare the original Joye's double-add algorithm (Algorithm 5) and the corresponding algorithm we got using co-Z arithmetic (Algorithm 8), this is actually what is achieved. We can compute $(\boldsymbol{T}, \boldsymbol{P}) \leftarrow \mathrm{ZADDU}(\boldsymbol{P}, \boldsymbol{Q})$ followed by $(\boldsymbol{R}, \boldsymbol{Q}) \leftarrow \mathrm{ZADDC}(\boldsymbol{T}, \boldsymbol{P})$. We let $(\boldsymbol{R}, \boldsymbol{Q}) \leftarrow \mathrm{ZDAU}(\boldsymbol{P}, \boldsymbol{Q})$ denote the corresponding operation (ZDAU stands for *co-Z double-add with update*).

Algorithmically, we have:

1: $C' \leftarrow (X_1 - X_2)^2$
2: $W_1' \leftarrow X_1 C'; W_2' \leftarrow X_2 C'$
3: $D' \leftarrow (Y_1 - Y_2)^2; A_1' \leftarrow Y_1(W_1' - W_2')$
4: $X_3' \leftarrow D' - W_1' - W_2'; Y_3' \leftarrow (Y_1 - Y_2)(W_1' - X_3') - A_1'; Z_3' \leftarrow Z(X_1 - X_2)$
5: $X_1 \leftarrow W_1'; Y_1 \leftarrow A_1'; Z_1 \leftarrow Z_3'$
6: $C \leftarrow (X_3' - X_1)^2$
7: $W_1 \leftarrow X_3' C; W_2 \leftarrow X_1 C$
8: $D \leftarrow (Y_3' - Y_1)^2; A_1 \leftarrow Y_3'(W_1 - W_2)$
9: $X_3 \leftarrow D - W_1 - W_2; Y_3 \leftarrow (Y_3' - Y_1)(W_1 - X_3) - A_1; Z_3 \leftarrow Z_3'(X_3' - X_1)$
10: $\overline{D} \leftarrow (Y_3' + Y_1)^2$
11: $X_2 \leftarrow \overline{D} - W_1 - W_2; Y_2 \leftarrow (Y_3' + Y_1)(W_1 - X_2) - A_1; Z_2 \leftarrow Z_3$

A close inspection of the above algorithm shows that two (field) multiplications can be traded against two (field) squarings. Indeed, with the same notations, we have:

$$2Y_3' = (Y_1 - Y_2 + W_1' - X_3')^2 - D' - C - 2A_1' \ .$$

Also, we can skip the intermediate computation of $Z_3' = Z(X_1 - X_2)$ and obtain directly $2Z_3 = 2Z(X_1 - X_2)(X_3' - X_1)$ as

$$2Z_3 = Z\big((X_1 - X_2 + X_3' - X_1)^2 - C' - C\big) \ .$$

These modifications (in Lines 4 and 9) require some rescaling. For further optimization, some redundant or unused variables are suppressed. The resulting algorithm is detailed hereafter (Algorithm 9). It clearly appears that the ZDAU operation only requires $\underline{9M + 7S}$.

Algorithm 9. Co-Z point doubling-addition with update (ZDAU)

Require: $P = (X_1 : Y_1 : Z)$ and $Q = (X_2 : Y_2 : Z)$
Ensure: $(R, Q) \leftarrow \text{ZDAU}(P, Q)$ where $R \leftarrow 2P + Q = (X_3 : Y_3 : Z_3)$ and $Q \leftarrow (\lambda^2 X_2 : \lambda^3 Y_2 : Z_3)$ with $Z_3 = \lambda Z$ for some $\lambda \neq 0$

---

  function ZDAU(P, Q)
    $C' \leftarrow (X_1 - X_2)^2$
    $W_1' \leftarrow X_1 C'; W_2' \leftarrow X_2 C'$
    $D' \leftarrow (Y_1 - Y_2)^2; A_1' \leftarrow Y_1(W_1' - W_2')$
    $\hat{X}_3' \leftarrow D' - W_1' - W_2'$
    $C \leftarrow (\hat{X}_3' - W_1')^2$
    $Y_3' \leftarrow [(Y_1 - Y_2) + (W_1' - \hat{X}_3')]^2 - D' - C - 2A_1'$
    $W_1 \leftarrow 4\hat{X}_3' C; W_2 \leftarrow 4W_1' C$
    $D \leftarrow (Y_3' - 2A_1')^2; A_1 \leftarrow Y_3'(W_1 - W_2)$
    $X_3 \leftarrow D - W_1 - W_2; Y_3 \leftarrow (Y_3' - 2A_1')(W_1 - X_3) - A_1$
    $Z_3 \leftarrow Z\big((X_1 - X_2 + \hat{X}_3' - W_1')^2 - C' - C\big)$
    $\overline{D} \leftarrow (Y_3' + 2A_1')^2$
    $X_2 \leftarrow \overline{D} - W_1 - W_2; Y_2 \leftarrow (Y_3' + 2A_1')(W_1 - X_2) - A_1; Z_2 \leftarrow Z_3$
  end function

The combined ZDAU operation immediately gives rise to an alternative implementation of Joye's double-add algorithm (Algorithm 5). Compared to our first implementation (Algorithm 8), the cost per bit amounts to $\underline{9M + 7S}$ (instead of $11M + 5S$).

Algorithm 10. Joye's double-add algorithm with co-Z addition formulæ (II)

Input: $P \in E(\mathbb{F}_q)$ and $k = (k_{n-1}, \ldots, k_0)_2 \in \mathbb{N}$ with $k_0 = 1$
Output: $Q = kP$

1: $b \leftarrow k_1; R_b \leftarrow P; (R_{1-b}, R_b) \leftarrow \text{TPLU}(R_b)$
2: **for** $i = 2$ to $n - 1$ **do**
3:    $b \leftarrow k_i$
4:    $(R_{1-b}, R_b) \leftarrow \text{ZDAU}(R_{1-b}, R_b)$
5: **end for**
6: **return** R_0

Similar savings can be obtained for our implementation of the Montgomery ladder (i.e., Algorithm 7). However, as the ZADDU and ZADDC operations appear in reverse order, it is more difficult to handle. It is easy to trade 1M against 1S. In order to trade 2M against 2S, one has to consider two bits of scalar k at a time so as to allow to have the ZADDC operation performed prior to the ZADDU operation. The two previous M/S trade-offs can then be applied.

5   Discussion

5.1   Performance Analysis

Table 1 summarizes the cost of different types of addition and doubling-addition formulæ on elliptic curves. Each type of formula presents its own advantages depending on the coordinate system and the underlying scalar multiplication algorithm. Symbols $\mathcal{J}$ and $\mathcal{A}$ respectively stand for Jacobian coordinates and affine coordinates.

Table 1. Performance comparison of addition and doubling-addition formulæ

Operation	Notation	System	Cost
Point addition:			
– General addition [2]	ADD	$(\mathcal{J},\mathcal{J}) \to \mathcal{J}$	$11M + 5S$
– Co-Z addition [22]	ZADD	$(\mathcal{J},\mathcal{J}) \to \mathcal{J}$	$5M + 2S$
– Co-Z addition with update [22][a]	ZADDU	$(\mathcal{J},\mathcal{J}) \to \mathcal{J}$	$5M + 2S$
– General conjugate addition [19]	ADDC	$(\mathcal{J},\mathcal{J}) \to \mathcal{J}$	$12M + 6S$
– Conjugate co-Z addition (Alg. 6)	ZADDC	$(\mathcal{J},\mathcal{J}) \to \mathcal{J}$	$6M + 3S$
Point doubling-addition:			
– General doubling-addition [18]	DA	$(\mathcal{J},\mathcal{J}) \to \mathcal{J}$	$13M + 8S$
– Mixed doubling-addition [20]	mDA	$(\mathcal{J},\mathcal{A}) \to \mathcal{J}$	$11M + 7S$
– Co-Z doubling-addition with update (Alg. 9)	ZDAU	$(\mathcal{J},\mathcal{J}) \to \mathcal{J}$	$9M + 7S$

[a] See also Algorithm 1.

For the sake of comparison, we consider the typical ratio $S/M = 0.8$. Similar results can easily be derived for other ratios. We see that the co-Z addition (with or without update) improves the general addition by a speed-up factor of 56%. Almost as well, our conjugate co-Z addition formula improves the general conjugate addition by a factor of 50%. For the doubling-addition operations, our co-Z formula (including the update) is always faster; it is even faster than the best mixed doubling-addition formula. It yields a respective speed-up factor of 25% and of 12% compared to the general doubling-addition and to the mixed doubling-addition. In addition to speed, our new formulæ are also very efficient memory-wise. See [12, Appendix A] for detailed register allocations.

Table 2 compares the performance of our co-Z implementations with previous ones. Our improved right-to-left co-Z scalar multiplication algorithm (i.e., Algorithm 10) requires $9M + 7S$ per bit of scalar k. An application of Joye's double-add algorithm with the best doubling-addition (DA) formula [18] requires $13M + 8S$ per bit. Hence, with the usual ratio $S/M = 0.8$, our co-Z version of Joye's double-add algorithm yields a speed-up factor of 25%.

Furthermore, our left-to-right co-Z algorithm (i.e., Algorithm 7 as modified in §4.4) offers a speed competitive with known implementations of Montgomery ladder for

Table 2. Performance comparison of scalar multiplication algorithm

Algorithm	Operations	Cost per bit
Joye's double-add algorithm [14]: R → L		
− Basic version	DA	13M + 8S
− Co-Z version (Algorithm 10)	ZDAU	9M + 7S
Montgomery ladder [24]: L → R		
− Basic version	DBL and ADD	14M + 10S[a]
− X-only version [5,9,13]	XDBL and XADD	9M + 7S[b]
− Co-Z version (Algorithm 7)	ZADDC and ZADDU	9M + 7S[c]

[a] The cost assumes that curve parameter a is equal to -3. This allows the use of the faster point doubling formula: 3M + 5S instead of 1M + 8S + 1c; cf. Section 2.

[b] The cost assumes that multiplications by curve parameter a are negligible; e.g., $a = -3$. It also assumes that input point P is given in affine coordinates; i.e., $Z(P) = 1$. See [12, Appendix B] for a detailed implementation.

[c] With the improvements mentioned in § 4.4. The direct implementation of Algorithm 7 has a cost of 11M + 5S per bit.

general[2] elliptic curves. It only requires 9M + 7S per bit of scalar k. Moreover, we note that this cost is *independent* of the curve parameters.

5.2 Security Considerations

As explained in Section 3, Montgomery ladder and Joye's double-add algorithm are naturally protected against SPA-type attacks and safe-error attacks. Since our implementations are built on them and maintain the same regular pattern of instructions without using dummy instructions, they inherit of the same security features. Moreover, our proposed co-Z versions (i.e., Algorithms 7, 8 and 10) can be protected against DPA-type attacks; cf. [1, Chapter 29] for several methods.

Yet another important class of attacks against implementations are the fault attacks [3,4]. An additional advantage of Algorithm 7 (and of Algorithms 8 and 10) is that it is easy to assess the correctness of the computation by checking whether the output point belongs to the curve. We remark that the X-only versions of Montgomery ladder ([5,9,13]) do not permit it and so may be subject to (regular) fault attacks, as was demonstrated in [10].

6   Conclusion

Co-Z arithmetic as developed by Meloni provides an extremely fast point addition formula. So far, their usage for scalar multiplication algorithms was confined to Euclidean addition chains and the Zeckendorf's representation. In this paper, we developed new

---

[2] Montgomery introduced in [24] a curve shape that nicely combines with the X-only ladder, leading to a better cost per bit. But this shape does not cover all classes of elliptic curves. In particular, it does not apply to NIST recommended curves [25, Appendix D].

strategies and proposed a co-Z conjugate point addition formula as well as other companion co-Z formulæ. The merit of our approach resides in that the fast co-Z arithmetic nicely combines with certain binary ladders. Specifically, we applied co-Z techniques to Montgomery ladder and Joye's double-add algorithm. The so-obtained implementations are efficient and protected against a variety of implementation attacks. All in all, the implementations presented in this paper constitute a method of choice for the efficient yet secure implementation of elliptic curve cryptography in embedded systems or other memory-constrained devices.

As a side result, this paper also proposed the fastest point doubling-addition formula.

Acknowledgments. The authors would like to thank Jean-Luc Beuchat, Francisco Rodrìguez Henrìquez, Patrick Longa, and Francesco Sica for helpful discussions. We would also like to thank the anonymous referees for their useful comments. In particular, we thank the referee pointing out that the cost with the Montgomery ladder can be reduced to $9M + 7S$ per bit for general elliptic curves.

References

1. Avanzi, R., Cohen, H., Doche, C., Frey, G., Lange, T., Nguyen, K., Vercauteren, F.: Handbook of Elliptic and Hyperelliptic Curve Cryptography. CRC Press, Boca Raton (2005)
2. Bernstein, D.J., Lange, T.: Explicit-formulas database,
   http://www.hyperelliptic.org/EFD/jacobian.html
3. Biehl, I., Meyer, B., Müller, V.: Differential fault attacks on elliptic curve cryptosystems. In: Bellare, M. (ed.) CRYPTO 2000. LNCS, vol. 1880, pp. 131–146. Springer, Heidelberg (2000)
4. Boneh, D., DeMillo, R.A., Lipton, R.J.: On the importance of eliminating errors in cryptographic computations. Journal of Cryptology 14(2), 110–119 (2001); Extended abstract in Proc. of EUROCRYPT'97 (1997)
5. Brier, E., Joye, M.: Weierstraß elliptic curves and side-channel attacks. In: Naccache, D., Paillier, P. (eds.) PKC 2002. LNCS, vol. 2274, pp. 335–345. Springer, Heidelberg (2002)
6. Chudnovsky, D.V., Chudnovsky, G.V.: Sequences of numbers generated by addition in formal groups and new primality and factorization tests. Advances in Applied Mathematics 7(4), 385–434 (1986)
7. Cohen, H., Miyaji, A., Ono, T.: Efficient elliptic curve exponentiation using mixed coordinates. In: Ohta, K., Pei, D. (eds.) ASIACRYPT 1998. LNCS, vol. 1514, pp. 51–65. Springer, Heidelberg (1998)
8. Coron, J.-S.: Resistance against differential power analysis for elliptic curve cryptosystems. In: Koç, Ç.K., Paar, C. (eds.) CHES 1999. LNCS, vol. 1717, pp. 292–302. Springer, Heidelberg (1999)
9. Fischer, W., Giraud, C., Knudsen, E.W., Seifert, J.-P.: Parallel scalar multiplication on general elliptic curves over $\mathbb{F}_p$ hedged against non-differential side-channel attacks. Cryptology ePrint Archive, Report 2002/007 (2002), http://eprint.iacr.org/
10. Fouque, P.-A., Lercier, R., Réal, D., Valette, F.: Fault attack on elliptic curve Montgomery ladder implementation. In: Breveglieri, L., et al. (eds.) Fault Diagnosis and Tolerance in Cryptography (FDTC 2008), pp. 92–98. IEEE Computer Society, Los Alamitos (2008)

11. Galbraith, S., Lin, X., Scott, M.: A faster way to do ECC. Presented at 12th Workshop on Elliptic Curve Cryptography (ECC 2008), Utrecht, The Netherlands, September 22–24 (2008), Slides available at, http://www.hyperelliptic.org/tanja/conf/ECC08/slides/Mike-Scott.pdf

12. Goundar, R.R., Joye, M., Miyaji, A.: Co-Z addition formulæ and binary ladders on elliptic curves. Cryptology ePrint Archive, Report 2010/309 (2010), http://eprint.iacr.org/

13. Izu, T., Takagi, T.: A fast parallel elliptic curve multiplication resistant against side channel attacks. In: Naccache, D., Paillier, P. (eds.) PKC 2002. LNCS, vol. 2274, pp. 280–296. Springer, Heidelberg (2002)

14. Joye, M.: Highly regular right-to-left algorithms for scalar multiplication. In: Paillier, P., Verbauwhede, I. (eds.) CHES 2007. LNCS, vol. 4727, pp. 135–147. Springer, Heidelberg (2007)

15. Joye, M., Yen, S.-M.: The Montgomery powering ladder. In: Kaliski Jr., B.S., Koç, Ç.K., Paar, C. (eds.) CHES 2002. LNCS, vol. 2523, pp. 291–302. Springer, Heidelberg (2003)

16. Koblitz, N.: Elliptic curve cryptosystems. Mathematics of Computation 48(177), 203–209 (1987)

17. Kocher, P.C., Jaffe, J., Jun, B.: Differential power analysis. In: Wiener, M. (ed.) CRYPTO 1999. LNCS, vol. 1666, pp. 388–397. Springer, Heidelberg (1999)

18. Longa, P.: ECC Point Arithmetic Formulae (EPAF), http://patricklonga.bravehost.com/Jacobian.html

19. Longa, P., Gebotys, C.H.: Novel precomputation schemes for elliptic curve cryptosystems. In: Abdalla, M., Pointcheval, D., Fouque, P.-A., Vergnaud, D. (eds.) ACNS 2009. LNCS, vol. 5536, pp. 71–88. Springer, Heidelberg (2009)

20. Longa, P., Miri, A.: New composite operations and precomputation for elliptic curve cryptosystems over prime fields. In: Cramer, R. (ed.) PKC 2008. LNCS, vol. 4939, pp. 229–247. Springer, Heidelberg (2008)

21. López, J., Dahab, R.: Fast multiplication on elliptic curves over $GF(2^m)$ without precomputation. In: Koç, Ç.K., Paar, C. (eds.) CHES 1999. LNCS, vol. 1717, pp. 316–327. Springer, Heidelberg (1999)

22. Meloni, N.: New point addition formulæ for ECC applications. In: Carlet, C., Sunar, B. (eds.) WAIFI 2007. LNCS, vol. 4547, pp. 189–201. Springer, Heidelberg (2007)

23. Miller, V.S.: Use of elliptic curves in cryptography. In: Williams, H.C. (ed.) CRYPTO 1985. LNCS, vol. 218, pp. 417–426. Springer, Heidelberg (1986)

24. Montgomery, P.L.: Speeding up the Pollard and elliptic curve methods of factorization. Mathematics of Computation 48(177), 243–264 (1987)

25. National Institute of Standards and Technology. Digital Signature Standard (DSS). Federal Information Processing Standards Publication, FIPS PUB 186-3 (June 2009)

26. Yen, S.-M., Joye, M.: Checking before output may not be enough against fault-based cryptanalysis. IEEE Transactions on Computers 49(9), 967–970 (2000)

27. Yen, S.-M., Kim, S., Lim, S., Moon, S.-J.: A countermeasure against one physical cryptanalysis may benefit another attack. In: Kim, K.-c. (ed.) ICISC 2001. LNCS, vol. 2288, pp. 414–427. Springer, Heidelberg (2002)

Efficient Techniques for High-Speed Elliptic Curve Cryptography

Patrick Longa and Catherine Gebotys

Department of Electrical and Computer Engineering,
University of Waterloo, Canada
{plonga,cgebotys}@uwaterloo.ca

Abstract. In this paper, a thorough bottom-up optimization process (field, point and scalar arithmetic) is used to speed up the computation of elliptic curve point multiplication and report new speed records on modern x86-64 based processors. Our different implementations include elliptic curves using Jacobian coordinates, extended Twisted Edwards coordinates and the recently proposed Galbraith-Lin-Scott (GLS) method. Compared to state-of-the-art implementations on identical platforms the proposed techniques provide up to 30% speed improvements. Additionally, compared to the best previous published results on similar platforms improvements up to 31% are observed. This research is crucial for advancing high speed cryptography on new emerging processor architectures.

Keywords: Elliptic curve cryptosystem, point multiplication, point operation, field arithmetic, incomplete reduction, software implementation.

1 Introduction

Elliptic curve point multiplication, defined as $[k]P$, where P is a point with order r on an elliptic curve $E(\mathbb{F}_p)$ and $k \in [1, r-1]$ is an integer, is the central and most time-consuming operation in Elliptic Curve Cryptography (ECC). Hence, its efficient realization on commodity processors, such as the new generation based on the x86-64 ISA, has gained increasing importance in recent years.

In this work, we combine several efficient techniques at the different computational levels of point multiplication to achieve significant speed improvements on x86-64 based CPUs:

- At the field arithmetic, code scheduling on hand-written assembly modules is carefully tuned for high performance field operations. Furthermore, optimal combination of well-known techniques such as incomplete reduction (IR) [21] and elimination of conditional branches is performed.
- At the point arithmetic, the cost of explicit formulas is reduced further by minimizing the number of additions/subtractions and small constants and maximizing the use of operations exploiting IR. Also, we study the negative effect of (true) data dependencies between "close" field operations and propose *three* techniques to reduce their effect: field arithmetic scheduling, merging of point operations and merging of field operations.
- At the scalar arithmetic, we discuss our choice of recoding method and precomputation scheme and describe their efficient implementation.

S. Mangard and F.-X. Standaert (Eds.): CHES 2010, LNCS 6225, pp. 80–94, 2010.

Our implementations are carried out on elliptic curves using Jacobian and (extended) Twisted Edwards coordinates [13]. We also present results when applying the GLS method [8] that exploits an efficiently computable endomorphism to speed up the point multiplication over a quadratic extension field.

By efficiently combining the aforementioned techniques and other optimizations, we are able to compute a 256-bit point multiplication for the case of Jacobian and (extended) Twisted Edwards coordinates in only 337000 and 281000 cycles, respectively, on one core of an Intel Core 2 Duo processor. Compared to the previous results of 468000 and 362000 cycles (respect.) by Hisil et al. [14], our results achieve an improvement of about 28% and 22% (respect.). In the case of the GLS method, for Jacobian and (extended) Twisted Edwards coordinates, we compute one point multiplication in about 252000 and 229000 cycles (respect.) on the same processor, which compared to the best previous results by Galbraith et al. [7,8] (326000 and 293000 cycles, respect.) translate to improvements of about 23% and 22%, respectively.

Our implementations use the well-known MIRACL library by M. Scott [20], which contains an extensive set of cryptographic functions that simplified the development/optimization process of our crypto routines. Our programs, based on M. Scott's software, are faster due to several improvements discussed in this paper. We greatly thank M. Scott for making his software freely available for educational purposes.

Although our programs are portable to any x86-64 based CPU, in this work we present test results on *three* processors: 2.66GHz Intel Core 2 Duo E6750, 2.83GHz Intel Xeon E5440 and 2.6GHz AMD Opteron 252.

Our work is organized as follows. In Section 2, we briefly introduce ECC over prime fields and the GLS method, and summarize the most relevant features of x86-64 based processors. In Sections 3, 4 and 5 we describe the different techniques employed for the speed-up of point multiplication at the field, point and scalar arithmetic levels. In Section 6, we discuss how our optimizations apply to implementations using GLS. Finally, in Section 7, we present our timings for point multiplication and compare them to the best previous results.

2   Preliminaries

For a background in elliptic curves, the reader is referred to [12]. In this work, we consider the standard elliptic curve equation E: $y^2 = x^3 + ax + b$ (also known as short Weierstrass equation) over a prime field $\mathbb{F}_p$, where $a, b \in \mathbb{F}_p$.

Representation of points using (x, y) is known as affine coordinates ($\mathcal{A}$). It is common practice to replace this representation with projective coordinates since affine coordinates are expensive over prime fields due to costly field inversions. In this work, we use Jacobian coordinates ($\mathcal{J}$), where each projective point $(X : Y : Z)$ corresponds to the affine point $(X/Z^2, Y/Z^3)$, $Z \neq 0$. The negative of $(X : Y : Z)$ is $(X : -Y : Z)$, and $(X : Y : Z) = \{(\lambda^2 X, \lambda^3 Y, \lambda Z) : \lambda \in \mathbb{F}_p^*\}$.

The central operation, namely point multiplication (denoted by $[k]P$, for a point $P \in E(\mathbb{F}_p)$), is traditionally carried out through a series of point doublings and additions using some algorithm such as double-and-add. More efficiently, a doubling followed by another doubling can be computed as $\mathcal{J} \leftarrow 2\mathcal{J}$ and every doubling followed by an addition can utilize the new doubling-addition by [15,19] to compute $\mathcal{J} \leftarrow 2\mathcal{J}+\mathcal{A}$ or $\mathcal{J} \leftarrow 2\mathcal{J}+\mathcal{J}$. All these formulas can also be found in our database of state-of-the-art formulas using Jacobian coord. [16].

Different curve forms exhibiting faster group arithmetic have been studied during the last few years. A good example is given by Twisted Edwards. This curve form, proposed in [2], is a generalization of Edwards curves [3] and has the equation $ax^2+y^2 = 1+dx^2y^2$, where $a, d \in \mathbb{F}_p$ are distinct nonzero elements. For this case, each triplet $(X : Y : Z)$ corresponds to the affine point $(X/Z, Y/Z)$, $Z \neq 0$, in homogeneous projective coordinates (denoted by $\mathcal{E}$). Later, Hisil et al. [13] introduced an extended system (called extended Twisted Edwards coord.; denoted by $\mathcal{E}^e$), where each point $(X : Y : Z : T)$ corresponds to $(X/Z, Y/Z, 1, T/Z)$ in affine, $T = XY/Z$ and $(X : Y : Z : T) = \{(\lambda X, \lambda Y, \lambda Z, \lambda T) : \lambda \in \mathbb{F}_p^*\}$.

Hisil et al. [13] also suggest the map $(x, y) \mapsto (x/\sqrt{-a}, y)$ to convert the previous curve to $-x^2+y^2 = 1+d'x^2y^2$, where $d' = -d/a$, allowing further reductions in the cost of point operations. For the point multiplication, they ultimately propose to compute a doubling followed by an addition as $\mathcal{E}^e \leftarrow 2\mathcal{E}$ and $\mathcal{E} \leftarrow \mathcal{E}^e+\mathcal{E}^e$ or $\mathcal{E} \leftarrow \mathcal{E}^e+\mathcal{A}$ (which can be unified into a doubling-addition operation with the form $\mathcal{E} \leftarrow (2\mathcal{E})^e+\mathcal{E}^e$ or $\mathcal{E} \leftarrow (2\mathcal{E})^e+\mathcal{A}$), and the remaining doublings as $\mathcal{E} \leftarrow 2\mathcal{E}$.

In Table 1, we have summarized the cost of formulas[1] using $\mathcal{J}$ and $\mathcal{E}/\mathcal{E}^e$. Although variations to these formulas exist [16], these sometimes involve an increased number of "small" operations such as additions/subtractions. On some platforms, the extra cost may not be negligible. Formulas in Table 1 have been selected so that the *overall* cost is minimal on the targeted platforms. In Section 4, we apply some techniques to minimize the number of such "small" operations and, thus, to reduce the cost of point operations further.

Table 1. Cost of point operations on Weierstrass and Twisted Edwards curves

Point Operation	Coord.	Weierstrass ($a = -3$)	Coord.	Twisted Edw. ($a = -1$)
Doubling	$\mathcal{J} \leftarrow 2\mathcal{J}$	4M+4S	$\mathcal{E} \leftarrow 2\mathcal{E}$	4M+3S
Mixed addition	$\mathcal{J} \leftarrow \mathcal{J}+\mathcal{A}$	8M+3S	$\mathcal{E} \leftarrow \mathcal{E}^e+\mathcal{A}$	7M
General addition	$\mathcal{J} \rightarrow \mathcal{J}+\mathcal{J}$	11M+3S [1]	$\mathcal{E} \leftarrow \mathcal{E}^e+\mathcal{E}^e$	8M
Mixed doubling-addition	$\mathcal{J} \leftarrow 2\mathcal{J}+\mathcal{A}$	13M+5S	$\mathcal{E} \leftarrow (2\mathcal{E})^e+\mathcal{A}$	11M+3S
General doubling-addition	$\mathcal{J} \rightarrow 2\mathcal{J}+\mathcal{J}$	16M+5S [1]	$\mathcal{E} \leftarrow (2\mathcal{E})^e+\mathcal{E}^e$	12M+3S

(1) Using cached values.

A recent method to improve the computation of point multiplication was proposed by Galbraith et al. [8], in which the computation is performed on a quadratic twist of a curve E over $\mathbb{F}_{p^2}$ with an efficiently computable homomorphism $\psi(x, y) \rightarrow (\alpha x, \alpha y)$, $\psi(P) = \lambda P$. Then, following [9], $[k]P$ can be computed as a multiple point multiplication with the form $[k_0]P + [k_1](\lambda P)$, where k_0 and k_1 have approximately half the bitlength of k. See [7,8] for complete details.

---

[1] Field operations: M = multiplication, S = squaring, Add = addition, Sub = subtraction, Mulx = multiplication by x, Divx = division by x, Neg = negation.

In this work, we present two "traditional" implementations (on Weierstrass and Twisted Edwards curves) and another two using the GLS method (again, one per curve). For the traditional case (and to be competitive with other implementations in the literature), we have written the underlying field arithmetic over $\mathbb{F}_p$ using assembly language. On the other hand, for the GLS method we reuse the efficient modules for $\mathbb{F}_{p^2}$ field arithmetic provided with MIRACL.

For $\mathbb{F}_p$, we consider for maximal speed-up a pseudo-Mersenne prime with the form $p = 2^m - c$, where $m = n.w$ on an w-bit platform, $n \in \mathbb{Z}^+$, and c is a "small" integer (i.e., $c < 2^w$). These primes are highly efficient for performing modular reduction and support other optimizations such as elimination of conditional branches. Similarly, for the GLS method, field arithmetic over $\mathbb{F}_{p^2}$ provided by MIRACL considers a Mersenne prime $p = 2^t - 1$ (i.e., t is prime).

For a more in-depth treatment of the techniques exploited in our implementations, the reader is referred to the extended paper version [18].

2.1   The x86-64 Based Processor Family

Modern CPUs from the desktop and server classes have decisively adopted the 64-bit x86 ISA (a.k.a. x86-64). This new instruction set expands general-purpose registers (GPRs) from 32 to 64 bits, allows arithmetic and logical operations on 64-bit integers and increments the number of GPRs, among other enhancements.

It seems that the move to 64 bits, with the inclusion of a powerful 64-bit integer multiplier, favors prime fields. Although the analysis becomes complex and processor dependent, our tests on the targeted processors suggest that SSE2 and its extensions seem not to be advantageous by themselves for the $\mathbb{F}_p$ arithmetic. This is probably due to the lack of carry handling and the fact that SSE2 multipliers can perform vector-multiplication with operands up to 32 bits only [11]. However, this outcome could change with improved SIMD extensions.

Another relevant feature of modern CPUs is their highly pipelined architectures. For instance, experiments by [6] suggest that Core 2 Duo and AMD architectures have pipelines with 15 and 12 stages, respectively. Although sophisticated branch prediction techniques exist, it is expected that the "random" nature of crypto computations, specifically of modular reduction, causes expensive mispredictions that force the pipeline to flush. In this work, we present experimental data quantifying the performance improvement obtained by eliminating branches in the field arithmetic (see Section 3).

Another direct consequence of highly pipelined architectures is that data dependencies between "close" instructions may insert a high penalty. Data dependencies that are relevant to our application are read-after-write (RAW), which can be found between a considerable number of field operations when the result of a previous operation is required as input by the next operation. Our tests show that, if field operations are not scheduled properly, RAW dependencies can cause the pipeline to stall for several cycles degrading the performance significantly. In this work, we propose several techniques that help to minimize this problem, enhancing the performance of point multiplication (see Section 4).

3   Optimizations at the Field Arithmetic Level

In this section, we discuss the algorithms and optimizations that were applied to modular operations. All tests described were performed on our assembly language module implementing the field arithmetic over $\mathbb{F}_p$.

3.1   Field Multiplication

Schoolbook and Comba are the methods of choice for performing this operation on general purpose processors (GPPs). Methods such as Karatsuba multiplication theoretically reduce the number of integer multiplications but increase the number of other (cheaper) operations, which are not inexpensive in our case.

In x86-64 based CPUs, integer multiplication is relatively expensive. For instance, on an Intel Core 2 Duo, 64-bit multiplications can be executed every 5 clock cycles in a dependence chain [5]. A strategy to reduce costs is to interleave other (cheaper) operations with integer multiplications to exploit the *instruction-level parallelism* (ILP) found in modern processors. Precisely, both schoolbook (also known as operand scanning) and Comba's method (also known as product scanning) exhibit this attractive feature. Both methods require n^2 w-bit multiplications when multiplying two n-digit numbers. However, we choose to implement Comba's method since it requires approximately $3n^2$ w-bit additions, whereas schoolbook requires $4n^2$ (see Section 5.3.1 of [4]).

3.2   Other "Cheaper" Operations

There are *two* key techniques that we exploit to reduce the cost of additions, subtractions, and divisions/multiplications by small constants:

Incomplete Reduction (IR). This technique was introduced by Yanik et al. [21]. Given two numbers in the range $[0, p-1]$, it consists of allowing the result of an operation to stay in the range $[0, 2^s-1]$ instead of executing a complete reduction, where $p < 2^s < 2p-1$, $s = n.w$, w is the wordlength (e.g., $w = 64$) and n is the number of words. If the modulus is a pseudo-Mersenne prime with form $2^m - c$ such that $m = s$ and $c < 2^w$, the method gets even more advantageous. For example, in the case of addition the result can be reduced by first discarding the carry bit in the most significant word and then adding the correction value c.

In Table 2, we summarize the cost of field operations and the gain in performance when exploiting IR. As can be seen, in our experiments using $p = 2^{256} - 189$ we obtain significant reductions in cost ranging from 20% to up to 41%.

It is important to note that, because multiplication and squaring accept inputs in the range $[0, 2^s - 1]$, an operation using IR can precede any of these two operations. Then it turns out that virtually all additions and multiplications/divisions by small constants can be implemented with IR in our software.

Elimination of Conditional Branches. Following the trend of other crypto implementations [10,20] and to avoid the high cost of branch misprediction on highly pipelined processors, we have implemented field addition, subtraction and multiplication/division by small constants without conditional branches [18].

Table 2. Cost (in cycles) of modular operations when using incomplete reduction (IR) against complete reduction (CR) ($p = 2^{256} - 189$)

Modular Operation	Core 2 Duo E6750			Opteron 252		
	IR	CR	Cost reduction (%)	IR	CR	Cost reduction (%)
Addition	20	25	20%	13	20	35%
Multiplication by 2	19	24	21%	10	17	41%
Multiplication by 3	28	43	35%	15	23	35%
Division by 2	20	25	20%	11	18	39%

In Table 3, we present the difference in performance for several field operations. In our tests using the prime $p = 2^{256} - 189$, we observed cost reductions as high as 50%. Remarkably, it can be seen that the greatest performance gains are obtained for operations exploiting IR. In conclusion, elimination of conditional branches favors more strongly our implementations, which are based on IR.

Table 3. Cost (in cycles) of modular operations without conditional branches (w/o CB) against operations using conditional branches (with CB) ($p = 2^{256} - 189$)

Modular Operation	Core 2 Duo E6750			Opteron 252		
	w/o CB	with CB	Cost reduction (%)	w/o CB	with CB	Cost reduction (%)
Subtraction	21	37	43%	16	23	30%
Addition with IR	20	37	46%	13	21	38%
Addition	25	39	36%	20	23	13%
Mult. by 2 with IR	19	38	50%	10	19	47%
Multiplication by 2	24	38	37%	17	20	15%

Table 4. Cost (in cycles) of modular operations

Modular Operation	Intel Core 2 Duo		AMD Opteron	
	This work $p = 2^{256} - 189$	mpFq [10] $p = 2^{255} - 19$	This work $p = 2^{256} - 189$	mpFq [10] $p = 2^{255} - 19$
Addition	20 [1]	21	13 [1]	19
Subtraction	21	24	16	22
Multiplication by 2	19 [1]	N/A	10 [1]	N/A
Division by 2	20 [1]	N/A	11 [1]	N/A
Squaring	101	107	65	72
Multiplication	110	141	80	108

(1) Using incomplete reduction.

Finally, Table 4 summarizes the cost of field operations optimized with the techniques discussed above and used in our implementations, and compare them with mpFq [10], a well-known and highly-efficient crypto library. Note that timings for mpFq are reported for Intel Core 2 Duo 6700 and AMD Opteron 250 [10], which have very similar architectures to those used for our tests. Although our modular operations and those from mpFq are based on a different modulus

p, comparisons in Table 4 are useful to explain part of the performance improvement obtained by our implementations in comparison with the implementation of *curve25519* using `mpFq` (see comparisons in Section 7).

4   Optimizations at the Point Arithmetic Level

In this section, we describe our choice of point formulas and some techniques to reduce their costs further. Also, we analyze how to reduce the computing cost of point multiplication by minimizing the number of pipeline stalls caused by contiguous field operations holding (true) data dependencies.

4.1   Our Choice of Explicit Formulas

For our programs, we choose the execution patterns based on doublings and doubling-additions proposed by Longa [15] and Hisil et al. [13] for $\mathcal{J}$ and $\mathcal{E}/\mathcal{E}^e$, respectively (see Section 2). For $\mathcal{J}$, we take as starting points the doubling formula from pp. 90 of [12] that costs 4M+4S, and the doubling-addition formula (3.5), pp. 37 of [15], that costs 13M+5S (16M+5S in the general case [16]). For $\mathcal{E}/\mathcal{E}^e$, we choose the doubling formula on pp. 400 of [2] that costs 4M+3S and the (dedicated) doubling-(dedicated) addition formulas from pp. 332-333 of [13] which cost in total 11M+3S (12M+3S in the general case). The previous costs (which are minimal on the targeted platforms in terms of mults and squarings) are obtained by setting $a = -3$ on $\mathcal{J}$ and $a = -1$ on $\mathcal{E}/\mathcal{E}^e$ [13] and avoiding the S-M tradings. Moreover, following [13], we precalculate (X_2+Y_2), (X_2-Y_2), $2Z_2$ and $2T_2$ to save *two* Adds and *two* Mul2 in the (dedicated) addition formula.

4.2   Minimizing the Cost of Point Operations

Further cost reduction of point operations can be achieved by exploiting the equivalence relation of projective coordinates. Consider, for example, the doubling formula using $\mathcal{J}$ in pp. 90-91 of [12] that has an overall cost of 4M+4S+ 1Add+4Sub+2Mul2+1Mul3+1Div2. If we fix $\lambda = 2^{-1} \in \mathbb{F}_p^*$ that formula can be modified to the following

$$X_2 = A^2 - 2B, \quad Y_2 = A(B - X_2) - Y_1^4, \quad Z_2 = Y_1 Z_1 \qquad (1)$$

where $A = 3(X_1 + Z_1^2)(X_1 - Z_1^2)/2$, $B = X_1 Y_1^2$. With formula (1), the operation count is reduced to 4M+4S+1Add$_{IR}$+5Sub+1Mul3$_{IR}$+1Div2$_{IR}$ (where *operation*$_{IR}$ represents an operation using incomplete reduction), replacing *two* multiplications by 2 with *one* subtraction and allowing the optimal use of incomplete reductions (every addition and multiplication/division by constants precedes a multiplication or squaring).

Additionally, depending on the relative cost of additions and subtractions (and the feasibility of using efficient "fused" subtractions such as $a - 2b \pmod{p}$; see Section 4.3) one may "convert" additions to subtractions (or vice versa) by applying $\lambda = -1 \in \mathbb{F}_p^*$ to a given formula. Refer to Appendix A for the details of the revised formulas exploiting these techniques.

4.3   Minimizing the Effect of Data Dependencies

Next, we present *three* techniques that help to reduce the number of memory stalls caused by RAW dependencies between successive field operations. For the remainder (and abusing notation), we define as *contiguous data dependence* if the output of a field operation is required as input by the immediately following operation causing the pipeline to stall in certain processor architecture.

Scheduling of Field Operations. The simplest solution to eliminate contiguous data dependencies is to perform a careful scheduling of the field operations inside point formulas. However, there is no unique solution and finding the optimal "arrangement" could be quite difficult and compiler/platform dependent. Instead, we demonstrate that some effort minimizing the number of these dependencies increases the overall performance significantly.

We tested several field operation "arrangements" to observe the potential impact of scheduling field operations. We detail here a few of our tests with field multiplication on an Intel Core 2 Duo. For example, let us consider the operation sequences given in Table 5. As can be seen, Sequence 1 involves a series of "ideal" data-independent multiplications, where the output of a given operation is not an input to the immediately following operation. In this case, the execution reaches its maximal performance with approx. 110 cycles/multiplication (see Table 4). Contrarily, the second sequence is highly-dependent because each output is required as input in the following operation. This is the worst-case scenario with an average of 128 cycles/mult., which is about 14% less efficient than the "ideal" case. We also studied other possible arrangements such as Sequence 3, in which operands of Sequence 2 have been reordered. This slightly amortizes the impact of contiguous data dependencies, improving the performance to 125 cycles/mult.

Table 5. Various sequences of field operations with different levels of contiguous data dependence. `Mult(opi,opj,resk)` denotes the field operation $resk \leftarrow opi * opj$

Sequence 1	Sequence 2	Sequence 3
> Mult(op1,op2,res1)	> Mult(op1,op2,res1)	> Mult(op1,op2,res1)
> Mult(op3,op4,res2)	> Mult(res1,op4,res2)	> Mult(op4,res1,res2)
> Mult(op5,op6,res3)	> Mult(res2,op6,res3)	> Mult(op6,res2,res3)
> Mult(op7,op8,res4)	> Mult(res3,op8,res4)	> Mult(op8,res3,res4)

Similarly, we have also tested the effect of contiguous data dependencies on other field operations, and detected that the cost reduction obtained by switching from an execution with strong contiguous data dependence (worst-case scenario, Sequence 2) to an execution with no contiguous data dependencies (best-case scenario, Sequence 1) ranges from approx. 9% to up to 33% on an Intel Core 2 Duo.

Merging point operations. This technique complements and increases the gain obtained by scheduling field operations. As expected, in some cases it is not possible to eliminate all contiguous data dependencies in a point formula by simple rescheduling. A clever way to increase the chances of eliminating more of these dependencies is by "merging" successive point operations.

It appears natural to merge successive doublings or a doubling and an addition. For our implementations, we use wNAF with window size $w = 5$ to recode the scalar (see Section 5). Then, at least *five* successive doublings between additions are expected. An efficient solution is to merge *four* consecutive doublings in a separate function and merge each addition with the precedent doubling in another function. In this way, we have been able to minimize most contiguous data dependencies and improve the overall performance further. As a side-effect, the number of function calls to point formulas is also reduced dramatically.

Merging field operations. If certain field operations are merged (and there are enough registers available) one can directly avoid memory stalls caused by dependencies between the writing to memory of the result and its posterior reading in the following field operation. A positive side-effect of this approach is that memory accesses (and potential cache misses) are also minimized.

Some crypto libraries have already experimented with this approach to certain extent. For example, MIRACL includes a double subtraction operation that executes $a-b-c \,(\mathrm{mod}\,p)$ and a multiplication by 3 executed as $a+a+a \,(\mathrm{mod}\,p)$. However, in this work we have maximized the use of registers and included other combinations such as $a-2b \,(\mathrm{mod}\,p)$ and the merging of $a-b \,(\mathrm{mod}\,p)$ and $(a-b)-2c \,(\mathrm{mod}\,p)$. We remark that this list is not exhaustive. Different platforms with more registers or different coordinate systems/underlying fields may enable a much wider range of merging options (for instance, see Section 6 for the merging options suggested for quadratic extension fields).

To illustrate the impact of scheduling field operations, merging point operations and merging field operations, we show in Table 6 the cost of a point doubling when using these techniques in comparison with a naïve implementation with a high number of dependencies.

Table 6. Cost (in cycles) of point doubling with different number of contiguous data dependencies (Jacobian coordinates, $p = 2^{256} - 189$)

Technique	# contiguous data-depend. per doubling	Core 2 Duo E6750		Opteron 252	
		Cost per doubling	Relative reduction (%)	Cost per doubling	Relative reduction (%)
"Unscheduled"	10	1115	–	786	–
Scheduled/merged	1.25	979	12%	726	8%

As shown in Table 6, by reducing the number of dependencies from *ten* to about *one* per doubling, minimizing function calls and reducing the number of memory reads/writes, we are able to reduce the cost of a doubling by 12% and 8% on Core 2 Duo and Opteron processors, respectively.

Following the strategies presented in this section, we have first minimized the cost of point operations (cf. §4.2) and then carefully scheduled (merged) field operations inside (merged) point operations so that memory stalls and memory accesses are minimized. See Appendix A for costs and scheduling details of most relevant point operations used in our implementations and discussed in §4.1.

5 Optimizations at the Scalar Arithmetic Level

In this section, we describe our choice of algorithms for the computation of point multiplication and precomputation.

For scalar recoding we use width-w Non-Adjacent Form (wNAF), which offers minimal nonzero density among signed binary representations for a given window width [1]. In particular, we use Alg. 3.35 of [12] for conversion from integer to wNAF representation. Although left-to-right conversion algorithms exist [1], which save memory and allow on-the-fly computation of point multiplication, they are not advantageous on the targeted CPUs. In fact, our tests show that converting the scalar to wNAF and then executing the point multiplication achieves higher performance than interleaving both stages. This could be explained by the fact that the latter approach "interrupts" the otherwise smooth flow of point multiplication by calling the conversion function at every iteration of the double-and-add algorithm.

For precomputation on $\mathcal{J}$, we have chosen a variant of the LM scheme [19] that does not require inversions (see Section 7.1 of [17]). This method achieves the lowest precomputing cost, given by $(5L+2)M+(2L+4)S$, where L represents the number of non-trivial points (note that we avoid here the S-M trading in the first doubling). On $\mathcal{E}/\mathcal{E}^e$ coordinates, we precompute points using the traditional sequence $P + 2P + \ldots + 2P$, adding $2P$ with general additions. Because precomputed points are left in projective coordinates no inversion is required and the cost is given by $(8L + 4)M+2S$. For both $\mathcal{J}$ and $\mathcal{E}/\mathcal{E}^e$, we have chosen a window with size $w = 5$ (i.e., precomputing $\{P, [3]P, \ldots, [15]P\}, L = 7$), which is optimal and slightly better than fractional windows using $L = 6$ or $L = 8$.

6 Implementation Using GLS

For our implementations using GLS, we apply similar techniques to those described in Sections 4 and 5 for the elliptic curve arithmetic. As mentioned previously, we use the optimized assembly implementation of the field arithmetic over $\mathbb{F}_{p^2}$ by M. Scott [20]. This library exploits the "nice" Mersenne prime $2^{127} - 1$, which allows a very simple reduction step with no conditional branches.

Note that the field arithmetic over $\mathbb{F}_{p^2}$ in fact translates to a bunch of $\mathbb{F}_p$ operations, where p has 127 bits in our case. For instance, each multiplication using Karatsuba (as implemented in [20]) involves 3 $\mathbb{F}_p$ multiplications and 5 $\mathbb{F}_p$ additions/subtractions. Thus, the scheduling and merging of field operations described in Section 4.3 are first applied to this underlying layer over $\mathbb{F}_p$ and then extended to the upper layer over $\mathbb{F}_{p^2}$.

For the point arithmetic, we slightly modify formulas described in Section 4 and Appendix A since in this case these require a few extra multiplications with the twisted curve parameter μ (see Appendix B). For example, the (dedicated) addition using $\mathcal{E}/\mathcal{E}^e$ with cost 8M has to be replaced with a formula that costs 9M (discussed in pp. 332 of [13] as "9M+1D"). Moreover, field arithmetic over $\mathbb{F}_{p^2}$ enables a much richer opportunity for merging field operations. In our implementations, we include $a - 2b \,(\mathrm{mod}\,p)$, $(a + a + a)/2 \,(\mathrm{mod}\,p)$, $a + b - c \,(\mathrm{mod}\,p)$, the merging of $a+b\,(\mathrm{mod}\,p)$ and $a-b\,(\mathrm{mod}\,p)$, the merging of $a-b\,(\mathrm{mod}\,p)$ and $c-d\,(\mathrm{mod}\,p)$, and the merging of $a+a\,(\mathrm{mod}\,p)$ and $a+a+a\,(\mathrm{mod}\,p)$. For complete details about point formulas and their implementation for the GLS method, the reader is referred to Appendix B in the extended paper version [18].

For the multiple point multiplication $[k_0]P+[k_1](\lambda P)$, each of the two scalars k_0 and k_1 is converted using fractional wNAF, and then the evaluation stage is executed using interleaving (see Alg. 3.51 of [12]). Again, we remark that the separation of the conversion and evaluation stages yields better performance in our case. For precomputation on $\mathcal{J}$, we use the LM scheme (see Section 4 of [19]) that has minimal cost among methods using only *one* inversion, i.e., $1\mathrm{I}+(9L+1)\mathrm{M}+(2L+5)\mathrm{S}$ (we avoid here the S-M trading in the first doubling). A fractional window with $L = 6$ achieves the optimal performance in our case. Again, on $\mathcal{E}/\mathcal{E}^e$ we precompute points using general additions in the sequence $P+2P+\ldots+2P$. Precomputed points are better left in projective coordinates, in which case the cost is given by $(9L+4)\mathrm{M}+2\mathrm{S}$. In this case, an integral window $w = 5$ (i.e., $L = 7$) achieves optimal performance. As pointed out by [8], precomputing $\{P,[3]\psi(P),\ldots,[2L+1]\psi(P)\}$ can be done on-the-fly at low cost.

7   Implementation Results

In this section, we summarize the timings obtained by our "traditional" implementations using $\mathcal{E}/\mathcal{E}^e$ and $\mathcal{J}$ (called *ted256189* and *jac256189*, respect.), and our implementations using GLS (called *ted1271gls* and *jac1271gls*, respect.), when running them on a single core of the targeted x86-64 based CPUs. The curves used in these implementations are described in detail in Appendix B. For verification of each implementation, the results of 10^4 point multiplications with "random" scalars were all validated using MIRACL. Several "random" point multiplications were also verified with Magma.

All the tested programs were compiled with gcc v4.4.1 on the Intel Core 2 Duo E6750 and with gcc v4.3.4 on the Intel Xeon E5440 and Opteron 252 processors. For measuring computing time, we follow [10] and use a method based on cycle counts. To obtain our timings, we ran each implementation 10^5 times with randomly generated scalars, averaged and approximated the results to the nearest 1000 cycles. Table 7 summarizes our results, labeled as *ted1271gls*, *jac1271gls*, *ted256189* and *jac256189*. All costs include scalar conversion, the point multiplication computation (precomputation and evaluation stages) and the final normalization step to affine. Table 7 also shows the cycle counts that we obtained when running the implementations by M. Scott (displayed as *gls1271-ref4* and *gls1271-ref3* [20]) on exactly the same platforms. Finally, the last 5 rows of the table detail cycle counts of several state-of-the-art implementations as reported in the literature. However, these referenced results are used only to provide an approximate comparison since the processor platforms are not identical (though they use very similar processors).

As can be seen in Table 7, our fastest implementation on the targeted platforms is *ted1271gls*, using $\mathcal{E}/\mathcal{E}^e$ with the GLS method. This implementation is about 22% faster than the previous record set by *gls1271-ref4* [7] on a slightly

Table 7. Cost (in cycles) of point multiplication

Implementation	Coord.	Field Arithm.	Core 2 Duo E6750	Xeon E5440	Opteron 252
ted1271gls	$\mathcal{E}/\mathcal{E}^e$	$\mathbb{F}_{p^2}$, 127-bit	**229000**	**230000**	**211000**
jac1271gls	$\mathcal{J}$	$\mathbb{F}_{p^2}$, 127-bit	252000	255000	238000
ted256189	$\mathcal{E}/\mathcal{E}^e$	$\mathbb{F}_p$, 256-bit	281000	289000	232000
jac256189	$\mathcal{J}$	$\mathbb{F}_p$, 256-bit	337000	343000	274000
gls1271-ref4 [20]	$\mathcal{E}^{inv}$	$\mathbb{F}_{p^2}$, 127-bit	295000	296000	295000
gls1271-ref3 [20]	$\mathcal{J}$	$\mathbb{F}_{p^2}$, 127-bit	332000	332000	341000
gls1271-ref4 [7]	$\mathcal{E}^{inv}$	$\mathbb{F}_{p^2}$, 127-bit	293000 [1]	–	–
gls1271-ref3 [8]	$\mathcal{J}$	$\mathbb{F}_{p^2}$, 127-bit	326000 [1]	–	–
curve25519 [10]	Montgomery	$\mathbb{F}_p$, 255-bit	386000 [2]	–	307000 [4]
Hisil et al. [14]	$\mathcal{E}/\mathcal{E}^e$	$\mathbb{F}_p$, 256-bit	362000 [3]	–	–
Hisil et al. [14]	$\mathcal{J}$	$\mathbb{F}_p$, 256-bit	468000 [3]	–	–

(1) On a 1.66GHz Intel Core 2 Duo. (2) On a 2.66GHz Intel Core 2 Duo E6700.
(3) On a 2.66GHz Intel Core 2 Duo E6550. (4) On a 2.4GHz AMD Opteron 250.

different processor (1.66GHz Intel Core 2 Duo). A more precise comparison, however, would be between measurements on identical processor platforms. In this case, *ted1271gls* is approx. 22%, 22% and 28% faster than *gls1271-ref4* [20] on Intel Core 2 Duo E6750, Intel Xeon E5440 and AMD Opteron 252, respectively. Although [20] uses inverted Twisted Edwards coordinates ($\mathcal{E}^{inv}$), the improvement with the change of coordinates only explains a small fraction of the speed-up. Similarly, in the case of $\mathcal{J}$ combined with GLS, *jac1271gls* is about 23% faster than the record set by *gls1271-ref3* [8] on a 1.66GHZ Intel Core 2 Duo. When comparing cycle counts on identical processor platforms, *jac1271gls* is 24%, 23% and 30% faster than *gls1271-ref3* [20] on Intel Core 2 Duo E6750, Intel Xeon E5440 and AMD Opteron 252, respect. Our implementations are also significantly faster than the implementation of Bernstein's *curve25519* by Gaudry and Thomé [10]. For instance, *ted1271gls* is 41% faster than *curve25519* [10] on a 2.66GHz Intel Core 2 Duo.

If GLS is not considered, the fastest implementations using $\mathcal{E}/\mathcal{E}^e$ and $\mathcal{J}$ are *ted256189* and *jac256189*, respectively. In this case, *ted256189* and *jac256189* are 22% and 28% faster than the previous best cycle counts due to Hisil et al. [14] using also $\mathcal{E}/\mathcal{E}^e$ and $\mathcal{J}$, respect., on a 2.66GHz Intel Core 2 Duo.

It is also interesting to note that the performance boost given by the GLS method depends on the characteristics of a given platform. For instance, *ted1271gls* and *jac1271gls* are about 19% and 25% faster than their "counterparts" over $\mathbb{F}_p$, namely *ted256189* and *jac256189*, respect., on a Core 2 Duo E6750. However, on the AMD Opteron processor the gap between the costs of field operations over $\mathbb{F}_p$ and $\mathbb{F}_{p^2}$ is shorter. As consequence, on Opteron 252 *ted1271gls* and *jac1271gls* only achieve a reduction of approx. 9% and 13% with respect to *ted256189* and *jac256189*, respectively. For the record, *ted1271gls* achieves the best cycle count on an AMD Opteron with an advantage of about 31% over the best previous result in the literature, i.e., *curve25519* [10].

In summary, this paper has illustrated that a significant speed-up can be achieved using a combination of optimizing techniques applied to all levels of the ECC computation and adapted to the architectural features of modern processors. This research is crucial for advancing the state-of-the-art crypto implementations in present and future platforms. Also, although our implementations (in their current form) only compute $[k]P$ where k and P vary, several of the optimizations discussed in this work are generic and can be easily adapted to speed up other implementations using a fixed point P, digital signatures and different coordinate systems/curve forms/underlying fields.

Acknowledgments. This work was made possible by the facilities of the Shared Hierarchical Academic Research Computing Network (SHARCNET) and Compute/Calcul Canada. We would like to thank the Natural Sciences and Engineering Research Council of Canada (NSERC) and the Ontario Centres of Excellence (OCE) for partially supporting this work. We would also like to thank Mike Scott, Hiren Patel and the reviewers for their useful comments.

References

1. Avanzi, R.: A Note on the Signed Sliding Window Integer Recoding and its Left-to-Right Analogue. In: Handschuh, H., Hasan, M.A. (eds.) SAC 2004. LNCS, vol. 3357, pp. 130–143. Springer, Heidelberg (2004)
2. Bernstein, D., Birkner, P., Joye, M., Lange, T., Peters, C.: Twisted Edwards Curves. In: Vaudenay, S. (ed.) AFRICACRYPT 2008. LNCS, vol. 5023, pp. 389–405. Springer, Heidelberg (2008)
3. Edwards, H.: A Normal Form for Elliptic Curves. Bulletin of the American Mathematical Society 44, 393–422 (2007)
4. Erdem, S.S., Yanik, T., Koç, Ç.K.: Fast Finite Field Multiplication. In: Koç, Ç.K. (ed.) Cryptographic Engineering, ch. 5. Springer, Heidelberg (2009)
5. Fog, A.: Instruction Tables: Lists of Instruction Latencies, Throughputs and Micro-operation Breakdowns for Intel, AMD and VIA CPUs (2009), http://www.agner.org/optimize/#manuals (accessed, January 2010)
6. Fog, A.: The Microarchitecture of Intel, AMD and VIA CPUs (2009), http://www.agner.org/optimize/#manuals (accessed, January 2010)
7. Galbraith, S., Lin, X., Scott, M.: Endomorphisms for Faster Elliptic Curve Cryptography on a Large Class of Curves. Cryptology ePrint Archive, Report 2008/194 (2008)
8. Galbraith, S., Lin, X., Scott, M.: Endomorphisms for Faster Elliptic Curve Cryptography on a Large Class of Curves. In: Joux, A. (ed.) EUROCRYPT 2009. LNCS, vol. 5479, pp. 518–535. Springer, Heidelberg (2010)
9. Gallant, R., Lambert, R., Vanstone, S.: Faster Point Multiplication on Elliptic Curves with Efficient Endomorphisms. In: Kilian, J. (ed.) CRYPTO 2001. LNCS, vol. 2139, pp. 190–200. Springer, Heidelberg (2001)
10. Gaudry, P., Thomé, E.: The mpFq Library and Implementing Curve-Based Key Exchanges. In: SPEED 2007, pp. 49–64 (2007)
11. Hankerson, D., Menezes, A., Scott, M.: Software Implementation of Pairings. In: Joye, M., Neven, G. (eds.) Identity-Based Cryptography, ch. 12. IOS Press, Amsterdam (2009)
12. Hankerson, D., Menezes, A., Vanstone, S.: Guide to Elliptic Curve Cryptography. Springer, Heidelberg (2004)

13. Hisil, H., Wong, K., Carter, G., Dawson, E.: Twisted Edwards Curves Revisited. In: Pieprzyk, J. (ed.) ASIACRYPT 2008. LNCS, vol. 5350, pp. 326–343. Springer, Heidelberg (2008)
14. Hisil, H., Wong, K., Carter, G., Dawson, E.: Jacobi Quartic Curves Revisited. Cryptology ePrint Archive, Report 2009/312 (2009)
15. Longa, P.: Accelerating the Scalar Multiplication on Elliptic Curve Cryptosystems over Prime Fields. Master's Thesis, University of Ottawa (2007), http://patricklonga.bravehost.com/publications.html#thesis
16. Longa, P.: ECC Point Arithmetic Formulae, EPAF (2008), http://patricklonga.bravehost.com/jacobian.html
17. Longa, P., Gebotys, C.: Setting Speed Records with the (Fractional) Multibase Non-Adjacent Form Method for Efficient Elliptic Curve Scalar Multiplication. CACR technical report, CACR 2008-06 (2008)
18. Longa, P., Gebotys, C.: Analysis of Efficient Techniques for Fast Elliptic Curve Cryptography on x86-64 based Processors (2010), http://patricklonga.bravehost.com/publications.html
19. Longa, P., Miri, A.: New Composite Operations and Precomputation Scheme for Elliptic Curve Cryptosystems over Prime Fields. In: Cramer, R. (ed.) PKC 2008. LNCS, vol. 4939, pp. 229–247. Springer, Heidelberg (2008)
20. Scott, M.: MIRACL - Multiprecision Integer and Rational Arithmetic C/C++ Library (1988-2007), ftp://ftp.computing.dcu.ie/pub/crypto/miracl.zip
21. Yanik, T., Savaş, E., Koç, Ç.K.: Incomplete Reduction in Modular Arithmetic. IEE Proc. of Computers and Digital Techniques 149(2), 46–52 (2002)

A   Point Operations Using $\mathcal{J}$ and $\mathcal{E}/\mathcal{E}^e$ Coordinates

The Maple scripts below verify most representative formulas used in our "traditional" implementations. Revised formulas for the GLS method can be found in the extended paper version [18]. Note that field operations are carefully merged and scheduled to reduce pipeline stalls and memory reads/writes. Temporary registers are denoted by t_i, DblSub represents $a-2b \,(\mathrm{mod}\,p)$ and SubDblSub merges $a-b \,(\mathrm{mod}\,p)$ and $(a-b)-2c \,(\mathrm{mod}\,p)$. Underlined operations are merged.

```
# Weierstrass curve (for verification):
x1:=X1/Z1^2; y1:=Y1/Z1^3; x2:=X2/Z2^2; y2:=Y2/Z2^3; ZZ2:=Z2^2; ZZZ2:=Z2^3; a:=-3;
x3:=((3*x1^2+a)/(2*y1))^2-2*x1; y3:=((3*x1^2+a)/(2*y1))*(x1-x3)-y1;
x4:=((y1-y2)/(x1-x2))^2-x2-x1; y4:=((y1-y2)/(x1-x2))*(x2-x4)-y2;
x5:=((y1-y4)/(x1-x4))^2-x4-x1; y5:=((y1-y4)/(x1-x4))*(x4-x5)-y4;
```

DBL, $\mathcal{J} \leftarrow 2\mathcal{J}$: $(X_{out}, Y_{out}, Z_{out}) \leftarrow 2(X_1, Y_1, Z_1)$. Cost $= 4\mathrm{M}+4\mathrm{S}+3\mathrm{Sub}+1\mathrm{DblSub}+1\mathrm{Add}_{\mathrm{IR}}+1\mathrm{Mul}3_{\mathrm{IR}}+1\mathrm{Div}2_{\mathrm{IR}}$; 5 contiguous data depend.

```
# In practice, Xout,Yout,Zout reuse the registers X1,Y1,Z1 for all cases below.
t4:=Z1^2; t3:=Y1^2; t1:=X1+t4; t4:=X1-t4; t0:=3*t4; t5:=X1*t3; t4:=t1*t0; t0:=t3^2; t1:=t4/2;
t3:=t1^2; Zout:=Y1*Z1; Xout:=t3-2*t5; t3:=t5-Xout; t5:=t1*t3; Yout:=t5-t0;
simplify([x3-Xout/Zout^2]), simplify([y3-Yout/Zout^3]); # Check
```

4DBL, $\mathcal{J} \leftarrow 8\mathcal{J}$: $(X_{out}, Y_{out}, Z_{out}) \leftarrow 8(X_1, Y_1, Z_1)$. Cost $= 4*(4\mathrm{M}+4\mathrm{S}+3\mathrm{Sub}+1\mathrm{DblSub}+1\mathrm{Add}_{\mathrm{IR}}+1\mathrm{Mul}3_{\mathrm{IR}}+1\mathrm{Div}2_{\mathrm{IR}})$; 1.25 contiguous data depend./doubling

```
t4:=Z1^2; t3:=Y1^2; t1:=X1+t4; t4:=X1-t4; t2:=3*t4; t5:=X1*t3; t4:=t1*t2; t0:=t3^2; t1:=t4/2;
Zout:=Y1*Z1; t3:=t1^2; t4:=Z1^2; Xout:=t3-2*t5; t3:=t5-Xout; t2:=Xout+t4; t5:=t1*t3; t4:=Xout-
t4; Yout:=t5-t0; t1:=3*t4; t3:=Yout^2; t4:=t1*t2; t5:=Xout*t3; t1:=t4/2; t0:=t3^2; t3:=t1^2;
```

```
Zout:=Yout*Zout; Xout:=t3-2*t5; t4:=Zout^2; t3:=t5-Xout; t2:=Xout+t4; t5:=t1*t3; t4:=Xout-t4;
Yout:=t5-t0; t1:=3*t4; t3:=Yout^2; t4:=t1*t2; t5:=Xout*t3; t1:=t4/2; t0:=t3^2; t3:=t1^2;
Zout:=Yout*Zout; Xout:=t3-2*t5; t4:=Zout^2; t3:=t5-Xout; t2:=Xout+t4; t5:=t1*t3; t4:=Xout-t4;
Yout:=t5-t0; t1:=3*t4; t3:=Yout^2; t4:=t1*t2; t5:=Xout*t3; t1:=t4/2; t0:=t3^2; t3:=t1^2;
Zout:=Yout*Zout; Xout:=t3-2*t5; t3:=t5-Xout; t5:=t1*t3; Yout:=t5-t0;
```

DBLADD, $\mathcal{J} \leftarrow 2\mathcal{J}+\mathcal{J}$: $(X_{out}, Y_{out}, Z_{out}) \leftarrow 2(X_1, Y_1, Z_1)+(X_2, Y_2, Z_2, Z_2^2, Z_2^3)$.
Cost = 16M+5S+7Sub+2DblSub+1Add$_{IR}$+1Mul2$_{IR}$; 3 contiguous data depend.

```
t0:=X1*ZZ2; t5:=Z1^2; t7:=Y1*ZZZ2; t4:=X2*t5; t6:=t5*Z1; t1:=t4-t0; t5:=Y2*t6; t6:=t1^2; t2:=
t5-t7; t4:=t2^2; t5:=t6*t0; t0:=t1*t6; t3:=t4-2*t5; t6:=Z1*t1; t3:=t3-t5; t4:=Z2*t6; t3:=t3-t0;
t6:=t7*t0; Zout:=t4*t3; t4:=t2*t3; t1:=2*t6; t0:=t3^2; t1:=t1+t4; t4:=t0*t5; t7:=t1^2; t5:=t0*t3;
Xout:=t7-2*t4; Xout:=Xout-t5; t3:=Xout-t4; t0:=t5*t6; t4:=t1*t3; Yout:=t4-t0;
simplify([x5-Xout/Zout^2]), simplify([y5-Yout/Zout^3]); # Check
```

```
# Twisted Edwards curve (for verification):
x1:=X1/Z1; y1:=Y1/Z1; x2:=X2/Z2; y2:=Y2/Z2; T2:=X2*Y2/Z2; a:=-1;
x3:=(2*x1*y1)/(y1^2+a*x1^2); y3:=(y1^2-a*x1^2)/(2-y1^2-a*x1^2);
x4:=(x3*y3+x2*y2)/(y3*y2+a*x3*x2); y4:=(x3*y3-x2*y2)/(x3*y2-y3*x2);
```

DBL, $\mathcal{E} \leftarrow 2\mathcal{E}$: $(X_{out}, Y_{out}, Z_{out}) \leftarrow 2(X_1, Y_1, Z_1)$. Cost = 4M+3S+1SubDblSub+1Add$_{IR}$+1Mul2$_{IR}$+1Neg; no contiguous data dependencies

```
t1:=2*X1; t2:=X1^2; t4:=Y1^2; t3:=Z1^2; Xout:=t2+t4; t4:=t4-t2; t3:=t4-2*t3; t2:=t1*Y1; Yout:=
-t4; Zout:=t4*t3; Yout:=Yout*Xout; Xout:=t3*t2;
simplify([x3-Xout/Zout]), simplify([y3-Yout/Zout]); # Check
# Iterate this code n times to obtain nDBL with cost n(4M+3S+1SubDblSub+1AddIR+1Mul2IR+1Neg)
```

Merged DBL–ADD, $\mathcal{E} \leftarrow (2\mathcal{E})^e + \mathcal{E}^e$: $(X_{out}, Y_{out}, Z_{out}) \leftarrow 2(X_1, Y_1, Z_1)+((X_2+Y_2),$
$(X_2 - Y_2), 2Z_2, 2T_2)$. Cost = 12M+3S+3Sub+1SubDblSub+4Add$_{IR}$+ 1Mul2$_{IR}$;
no contiguous data dependencies

```
t1:=2*X1; t5:=X1^2; t7:=Y1^2; t6:=Z1^2; Xout:=t5+t7; t7:=t7-t5; t6:=t7-2*t6; t5:=t1*Y1; t8:=t7*
Xout; t0:=t7*t6; t7:=t6*t5; t6:=Xout*t5; Xout:=t7+t8; t1:=t7-t8; t7:=(2*T2)*t0; t5:=(2*Z2)*t6;
t0:=(X2-Y2)*t1; t1:=t5+t7; t6:=(X2+Y2)*Xout; Xout:=t5-t7; t7:=t0-t6; t0:=t0+t6; Xout:=Xout*t7;
Yout:=t1*t0; Zout:=t0*t7;
simplify([x4-Xout/Zout]), simplify([y4-Yout/Zout]); # Check
```

B The Curves

The curves below provide approximately 128-bit level of security and were found by using a modified version of the Schoof's algorithm provided with MIRACL.

- *Jac256189* uses the Weierstrass curve $E_w : y^2 = x^3 - 3x + B$ over $\mathbb{F}_p$ with $\mathcal{J}$, where $p = 2^{256} - 189$, $B =$ 0xfd63c3319814da55e88e9328e96273c483dca6cc84 df53ec8d91b1b3e0237064 and $\#E_w(\mathbb{F}_p) = 10r$ (r is a 253-bit prime).
- *Ted256189* uses the Twisted Edwards curve $E_{tedw}: -x^2 + y^2 = 1 + 358\,x^2 y^2$ over $\mathbb{F}_p$ with $\mathcal{E}/\mathcal{E}^e$, where $p = 2^{256} - 189$ and $\#E_{tedw}(\mathbb{F}_p) = 4r$ (r is a 255-bit prime).
- *Jac1271gls* uses the quadratic twist $E'_{w-gls}: y^2 = x^3 - 3\mu x + 44\mu$ of the Weierstrass curve $E_{w-gls}(\mathbb{F}_{p^2})$, where $\mu = 2 + i \in \mathbb{F}_{p^2}$ is non-square, $E_{w-gls}/\mathbb{F}_p$: $y^2 = x^3 - 3x + 44$ and $p = 2^{127} - 1$. In this case, $\#E'_{w-gls}(\mathbb{F}_{p^2})$ is a 254-bit prime. The same curve is also used in [8].
- *Ted1271gls* uses the quadratic twist $E'_{tedw-gls} : -\mu x^2 + y^2 = 1 + 109\mu x^2 y^2$ of the Twisted Edwards curve $E_{tedw-gls}(\mathbb{F}_{p^2})$, where $\mu = 2 + i \in \mathbb{F}_{p^2}$ is non-square, $E_{tedw-gls}/\mathbb{F}_p : -x^2 + y^2 = 1 + 109 x^2 y^2$ and $p = 2^{127} - 1$. In this case, $\#E'_{tedw-gls}(\mathbb{F}_{p^2}) = 4r$ where r is a 252-bit prime.

Analysis and Improvement of the Random Delay Countermeasure of CHES 2009

Jean-Sébastien Coron and Ilya Kizhvatov

Université du Luxembourg
6, rue Richard Coudenhove-Kalergi
L-1359 Luxembourg
{jean-sebastien.coron,ilya.kizhvatov}@uni.lu

Abstract. Random delays are often inserted in embedded software to protect against side-channel and fault attacks. At CHES 2009 a new method for generation of random delays was described that increases the attacker's uncertainty about the position of sensitive operations. In this paper we show that the CHES 2009 method is less secure than claimed. We describe an improved method for random delay generation which does not suffer from the same security weakness. We also show that the paper's criterion to measure the security of random delays can be misleading, so we introduce a new criterion for random delays which is directly connected to the number of acquisitions required to break an implementation. We mount a power analysis attack against an 8-bit implementation of the improved method verifying its higher security in practice.

Keywords: Side channel attacks, DPA, countermeasures, random delays.

1 Introduction

Embedded software implementations of cryptographic algorithms are threatened by physical attacks like Differential Power Analysis (DPA) or fault injection. The simplest method of protection against such attacks consists in randomizing the flow of the operations by shuffling the order of the operations or inserting random delays composed of dummy operations. These *hiding* countermeasures offer less security than *masking* countermeasures but have smaller implementation and performance costs. For the general background on physical attacks and countermeasures we refer the reader to the book [6].

Random delays in software. Software random delays are implemented as loops of "dummy" operations that are inserted in the execution flow of an algorithm being protected. A single delay can be removed relatively easy by static alignment of side-channel traces, *e.g.* with cross-correlation techniques [4]. Therefore, the execution should be interleaved with delays in multiple places. To minimize the performance overhead in this setting, individual delays should be possibly shorter. An attacker would typically face a cumulative sum of the delays

S. Mangard and F.-X. Standaert (Eds.): CHES 2010, LNCS 6225, pp. 95–109, 2010.

between the synchronization point (which would usually be at the start or at the end of the execution) and the target event. So the cumulative delay should increase the uncertainty of an attacker about the location of the target event in time. For further discussion on random delays in software we refer the reader to [3] and [8]. We also note that elastic alignment techniques were reported in [9] to be able to reduce the effect of multiple delays. Within the scope of this paper we do not verify these techniques, assuming an attacker without elastic alignment.

Previous work. An efficient method for random delay generation in embedded software was suggested in CHES 2009 [3] under the name of Floating Mean. The central idea of the method is to generate the delays non-independently within one run of a protected algorithm. In this way, the adversary facing the effect of the cumulative sum of the delays will have to cope with much larger variance and thus will require significantly more side channel measurements compared to other methods with independently generated random delays such as [8]. We recall the Floating Mean method in Sect. 2.

Our contributions. Here we discover that the Floating Mean method of [3] can suffer from an improper parameter choice, offering less security than expected. We perform a detailed statistical analysis of the Floating Mean method of [3] and we show how to choose correct parameters (see Sect. 3).

However these new parameters require longer delays, which means the number of delays should be relatively small to keep a reasonable performance overhead. This is not good for security because, as discussed above, in general few long delays are easier to detect and remove than multiple short delays. Therefore we propose an improved method for random delay generation which can work with short delays (Sect. 4). Our new method is easy to implement; we describe a concrete implementation in assembly language (Appendix B).

We also show that the criterion in [3] to measure the efficiency of random delays can be misleading and derive a new efficiency criterion that is information-theoretically sound (Sect. 5). Finally, we mount a practical DPA attack against the implementation of our improved method on an 8-bit AVR microcontroller to verify its higher security. With these results, we target practical designers who implement the timing randomization countermeasures for protection of their embedded software implementations.

2   The Floating Mean Method

Here we recall the Floating Mean method introduced in [3]. Most methods for random delay generation use independent individual delays; in this way, the cumulative sum of many delays which an adversary is facing in an attack tends to normal distribution with the variance being the sum of variances of the individual delays. Instead the core idea of the Floating Mean method from

[3] is to generate random delays non-independently, in order to increase the variance of the cumulative sum. More precisely, the delays are generated as follows:

1. Initially the integer parameters a and b are chosen so that $b < a$, where a determines the worst-case delay and b determines the variance of the delays within a single run of a protected algorithm;
2. Before each run of a protected algorithm, a integer value m is generated independently and uniformly on $[0, a - b]$;
3. Within each run, the integer lengths of individual delays are generated independently and uniformly on $[m, m + b]$.

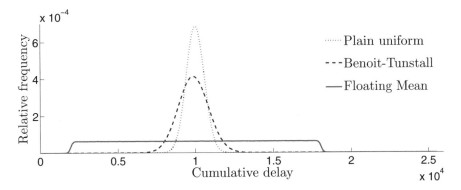

Fig. 1. Empirical distributions of the sum of 100 delays for random delay generation algorithms, for the case of equal means (based on [3]; delay length counted in atomic delay units)

As shown in [3] the variance of the cumulative sum of the delays in an execution becomes quadratic in the number N of delays instead of linear when the delays are generated independently. This is a significant improvement over the plain independent uniform delays or the table-based method of Benoit and Tunstall [8]. An illustrative example of the distribution of the cumulative sum of 100 delays for the Floating Mean compared to other methods is illustrated by Figure 1 based on [3]. The parameters for the Floating Mean here are $a = 200$, $b = 40$; the parameters for other methods are chosen so that all the methods yield the same mean cumulative delay.

However in practice it is better to have many short random delays rather than long delays, as recommended in [3]; this is because it is more complex to distinguish and remove many short delays than just few long delays. Therefore under this recommendation one should choose smaller values of a and b, for example $a = 18$ and $b = 3$ as used in [3] for the practical implementation. However in the next section we show that for such range of parameters the Floating Mean method provides less security than expected.

3   The Real Behavior of Floating Mean

In this section, we show that the Floating Mean method from [3] provides less security than expected for small a and b.

We begin with taking a detailed look at the distributions of different methods by simulating them with the exact experimental parameters used in the implementation of [3]. In Figure 2 we present histograms of the distributions for different methods. Namely, the number of delays in the sum is $N = 32$ and the parameters of the Floating Mean method are $a = 18$, $b = 3$. The histograms present the relative frequency of the cumulative delay against its duration[1]. We clearly see a multimodal distribution for the Floating Mean method: the histogram has a distinct shape of a saw with 16 cogs, and not a flat plateau as one would expect from [3].

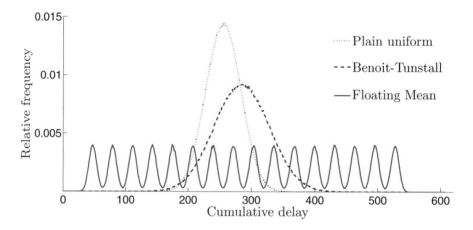

Fig. 2. Empirical distributions of the sum of 32 delays in the experiment of [3]

These cogs are not good for security since they make it easier for an attacker to mount an attack. The classical DPA will be more efficient since the signal is concentrated on the top of the 16 cogs instead of being spread over the clock cycles. In case of an attack with windowing and integration [2], the attacker would integrate the values around cog maximums, omit the minimums to reduce the noise (assuming noise is the same in all the points of the trace) and thus gain a reduction in the number of traces required for an attack.

3.1   Explaining the Cogs

Here we explain how cogs arise in the distribution of the Floating mean and we show how to choose the parameters to avoid the cogs.

---

[1] As in Figure 1, the duration in Figure 2 is expressed in atomic delay units, *i.e.* as the total number of delay loop iterations. To obtain the value in clock cycles one should multiply this by the length of a single delay loop, which is 3 clock cycles in the implementation of [3].

The distribution for the Floating Mean is in fact a *finite mixture* [7] of $a - b + 1$ components with equal weights. Every component corresponds to a given integer value m in $[0, a-b]$. For a given m, the cumulative sum of random delay durations is the sum of N random variables uniformly and independently distributed in $[m, m + b]$. Therefore it can be approximated by a Gaussian distribution with mean

$$\mu_m = N \cdot (m + b/2)$$

and variance

$$V = N \frac{(b + 1)^2 - 1}{12}.$$

Therefore the probability density of the distribution for random integer $m \in [0, a - b]$ can be approximated by:

$$f(x) = \sum_{m=0}^{a-b} \frac{1}{(a - b + 1)\sigma\sqrt{2\pi}} \exp\left(-\frac{(x - \mu_m)^2}{2\sigma^2}\right)$$

where all components have the same standard deviation $\sigma = \sqrt{V}$:

$$\sigma = \sqrt{N} \cdot \sqrt{\frac{(b + 1)^2 - 1}{12}}.$$

The cog peaks are the modes of the components, located in their means μ_m. The distance between the means of successive components is $\mu_{m+1} - \mu_m = N$. We can consider the cogs distinguishable by comparing the standard deviation σ of the components to the distance N between the means. Namely, the distribution becomes multimodal whenever $\sigma \ll N$. In the case of practical implementation in [3], we have $a = 18$, $b = 3$ and $N = 32$, which gives $\sigma = 6.3$; therefore we have $\sigma < N$ which explains why the 16 cogs are clearly distinguishable in the distribution in Figure 2. However for $a = 200$, $b = 40$ and $N = 100$ we get $\sigma = 118$ so $\sigma > N$ which explains why the cogs are indistinguishable in Figure 1 and a flat plateau is observed instead.

3.2 Choosing Correct Parameters

From the above we derive the simple rule of thumb for choosing Floating Mean parameters. To ensure that no distinct cogs arise, parameter b should be chosen such that $\sigma \gg N$. For sufficiently large b we can approximate σ by:

$$\sigma \simeq \frac{\sqrt{3}}{6} \cdot b \cdot \sqrt{N}$$

Therefore this gives the condition:

$$b \gg \sqrt{N}. \tag{1}$$

However as observed in [3] it is better to have a large number of short random delays rather than a small number of long delays; this is because rare longer delays are a priori easier to detect and remove than multiple short delays. But we see that condition (1) for Floating Mean requires longer delays since by definition the length of random delays is between $[m, m + b]$ with $m \in [0, a - b]$. In other words, condition (1) contradicts the requirement of having many short random delays.

In the next section we describe a variant of the Floating Mean which does not suffer from this contradiction, *i.e.* we show how to get short individual random delays without having the cogs in the cumulative sum.

4   Improved Floating Mean

In the original Floating Mean method an integer m is selected at random in $[0, a - b]$ before each new execution and the length of individual delays is then a random integer in $[m, m + b]$. The core idea of the new method is to improve the granularity of random delay generation by using a wider distribution for m. More precisely the new method works as follows:

1. Initially the integer parameters a and b are chosen so that $b < a$; additionally we generate a non-negative integer parameter k.
2. Prior to each execution, we generate an integer value m' in the interval $[0, (a - b) \cdot 2^k[$.
3. Throughout the execution, the integer length d of an individual delay is obtained by first generating a random integer $d' \in [m', m' + (b + 1) \cdot 2^k[$ and then letting $d \leftarrow \lfloor d' \cdot 2^{-k} \rfloor$.

4.1   Analysis

We see that as in the original Floating Mean method, the length of individual delays is in the interval $[0, a]$. Moreover if the integer m' generated at step 2 is such that $m' = 0 \mod 2^k$, then we can write $m' = m \cdot 2^k$ and the length d of individual delays is uniformly distributed in $[m, m+b]$ as in the original Floating Mean method.

When $m' \neq 0 \mod 2^k$, writing $m' = m \cdot 2^k + u$ with $0 \leq u < 2^k$, the delay's length d is distributed in the interval $[m, m + b + 1]$ with a slightly non-uniform distribution:

$$\Pr[d = i] = \begin{cases} \frac{1}{b+1} \cdot (1 - u \cdot 2^{-k}) & \text{for } i = m \\ \frac{1}{b+1} & \text{for } m + 1 \leq i \leq m + b \\ \frac{1}{b+1} \cdot u \cdot 2^{-k} & \text{for } i = m + b + 1 \end{cases}$$

Therefore when m' increases from $m \cdot 2^k$ to $(m + 1) \cdot 2^k$ the distribution of the delay length d moves progressively from uniform in $[m, m + b]$ to uniform in $[m + 1, m + b + 1]$. In Appendix A we show that for a fixed m':

$$E[d] = m' \cdot 2^{-k} + \frac{b}{2}$$

$$\text{Var}[d] = E[d^2] - E[d]^2 = \frac{(b + 1)^2 - 1}{12} + 2^{-k} \cdot u \cdot (1 - 2^{-k} \cdot u)$$

where $u = m' \mod 2^k$.

For a fixed m' the cumulative sum of random delay durations is the sum of N independently distributed random variables. Therefore it can be approximated by a Gaussian distribution with mean:

$$\mu_{m'} = N \cdot \mathrm{E}[d] = N \cdot \left(m' \cdot 2^{-k} + \frac{b}{2}\right) \tag{2}$$

and variance

$$V_{m'} = N \cdot \mathrm{Var}[d] = N \cdot \left(\frac{(b+1)^2 - 1}{12} + 2^{-k} \cdot u \cdot (1 - 2^{-k} \cdot u)\right)$$

For random $m' \in [0, (a-b) \cdot 2^k[$ the probability density of the cumulative sum can therefore be approximated by:

$$f(x) = \sum_{m'=0}^{(a-b) \cdot 2^k - 1} \frac{1}{(a-b)2^k \sigma_{m'} \sqrt{2\pi}} \exp\left(-\frac{(x - \mu_{m'})^2}{2\sigma_{m'}^2}\right)$$

where $\sigma_{m'} = \sqrt{V_{m'}}$. We have:

$$\sigma_{m'} > \sigma = \sqrt{N} \cdot \sqrt{\frac{(b+1)^2 - 1}{12}}$$

where σ is the same as for the original Floating Mean method.

As previously we obtain a multimodal distribution. The distance between the means of successive components is $\mu_{m'+1} - \mu_{m'} = N \cdot 2^{-k}$ and the standard deviation of a component is at least σ. Therefore the cogs become indistinguishable when $\sigma \gg N \cdot 2^{-k}$ which gives the condition:

$$b \gg \sqrt{N} \cdot 2^{-k}$$

instead of $b \gg \sqrt{N}$ for the original Floating Mean. Therefore by selecting a sufficiently large k we can accommodate a large number of short random delays (large N and small b). In practice, already for k as small as 3 the effect is considerable; we confirm this practically in Sect. 5.3.

We now proceed to compute the mean and variance of the cumulative sum for random m'. Let denote by S_N the sum of the N delays. We have from (2):

$$\mathrm{E}[S_N] = \mathrm{E}[\mu_{m'}] = N \cdot \left(\frac{a}{2} - 2^{-k-1}\right)$$

which is the same as the original Floating Mean up to the 2^{-k-1} term.

To compute the standard deviation of S_N, we represent an individual delay as a random variable $d_i = m + v_i$ where $m = \lfloor m' \cdot 2^{-k} \rfloor$ and v_i is a random variable in the interval $[0, b+1]$. Since m' is uniformly distributed in $[0, (a-b) \cdot 2^k[$, the integer m is uniformly distributed in $[0, a-b[$; moreover the distribution of v_i is independent of m and the v_i's are identically distributed. From

$$S_N = \sum_{i=1}^{N} d_i = Nm + \sum_{i=1}^{N} v_i \,.$$

we get:

$$\mathrm{Var}(S_N) = N^2 \cdot \mathrm{Var}(m) + N \cdot \mathrm{Var}(v_1)$$

For large N we can neglect the term $N \cdot \mathrm{Var}(v_1)$ which gives:

$$\mathrm{Var}(S_N) \simeq N^2 \cdot \mathrm{Var}(m) = N^2 \cdot \frac{(a-b)^2 - 1}{12}$$

which is approximately the same as for the original Floating Mean method. As for the original Floating Mean the variance of the sum of N delays is in $\Theta\left(N^2\right)$ in comparison to plain uniform delays and the method of [8] that both have variances in $\Theta\left(N\right)$.

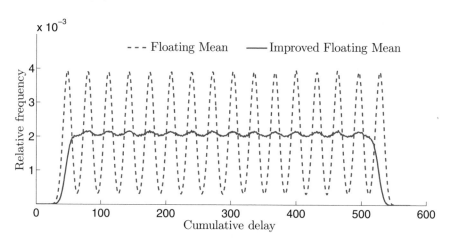

Fig. 3. Improved Floating Mean with $k = 3$ compared to the original method of [3]; $a = 18$, $b = 3$ and $N = 32$ for both methods

4.2   Illustration

The result is shown in Figure 3 compared to the original Floating Mean. The parameters of both methods were the same as in Figure 2: $a = 18$, $b = 3$ and $N = 32$. We take $k = 3$ for our Improved Floating Mean method. We can see that we have flattened out the cogs while keeping the same mean. This is because we still have $\sigma \simeq 6.3$ but the distance between successive cogs is now $N \cdot 2^{-k} = 32 \cdot 2^{-3} = 4$ instead of 32 so the cogs are now almost indistinguishable.

4.3   Full Algorithm

The Improved Floating Mean method is formally defined by Algorithm 1. Following [3], by $\mathcal{DU}[y, z[$ we denote discrete uniform distribution on $[y, z[$, $y, z \in \mathbb{Z}$, $y < z$. Note that as in [3] we apply the technique of "flipping" the mean in the middle of the execution to make the duration of the entire execution independent of m'. In Appendix B we show that Improved Floating Mean can be efficiently implemented on a constrained platform by describing an implementation in assembly language.

Algorithm 1. Improved Floating Mean

Input: $a, b, k, M \in \mathbb{N}, b \leq a, N = 2M$

$\quad m' \leftarrow \mathcal{DU}[0, (a - b) \cdot 2^k[$

$\quad$ **for** $i = 1$ **to** $N/2$ **do**

$\qquad d_i \leftarrow \lfloor (m' + \mathcal{DU}\left[0, (b+1) \cdot 2^k\right[) \cdot 2^{-k} \rfloor$

$\quad$ **end for**

$\quad$ **for** $i = N/2 + 1$ **to** N **do**

$\qquad d_i \leftarrow \lfloor (a \cdot 2^k - m' - \mathcal{DU}\left[0, (b+1) \cdot 2^k\right[) \cdot 2^{-k} \rfloor$

$\quad$ **end for**

Output: $d_1, d_2, \ldots, d_N$

---

5 The Optimal Criterion of Efficiency

In [3], the ratio σ/μ called coefficient of variation was suggested as a criterion for measuring the efficiency of the random delays countermeasure, where σ is the standard deviation of the cumulative sum of the delays, and μ the mean of the cumulative sum[2], where a higher ratio meant a better efficiency. Here we argue that this measure is misleading and suggest a new criterion.

5.1 Drawbacks of the Coefficient of Variation

We first take a closer look at the experimental data from [3] and establish consistency with the theoretical expectations. In Figure 4(a) we present the relation between the standard deviation σ of the cumulative sum of 32 delays and the number T_{CPA} of traces required for a successful CPA attack on the implementation without the random delays countermeasure (no delays, ND) and with different random delay generation methods: plain uniform (PU), Benoit-Tunstall (BT), Floating Mean (FM). The data were taken from Table 2 of [3].

Following [2] and [5] one would expect that without integration, which was the case for the experiments in [3], the number of traces grows quadratically with the standard deviation σ. However, from Figure 4(a) one can see that T_{CPA} exhibits growth with σ that is almost linear and not quadratic as one would have expected.

The problem is that standard deviation σ is in general a very rough way to estimate the number of traces which only works for very similar distributions (like two normal distributions). If we look at the figures in Table 2 in [3], we will see that σ for Floating Mean is 5.3 times larger than that for the plain uniform delays. If one expects the number of traces to be in σ^2 in the attack without integration, then $5.3^2 = 28$ more traces are expected to attack the Floating Mean. But observed was only $45000/2500 = 18$ times increase. This means that by looking at the variance, one can overestimate the security level of the countermeasure.

---

[2] Here σ is the standard deviation of the cumulative sum across various executions, as opposed to Sections 2 and 4 where σ was the standard deviation for a single execution with a fixed m.

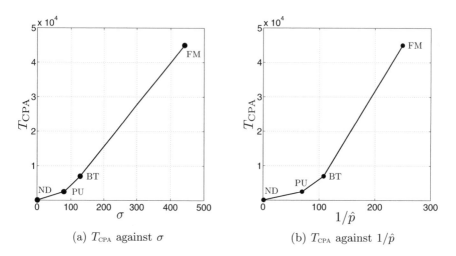

(a) T_{CPA} against σ (b) T_{CPA} against $1/\hat{p}$

Fig. 4. Attack complexity as a function of cumulative sum distribution parameters

We illustrate this with a simple example. Consider the uniform distribution U of integers on some interval $[a, b]$, $a, b \in \mathbb{Z}$, and the distribution X with $\Pr[X = a] = \Pr[X = b] = 1/2$. We have $\text{Var}(U) = ((a - b + 1)^2 - 1)/12$ and $\text{Var}(X) = (a-b)^2/4$, so $\text{Var}(X) > \text{Var}(U)$. Therefore the efficiency of X counted in σ/μ will be higher than for U. But with X the DPA signal is only divided by 2 instead of $(b - a + 1)$ with U, so the number of traces required to break an implementation with U will be smaller than with X. So in this case the criterion from [3] is misleading.

An accurate estimate is the maximum $\hat{p}$ of the probability mass function (p.m.f.)[3] of the distribution of the cumulative sum. From [2] and [5] we recall that the number of traces T required for a DPA attack is determined by the maximal correlation coefficient ρ_{max} observed in the correlation trace for the correct key guess. Namely, the number of traces can be estimated as

$$T = 3 + 8 \left(\frac{Z_\alpha}{\ln \left(\frac{1+\rho_{max}}{1-\rho_{max}} \right)} \right)^2 \tag{3}$$

where Z_α is a quantile of a normal distribution for the 2-sided confidence interval with error $1-\alpha$. For $\rho_{max} < 0.2$, $\ln \left(\frac{1+x}{1-x} \right) \approx 2x$ holds, so we can approximate (3) for $Z_{\alpha=0.9} = 1.282$ as $T \approx 3/\rho_{max}^2$. So the number of traces is in ρ_{max}^{-2}. In turn, the effect of the timing disarrangement on ρ_{max} is in $\hat{p}$ in case no integration is used. So the number of traces is in $1/\hat{p}^2$.

We now compute $\hat{p}$ for the distributions shown in Figure 2 and plot T_{CPA} from [3] against $1/\hat{p}$ in Figure 4(b). We can now see that quadratic dependency has become clear, which can be verified by the computations: $\hat{p}$ suggests that the

---

[3] And not p.d.f. since the distribution is discrete.

number of traces for the Floating Mean will be 13 times higher than for plain uniform delays. Now we have underestimation, but the relation of the number of traces to $\hat{p}$ is still more accurate than to σ. Note that such calculations will not hold for the case without delays since in this case ρ_{max} was about 0.6 (see Figure 5 in [3]), whereas in the other cases $\rho_{max} < 0.2$ holds.

5.2   The New Criterion

We propose a better criterion for the efficiency E of the random delays counter-measure:

$$E = 1/(2\hat{p} \cdot \mu)$$

where $\hat{p}$ is the maximum of the probability mass function and μ is the mean of the distribution of the cumulative sum of the delays. This criterion is normalized and *optimal* with respect to the desired properties of the countermeasure, as shown below.

With the countermeasure, we want to maximize the number of traces in an attack, *i.e.* minimize $\hat{p}$, while keeping the smallest possible overhead, *i.e.* smallest mean μ. One can see that from all distributions with the given $\hat{p}$, the one with the smallest μ is uniform on $[0, 1/\hat{p}]$. In this case, $\mu = 1/(2\hat{p})$ and the criterion E is equal to 1. In all other cases (same $\hat{p}$ but larger μ) the value of the criterion will be smaller, and the closer to zero – the farther is the distribution from the optimal one (*i.e.* the uniform one).

This tightly relates the criterion to the entropy of the distribution. Namely, the new criterion is directly linked to min-entropy, which is defined for a random variable S as

$$H_\infty(S) = -\log \max_i p_i = -\log \hat{p}.$$

Note that $H_\infty(S) \leq H(S)$, where $H(S) = -\sum_i p_i \log p_i$ is the Shannon entropy, so min-entropy can be considered as a worst-case measure of uncertainty. Now we have $\hat{p} = 2^{-H_\infty(S)}$ and the new efficiency criterion is expressed as

$$E = \frac{2^{H_\infty(S)-1}}{\mu}.$$

Indeed, for a fixed worst-case cumulative delay, the distributions with the higher entropy, *i.e.* maximizing uncertainty for the attacker, will have lower $\hat{p}$, larger number of traces to attack and thus more efficient as a countermeasure.

This criterion is easily computable once the designer have simulated the real distribution for the concrete parameters of a method (taking into consideration the number of clock cycles per delay loop) and obtained $\hat{p}$.

5.3   Comparing Efficiency

In our example in Sect. 4.2 with parameters $a = 18$, $b = 3$ and $N = 32$, with the Improved Floating Mean (IFM) method we have decreased $\hat{p}$ by a factor 2, as illustrated in Figure 3. So the number of traces for the successful straightforward

Table 1. New efficiency criterion for different methods

	ND	PU	BT [8]	FM [3]	IFM [this paper]
μ, cycles	0	720	860	862	953
$\hat{p}$	1	0.0144	0.0092	0.0040	0.0020
$E = 1/(2\hat{p}\mu)$	–	0.048	0.063	0.145	0.259
T_{CPA}, traces	50	2500	7000	45000	> 150000

DPA attack will be in principle almost 4 times larger (around 160000), and according to the optimal criterion the efficiency is almost 2 times higher. Table 1 below revises Table 2 of [3] with the new criterion and the new method.

We have performed a practical power analysis attack against an AES-128 implementation with the new IFM method running on ATmega16, an 8-bit AVR microcontroller. To be consistent with the previous results, the implementation and the measurement setup were as in [3]. Namely, there were 10 random delays per round, and 3 dummy rounds were added before and after the encryption, so $N = 32$ delays occur between the start of the execution (the synchronization point) and the 1-st S-Box lookup of the 1st encryption round the attack target. The parameters for IFM were $a = 19$, $b = 3$, $k = 3$. Note that a different value for a was chosen to ensure efficient implementation of the method as described in Appendix B, so the mean for IFM is larger, but still the efficiency is 1.8 times higher. We could not break this implementation by a CPA attack [1] with $150 \cdot 10^3$ traces, which corresponds to the theoretical expectations.

Due to its definition, the new criterion reflects well the number of traces observed in the experimental attack. For example, looking at the new criterion we expect the number of traces for the Floating Mean be $0.145^2/0.063^2 = 5.3$ times higher than for the table method of Benoit and Tunstall [8]. In the experiment, it was $45000/7000 = 6.4$ times higher.

6   Conclusion

We have shown that the Floating Mean method for random delay generation in embedded software [3] exhibits lower security if its parameters are improperly chosen. We have suggested how to choose the parameters of the method so that it generates a good distribution; however this requires to generate longer delays while in practice it is preferable to have multiple shorter delays. We have proposed an improved method that allows for a wider choice of parameters while having an efficient implementation. Finally, we have suggested an optimal criterion for measuring the efficiency of the random delays countermeasure.

References

1. Brier, E., Clavier, C., Benoit, O.: Correlation power analysis with a leakage model. In: Joye, M., Quisquater, J.-J. (eds.) CHES 2004. LNCS, vol. 3156, pp. 135–152. Springer, Heidelberg (2004)

2. Clavier, C., Coron, J.-S., Dabbous, N.: Differential power analysis in the presence of hardware countermeasures. In: Koç, Ç.K., Paar, C. (eds.) CHES 2000. LNCS, vol. 1965, pp. 252–263. Springer, Heidelberg (2000)
3. Coron, J.-S., Kizhvatov, I.: An efficient method for random delay generation in embedded software. In: Clavier, C., Gaj, K. (eds.) CHES 2009. LNCS, vol. 5747, pp. 156–170. Springer, Heidelberg (2009)
4. Homma, N., Nagashima, S., Sugawara, T., Aoki, T., Satoh, A.: A high-resolution phase-based waveform matching and its application to side-channel attacks. IEICE Trans. Fundam. Electron. Commun. Comput. Sci. E91-A(1), 193–202 (2008)
5. Mangard, S.: Hardware countermeasures against DPA – a statistical analysis of their effectiveness. In: Okamoto, T. (ed.) CT-RSA 2004. LNCS, vol. 2964, pp. 222–235. Springer, Heidelberg (2004)
6. Mangard, S., Oswald, E., Popp, T.: Power Analysis Attacks: Revealing the Secrets of Smart Cards. Springer, Heidelberg (2007)
7. McLachlan, G., Peel, D.: Finite Mixture Models. John Wiley & Sons, Chichester (2000)
8. Tunstall, M., Benoit, O.: Efficient use of random delays in embedded software. In: Sauveron, D., Markantonakis, K., Bilas, A., Quisquater, J.-J. (eds.) WISTP 2007. LNCS, vol. 4462, pp. 27–38. Springer, Heidelberg (2007)
9. van Woudenberg, J.G.J., Witteman, M.F., Bakker, B.: Improving Differential Power Analysis by elastic alignment (2009),
http://www.riscure.com/fileadmin/images/Docs/elastic_paper.pdf

A   Distribution of Delay's Length d

We have:

$$E[d] = \frac{1}{(b+1)2^k} \sum_{i=m'}^{m'+(b+1)\cdot 2^k -1} \lfloor i \cdot 2^{-k} \rfloor$$

Write $m' = m \cdot 2^k + u$ with $0 \leq u < 2^k$. This gives:

$$E[d] = \frac{1}{(b+1)2^k} \sum_{i=m2^k+u}^{(m+b+1)2^k+u-1} \lfloor i \cdot 2^{-k} \rfloor$$

$$= \frac{1}{(b+1)2^k} \left(\sum_{i=m2^k+u}^{(m+1)2^k-1} \lfloor i \cdot 2^{-k} \rfloor + \sum_{i=(m+1)2^k}^{(m+b+1)2^k-1} \lfloor i \cdot 2^{-k} \rfloor + \sum_{i=(m+b+1)2^k}^{(m+b+1)2^k+u-1} \lfloor i \cdot 2^{-k} \rfloor \right)$$

$$= \frac{1}{(b+1)2^k} \left(m \cdot (2^k - u) + 2^k \sum_{j=m+1}^{m+b} j + (m+b+1) \cdot u \right)$$

$$= \frac{1}{(b+1)2^k} \left(m \cdot 2^k + (b+1) \cdot u + b \cdot 2^k \cdot \left(m + \frac{b+1}{2} \right) \right)$$

$$= \frac{1}{(b+1)2^k} \left(m \cdot (b+1) \cdot 2^k + (b+1) \cdot u + b \cdot 2^k \cdot \frac{b+1}{2} \right)$$

$$= m + u \cdot 2^{-k} + \frac{b}{2} = m' \cdot 2^{-k} + \frac{b}{2}$$

Similarly we have:

$$E[d^2] = \frac{1}{(b+1)2^k} \sum_{i=m2^k+u}^{(m+b+1)2^k+u-1} \lfloor i \cdot 2^{-k} \rfloor^2$$

$$= \frac{1}{(b+1)2^k} \left(\sum_{i=m2^k+u}^{(m+1)2^k-1} \lfloor i \cdot 2^{-k} \rfloor^2 + \sum_{i=(m+1)2^k}^{(m+b+1)2^k-1} \lfloor i \cdot 2^{-k} \rfloor^2 + \sum_{i=(m+b+1)2^k}^{(m+b+1)2^k+u-1} \lfloor i \cdot 2^{-k} \rfloor^2 \right)$$

$$= \frac{1}{(b+1)2^k} \left(m^2 \cdot (2^k - u) + 2^k \sum_{j=m+1}^{m+b} j^2 + (m+b+1)^2 \cdot u \right)$$

After simplifications this gives:

$$\mathrm{Var}[d] = E[d^2] - E[d]^2 = \frac{(b+1)^2 - 1}{12} + 2^{-k} \cdot u(1 - 2^{-k}u)$$

B   Efficient Implementation of Improved Floating Mean

Here we show that our new Improved Floating Mean method can be efficiently implemented and introduces only a slight additional performance overhead compared to the original Floating Mean (*cf.* Appendix B of [3]).

In the Improved Floating Mean method one has to generate the mean and the individual delays in a broader range but then round them. The former is done by modifying the mask for truncating the random numbers so it is k bits longer, the latter – by shifting the register with the delay right by k bits.

As a reference, the new implementation of the delay loop in the 8-bit AVR assembly (*cf.* [3]) for the Improved Floating Mean is:

```
    rcall randombyte    ; obtain a random byte in RND
    and   RND, MASKBK   ; truncate to the desired length including k
    add   RND, FM       ; add 'floating mean'
    lsr   RND           ;
    ...                 ; logical shit right by k bits
    lsr   RND           ;
    tst   RND           ; balancing between zero and
    breq  zero          ;   non-zero delay values
    nop
    nop
dummyloop:
    dec   RND
    brne  dummyloop
zero:
    ret
```

and the generation of m in register FM in the beginning of the execution looks like:

```
    rcall randombyte    ; obtain a random byte in RND
    and RND, MASKMK     ; truncate to the desired length including k
    mov FM, RND         ; store 'floating mean' on register FM
```

Here, the masks have the following form:

$$\texttt{MASKBK} = 0\ldots0\underbrace{1\ldots1}_{t}\underbrace{1\ldots1}_{k}$$

$$\texttt{MASKMK} = 0\ldots0\underbrace{1\ldots1}_{s}\underbrace{1\ldots1}_{k}$$

where $2^t - 1 = b$ and $2^s = a - b$ (we note that this choice of parameters is slightly different from the one for efficient implementation of the Floating Mean, and therefore a was set to 19 for $b = 3$ in our experiments reported in Sect 5.3). To ensure that the operations are performed on a single register and no overflow occurs on an n-bit microcontroller, s, t, and k should be chosen such that $\max(s, t) + k + 1 \leq n$.

Note that the number of cycles per delay loop itself did not change. What changed is the additional overhead per delay. In the case of the 8-bit AVR implementation it is k additional cycles required for k-bit shift right. For a small k like $k = 3$ the impact is therefore insignificant.

New Results on Instruction Cache Attacks

Onur Acıiçmez[1], Billy Bob Brumley[2,*], and Philipp Grabher[3,**]

[1] Samsung Electronics, USA
o.aciicmez@samsung.com
[2] Aalto University School of Science and Technology, Finland
billy.brumley@tkk.fi
[3] University of Bristol, UK
grabher@cs.bris.ac.uk

Abstract. We improve instruction cache data analysis techniques with a framework based on vector quantization and hidden Markov models. As a result, we are capable of carrying out efficient automated attacks using live I-cache timing data. Using this analysis technique, we run an I-cache attack on OpenSSL's DSA implementation and recover keys using lattice methods. Previous I-cache attacks were proof-of-concept: we present results of an actual attack in a real-world setting, proving these attacks to be realistic. We also present general software countermeasures, along with their performance impact, that are not algorithm specific and can be employed at the kernel and/or compiler level.

1 Introduction

Cache-timing attacks are emerging attack vectors on security-critical software. They belong to a larger group of cryptanalysis techniques within side-channel analysis called Microarchitectural Attacks (MA). Microarchitectural Cryptanalysis focuses on the effects of common processor components and their functionalities on the security of software cryptosystems. The main characteristic of microarchitectural attacks, which sets them aside from classical side-channel attacks, is the simple fact that they exploit the microarchitectural behavior of modern computer systems. MA techniques have been shown to be effective and practical on real-world systems. For example, Osvik et. al. used cache attacks on dm-crypt application to recover AES keys [11]. Ristenpart et. al. successfully applied cache attacks in Amazon's EC2 cloud infrastructure and showed the information leakage from one virtualized machine to another [14]. Several studies showed the effectiveness of these attacks on various cryptosystems including AES [11,5], RSA [13,4,3], and ECC [6]. Popular cryptographic libraries such as OpenSSL have gone under several revisions to mitigate different MA attacks, c.f. e.g. [1].

---

* Supported in part by the European Commission's Seventh Framework Programme (FP7) under contract number ICT-2007-216499 (CACE).
** Supported in part by EPSRC grant EP/E001556/1.

S. Mangard and F.-X. Standaert (Eds.): CHES 2010, LNCS 6225, pp. 110–124, 2010.

There are usually two types of caches in today's processors, data cache and instruction cache, which have different characteristics, and hence we have two different types of cache-timing attacks. Our work presented in this paper deals only with instruction caches. I-cache attacks rely on the fact that instruction cache misses increase the execution time of a software. An adversary executes a so-called spy process on the same machine that his target software (e.g. an encryption application) is running on and this spy uses some techniques to keep track of the changes in the state of I-cache during the execution of the target software. Knowing the state changes in I-cache may allow the adversary to extract the instruction flow of the target software. Cipher implementations that have key-dependent instruction flows can be vulnerable to I-cache attacks unless effective countermeasures are in place. I-cache analysis technique was introduced in [2]. We have seen I-cache attack vulnerabilities in widely used RSA implementations [4]. Previous works on I-cache analysis were, in a sense, only proof-of-concept attacks. Spy measurements were either taken *within* the cipher process or in a simplified experimental setup.

In this paper, we present several contributions related to I-cache attacks, their data analysis, and countermeasures. We apply the templating cache-timing data analysis framework [6] to I-cache data. It makes use of Vector Quantization (VQ) and Hidden Markov Models (HMM) to automate the side-channel data analysis step. This allows us to mount a lattice attack on an unmodified OpenSSL-DSA implementation and successfully recover DSA keys. These are the first published results of a real-life I-cache attack on a cryptosystem. In a nut-shell, our contributions in this paper include:

- improving I-cache data analysis techniques,
- mounting a lattice attack on OpenSSL's DSA implementation using this improved analysis,
- presenting results of I-cache Analysis in a real-world attack settings,
- and outlining possible countermeasures to prevent I-cache attacks and measuring their performance impacts.

We give an overview of the original I-cache attack of [2] in Section 2 and present the details of our improved attack and our results on OpenSSL-DSA in Section 3. Our results prove the dangers of I-cache attacks and the necessity of employing appropriate countermeasures. We studied some of the possible countermeasures and analyzed their impacts on cipher performance and also on the performance of the entire system. We discuss these countermeasures and present our results in Sections 4 and 5.

2   I-Cache Attack Concept

I-cache analysis relies on the fact that instruction cache misses increase the execution time of software applications. Each I-cache miss mandates an access to a higher level memory, i.e., a higher level cache or main memory, and thus results in additional execution time delays. In I-cache analysis, an adversary runs a so-called spy process that monitors the changes in I-cache. They spy process continuously executes a set of "dummy" instructions in a loop in a particular order and measures how much time it takes to bring the I-cache to a predetermined state. Sec. 3.1 contains an example of such a spy routine.

If another process is running simultaneously with the spy on the same physical core of an SMT processor, the instructions executed by this process will alter the I-cache state and cause evictions of spy's dummy instructions. When the spy measures the time to re-execute its instructions, the latency will be higher for any evicted dummy instructions that must be fetched from a higher memory level. In this manner the spy detects changes in the I-cache state induced by the other (i.e., "spied-on") process and can follow the footprints of this process.

[2] shows an attack on OpenSSL's RSA implementation. They take advantage of the fact that OpenSSL employs sliding window exponentiation which generates a key dependent sequence of modular operations in RSA. Furthermore, OpenSSL uses different functions to compute modular multiplications and square operations that leaves different footprints on I-cache. Thus, a spy can monitor these footprints and can easily determine the operation sequence of RSA. [2] uses a different spy than the one we outline in Sec. 3.1. They try to extract the sequence of multiplication and square operations and thus their spy monitors only the I-cache sets related to these functions. Furthermore, their spy does not take timing measurements for each individual I-cache set, but instead considers a number of sets as a group and takes combined measurements. In our work, the spy takes individual measurements for each I-cache set so that we can monitor each set independently and devise template I-cache attacks.

3   Improved Attack Techniques

In this section, we present our improvements to I-cache timing data analysis and subsequently apply the results to run an I-cache attack on OpenSSL's DSA implementation (0.9.8l) to recover keys. We concentrate on Intel's Atom processor featuring Intel's HyperThreading Technology (HT).

3.1   Spying on the Instruction Cache

The templating framework in [6] used to analyze cache-timing data assumes vectors of timing data where each component is a timing measurement for a distinct cache set. We can realize this with the I-cache as well using a spy process that is essentially the I-cache analogue of Percival's D-cache spy process [13]. It pollutes the I-cache with its own data, then measures the time it takes to re-execute code that maps to a distinct set, then repeats this procedure indefinitely for any desired I-cache sets.

To this end, we outline a generic instruction cache spy process; the example here is for the Atom's 8-way associative 32KB cache, $c = 64$ cache sets, but is straightforwardly adaptable to other cache structures. We lay out contiguous 64-byte regions of code (precisely the size of one cache line) in labels

```
xor %edi, %edi           .endr                    .rept 49
mov <buffer addr>, %ecx  ...                      nop
rdtsc                   L64:                      .endr
mov %eax, %esi            jmp L128                 ...
jmp L0                    .rept 59               L511:
.align 4096               nop                      rdtsc
L0:                       .endr                    sub %esi, %eax
  jmp L64                 ...                      movb %al, (%ecx,%edi)
  .rept 59              L448:                      add %eax, %esi
  nop                      rdtsc                   inc %edi
  .endr                    sub %esi, %eax          cmp <buffer len>, %edi
L1:                        movb %al, (%ecx,%edi)   jge END
  jmp L65                  add %eax, %esi          jmp L0
  .rept 59                 inc %edi
  nop                      jmp L1
```

Fig. 1. Outline of a generic I-cache spy process

$\mathcal{L} = \{L_0, L_1, \ldots, L_{511}\}$. Denote subsets $\mathcal{L}_i = \{L_j \in \mathcal{L} : j \bmod c = i\}$ in this case each with cardinality eight, where all regions map to the same cache set yet critically do not share the same address tag. These subsets naturally partition $\mathcal{L} = \bigcup_{i=0}^{c-1} \mathcal{L}_i$. Observe that stepping through a given $\mathcal{L}_i$ pollutes the corresponding cache set i and repeating for all i completely pollutes the entire cache.

The spy steps iteratively through these $\mathcal{L}_i$ and measures their individual execution time. For example, it begins with regions that map to cache set zero: $\mathcal{L}_0 = \{L_0, L_{64}, L_{128}, \ldots, L_{448}\}$, stores the execution time, then continues with cache set one: $\mathcal{L}_1 = \{L_1, L_{65}, \ldots, L_{449}\}$ and so on through all $0 \leq i < c$. For each i we get a single latency measurement, and for all i a vector of measurements: repeating this process gives us the desired side-channel. For completeness, we provide a code snippet in Fig. 1. The majority of the code is **nop** instructions, but they are only used for padding and never executed. Note **rdtsc** is a clock cycle metric.

3.2 Realizing the DSA

We use the following notation for the DSA. The parameters include a hash function h and primes p, q such that $g \in \mathbb{F}_p^*$ generates a subgroup of order q. Currently, a standard choice for these would be a 1024-bit p and 160-bit q. Parties select a private key x uniformly from $0 < x < q$ and publish the corresponding public key $y = g^x \bmod p$. To sign a message m, parties select nonce k uniformly from $0 < k < q$ then compute the signature (r, s) by

$$r = g^k \bmod p \bmod q \tag{1}$$

$$s = (h(m) + xr)k^{-1} \bmod q \tag{2}$$

and note OpenSSL pads nonces to thwart traditional timing attacks by adding either q or $2q$ to k.

The performance bottleneck for the above signatures is the exponentiation in (1); extensive literature exists on speeding up said operation. Arguably the most widely implemented method in software is based on the basic left-to-right square-and-multiply algorithm and employs a standard sliding window (see [8, 14.85]).

It is a generalization where multiple bits of the exponent can be processed during a given iteration. This is done to reduce the total number of multiplications using a time-memory trade-off. With the standard 160-bit q, a reasonable choice (and what OpenSSL uses) is a window width $w = 4$.

The OpenSSL library includes an implementation of this algorithm, and uses it for DSA computations. Its speed is highly dependent on how the modular squaring and multiplication functions are implemented. Computations modulo p are carried out in a textbook manner using Montgomery reduction. Outside of the reduction step, the actual squaring and multiplication are implemented in separate functions; this is because we can square numbers noticeably faster than we can multiply them.

3.3   The Attack

We aim to determine partial nonce data during the computation of (1) by observing I-cache timings and use said partial data on multiple nonces to mount a lattice attack on (2) to recover the private key x.

In Sec. 3.2 we mention that squaring and multiplication are implemented as two distinct functions. In light of this, it is reasonable to assume that:

- All portions of these two sections of code are *unlikely* to map to the same I-cache sets;
- The load and consequentially execution time of (1) is dependent on their respective availability in the I-cache;
- An attacker capable of taking I-cache timings by executing their *own* code as outlined in Sec. 3.1 in parallel with the computation of (1) can deduce information about the state of the exponentiation algorithm—thus obtaining critical information about k.

The resulting side-channel is a list of vectors where each vector component is a timing measurement for a distinct cache set. We illustrate in Fig. 2, where we hand picked 16 of 64 possible I-cache sets that seemed to carry pertinent information.

Analyzing Timing Data. Next, we analyze this data to determine the sequence of states the exponentiation algorithm passed through. Just the sequence of squarings and multiplications that the sliding window algorithm passes through implies a significant amount of information about the exponent input. We utilize the framework of [6] to analyze the timing data, obtain a good guess at the algorithm state sequence, and infer a number of bits for each nonce. The steps include:

- For each operation we wish to distinguish (for example, squaring and multiplication), take a number of exemplar timing vectors that represent the I-cache behavior during said operation; [6, Sec. 4.2] calls this "templating".
- With these templates, create a Vector Quantization (VQ) codebook for each operation; this is done using a standard supervised learning method called LVQ. This can help eliminate noise and reduce the size of the codebook.

- Create a Hidden Markov Model (HMM) that accurately reflects the control flow of the considered algorithm. The observation input to the HMM is the output from VQ.
- Use the Viterbi algorithm to predict the most likely state sequence given a (noisy) observation sequence (VQ output of I-cache timing data).

Vector Quantization. We categorize timing vectors using VQ, which maps the input vectors to their closest (Euclidean distance-wise) representative vector in a fixed codebook. We obtain codebook vectors during a profiling stage of the attack, where we examine timing data from known input to classify the vectors in the codebook. Essentially, this means we setup an environment similar to the one under attack, obtain side-channel and DSA signatures with our own known key, then partition the obtained vectors into a number of sets with fixed labels. These sets represent the I-cache access behavior of the algorithm in different states, such as multiplication and squaring; these are the labels. When running the attack, we classify the incoming timing vectors using VQ. The algorithm state guess is the label of the closest vector in the codebook. To summarize, we guess at the algorithm state based on previously observed (known) algorithm state.

Hidden Markov Models. We also build and train the HMM during the profiling stage, using the classical Baum-Welch algorithm. The training data is the output from VQ above: the observation domain for the HMM is the range of VQ (the labels). As multiplication and squaring steps in the algorithm span multiple timing vectors in the trace, we consider these steps as meta-states, represented explicitly in the HMM by a number of sub-states corresponding to this span. When running the attack, we feed the trace through VQ and send the output to the HMM. The classical Viterbi algorithm outputs the state sequence that maximizes the probability of the observation sequence. To summarize, we guess the algorithm state sequence that best explains the side-channel observations.

Example. In addition to the timing data (rows 0-15) in Fig. 2, we give the VQ output (rows 16-17) and HMM state prediction (rows 18-19). Normally the purpose of any HMM in signal processing is to clean up a noisy signal, but in this case we are able to obtain extremely accurate results from VQ. This leaves little work in the end for the HMM. We chose to template squaring (the dark gray), multiplication (black), and what we can only assume is the Montgomery reduction step (light gray).

Using Partial Nonce Data. Having obtained a state sequence guess and thus partial information on nonces k for many signatures, the endgame is a lattice attack.

In such an attack it is difficult to utilize sparse key data, thus an attacker usually concentrates on a fairly long run of consecutive (un)known bits, and obtains more equations instead. Furthermore, we experienced that guesses on bits of k get less accurate the farther away they are from the LSB. We sidestep these issues by concentrating on signatures where we believe k has $\{0,1\}\{0\}^6$ in

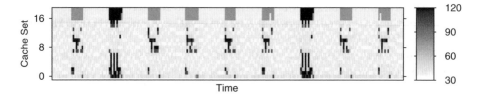

Fig. 2. Live I-cache timing data produced by a spy process running in parallel with an OpenSSL DSA sign operation; roughly 250 timing vectors (in CPU cycles), and time moves left-to-right. The bottom 16 rows are the timing vector components on 16 out of 64 possible cache sets. The top four are meta-data, of which the bottom two are the VQ classification and the top two the HMM state guess given the VQ output. Seven squarings are depicted in dark gray and two multiplications in black.

the LSBs—that is, six zeros followed by a zero or one. The top bit is fixed due to the padding, giving us a total of eight known bits separated by a single long run of unknown bits. Experiments suggest we need 37 such signatures to recover the long term key x.

Results. We obtained 17K signatures, messages, and corresponding I-cache timing data. Considering we expect the given bit pattern in k with probability 2^{-6}, this number seems unnecessarily high at first glance. Like many practical side-channel attacks, this is due to inherent issues such as noise, context switching, OS scheduling, and lack of synchronization. As our spy process is truly decoupled from the OpenSSL code, running as an independent process, we get absolutely no guarantee that they will execute simultaneously—or when they happen to, for how long.

After obtaining these 17K signatures, our analysis resulted in 75 signatures believed to match the pattern. We ran the lattice attack on five Intel Core2 quad core machines, taking random samples of size 37 until the result yielded a private key that corresponded to the given public key. The first core to succeed did so in 54 minutes, after roughly 3200 lattice attack iterations. Checking afterwards, 59 of these guesses were correct and 16 incorrect.

4   Closing the Instruction Cache Side-Channel

Countermeasures to mitigate the I-cache side-channel can be employed at a hardware and/or a software level. Hardware countermeasures require changes to the micro-architecture and it might take a while until such a new processor generation is available on the market. Previous work in this area proposed using alternative cache hardware, such as Partitioned Caches [12], Partition-locked Caches and Random-permutation Caches [16]. Most current processor designs are driven by performance and power criteria, leaving security as a secondary consideration; it is questionable whether this view will change in the foreseeable future. In this work, we focus solely on software techniques to address this vulnerability. Such countermeasures can be applied instantly by a software engineer as long as no hardware equivalents are present. In contrast to previously proposed software techniques, which are usually algorithm specific (e.g., Montgomery's powering ladder [9]), our aim is to provide generic methods be employed at the kernel and/or compiler level.

In the following discussion, we have to distinguish between countermeasures applicable to architectures which support SMT and conventional single-threaded processors. While in both architectures multiple threads can exist at the same time, there is a substantial difference in how they are scheduled. Processors with SMT support essentially split a single physical processor into two logical processors by duplicating some sections of the micro-architecture responsible for architectural state. In this way, the OS can schedule two threads/processes to be executed simultaneously on the same processor. These two threads execute literally simultaneously, not in a time-sharing fashion. Memory accesses of both execution threads alter the cache states at the same time.

In contrast, single-threaded processors are only capable to execute a single process/thread at any given point in time. In such architectures, execution time is allocated in time slices to the different processes/threads; by frequently switching between processes/threads, it gives an outward impression that multiple tasks are executed simultaneously. This type of execution is called quasi-parallel execution.

Cache attacks can work on both SMT processors and single-threaded processors. It is easier to run these attacks on SMT because spy and cipher can run simultaneously on different virtual cores in a single physical processor and spy can monitor cipher execution while cipher is performing its computations. Running cache attacks on single-threaded processors is more difficult. An attacker needs to use some tricks to have a "ping-pong" effect between the spy and cipher processes. [10] showed that it is possible to pause the cipher execution at a determined point and let a spy to examine the cache state. [10] exploited an OS scheduling trick to achieve this functionality and devised an attack on the last round of AES. A similar OS trick was shown in [15] to let a malicious party monopolize CPU cycles. [15] proposes to exploit OS scheduling mechanism to steal CPU cycles unfairly. Their cheating idea and the source code can easily be adapted to cache attacks on single-threaded processors.

Disable Multi-threading. In general, cache-based side-channel attacks take advantage of the fact that modern computer systems provide multi-threading capability. This fact allows an attacker to introduce an unprivileged spy process to run simultaneously with a security-critical code, thereby deriving secret key information from the state of the I-cache. A simple solution to eliminate this vulnerability is to turn off multi-threading when a security-critical process is scheduled to be executed: since it is the task of the OS to schedule processes/threads, it can simply decide to ignore all unprivileged processes/threads and not run them. On processors with SMT capability, the OS can adopt a scheduling policy that does not permit to execute another process in parallel with the crypto process. Alternatively, SMT can be turned off in the BIOS. According to Intel, SMT improves performance of multi-threaded applications by up to 30 %. Therefore it needs to be decided on a case-by-case basis if disabling

SMT for a more secure processing platform is acceptable from a performance point of view. Disabling multi-threading alone does not suffice to close I-cache side channel. I-cache attacks can be used on single-threaded processors without SMT capability as we explained above.

Fully Disable Caching. Another intuitively simple solution to close the information leakage through the I-cache is to disable the cache entirely. The Intel x86 architecture makes the cache visible to the programmer through the CD flag in the control register *cr0*: if said flag is set, caching is enabled otherwise it is disabled. However, such an approach severely affects the performance of the system as a whole. A more fine-grained control sees the cache only disabled when security-critical code is scheduled to be executed.

Partially Disable Caching. The x86 caches allow the OS to use a different cache management policy for each page frame. Of particular interest in this context is the PCD flag in control register *cr0* which determines whether the accessed data included in the page frame is stored in the cache or not. In other words, by setting the PCD flag of the page frames containing security-critical code it is possible to partially disable the caching mechanism. While such an approach successfully eliminates the I-cache side-channel we argue that is has a considerable negative impact on performance (albeit not as severe as with completely turning off the cache). The reason is that most cryptographic primitives spend the vast majority of the execution time in some small time-critical code sections; hence, not caching parts of these sections will be reflected in longer execution times.

Cache Flushing. Ideally, the processor would provide an instruction to flush the content of the L1 I-cache only. Unfortunately, such an instruction is not yet available on Intel's x86 range of processors. Instead, the WBINVD instruction [7] can be executed during context switches to flush the L1 I-cache. Note, that this instruction invalidates all internal caches, i.e., the instruction cache as well as the data cache; modified cache lines in the data cache are written back to main memory. After that, the instruction signals the external caches, i.e., the L2 and L3 cache to be invalidated. Invalidation and writing back modified data from the external caches proceeds in the background while normal program execution resumes, which partly mitigates the associated performance overhead. OS can flush the cache when a security-critical process such as a cipher switched out and thus the next process scheduled right after the cipher cannot extract any useful information from the cache state. This countermeasure is not effective on SMT systems because flushing happens during context switch and spy that runs simultaneously with a cipher on SMT can still monitor cipher's execution.

Partial Cache Flushing. Flushing the entire L1 I-cache negatively affects performance of both the security application as well as of all the other existing threads. This performance impact can be reduced when following a more fine-grained approach: instead of flushing the entire I-cache we propose to invalidate only those cache sets that contain security-critical instructions via some kind of OS support.

The x86 processor does not include such a mechanism that allows flushing of specific cache sets. Instead, some architectures provide the CLFLUSH instruction [7] capable of invalidating a cache line from the cache hierarchy. This instruction takes the linear address of the cache line to be invalidated as an argument. Consequently, flushing an entire cache set with this instruction would require the knowledge of both the linear address space of the spy process as well as of the security-critical code sections of the crypto process. While the later can be made easily available to the OS, it is much more difficult to reason about the linear address space of the spy process. This instruction is not suitable for our purposes as a result.

However, flushing of specific cache sets on x86 processors can still be accomplished by beating an attacker at his own game. The simple idea is to divert the spy process from its intended use by employing it as defence mechanism; essentially, the kernel integrates a duplicate spy process into the context switch. This permits the eviction of security-critical code sections from the I-cache each time security-critical code is switched out.

At first glance, it might seem that invalidating only those cache lines containing security-critical code before giving control to another process (possibly the spy process) can defeat the I-cache attack. However, from the spy's point of view, it makes no difference whether lines with security-critical code have been invalidated or not: in any case, the spy process will measure a longer execution time since the crypto process has evicted a cache line belonging to the spy. Therefore, invalidating only cache lines with security-critical code is not sufficient and the entire sets that hold them need to be invalidated. Similar to flushing the entire cache, partial flushing is not effective on SMT processors as explained above.

Cache-conscious Memory Layout. Fundamentally, I-cache attacks rely on the premise that the security-critical code sections or parts of them map to different regions in the I-cache. By mapping these security-critical code sections exactly to the same regions in the cache, the I-cache attacks can no longer recover the operation sequence. However, in some cases this approach might not be sufficient. For example, consider the case of two security-critical code sections that are of equal size and map to the same sets, where the majority of execution time of the two security-critical code sections is spent in disjoint cache sets. In such a scenario, it is still highly likely that the spy observes distinct traces despite the appropriate alignment in memory. Cache-conscious memory layout can be accomplished either by a compiler or via OS support. Given the I-cache parameters and the security-critical code sections, a compiler can generate an executable resistant against I-cache attacks by appropriately aligning said sections. To balance the sizes of these sections, it might be necessary to add some dummy instructions, e.g., NOPs, before and/or after the sections; this padding with dummy operations implies some performance penalty and results in an increase in the size of the executable. Alternatively, the OS can be in charge of placing the security-critical code sections in such a way in memory that they map to the same regions in the cache if the cache is physically addressed. For that, the executable needs to specify the memory sections with security-critical information. Similar to the compiler approach, additional dummy operations might be required to make the security-critical code sections equal in size. None of the above countermeasures provide an effective yet practical mechanism for SMT systems, except cache-conscious memory layout . This countermeasure incurs very low overhead as we will explain in the next section and it is also effective on SMT systems.

5   Performance Evaluation

All our practical experiments were conducted on a Intel Core Duo machine running at 2.2 GHz with a Linux (Ubuntu) Operating System. To minimize the variations in our timing measurements due to process-interference we used the process affinity settings to bind the crypto process to one core and assigned all the other processes to the other core.

Performance impact on the crypto process. For the performance evaluation of our proposed software countermeasures we used the RSA decryption function of OpenSSL (version 0.9.8d) as a baseline. Table 1 summarizes the performance impact of our proposed countermeasures upon OpenSSL/RSA in comparison to the baseline implementation; results are given for different key lengths, i.e., 1024-bits, 2048-bits and 4096-bits.

Table 1. Performance evaluation of the proposed countermeasures

Implementation		1024-bit	2048-bit	4096-bit
Baseline OpenSSL/RSA	Execution time (in ms)	1.735	9.606	57.9
	Decryptions/s	576.3	104	17.3
OpenSSL/RSA with	Execution time (in ms)	1273	7204	45060
cache disabled	Decryptions/s	0.8	0.1	0.02
OpenSSL/RSA with	Execution time (in ms)	1.888	11.192	60.6
cache flushing	Decryptions/s	530	89.3	16.5
OpenSSL/RSA with	Execution time (in ms)	1.734	9.535	58.2
partial flushing	Decryptions/s	576.8	104.9	17.1
OpenSSL/RSA with	Execution time (in ms)	1.755	9.727	58.2
cache-conscious layout	Decryptions/s	570	102.8	17.2

The performance evaluation supports our claim that turning the cache off results in an immense performance overhead. For instance, execution of a 1024-bit OpenSSL/RSA with a disabled cache leads to a 3-orders of magnitude degradation in performance. This experiment was conducted with the help of a simple kernel module which turns the cache off when loaded into the kernel and turns the cache on again when unloaded. Similarly, we expect an unacceptable impact on performance when just partially disabling the cache since this forces the processor to repeatedly fetch instructions from the slow main memory; for that reason we refrained from investigating this approach in more detail. Flushing the cache hierarchy during a context switch incurs a performance overhead of about

$5 - 15$ % for the different key sizes. Even more severe than this non-trivial performance penalty is the significant increase in context switch time: using Intel's RDTSC instruction, we measured a 10-fold increase. The performance overhead from both an application as well as OS point of view can be significantly reduced when only invalidating the cache sets containing security-critical code. For that, we aligned the spy process appropriately in the context switch routine to evict the cache sets that hold data of both the OpenSSL/RSA multiplication and squaring routines; in total, it was necessary to evict 29 sets (i.e., 18 sets are occupied by multiplication instructions and 11 sets by squaring instructions) from the instruction cache.

From Table 1 it appears that no noticeable performance overhead is associated with this countermeasure. This result is somewhat expected since the overhead of bringing the evicted instructions from the L2 cache back into the instruction cache is negligible.

Finally, we investigated the cache-conscious memory layout approach. It was necessary to pad the OpenSSL/RSA squaring routine with 406 NOP instructions in total so that it matches the size of the OpenSSL/RSA multiplication. However, having the same code size alone does not prevent information leakage; the security-critical code sections also need to be aligned in memory in such a way that they map into the same cache sets. This can be done by rewriting the linker script to control the memory layout of the output file. In more detail, we first used the gcc compiler option "-ffunction-sections" to place the two security-critical code sections in a separate ELF section each. Then, we redefined the memory model so that each section is placed at a known address such that they will be placed in the same sets in the cache. The performance overhead associated with this countermeasure is so minimal that in practice it can be regarded as negligible.

Performance impact on the system caused by the countermeasures. Our proposed software countermeasures may have a negative impact on other processes that are running concurrently with a security-critical application. This impact might be in particular noticeable for the solution where the entire cache content is flushed during a context switch.

To estimate the impact on the system as a whole, we ran the SPEC2000int benchmark simultaneously with a security-critical application; this means the processor's entire cache hierarchy is invalidated at regular intervals, i.e., every time the security-critical process is switched out. Figure 3 illustrates this performance impact on the SPEC benchmark in presence of this countermeasure.

On average, invalidation of the cache during context switches results in a 10 % degradation in performance. This decline is caused by bringing data back into the cache after it has been discarded during the context switch. Note, this overhead gives an estimation for the worst-case scenario and the impact will typically be less severe on systems where security-critical applications are executed less frequently. Figure 3 also shows the run time of the SPEC benchmark in presence of partial cache eviction as a countermeasure instead. Essentially, the performance impact on the system as a whole is negligible in this case. Further, some of our

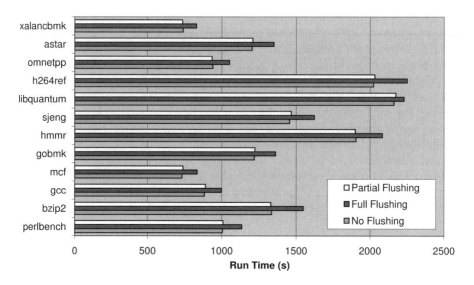

Fig. 3. Performance impact of cache flushing on the SPEC2000int benchmark

countermeasures can influence the time it takes to perform a context switch. If
there is no process using our countermeasures, which will probably be the case
most of the time, the only extra work that needs to be done is to check a flag (i.e.,
the flag that indicates whether the current process needs to be protected). However, switching out a security-critical process possibly requires some additional
work for the scheduler which results in a longer execution time of the context
switch. This behaviour is in particular apparent when using the cache flushing
approach since the scheduler needs to wait until all dirty data cache lines have
been written back to maintain memory coherence. Consequently, depending on
the type of process that is switched out, a different amount of time is spent in
the context switch routine. This can pose a serious problem to real-time systems,
where highly deterministic behaviour is required. Note that partial eviction has
a considerably smaller impact on the context switch; in theory, for each cache
set that contains security-critical instructions, the OS simply needs to execute
a small number of appropriately aligned dummy instructions and this number
needs to be equal or larger than the associativity of the I-cache.

6   Conclusions

We presented improved I-cache analysis techniques based on vector quantization
and hidden Markov models. The analysis is automated and fast, capable of
analyzing large volumes of concrete I-cache timing data. This can be used to
perform automated I-cache attacks.

We demonstrated its effectiveness by carrying out an I-cache attack on an unmodified version of OpenSSL's DSA implementation (0.9.8l). We used the framework to process the timing data from thousands of signatures and subsequently

recovered keys using lattice methods. This attack is automated, recovering a DSA private key within an hour.

Our study clearly proves that I-cache cryptanalysis is realistic, practical, and a serious security threat for software systems. We believe it is necessary to conduct a thorough analysis on current software cryptosystems to detect I-cache analysis (more generally Microarchitectural Analysis) vulnerabilities. We already saw several MA vulnerabilities in cryptographic software like OpenSSL and they were fixed by specific algorithm-level solutions such as removing extra reduction step from Montgomery multiplication. However, it is crucial to design generic algorithm-agnostic mitigation mechanisms.

Mitigation mechanisms can be employed at a hardware and/or a software level. Hardware countermeasures require changes to the micro-architecture and much longer time to hit the market compared to software countermeasures. Thus, we focused solely on generic software-level mitigations in our work and presented some countermeasures to close I-cache side channel. We studied their impacts on cipher performance and also on the performance of the overall system. Naive approaches such as disabling cache or flushing the entire cache before or after the execution of security critical software have high performance overheads associated with them. Thus, such approaches are far from gaining wide usage due to their low practicality even though they can eliminate I-cache side channel leakage. However, we presented two practical approaches, "partial flushing" and "cache conscious memory layout", that have very low performance overheads.

We realized that even very primitive support from the hardware can be very helpful towards designing and developing low-cost mitigations. For instance, if a processor's ISA includes an instruction permitting to flush cache sets, mitigations like partial flushing become much easier to implement and have lower performance overheads.

As our final remark, we want to emphasize that our results stress the significance of considering security as a dimension in processor design space and paying it the same level of attention as cost, performance, and power.

Acknowledgments. The authors would like to thank Dan Page for his input throughout the duration of this work.

References

1. http://cvs.openssl.org/chngview?cn=16275
2. Acıiçmez, O.: Yet another microarchitectural attack: Exploiting I-cache. In: Proceedings of the 1st ACM Workshop on Computer Security Architecture (CSAW 2007), pp. 11–18. ACM Press, New York (2007)
3. Acıiçmez, O., Koç, Ç.K., Seifert, J.-P.: On the power of simple branch prediction analysis. In: Proceedings of the 2nd ACM Symposium on Information, Computer and Communications Security (ASIACCS 2007), pp. 312–320. ACM Press, New York (2007)
4. Acıiçmez, O., Schindler, W.: A vulnerability in rsa implementations due to instruction cache analysis and its demonstration on openssl. In: Malkin, T.G. (ed.) CT-RSA 2008. LNCS, vol. 4964, pp. 256–273. Springer, Heidelberg (2008)

5. Acıiçmez, O., Schindler, W., Koç, Ç.K.: Cache based remote timing attacks on the AES. In: Abe, M. (ed.) CT-RSA 2007. LNCS, vol. 4377, pp. 271–286. Springer, Heidelberg (2006)
6. Brumley, B.B., Hakala, R.M.: Cache-timing template attacks. In: Matsui, M. (ed.) ASIACRYPT 2009. LNCS, vol. 5912, pp. 667–684. Springer, Heidelberg (2009)
7. Intel Corporation: Intel(R) 64 and IA-32 Architectures Software Developer's Manual, http://developer.intel.com/Assets/PDF/manual/253667.pdf
8. Menezes, A., Vanstone, S., van Oorschot, P.: Handbook of Applied Cryptography. CRC Press, Inc., Boca Raton (1996)
9. Montgomery, P.L.: Speeding the Pollard and elliptic curve methods of factorization. Mathematics of Computation 48(177), 243–264 (1987)
10. Neve, M., Seifert, J.P.: Advances on access-driven cache attacks on AES. In: Biham, E., Youssef, A.M. (eds.) SAC 2006. LNCS, vol. 4356, pp. 147–162. Springer, Heidelberg (2007)
11. Osvik, D.A., Shamir, A., Tromer, E.: Cache attacks and countermeasures: The case of AES. In: Pointcheval, D. (ed.) CT-RSA 2006. LNCS, vol. 3860, pp. 1–20. Springer, Heidelberg (2006)
12. Page, D.: Partitioned cache architecture as a side-channel defense mechanism. Cryptology ePrint Archive, Report 2005/280 (2005), http://eprint.iacr.org
13. Percival, C.: Cache missing for fun and profit. In: Proceedings of BSDCan 2005 (2005), http://www.daemonology.net/papers/htt.pdf
14. Ristenpart, T., Tromer, E., Shacham, H., Savage, S.: Hey, you, get off of my cloud: Exploring information leakage in third-party compute clouds. In: Proceedings of the 16th ACM Conference on Computer and Communications Security (CCS 2009), pp. 199–212. ACM Press, New York (2009)
15. Tsafrir, D., Etsion, Y., Feitelson, D.G.: Secretly monopolizing the CPU without superuser privileges. In: Proceedings of the 16th USENIX Security Symposium (SECURITY 2007), pp. 239–256. USENIX Association (2007)
16. Wang, Z., Lee, R.B.: New cache designs for thwarting software cache-based side-channel attacks. In: Proceedings of the 34th Annual International Symposium on Computer Architecture (ISCA 2007), pp. 494–505. ACM Press, New York (2007)

Correlation-Enhanced Power Analysis Collision Attack

Amir Moradi[1], Oliver Mischke[1], and Thomas Eisenbarth[2]

[1] Horst Görtz Institute for IT Security, Ruhr University Bochum, Germany
[2] Department of Mathematical Sciences, Florida Atlantic University, FL, USA
{moradi,mischke}@crypto.rub.de, teisenba@fau.edu

Abstract. Side-channel based collision attacks are a mostly disregarded alternative to DPA for analyzing unprotected implementations. The advent of strong countermeasures, such as masking, has made further research in collision attacks seemingly in vain. In this work, we show that the principles of collision attacks can be adapted to efficiently break some masked hardware implementation of the AES which still have first-order leakage. The proposed attack breaks an AES implementation based on the corrected version of the masked S-box of Canright and Batina presented at ACNS 2008. The attack requires only six times the number of traces necessary for breaking a comparable unprotected implementation. At the same time, the presented attack has minimal requirements on the abilities and knowledge of an adversary. The attack requires no detailed knowledge about the design, nor does it require a profiling phase.

1 Introduction

Ten years after the introduction of side-channel attacks [2,10,13,15,23], the creation of a DPA-resistant cryptographic hardware implementation remains a challenge. During the last years several countermeasures to prevent power and EM-analysis have been proposed [12,20,21,29,30]. One of the main targets of the side-channel community are implementations of the AES. AES [19], having been the NIST symmetric encryption standard for about 10 years, is probably the most widely used cipher in practical applications. Despite of its high cryptographic security in a black box scenario, implementations of AES are a popular and easy target for side-channel attacks such as DPA and SPA. Correspondingly, the efficient and leakage-minimized implementation of AES is a well-studied problem [8,9,21,24,25].

At the same time attacking techniques have been improved and defeated many of these countermeasures. The first practical evaluation was performed on one additive and one multiplicative masking scheme of AES [16]. It has been shown that though they are resistant to classical DPA attacks considering standard Hamming Weight (HW) and Hamming Distance (HD) models, more sophisticated attacks using more precise power models, e.g., the toggle count model [16], are capable of overcoming the masking countermeasure. However, these attacks usually require detailed information about the implementation such as the netlist

S. Mangard and F.-X. Standaert (Eds.): CHES 2010, LNCS 6225, pp. 125–139, 2010.

of the target device. Later it was shown in [17] that XOR gates of the mask multipliers of the masked S-box play the most significant role in the susceptibility of the evaluated schemes, but to our knowledge the proposed solutions have not been practically evaluated.

Another approach for attacking implementations using a power or EM side-channel are collision attacks. Here, the attacker concludes from the leakage that two identical intermediate values have been processed and uses this information to cryptanalize the encryption scheme. The practicability of these attacks has been shown against DES in [14,27]. Successful attacks against AES have been presented in [6,26]. Collision attacks remain less popular than DPA-like attacks because of their sometimes complicated setup, their strong dependence on noise, and the more complex key recovery phase. Although the number of traces actually used in an attack is usually lower than that of classical DPAs, the number of traces needed to generate a collision normally makes the attacks less efficient than, e.g., correlation DPAs. Finally, with the advent of randomizing-like countermeasures, collision attacks seem to be infeasible against protected implementations.

Our Contribution. In this work we present a method to identify and exploit collisions between masked S-boxes in a very efficient manner. In fact, we use correlation to combine the leakage of all possible collisions and thereby including the full set of obtained measurements in the attack. Since practical evaluation of attacks and countermeasures by means of making a state of the art ASIC chip is not a time- and cost-effective approach, we have applied our attack on a masked version of the AES, implemented on a Xilinx Virtex-II Pro chip mounted on the Side-channel Attack Standard Evaluation Board (SASEBO) [1]. Our implementation generates all masks for each plaintext byte uniformly at random and none of the mask bytes is reused in later encryptions. Our investigation shows that the applied masking scheme is capable of resisting against those first-order DPA attacks which use common and well-known power models, e.g., HD and HW. From the results of [16] it can be expected that the masking scheme can be overcome when using a more accurate power model, e.g., toggle count, or when applying template-based DPA attacks. These attacks, however, assume a powerful adversary, because detailed knowledge such as a back annotated netlist of the layout is needed, or a profiling phase using a controllable target implementation has to be performed. None of these requirements have to be met to perform the attack presented in this article.

Our proposed attack reduces the effect of randomness by means of averaging over known (not chosen) inputs, and detects the collisions on the S-box input/output by examining the leakage of averaged power traces. In fact, our attack reveals that in our target implementation even uniformly distributed masks cannot prevent a first-order leakage depending on the unmasked values. It should be noted that our attack does not depend on a specific leakage model. The experimental results show that our attack is able to recover the key by means of less than 20 000 traces while the secret starts leaking out by a zero value attack using at least 1 000 000 traces of the same implementation. For a second-order zero-offset DPA, even around 8 000 000 traces are needed to recover the secret key.

Organization. The remainder of this article is organized as follows: Section 2 describes the target implementation of the AES and sets it into the context of related work. In Section 3 we analyze the AES implementation with classical methods, before we detail on the proposed collision attack in Section 4. A conclusion is given in Section 5.

2   Hardware Implementation of the AES

Several optimizations for hardware implementations of the AES have been proposed. To minimize circuit area consumption of the AES, Rijmen [24] suggested the use of subfield arithmetic in $GF(2^4)$ to compute the inverse in $GF(2^8)$. The idea was taken further by Satoh et al. [25] using the "composite-field" approach/"tower-field" representation by Paar [22] to implement the inversion in $GF(2^4)$ by the use of sub-subfield arithmetic in $GF(2^2)$. Along with other innovations this resulted in a very compact AES S-box, which was further improved by Canright [8] through choosing the normal bases which yielded the smallest circuit size.

Several masking schemes have been proposed to create a masked AES S-box using either multiplicative or additive methods. Unfortunately, multiplicative ones [4,11] are vulnerable to certain attacks, especially the so-called zero-value attack, because a zero input value does not get masked by multiplication. The solution is to use the tower-field representation for an additive masking scheme because the inversion in $GF(2^2)$ is equivalent to squaring which is linear. The first at least algorithmically provable secure additive masking scheme was proposed by Blömer et al. [5]. Later Oswald et al. [21] proposed a more efficient scheme by using different bases and reusing some mask parts. Canright et al. [9] applied this idea to his very compact S-box resulting in the most compact masked S-box to date. [33] is also another design showing the interest of the research community on this topic.

2.1   Our Implementation

Our goal is the evaluation of a hardware implementation of the AES that is supposed to be secure against first-order side-channel attacks. To cover a wide range of possible implementations, we decided to implement two different architectures of the AES. The first one is designed to achieve low power consumption and has a low area requirement. This is achieved by choosing an 8-bit data path and features a single S-box that is sequentially used for SubBytes operations and the key scheduling. All registers are implemented as byte-wise shift-registers which can be clocked independently. The full data path of the complete AES engine excluding the key registers is masked. The mask values are generated internally by means of a PRNG, and the (uniform) distribution of the generated random values have been verified. The masks are different for each plaintext byte and differ in each execution of the encryption. The high level architecture of our AES

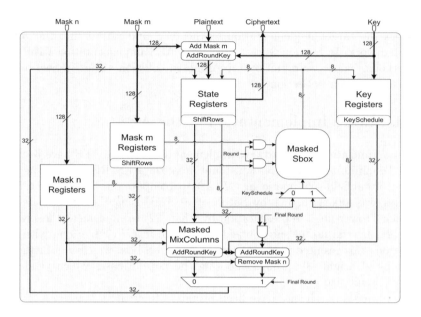

Fig. 1. Architecture of the AES design

design is depicted in Fig. 1. Unless stated otherwise, our analysis focuses on this implementation. To verify that our attack also works in the presence of noise, we implemented a second AES engine that has a 32-bit data path and features four parallel S-boxes. Details on this engine can be found in Appendix B of [18].

The design of the masked S-box is identical to [9] which uses two independent masks, m and n to randomize the input and the output. We retrieved the Verilog code from author's website and paid special attention that the order of the operations and other suggestions to maintain the masking scheme have been strictly kept by the synthesis tool.

Encryption starts by providing 128-bit plaintext, key, and masks m and n. The masks are independent and uniformly distributed and differ for each plaintext byte and each encryption execution. At the beginning of each round ShiftRows is performed on both the masked data state and the input mask m. The S-box is then first used by the key schedule unit to compute the first 32-bit part of the next round key without using any masks. In the following four clock cycles the masked S-box performs the SubBytes transformation on the first column. The consecutive masked MixColumns and AddRoundKey transformations are performed using a 32-bit wide datapath. During this operation the mask of the state is also changed back to mask m because during SubBytes the input mask m is replaced by the output mask n. This sequence of four times SubBytes followed by MixColumns and AddRoundkey is repeated four times to complete the round function.

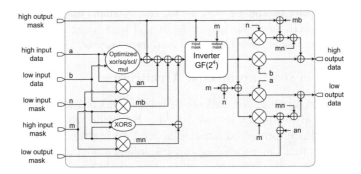

Fig. 2. Block diagram of the used masked $GF(2^8)$ inverter

2.2   Details on the Masked AES S-Box

The general structure of the used masked S-box is depicted in Fig. 2 omitting the tower-field conversion. While only the $GF(2^8)/GF(2^4)$ module is shown, the $GF(2^4)/GF(2^2)$ module uses the same structure the only difference being that instead of an $GF(2^2)$ inversion module, this step is merged as squaring to the overall design. As can be seen the additional elements in the datapath are all additive (XORs). It is important to introduce a new mask before adding the masked products since the distribution of the sum of two masked products is otherwise not uniformly distributed as explained in [9]. By doing all summations in the correct order the result of every computation is either uniformly distributed or has the random product distribution independent of the used plaintext and key. Therefore, as stated in [9], the scheme is considered to theoretically achieve perfect masking on an algorithmic level by the definition of [5].

3   Analysis of the AES Implementation

The whole design has been implemented on a Xilinx Virtex-II Pro FPGA (xc2vp7) of a SASEBO circuit board which is particularly designed for side-channel attack experiments [1]. To better understand the leakage of our implementation we performed several tests of our platform. We performed tests to identify when certain leakages occur. Subsequently we analyzed the vulnerability of our implementation to first-order DPA attacks based on correlation, both in the unmasked and in the masked case.

All tests are performed on the power consumption of the Virtex-II FPGA containing our implementation. Measurements are performed using a LeCroy WP715Zi 1.5GHz oscilloscope at a sampling rate of 5GS/s[1] and by means of a differential probe which captures the voltage drop over an 8Ω resistor in the

---

[1] This oversampling is not essential here; however, since glitches and toggles in hardware occur at very high frequencies, we decided to keep a high sampling rate, but we have confirmed the feasibility of the attacks using lower sampling rates, e.g., 1GS/s.

VDD (3.3V) supply of the FPGA. In all the experiments the clock signal is provided by a 24MHz oscillator which is divided by 8 using a frequency divider, i.e., our cryptographic engine is clocked at a frequency of 3MHz.

3.1   Analysis of the Unprotected Architecture

In a first step we analyze the leakage of an unprotected implementation that employs the highly compact unmasked AES S-box design of Canright [8]. A power trace of this unprotected implementation during the first 12 clock cycles is shown in Fig. 3(a). The processing order and hence the occurrence of leakages over clock cycles will pretty much stay the same for the masked implementation, as the high level architecture remains constant.

Similarly to what was observed in [16], DPA attacks using the HW of the S-box input/output are not successful. We get a good estimation about the leakage strength of the implementation platform performing a DPA attack predicting the HD of 8 bits of the state register[2]. The result of this HD-based DPA is shown by Fig. 3(b). As shown in Fig. 3(d), the leakage of approx. 3 000 traces suffices to perform a successful attack using a HD model.

As explained in [15], most of the time implementations of the AES S-box consume less power for the zero input value than for the other cases. It holds here as well, and an attack using the zero value model is possible which is shown by Fig. 3(c). Moreover, according to Fig. 3(d) 4 000 measurements are required for succeeding with the zero value attack.

3.2   Analysis of the Masked Architecture

Moving towards the masked version of the implementation, we should emphasize that neither the attacks using the HW model predicting S-box input/output nor those which use the HD model on the state register are expectedly able to reveal the secrets. Since in our architecture the state and the mask registers are shifted in the same fashion, both masked values and the masks are processed at the same time. Therefore, one can perform a zero-offset second-order DPA attack [31] by squaring the power values and by means of a HD model to predict the transitions in the state register. In practice higher-oder attacks usually require much more traces in comparison to the first-order ones, and we collected 10 000 000 traces to clearly distinguish the correct key guess amongst the others. The result of such an attack is shown by Fig. 4 indicating that around 8 000 000 traces are needed to have a successful attack. In our experiments we have examined several possible power models in first-order attacks, and interestingly the secret starts leaking by a zero value DPA attack using 1 000 000 traces. The relevant result is shown by Fig. 5. It shows that power consumption of the target implementation is not really independent of the unmasked values, and this issue motivated us to try for an alternative approach in order to decrease the number of measurements and to distinguish the secret more clearly. It should be emphasized that in our

---

[2] Note that to predict the HD of the state register in our target architecture, two key bytes amongst 2^{16} hypotheses should be guessed at the same time.

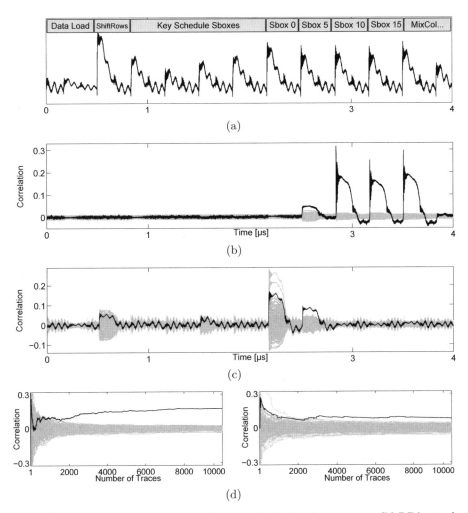

Fig. 3. (a) A measured power traces of an undefended implementation, (b) DPA attack result predicting toggles in the state register, (c) DPA attack using zero value model predicting the S-box input, and (d) the required number of traces in attacks using (left) HD model and (right) zero value model

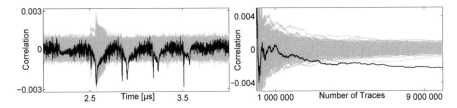

Fig. 4. Result of a zero-offset second-order DPA attack on the masked implementation using a HD model (left) by $10\,000\,000$ traces and (right) at point $2.9\mu s$ over the number of traces

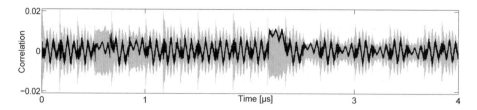

Fig. 5. Result of a zero value DPA on the masked implementation by 1 million traces

target implementation the mask values are internally generated by means of a PRNG, and the (uniform) distribution of the generated random values has been verified. Furthermore, the masks are different for each plaintext byte and differ in each execution of the encryption.

Before introducing our collision attack, we first explain some issues which we observed during practical experiments. As mentioned before, we acquired millions of traces for the aforementioned attacks. It should be noted that these traces have been recorded on randomly chosen plaintext bytes. To learn about the behavior of the implementation, we computed the average over the traces (measured from the masked implementation) based on a plaintext byte and thereby obtained 256 mean traces. By examining the variance of the mean traces we can detect in which clock cycle a function, e.g. the S-box, relevant to the selected plaintext byte is computed, if the mean traces are not ideally close to each other. If such features are detectable in the power traces, the mean traces are not independent of the unmasked values. In Fig. 6 two variance traces over the mean traces of plaintext byte 0 and byte 5 are shown[3]. The figure shows that a function over these two plaintext bytes is computed in two consecutive clock cycles, which fits to the target architecture. We have used 1 000 000 measurements to generate the variance traces shown in Fig. 6, but we have examined using less number of traces, e.g., 50 000, and the result had the same shape and the same feature.

Since the mean traces depend on the unmasked values, a couple of attacks are possible. For example, as expected by the authors of [9], a DPA using the toggle-count model should work here. Yet, for that attack the adversary needs to have access to the target netlist or layout to simulate and extract the toggle-count model. Moreover, a template-based DPA attack also might work, but the adversary needs to first create profiles for a known key. The aim of our attack is to avoid such limitations and strong assumptions.

4   Correlation-Enhanced Collision Attack

Based on the observations described in Section 3, we adapt collision attacks to be able to exploit any first-order leakage without knowing the precise hypothetical power model. The attack targets collisions in the full S-box computation. Detected collisions have the same 8-bit input and consequently the same 8-bit

---

[3] Note that the mean traces are computed based on each plaintext byte independently.

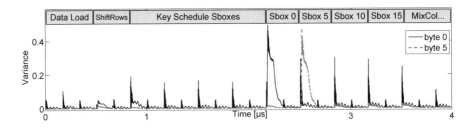

Fig. 6. Variance of mean traces for plaintext byte 0 and 5

output value. Please keep in mind that these values are always masked, with different masks for each measurement and each S-box. The developed attack does not require any sort of profiling phase with a known-key device. Of course, knowledge about the position of the execution of the S-box computation are helpful, but all information needed can be extracted from the measurements of the device under attack. As described in section 3.2, one way would be to compute the mean traces and perform a variance check. Alternatively, such information can be gained by *combing* through the power traces with an offset of e.g. one clock cycle [28]. We split the attack into a measurement phase, which is comparable to previous collision attacks against unmasked implementations, and an enhanced detection phase.

During the *measurement phase* we record the power consumption traces T_i of the encryption of random known plaintexts $P_i = \{p_j^i\}_{j=0}^{15}$. We know that each trace T_i contains the leakage of every S-box computation of the first round $s_j^i = S(p_j^i \oplus k_j)$, which we target in our collision attack. In our model, a collision occurs when two S-box computations at the byte position $j_1 = a$ and $j_2 = b$ collide, i.e., have equal output $s_a^{i_1} = s_b^{i_2}$ and due to the bijectivity of the S-box also equal input $p_a^{i_1} \oplus k_a = p_b^{i_2} \oplus k_b$. We can define the input difference $\Delta_{a,b}$ as

$$\Delta_{a,b} = p_a^{i_1} \oplus p_b^{i_2} = k_a \oplus k_b$$

Hence, this type of collision reveals a linear relation between two key bytes, depending only on the known difference $\Delta_{a,b}$. By finding more first-round collisions, eventually we will have relations for all 16 key bytes k_i, reducing the key entropy to 8 bits (i.e. 256 key candidates for the full 128-bit AES key), which can easily be recovered by trial and error. This attack is labeled *linear collision attack on AES* in [6]. In theory, this attack is prevented by masking, since both input and output of the S-box are masked, destroying any relation between input difference Δ and the plaintexts p^{i_1} and p^{i_2}. Yet, we show that there is a remaining leakage in the masked Canright/Batina S-box that can be exploited by an adaption of the linear collision attack we describe in the following.

The measurement phase is the same as for the normal linear collision attack. Yet, we apply a different *detection phase* to identify many collisions at once. As described above, we first have to detect where the leakage of the individual S-boxes occurs. To reduce the influence of the masks, we average the power consumption for equal input bytes p_j. We do this by browsing all of our traces T and averaging only those traces where the jth plaintext byte equals a certain value $\alpha \in GF(2^8)$. Hence, we get 2^8 average traces M_j^α for each plaintext byte position j, where M_j^α is the average of all traces T_i where $p_j^i = \alpha$.

$$M_j^\alpha = \overline{T_i \cdot \delta(p_j^i = \alpha)}$$

Unlike the classical *linear collision attack*, we do not try to detect a single collision, but directly include all possible collisions between two byte positions $j = a$ and $j = b$. We know that for one particular key pair k_a and k_b, the difference $\Delta_{a,b} = k_a \oplus k_b$ is constant. Hence, a collision occurs whenever the plaintexts at position a and b show the same difference, i.e., $p_a = \alpha$ and $p_b = \alpha \oplus \Delta_{a,b}$. Our approach is to guess the difference $\Delta_{a,b}$ and verify our guess by detecting all resulting collisions $p_a = \alpha$ and $p_b = \alpha \oplus \Delta_{a,b}$ for all $\alpha \in GF(2^8)$ at the same time. For detecting the correct $\Delta_{a,b}$, we correlate the averaged power consumption M_a^α of the S-box lookup of $p_a = \alpha$ to the averaged power consumption $M_b^{\alpha \oplus \Delta_{a,b}}$ of the S-box lookup of $p_b = \alpha \oplus \Delta_{a,b}$ for all $\alpha \in GF(2^8)$. The correct difference $\Delta_{a,b}$ of the two key bytes k_a and k_b is then given by:

$$\underset{\Delta_{a,b}}{\arg\max}\, \rho\left(M_a^\alpha, M_b^{\alpha \oplus \Delta_{a,b}}\right)$$

The correlation $\rho\left(M_a^\alpha, M_b^{\alpha \oplus \Delta_{a,b}}\right)$ is computed over all $\alpha \in GF(2^8)$ and can be computed for every point in time. It is maximum if $\Delta_{a,b}$ is correct. For wrong differences Δ, the correlation approaches zero. Hence, this attack behaves similar to a correlation attack. Unlike correlation based DPA, which correlates the power consumption to a power model that will never truly represent the real power consumption, our attack correlates the power consumption of one S-box computation to the power consumption of a different instantiation of the same S-box (processing the same value). Compared to classical collision attacks, our attack is stronger because all traces are included in calculating the correlation coefficient ρ, i.e. leakage from all traces T_i is used to recover the key relations.

If we go back to the last experimental results described in section 3.2, where millions of traces have been collected from the masked implementation while all 256 mask bits are randomly generated for each measured encryption, we had the mean traces for plaintext byte 0, M_0, and for plaintext byte 5, M_5. Also, we have shown that a function, here the S-box, exists, which processes a value depending on these two plaintext bytes at two consecutive clock cycles. Then we shift for example M_5 with the length of a clock cycle to the left to align the traces and perform the attack algorithm. The result of the attack is shown by Fig. 7 in which the correct hypothesis is obviously distinguishable amongst others. As mentioned before, our attack computes the correlation between the power consumption of two instances of the S-box computation; this explains why we get such a high correlation value for a correct guess of Δ. It should be noted that while initially 1 000 000 traces have been used to compute the mean traces, around 20 000 traces are enough to distinguish the correct key relation. The relevant figure for the number of required traces is also shown in Fig. 7.

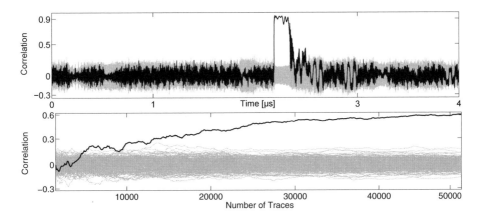

Fig. 7. (top) The result of collision attack using the mean traces of byte 0 and byte 5, (bottom) the result of the attack over the number of used measurements

Repeating this attack on other sets of mean traces, e.g., between M_0 and M_1, M_0 and M_2, and so on, will reveal 15 relations between all key bytes[4]. The correct AES key can easily be distinguished from the remaining 256 candidates by simple trial and error. In cases where this might not be feasible or to recover the full key of an implementation of AES-256, we simply extend our attack to the second round. An alternative for the case when ciphertexts are unknown is to compute the output of MixColumns of the first round (of AES) for each of the 256 possible key candidates and perform only one (collision) attack using the leakages of the second round for each of the 256 128-bit key candidates separately. The attack would be possible for only one of the key candidates since the use of others will lead to wrong input bytes for the AddRoundKey and therefore to incorrect averages (to make the mean traces). In fact, a variance test approach on mean traces of the second round can reveal the correct 128-bit key candidate, and performing the attack on the second round is not even necessary to recover the full key of AES-128. No new measurements are needed for this extension.

Knowing only the ciphertext bytes the same attack is possible using the power traces covering the last round of the encryption because of the absence of Mix-Columns at the last round. Similarly to the attack on the start of the encryption, 256 128-bit candidates will remain as the last round key. Then, for each of them the input of MixColumns of the 9th round can be computed which is also the output of the S-box of the same round. Therefore, the variance check approach as mentioned above will reveal the correct key. We practically performed the afore-mentioned attack, and were able to extract the secret using the same number of traces as in the known-plaintext attack.

Resemblance to Template DPA: A better understanding of how and why the attack works can be gained by a comparison to a template-based DPA as described

---

[4] In fact, 120 key byte relations can be computed and possibly all be evaluated by voting techniques [6,32].

in [3,15]. Since we do not have a profiling phase, the creation of templates is different. We create templates M_j^α for each input (or output) value $p_j = \alpha$ (like in some template attacks) and also for each input (or output) byte position j. The separate templates for different byte positions j are necessary, as we cannot match our templates M_j^α to specific states (or input-output combinations) of the S-boxes, because we lack knowledge about the key. In the next step, we compare pairs of these templates for two positions $j_1 = a$ and $j_2 = b$ by correlating them to each other. A template-based DPA attack, as described in [15], instead uses the template as power model which is correlated to each individual power trace. Due to the noise in each trace, the resulting correlation values are much lower when compared to our case. Our attack correlates two sets of templates, which have a much lower noise due to the averaging process. The relative distance between the correlation of the right and the wrong key hypotheses is quite similar in both attacks. Compared to a template-based DPA, our attack assumes a much weaker adversary that neither needs access to a known key implementation nor requires a profiling phase.

Attack on parallel architectures/Influence of noise: All the practical results shown are for an 8-bit architecture, and each S-box is computed in a separate clock cycle. In order to investigate the feasibility of the proposed attack in the presence of increased (switching) noise, we have examined the same attack on a 32-bit architecture where four S-boxes are executed in parallel at each clock cycle. The power consumption of the three unpredicted S-boxes enters the mean traces as noise, which can be reduced by increasing the number of measurements. According to the experimental results shown in the Appendix B of [18], the secret is revealed in the same way using around 300 000 measurements which shows the strength of the attack. We expect a similar behavior if the shuffling countermeasure [12,15] would be applied to the serial 8-bit architecture. For the case where just 4 S-boxes are shuffled (to avoid conflicts with MixColumns), we expect a similar behavior as in the 32-bit implementation, if the attacker applies combing [28]. If a full shuffling on all 16 S-boxes is applied or the 32-bit architecture is shuffled, the number of traces would accordingly increase further.

The proposed attack is not specific to the applied masking scheme which still has a first-order leakage. It should be efficient for any case where the mean traces are slightly different. For example, the attack works on an unprotected implementation as well, i.e., the adversary does not even need to know whether a countermeasure has been applied in the target device. We have practically evaluated this issue as well; as a result around 3 000 traces are required for the attack on an unprotected implementation using an 8-bit architecture.

Leakage due to an implementation error? One may ask about the source of the leakage which we found here since we have presented the practical result on a whole AES implementation, and if some flaws in the design architecture have caused the observed strong leakage. We should emphasize that as expressed before in Section 2, we have made sure to keep all necessary requirements suggested in the original design of the masked S-box [9], like the correct order of the product additions and the masked summation of these. Keep in mind that this design does not take glitches and their effect on the DPA leakage into account.

Moreover, we have implemented only one masked S-box on the same platform and have examined its leakage when the S-box input (including masked input and masks) solely change. The relevant results are shown in Appendix A of [18], and confirm that the S-box computation is the source of information leakage which caused the observed vulnerability. Although we have not performed a simulation to extract the toggle-count model, we believe that the source of the observed first-order leakage also is toggles and glitches of the combinational circuits similar to [16], which was already predicted by [9].

Applicability to other Algorithms: On other algorithms exhibiting a similar structure of a key addition operation followed by some kind of S-box operation (e.g. typical SPN structures), the attack can be applied in a similar fashion. One important property of the attack algorithm is the number of values which contribute to the computation of the correlation coefficient. In the case of AES, α can take 256 different values such that the correlation is computed over 256 points, which is not too high, but still yields a suitable estimation of the correlation. The estimation gets less reliable for target algorithms with smaller S-boxes, e.g., PRESENT [7] with $\alpha \in GF(2^4)$. To solve this problem one can define a window and perform the attack not only on a single point in time, but also using other adjacent power points, i.e., to compute the correlation in a 2-dimensional domain which equals to make vectors from matrices and get the correlation over two vectors. Alternatively, two S-boxes can be attacked in parallel by viewing them as a single one and predicting the difference Δ on two keys at the same time, if the S-boxes are processed at the same time.

5   Conclusion

In this work we have presented a collision attack that efficiently breaks a masked implementation with a remaining first-order leakage. We have further shown that combining all possible collisions via the correlation coefficient generates a highly efficient attack. The number of traces needed to overcome an implementation of the masking countermeasure of [9] only increases by a small factor of six when compared to a DPA on an unprotected implementation. Unlike other advanced attacks, the described attack is as general as a classical DPA attack, because it makes minimal assumptions about the adversary. In fact, the attack makes almost no assumptions about the leakage and does not require any detailed knowledge about the implementation (such as general architecture, layout, and netlist). Furthermore, the attack works out-of-the-box without requiring a profiling phase. The attack succeeds on any implementation as long as a leakage yields distinguishable differences in the means of the power consumption traces for certain inputs. To the best of our knowledge the presented attack is the first successful collision-based attack on a masked implementation. We have practically confirmed that not considering glitches in the implementation of algorithmic masking schemes leads to an exploitable side-channel leakage.

Acknowledgment. The authors would like to thank Akashi Satoh and RCIS for the prompt and kind help in obtaining SASEBOs.

References

1. Side-channel Attack Standard Evaluation Board (SASEBO). Further information are available via, http://www.rcis.aist.go.jp/special/SASEBO/index-en.html
2. Agrawal, D., Archambeault, B., Rao, J.R., Rohatgi, P.: The EM Side-Channel(s). In: Kaliski Jr., B.S., Koç, Ç.K., Paar, C. (eds.) CHES 2002. LNCS, vol. 2523, pp. 29–45. Springer, Heidelberg (2003)
3. Agrawal, D., Rao, J.R., Rohatgi, P.: Multi-channel Attacks. In: Walter, C.D., Koç, Ç.K., Paar, C. (eds.) CHES 2003. LNCS, vol. 2779, pp. 2–16. Springer, Heidelberg (2003)
4. Akkar, M.-L., Giraud, C.: An Implementation of DES and AES, Secure against Some Attacks. In: Koç, Ç.K., Naccache, D., Paar, C. (eds.) CHES 2001. LNCS, vol. 2162, pp. 309–318. Springer, Heidelberg (2001)
5. Blömer, J., Guajardo, J., Krummel, V.: Provably Secure Masking of AES. In: Handschuh, H., Hasan, M.A. (eds.) SAC 2004. LNCS, vol. 3357, pp. 69–83. Springer, Heidelberg (2004)
6. Bogdanov, A.: Multiple-Differential Side-Channel Collision Attacks on AES. In: Oswald, E., Rohatgi, P. (eds.) CHES 2008. LNCS, vol. 5154, pp. 30–44. Springer, Heidelberg (2008)
7. Bogdanov, A., Leander, G., Knudsen, L., Paar, C., Poschmann, A., Robshaw, M., Seurin, Y., Vikkelsoe, C.: PRESENT - An Ultra-Lightweight Block Cipher. In: Paillier, P., Verbauwhede, I. (eds.) CHES 2007. LNCS, vol. 4727, pp. 450–466. Springer, Heidelberg (2007)
8. Canright, D.: A Very Compact S-Box for AES. In: Rao, J.R., Sunar, B. (eds.) CHES 2005. LNCS, vol. 3659, pp. 441–455. Springer, Heidelberg (2005), The HDL specification is available at author's official webpage, http://faculty.nps.edu/drcanrig/pub/index.html
9. Canright, D., Batina, L.: A Very Compact "Perfectly Masked" S-Box for AES. In: Bellovin, S.M., Gennaro, R., Keromytis, A.D., Yung, M. (eds.) ACNS 2008. LNCS, vol. 5037, pp. 446–459. Springer, Heidelberg (2008), the corrected version is available at Cryptology ePrint Archive, Report 2009/011 (2009), http://eprint.iacr.org/2009/011
10. Gandolfi, K., Mourtel, C., Olivier, F.: Electromagnetic Analysis: Concrete Results. In: Koç, Ç.K., Naccache, D., Paar, C. (eds.) CHES 2001. LNCS, vol. 2162, pp. 251–261. Springer, Heidelberg (2001)
11. Golić, J.D., Tymen, C.: Multiplicative Masking and Power Analysis of AES. In: Kaliski Jr., B.S., Koç, Ç.K., Paar, C. (eds.) CHES 2002. LNCS, vol. 2523, pp. 198–212. Springer, Heidelberg (2003)
12. Herbst, C., Oswald, E., Mangard, S.: An AES Smart Card Implementation Resistant to Power Analysis Attacks. In: Zhou, J., Yung, M., Bao, F. (eds.) ACNS 2006. LNCS, vol. 3989, pp. 239–252. Springer, Heidelberg (2006)
13. Kocher, P.C., Jaffe, J., Jun, B.: Differential Power Analysis. In: Wiener, M. (ed.) CRYPTO 1999. LNCS, vol. 1666, pp. 388–397. Springer, Heidelberg (1999)
14. Ledig, H., Muller, F., Valette, F.: Enhancing Collision Attacks. In: Joye, M., Quisquater, J.-J. (eds.) CHES 2004. LNCS, vol. 3156, pp. 176–190. Springer, Heidelberg (2004)
15. Mangard, S., Oswald, E., Popp, T.: Power Analysis Attacks: Revealing the Secrets of Smart Cards. Springer, Heidelberg (2007)
16. Mangard, S., Pramstaller, N., Oswald, E.: Successfully Attacking Masked AES Hardware Implementations. In: Rao, J.R., Sunar, B. (eds.) CHES 2005. LNCS, vol. 3659, pp. 157–171. Springer, Heidelberg (2005)

17. Mangard, S., Schramm, K.: Pinpointing the Side-Channel Leakage of Masked AES Hardware Implementations. In: Goubin, L., Matsui, M. (eds.) CHES 2006. LNCS, vol. 4249, pp. 76–90. Springer, Heidelberg (2006)
18. Moradi, A., Mischke, O., Eisenbarth, T.: Correlation-Enhanced Power Analysis Collision Attack. Cryptology ePrint Archive, Report 2010/297 (2010), http://eprint.iacr.org/2010/297
19. National Institute of Standards and Technology (NIST). Announcing the Advanced Encryption Standard (AES) (November 2001), http://www.nist.gov/
20. Nikova, S., Rijmen, V., Schläffer, M.: Secure Hardware Implementation of Nonlinear Functions in the Presence of Glitches. In: Lee, P.J., Cheon, J.H. (eds.) ICISC 2008. LNCS, vol. 5461, pp. 218–234. Springer, Heidelberg (2009)
21. Oswald, E., Mangard, S., Pramstaller, N., Rijmen, V.: A Side-Channel Analysis Resistant Description of the AES S-Box. In: Gilbert, H., Handschuh, H. (eds.) FSE 2005. LNCS, vol. 3557, pp. 413–423. Springer, Heidelberg (2005)
22. Paar, C.: Efficient VLSI Architectures for Bit-Parallel Computation in Galois Fields. PhD thesis, Institure for Experimental Mathematics, University of Essen, Germany (1994)
23. Quisquater, J.-J., Samyde, D.: ElectroMagnetic Analysis (EMA): Measures and Counter-Measures for Smart Cards. In: Attali, S., Jensen, T. (eds.) E-smart 2001. LNCS, vol. 2140, pp. 200–210. Springer, Heidelberg (2001)
24. Rijmen, V.: Efficient Implementation of the Rijndael S-box (2000)
25. Satoh, A., Morioka, S., Takano, K., Munetoh, S.: A Compact Rijndael Hardware Architecture with S-Box Optimization. In: Boyd, C. (ed.) ASIACRYPT 2001. LNCS, vol. 2248, pp. 239–254. Springer, Heidelberg (2001)
26. Schramm, K., Leander, G., Felke, P., Paar, C.: A Collision-Attack on AES: Combining Side Channel- and Differential-Attack. In: Joye, M., Quisquater, J.-J. (eds.) CHES 2004. LNCS, vol. 3156, pp. 163–175. Springer, Heidelberg (2004)
27. Schramm, K., Wollinger, T., Paar, C.: A New Class of Collision Attacks and Its Application to DES. In: Johansson, T. (ed.) FSE 2003. LNCS, vol. 2887, pp. 206–222. Springer, Heidelberg (2003)
28. Tillich, S., Herbst, C.: Attacking State-of-the-Art Software Countermeasures - A Case Study for AES. In: Oswald, E., Rohatgi, P. (eds.) CHES 2008. LNCS, vol. 5154, pp. 228–243. Springer, Heidelberg (2008)
29. Tiri, K., Akmal, M., Verbauwhede, I.: A Dynamic and Differential CMOS Logic with Signal Independent Power Consumption to Withstand Differential Power Analysis on Smart Cards. In: European Solid-State Circuits Conference - ESSCIRC 2002, pp. 403–406 (2002)
30. Trichina, E., Korkishko, T., Lee, K.-H.: Small Size, Low Power, Side Channel-Immune AES Coprocessor: Design and Synthesis Results. In: Dobbertin, H., Rijmen, V., Sowa, A. (eds.) AES 2005. LNCS, vol. 3373, pp. 113–127. Springer, Heidelberg (2005)
31. Waddle, J., Wagner, D.: Towards Efficient Second-Order Power Analysis. In: Joye, M., Quisquater, J.-J. (eds.) CHES 2004. LNCS, vol. 3156, pp. 1–15. Springer, Heidelberg (2004)
32. Yu, P., Schaumont, P.: Secure FPGA Circuits using Controlled Placement and Routing. In: Hardware/Software Codesign and System Synthesis - CODES+ISSS 2007, pp. 45–50. ACM, New York (2007)
33. Zakeri, B., Salmasizadeh, M., Moradi, A., Tabandeh, M., Shalmani, M.T.M.: Compact and Secure Design of Masked AES S-Box. In: Qing, S., Imai, H., Wang, G. (eds.) ICICS 2007. LNCS, vol. 4861, pp. 216–229. Springer, Heidelberg (2007)

Side-Channel Analysis of Six SHA-3 Candidates

Olivier Benoît and Thomas Peyrin

Ingenico, France
forename.name@ingenico.com

Abstract. In this paper we study six 2nd round SHA-3 candidates from
a side-channel cryptanalysis point of view. For each of them, we give the
exact procedure and appropriate choice of selection functions to perform
the attack. Depending on their inherent structure and the internal primitives used (Sbox, addition or XOR), some schemes are more prone to
side channel analysis than others, as shown by our simulations.

Keywords: side-channel, hash function, cryptanalysis, HMAC, SHA-3.

1 Introduction

Hash functions are one of the most important and useful tools in cryptography. A n-bit cryptographic hash function H is a function taking an arbitrarily
long message as input and outputting a fixed-length hash value of size n bits.
Those primitives are used in many applications such as digital signatures or key
generation. In practice, hash functions are also very useful for building Message Authentication Codes (MAC), especially in a HMAC [5,33] construction.
HMAC offers a good efficiency considering that hash functions are among the
fastest bricks in cryptography, while its security can be proven if the underlying
function is secure as well [4].

In recent years, we saw the apparition of devastating attacks [38,37] that broke
many standardized hash functions [36,30]. The NIST launched the SHA-3 competition [32] in response to these attacks and in order to maintain an appropriate security margin considering the increase of the computation power or further
cryptanalysis improvements. The outcome of this competition will be a new hash
function security standard to be determined in 2012 and 14 candidates have been
selected to enter the 2nd round of the competition (among 64 submissions).

Differential and Simple Power Analysis (DPA and SPA) were introduced in
1998 by Kocher *et al.* [25] and led to a powerful class of attacks called side-
channel analysis. They consist in two main steps. First, the power consumption,
the electro-magnetic signal [1] or any others relevant physical leakage from an
integrated circuit is measured during multiples execution of a cryptographic
algorithm. Then, one performs a hypothesis test on subkeys given the algorithm
specification, the algorithm input and/or output values and the traces obtained
during the first step. This second step requires to compute an intermediary
results of the algorithm for a given key guess and all the input/output and
analyze correlation [13] with the actual experimental traces. The intermediary
result is the output of what we will call hereafter the "selection function".

S. Mangard and F.-X. Standaert (Eds.): CHES 2010, LNCS 6225, pp. 140–157, 2010.

Because of the widely developed utilization of HMAC (or any MAC built upon a hash function) in security applications, it makes sense to consider physical security of hash functions [27,17,28,19,34]. Indeed, those functions usually manipulate no secret and have been at little bit left apart from side-channel analysis for the moment. In practice, the ability to retrieve the secret key that generates the MACs with physical attacks is a real threat that needs to be studied and such a criteria is taken in account by the NIST for the candidates selections [23].

Our contributions. We present a side-channel analysis of six hash functions selected to the 2nd round of the SHA-3 competition: ECHO [7], Grøstl [18], SHAvite-3 [11] (three AES-based hash functions), BLAKE [3], CubeHash [9] and HAMSI [35]. This paper aims at finding the appropriate selection function for each SHA-3 candidates in a MAC setting and evaluating the relative efficiency through simulations of the corresponding attacks. Then, we draw conclusions concerning the relative complexity for protecting each candidate against first order side-channel cryptanalysis.

Of course, the intend of this paper is not to show that some particular hash functions can be broken with side-channel analysis, which should be easy in general when the implementation is not protected. However, we believe there are constructions that are naturally harder to attack or easier to implement in a secure and relatively efficient way.

2   How to Perform Side-Channel Attacks on Hash Functions

2.1   Message Authentication Codes with Hash Functions

A Message Authentication Code (MAC) is an algorithm that takes as input a arbitrary long message M and a secret key K and outputs a fixed-length value $V = MAC(M, K)$. One requires that it should be computationally impossible for an attacker to forge a valid MAC without knowing the secret key K, or to retrieve any information about K. This primitive allows the authentication of messages between two parties sharing the same secret key. MACs can be built upon block ciphers (i.e. CBC-MAC [6]) or hash functions in the case of HMAC [5,33]. HMAC instantiated with the hash function H is defined as follows:

$$HMAC(K, M) = H((K \oplus opad)||H((K \oplus ipad)||M))$$

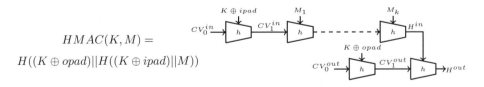

where $||$ denotes the concatenation of bit strings and $\oplus$ represents the bitwise exclusive or (XOR) boolean function. The two words *opad* and *ipad* are two constants of the size of a message block in the iterative hash function. This point is important: HMAC implicitly considers that H is an iterative hash function. Thus, for each iteration i we have an internal state (so-called chaining variable) CV_i that is updated thanks to a compression function h and a message block M_i : $CV_{i+1} = h(CV_i, M_i)$. The first chaining variable CV_0 is initialized with an initial vector IV and the hash output is the very last chaining variable or a truncated version of it.

It is easy to see that when computing the first hash function call of HMAC ($H^{in} = H((K \oplus ipad)\|M)))$, the first iteration is $CV_1 = h(IV, K \oplus ipad)$ and one only needs to guess CV_1 to complete the rest of this hash function computation, whatever the message M. Then, for the second hash function call ($H^{out} = H((K \oplus opad)\|H^{in}))$, the same remark applies: one only needs to guess CV_1 to complete the MAC computation whatever H^{in}. We denote by CV_i^{in} the chaining variables for the first hash call and CV_i^{out} the chaining variables for the second one. In practice, one can speed up the implementation by precomputing the CV_1^{in} and CV_1^{out} and starting the two hash processes with those two new initial chaining variables.

Therefore, when attacking HMAC with a side-channel technique, it is very interesting to recover CV_1^{in} and CV_1^{out}. We are now left with the problem of being able to retrieve a fixed unknown chaining variable with random message instances. This will have to be done two times, first for CV_1^{in} and then for CV_1^{out}. The attack process will be identical for each call, so in this article we only describe how to recover the unknown chaining variable with several known random message instances. However, note that attacking CV_1^{in} should be easier than attacking CV_1^{out} because in the former the attacker has full control over the message M, which is not the case in the latter (the incoming message for the second hash call is H^{in}, over which the attacker has no control). For some SHA-3 candidates, the ability to control the incoming message may reduce the number of power traces needed to recover CV_1^{in}. However, the maximal total improvement factor is at most 2 since the leading complexity phase remains the recovering of CV_1^{out}.

In the case of the so-called stream-based hash functions (for example Grindahl [24] or RadioGatún [10]), for which the message block size is rather small (smaller than the MAC key and hash output lengths), the HMAC construction is far less attractive and one uses in general the prefix-MAC construction: $MAC(K, M) = H(K\|M)$. In order to avoid trivial length-extension attacks (which makes classical Merkle-Damgård [29,14] based hash functions unsuitable for the prefix-MAC construction), the stream-based hash functions are usually composed of a big internal state (much bigger than the hash output length) and define an output function composed of blank rounds (iterations without message blocks incorporation) and a final truncation phase. However, the corresponding side-channel attack for breaking the prefix-MAC construction will not change here and our goal remains to recover the full internal state.

In this paper, we will study the 256-bit versions of the hash functions considered, but in most of the case the analysis is identical for the 512-bit versions. Moreover, when available, the salt input is considered as null. For all candidates reviewed, the message to hash is first padded and since we only give a short description of the schemes, we refer to the corresponding specification documents for all the details. Finally, since our goal is to recover the internal state before the message digesting phase, there is no need to consider potential output functions performed after all the message words have been processed.

2.2   Side-Channel Attacks

Regardless of the compression function $h(CV, M)$ considered, at some point in the algorithm (usually in the very first stage), the incoming chaining variable CV will be combined with the incoming message block M. The selection functions will be of the form:

$$w = f(cv, m)$$

where cv is a subset of CV and m is a subset from M. Usually w, cv and m have the same size (8 bits in the case of AES), but strongly compressing bricks (such as the DES Sbox) may impose a smaller w. Ideally the selection function must be non-linear and a modification of 1-bit in one of the input should potentially lead to multiple-bit modifications in the output. For example, the output of a substitution table (Sbox) is a good selection function: block cipher encryption algorithms such as DES [16] or AES [15] are very sensitive to side-channel analysis because they both use an Sbox in their construction (a $6 \mapsto 4$-bit Sbox for DES and a $8 \mapsto 8$-bit Sbox for AES).

Some algorithms or SHA-3 candidates (i.e. BLAKE or CubeHash) do not use such substitution table, while they rely exclusively on modular addition ⊞, rotation ⋘ and XOR ⊕ operations (so-called ARX constructions). In this case, side-channel analysis is still possible but the XOR or modular addition selection functions are less efficient than for the Sbox case. Moreover, it has been theoretically proven that the XOR selection function is less efficient that the modular addition operations [27]. Indeed, the propagation of the carry in the modular addition leads to some non-linearity whereas the XOR operation if completely linear. More precisely, we can quantify the efficiency difference between the AES Sbox, the HAMSI Sbox, the XOR and the modular addition selection functions by looking at the theoretical correlation results in the so-called hamming-weight model. The rest of this paper exclusively deals with the Hamming weight model since in practice this model leads to good results for the majority of the target devices.

In order to estimate the efficiency of a selection function $f(k, m)$, it is interesting to look at the theoretical correlation $c(j, r)$ between the data set x_i for a key guess j and the data set y_i for an arbitrary real key r. Where $x_i = HW(f(j, m_i))$ and $y_i = HW(f(r, m_i))$, with $i \in [0, \ldots, 2^N - 1]$, N being the number of bits of the selection function input message m and $HW(w)$ being the Hamming Weight of the word w. We also denote by $\overline{x}$ (respectively $\overline{y}$) the average value of the data set x_i (respectively y_i).

$$c(j, r) = \frac{\sum (x_i - \overline{x})(y_i - \overline{y})}{\sqrt{\sum (x_i - \overline{x})^2} . \sqrt{\sum (y_i - \overline{y})^2}}$$

Of course, when the key guess is equal to the real key ($j = r$), we have $c(j, r) = 1$.

The Figure 1 displays $c(j, 8)$ for $j \in [0, \ldots, 255]$ for the AES selection function $Sbox(k, m)$, for the XOR selection function $k \oplus m$ and for the modular addition selection function $k \boxplus m$. HAMSI is specific because the selection function is using a subset of the Sbox for a given key, therefore, the following table displays $c(j, r)$ for $j \in [0, \ldots, 3]$ and $r \in [0, \ldots, 3]$. Only two bits of the message and two bits of the key are handled in this selection function.

key guess / key value	$j = 0$	$j = 1$	$j = 2$	$j = 3$
$r = 0$	+1.00	−0.17	−0.56	−0.87
$r = 1$	−0.17	+1.00	+0.87	−0.09
$r = 2$	−0.56	+0.87	+1.00	+0.17
$r = 3$	−0.87	−0.09	+0.17	+1.00

The efficiency $E(f)$ of the selection function f is directly linked with the correlation contrast c_c between the correct key guess (correlation $= 1$) and the strongest wrong key guess (correlation $= c_w$). The higher this contrast, the more efficient the selection function will be to perform a side-channel analysis. Indeed, it will be able to sustain a much higher noise level.

$$c_c = \frac{1 - |c_w|}{|c_w|}$$

selection function	AES Sbox	modular addition	XOR[1]	HAMSI Sbox
c_w	0.23	0.75	−1	0.87
c_c	3.34	0.33	0	0.15

The values of c_w are extracted from Figure 1 by measuring the highest correlation peak (except the peak with correlation equal to 1 which corresponds to the correct guess). The result of this theoretical/simulation study is the following:

$$E(\text{AES Sbox}) > E(\text{modular addition}) > E(\text{HAMSI Sbox}) > E(\text{XOR})$$

In the rest of this article, we will search for the best selection function for each SHA-3 candidate analyzed with this conclusion in mind.

3   AES-Based SHA-3 Candidates

In this section, we analyze ECHO [7], Grøstl [18] and SHAvite-3 [11], three AES-based SHA-3 candidates. We recall that the round function of AES is composed of four layers (we use the order AddRoundKey, SubBytes, ShiftRows and Mix-columns) and we refer to [31] for a complete specification of this block cipher.

---

[1] In practice, if the attacker managed to characterize the chip leakage, he eventually can distinguish the wrong guess from the correct guess by taking in consideration the correlation sign (positive or negative). Note that a contrast of zero does not means that the XOR selection function is not yielding any information. Indeed, the attacker have reduced the subkey space from 256 to 2 values (with correlation 1 and -1).

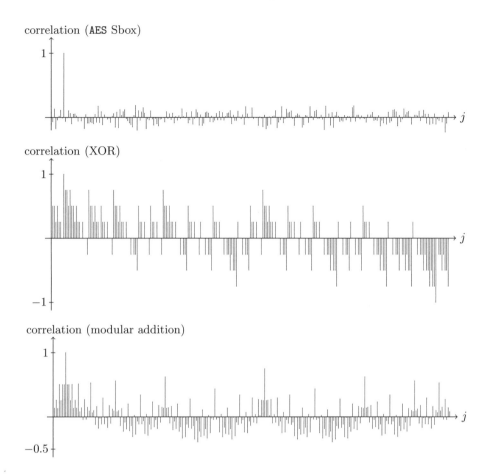

Fig. 1. Correlations $c(j, 8)$ in the Hamming Weight model for the AES Sbox, XOR and modular addition selection function respectively

3.1 ECHO

Description. ECHO [7] is an iterated hash function whose chaining variable CV_i is 512-bit long. Its compression function h maps CV_i and a 1536-bit message block M_i to a new chaining variable CV_{i+1}. More precisely, with CV_i and M_i the compression function h initializes a 2048-bit internal state which is viewed as a 4×4 matrix of 128-bit words (CV_i initializes the first column while M_i fills the three other ones). A fixed-key permutation P_E is applied to this internal state and the output chaining variable is built by calling the shrink$_{256}$ function that XORs all the 512-bit columns together after a feedforward:

$$CV_{i+1} = \mathsf{shrink}_{256}(P_E(CV_i||M_i) \oplus (CV_i||M_i)).$$

The permutation P_E contains 8 rounds, each composed of three functions very similar to the AES ones, but on 128-bit words instead of bytes. First, the BIG.SubBytes function mimics the application of 128-bit Sboxes on each state word by applying 2 AES rounds with fixed round keys (determined by the iteration and round numbers). Then, BIG.ShiftRows rotates the position in their matrix column of all the 128-bit words according to their row position (analog to the AES). Finally, BIG.MixColumns is a linear diffusion layer updating all the columns independently. More precisely, for one matrix column, it applies sixteen parallel AES MixColumns transformations (one for each byte position in a 128-bit word).

Side-channel analysis. The incoming chaining variable CV fills the first 128-bit words column (denoted cv_i in Figure 2) of the matrix representing the internal state, while the three other columns are filled with the known random incoming message (denoted m_i in Figure 2). The goal of the attacker is therefore to retrieve the words cv_i.

The first layer (BIG.SubBytes) handles each 128-bit word individually. The known and secret data are not mixed yet ($cv_i \mapsto cv_i'$ and $m_i \mapsto m_i'$) and therefore it is not possible to derive a selection function at this point. The same comment applies to the second layer (BIG.ShiftRows) and one has to wait for the third layer (BIG.MixColumns) to observe known and secret data mixing: each column will depend on one secret word cv_i' and three known words m_i' (see Figure 2). More precisely, for each 128-bit word column, BIG.MixColumns applies sixteen parallel and independent AES MixColumns transformations (one for each byte position) and each MixColumns call manipulates one byte of secret and three bytes of known data. Overall, in the end of the first round, every byte $w[b]$ of an internal state word w (we denote $w[b]$ the b-th byte of w) can be written as the following affine equation (see the AES MixColumns definition for the α, β, γ and δ values):

$$w_{i_0}[b] = \alpha \cdot cv_{i_1}'[b] \oplus \beta \cdot m_{i_2}'[b] \oplus \gamma \cdot m_{i_3}'[b] \oplus \delta \cdot m_{i_4}'[b]$$

with $b \in [0, \dots, 15]$, $i_0 \in [0, \dots, 15]$, $i_1 \in [0, \dots, 3]$ and $i_2, i_3, i_4 \in [0, \dots, 11]$. One could use those $w_i[b]$ as selection functions, but the mixing operation would be the exclusive or. As already explained, the selection function involving an XOR is the least efficient one. It seems much more promising to wait the first layer from the second round of the ECHO internal permutation.

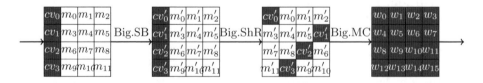

Fig. 2. Recovering the internal state for ECHO. The gray cells represent the words that depends on the initial secret chaining variable. Each cell represents a 128-bit word.

Indeed, the BIG.SubBytes transformation applies directly two AES rounds independently for each words w_i. The first function of the first AES round is the subkey incorporation and in the case of ECHO those subkeys are fully known constants (we denote them t_i). Then, the second function of the first AES round applies the AES Sbox to each byte of the internal state. Therefore, we obtain the words w_i' on the output:

$$w_i'[b] = Sbox(w_i[b] \oplus t_i[b]).$$

These equations can be used as selection functions manipulating only the AES Sbox which is much more efficient than the XOR case. Overall, one has to perform 64 AES Sbox side-channel attacks in order to guess all the words cv_i' byte per byte. By inverting the BIG.SubBytes layer from the words cv_i', one recovers completely CV. Note that for each byte of cv_i', one gets 4 selection functions involved. Thus, the overall number of curved can be reduced by a factor 4 at maximum by using and combining this extra information.

3.2 Grøstl

Description. Grøstl [18] is an iterated hash function whose compression function h maps a 512-bit chaining variable CV_i and a 512-bit message block M_i to a new chaining variable CV_{i+1}. More precisely, two fixed-key permutations P_G and Q_G, only differing in the constant subkeys used, are applied:

$$CV_{i+1} = P_G(CV_i \oplus M_i) \oplus Q_G(M_i) \oplus CV_i.$$

Each permutation is composed of 10 rounds very similar to the AES ones, except that they update a 512-bit state, viewed as a 8×8 matrix of bytes (instead of 4×4). Namely, for each round, constant subkeys are first XORed to the state (AddRoundConstant), then the AES Sbox is applied to each byte (SubBytes), the matrix rows are rotated with distinct numbers of positions (ShiftBytes) and finally a linear layer is applied to each byte column independently (MixBytes). This is depicted in Figure 3.

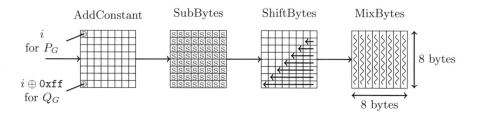

Fig. 3. One round of the internal permutation P_G of Grøstl. Each cell represents a byte.

Side-channel analysis. The Grøstl case is very simple. One can see that the incoming message block M is processed trough the Q_G permutation. Since the output of this permutation only depends on the message block and not on the incoming chaining variable CV, we can completely ignore this part of the compression function. Then, the permutation P_G takes as input $M \oplus CV$ and one may be tempted to perform the side-channel attack during this operation. As demonstrated earlier, it is much more convenient to wait for the opportunity to attack the AES Sbox instead. The first layer of the permutation P_G is the AddConstant function which XORs the round number i (the counting starting from 0) to the top left byte of the internal and therefore the operation is fully transparent. Then, the second layer of the first P_G round is the SubBytes function which applies the AES Sbox to every byte of the internal state $w = M \oplus CV$:

$$w[b] = m[b] \oplus CV[b]$$

with $b \in [0, \ldots, 63]$. The output state is denoted w' and we obtain the following selection function which recovers CV byte per byte:

$$w'[b] = Sbox(w[b]).$$

Note that it is possible to improve this attack when dealing with CV_1^{in} (the unknown chaining variable for the first hash call) by choosing appropriately the message. More precisely, one can divide the number of power traces by a factor 64 when choosing all $m[b]$ as equals. Indeed, this allows to perform in parallel the side-channel analysis of the 64 unknown bytes.

3.3   SHAvite-3

Description. SHAvite-3 [11] is an iterated hash function whose compression function h maps a 256-bit chaining variable CV_i and a 512-bit message block M_i to a new chaining variable CV_{i+1}. Internally, we have a block cipher E^S in classical Davies-Meyer mode

$$CV_{i+1} = CV_i \oplus E_{M_i}^S(CV_i).$$

This block cipher derives many subkeys thanks to a key schedule (all subkeys depending on the message M_i only) and is composed of 12 rounds of a 2-branch Feistel scheme (128 bits per branch). The basic primitive in the Feistel rounds is the application of 3 AES rounds with subkeys incoming from the key schedule.

Side-channel analysis. For SHAvite-3, we divide the attack in two phases (see Figure 4). In the first one, we recover the right part (in the Feistel separation) of the incoming chaining variable (CV^R) during the first round. Once this first phase succeeded, we recover the left part of the incoming chaining variable (CV^L) during the second round. The message expansion maps the incoming message M to three 128-bit message words (m_0^j, m_1^j, m_2^j) for each round j. One round j of SHAvite-3 consists in executing sequentially three AES round functions with as input one branch of the current SHAvite-3 state and (m_0^j, m_1^j, m_2^j) as subkeys. Consequently, for the first SHAvite-3 round, the secret vector (CV^R) is mixed with the known message word m_0^1 during the AddRoundKey layer of the first AES round and we note:

$$w[b] = CV^R[b] \oplus m_0^1[b].$$

with $b \in [0, \ldots, 15]$. One could use this equation as the selection function, but it is more appropriate to use the output of the very next transformation instead, i.e. the SubBytes layer:

$$w'[b] = Sbox(w[b]).$$

Before executing the second round of SHAvite-3, the left part of the chaining variable (CV^L) is XORed with the output w'' of the three AES rounds. Then, this word is mixed with $m_0^2[b]$ just before the first AES round of the second SHAvite-3 round and we note:

$$z[b] = CV^L[b] \oplus w''[b] \oplus m_0^2[b].$$

Obviously, after a successful first phase, it is possible to compute $w''[b]$ and therefore CV^L is the only unknown constant. Once again, one could use this equation as the selection function, but it is more appropriate to use the output of the very next transformation instead, i.e. the SubBytes layer:

$$z'[b] = Sbox(z[b]).$$

Overall, we recover byte per byte the CV^L and CV^R values.

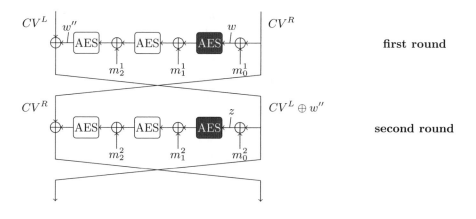

Fig. 4. Recovering the internal state for SHAvite-3. The AES rounds we use for recovering the internal state are depicted in black.

4   Other SHA-3 Candidates

In this section, we analyze BLAKE [3], CubeHash [9] and HAMSI [35], three 2nd round SHA-3 candidates.

4.1   BLAKE

Description. BLAKE [3] is an iterated hash function whose compression function h maps a 256-bit chaining variable CV_i and a 512-bit message block M_i to a new chaining variable CV_{i+1}. Internally, the update is done with a block cipher E^B, keyed with the message block (see Figure 5):

$$CV_{i+1} = final(E^B_{M_i}(init(CV_i)), CV_i).$$

where the *init* function initializes the 512-bit internal state with CV_i and constants. The *final* function computes the output chaining variables according to CV_i, constants and the internal state after the application of E^B. The internal state is viewed as a 4×4 matrix of 32-bit words and the block cipher E^B is composed of 10 rounds, each consisting of the application of eight 128-bit sub-functions G_i. Assume an internal state for BLAKE with v_{i+4j} representing the 32-bit word located on row j and column i, one round of E^B is:

$$G_0(v_0, v_4, v_8, v_{12}) \quad G_1(v_1, v_5, v_9, v_{13}) \quad G_2(v_2, v_6, v_{10}, v_{14}) \quad G_3(v_3, v_7, v_{11}, v_{15})$$
$$G_4(v_0, v_5, v_{10}, v_{15}) \quad G_5(v_1, v_6, v_{11}, v_{12}) \quad G_6(v_2, v_7, v_8, v_{13}) \quad G_7(v_3, v_4, v_9, v_{14})$$

A sub-function G_i incorporates 32-bit message chunks m_i and is itself made of additions, XORs and rotations. During the round r, the function $G_s(a, b, c, d)$ processes the following steps:

$$a \leftarrow (a \boxplus b) \boxplus (m_i \oplus k_j)$$
$$d \leftarrow (d \oplus a) \ggg 16$$
$$c \leftarrow (c \boxplus d)$$
$$d \leftarrow (b \oplus c) \ggg 12$$
$$a \leftarrow (a \boxplus b) \boxplus (m_j \oplus k_i)$$
$$d \leftarrow (d \oplus a) \ggg 8$$
$$c \leftarrow (c \boxplus d)$$
$$d \leftarrow (b \oplus c) \ggg 7$$

where $\ggg x$ denotes the right rotation of x positions, $i = \sigma_r(2s)$ and $j = \sigma_r(2s = 1)$. The notation σ_r represents a family of permutations on $\{0, \ldots, 15\}$ defined in the specifications document.

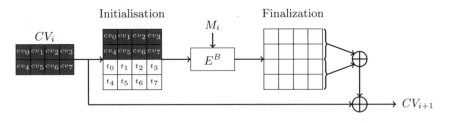

Fig. 5. The compression function of BLAKE

Side-channel analysis. The sixteen 32-bit internal state words are initialized with the eight secret chaining value CV words (denoted cv_i) and constants values t_i (see Figure 5). Then, the eight G_i functions during the first BLAKE round are applied to the internal state and one can check that the two first parameters of

G_0, G_1, G_2 and G_3 are (cv_0, cv_1), (cv_2, cv_3), (cv_4, cv_5) and (cv_6, cv_7) respectively. Our goal is therefore to recover a_0 and b_0 when applying $G_i(a_0, b_0, c_0, d_0)$ with $0 \leq i \leq 3$. The functions G_i consist in a sequence of eight transformations, the five first being:

$$a_1 = (a_0 \boxplus b_0) \boxplus m_k$$
$$d_1 = (d_0 \oplus a_1) \ggg 16$$
$$c_1 = c_0 \boxplus d_1$$
$$b_1 = (b_0 \oplus c_1) \ggg 12$$
$$a_2 = a_1 \boxplus b_1 \boxplus m_l$$

In practice, the first transformation will be computed in one of the three following way:

$$\text{first } a \leftarrow a \boxplus b \text{ then } a \leftarrow a \boxplus m_i$$
$$\text{first } a \leftarrow a \boxplus m_i \text{ then } a \leftarrow a \boxplus b$$
$$\text{first } x \leftarrow b \boxplus m_i \text{ then } a \leftarrow a \boxplus x$$

For the second and third case, a_0 and b_0 can be found by two side-channel analysis applied successively to the two modular addition selection function (working byte per byte):

$$w_i = cv_i \boxplus m_k \text{ and } z_i = cv_{i+1} \boxplus w_i.$$

For the first case, $a_0 \boxplus b_0$ can be recovered by performing the side-channel analysis on the second modular addition selection function. In order to solve the problem and estimate a_0 and b_0 the attacker has to target the output of the fourth transformation of G_i. However, in this case the selection function would be based on the XOR operation. Therefore it seems more interesting to aim for the fifth transformation of G_i, a modular addition.

4.2   CubeHash

Description. CubeHash [9] is an iterated hash function whose compression function h maps a 1024-bit chaining variable CV_i and a 256-bit message block M_i to a new chaining variable CV_{i+1}. Internally, the update is done with a permutation P_C:

$$CV_{i+1} = P_C(CV_i \oplus (M_i \| \{0\}^{768})).$$

The internal state is viewed as a table of 32 words of 32 bits each. The permutation P_C is composed of 16 identical rounds and one round is made of ten layers (see Figure 6): two intra-word rotation layers, two XOR layers, two 2^{32} modular addition layers and four word positions permuting layers.

Side-channel analysis. The attack is divided into 4 steps. The incoming chaining variable CV fills an internal state represented by a vector of four 256-bit words or eight 32-bit words (denoted cv_i). The known random incoming message M is first XORed with cv_0 and then starts the first round of permutation P_C. Thus, the first selection function is of XOR type and recovers cv_0 byte per byte:

$$w[b] = cv_0[b] \oplus M[b].$$

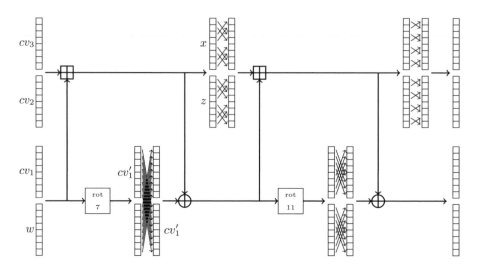

Fig. 6. Recovering the internal state of CubeHash during one round of the internal permutation P_C. Each cell represents a 32-bit word.

with $b \in [0, \ldots, 7]$. Once this first step successfully performed, the attacker fully knows w and the first layer of P_C adds each 32-bit words from w to those from cv_2 modulo 2^{32} (and each words from cv_1 to those from cv_3). The second selection function is therefore of modular addition type and recovers cv_2 byte per byte (starting from the LSB of the modular addition):

$$z[b] = cv_2[b] \boxplus w[b].$$

The second layer of P_C applies a rotation to each 32-bit word of w and cv_1 and then XORs z to this rotated version of cv_1 (denoted cv_1'). The selection function for the third step is then of XOR type and recovers cv_1' byte per byte:

$$y[b] = cv_1'[b] \oplus z[b].$$

Finally, going backward in the computation to the first P_C layer, the selection function for the fourth step is of modular addition type and recovers cv_3 byte per byte (starting from the LSB of the modular addition):

$$x[b] = cv_3[b] \boxplus cv_1[b].$$

Note that each step must be successful, otherwise it would compromise the results of the following steps.

4.3   HAMSI

Description. HAMSI [35] is an iterated hash function whose compression function h maps a 256-bit chaining variable CV_i and a 32-bit message block M_i to a new chaining variable CV_{i+1}. First, the message block is expanded into eight

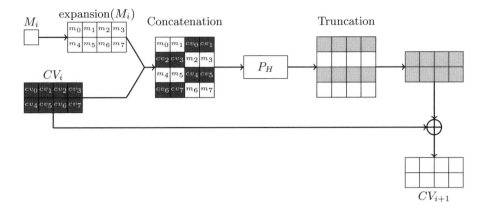

Fig. 7. The compression function of HAMSI. Each cell represents a 32-bit word.

32-bit words m_i that are concatenated to CV_i in order to initialize a 512-bit internal state (viewed as a 4×4 matrix of 32-bit words). Then a permutation P_H is applied to the state and a truncation allows to extract 256 bits. In the end, there is a feedforward of the chaining variable (see Figure 7):

$$CV_{i+1} = trunc(P_H(CV_i||expansion(M_i))) \oplus CV_i.$$

The permutation P_H contains three identical rounds. One round is composed of three layers: constants are first XORed to the internal state, then 4-bit Sboxes are applied to the whole state by taking one bit of each 32-bit word of the same column of the 4×4 matrix and repeating the process for all bit positions. Finally, a linear layer is applied to four 32-bit words diagonals of the 4×4 matrix independently (see Figure 8).

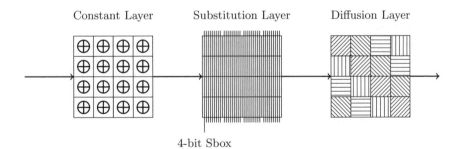

4-bit Sbox

Fig. 8. One round of the internal permutation P_H of HAMSI. Each cell represents a 32-bit word.

Side-channel analysis. The known random message M (after expansion) and the secret chaining variable CV fill the internal state matrix as shown in Figure 7

(32-bit words are denoted m_i and cv_i). The first layer of the permutation P_H XORs each matrix element with a constant t_i:

$$m_i' = m_i \oplus t_i \text{ and } cv_i' = cv_i \oplus t_{i+8}.$$

Then is applied the HAMSI Sbox layer. This Sbox is acting over 4-bits (one bit per word located in the same column of the state matrix). Therefore, the input of each Sbox is composed of 2 known message bits and 2 unknown chaining variable bits. The generic selection function for HAMSI can therefore be written as:

$$w = Sbox(m_i'[b]||cv_{i+2}'[b]||m_{i+4}'[b]||cv_{i+6}'[b])$$

or

$$w = Sbox(cv_i'[b]||m_{i+2}'[b]||cv_{i+4}'[b]||m_{i+6}'[b])$$

for $i \in [0, 1]$ and $b \in [0, \ldots, 127]$ where b is the bit index in a 128-bits word. Overall, one recovers two bits of cv_i' at a time with a total of 4 times 128 correlation computations (with 4 guess each).

5   Conclusion and Discussions

For each hash proposal considered in this article, we described an appropriate selection function for an efficient side-channel attack.

In the case of the AES-based SHA-3 candidates, we did not found significant differences of performance when choosing the selection function. Indeed, in ECHO, Grøstl and SHAvite-3, one has to compute several AES Sbox side-channel attacks in order to retrieve the full secret internal state. Up to a small complexity/number of power traces factor, the three schemes seem to provide the same natural vulnerability to side-channel cryptanalysis. As expected, their situation is therefore very close to the real AES block cipher.

Attacking BLAKE seems feasible in practice since we managed to derive a modular addition selection function for recovering the 256 bits of unknown chaining variable. The modular addition non-linearity is very valuable for the attacker as it increases the correlation contrast. Then, for CubeHash (a typical ARX function) we tried to force the selection function to be of modular addition type as much as possible. Overall, 512 bits can be recovered with modular addition selection function and 512 bits with XOR selection function. In practice, depending on the underlying hardware, it could be challenging to mount the attack. One must notice that the internal state is bigger for CubeHash than for other candidates. This is an additional strength since in practice, if the side-channel attack gives only probabilistic results, the final exhaustive search complexity will be higher. Finally, for HAMSI, the attack would be difficult to mount despite the fact that a substitution table is used. Indeed, the correlation contrast for this primitive is quite low compared to the AES Sbox. We believe that a better selection function involving a modular addition might possibly be found in the inner layer.

Of course, those results concern unprotected implementations and the ranking would be really different if we also considered methods for hardening the side-channel cryptanalysis. For example, in the case of AES-based hash functions, one could perform secure round computations and leverage all the research achieved so far on this subject [2,20]. Also, as ECHO processes parallel AES rounds, we believe it could benefit from secure bit-slice implementations regarding some side-channels attacks [26], while maintaining its normal use efficiency. Finally, ECHO and SHAvite-3 can take advantage of the natural protection inherited from the hardware implementations of the AES round such as the new AES NI instruction set on Intel microprocessor [8].

Side-channel countermeasures for ARX constructions such as BLAKE or CubeHash are of course possible, but they will require to constantly switch from boolean to arithmetic masking. As a consequence, one will observe an important decrease of the speed performance for secure implementations. AES-based hash functions seem naturally easier to attack regarding side-channel cryptanalysis, but are also easier to protect.

References

1. Agrawal, D., Archambeault, B., Rao, J.R., Rohatgi, P.: The EM Side-Channel(s). In: Jr., et al. (eds.) [22], pp. 29–45.
2. Akkar, M.-L., Giraud, C.: An Implementation of DES and AES, Secure against Some Attacks. In: Koç, Ç.K., Naccache, D., Paar, C. (eds.) CHES 2001. LNCS, vol. 2162, pp. 309–318. Springer, Heidelberg (2001)
3. Aumasson, J.-P., Henzen, L., Meier, W., Phan, R.C.-W.: SHA-3 proposal BLAKE. Submission to NIST (2008)
4. Bellare, M.: New Proofs for NMAC and HMAC: Security Without Collision-Resistance. Cryptology ePrint Archive, Report 2006/043 (2006), http://eprint.iacr.org/
5. Bellare, M., Canetti, R., Krawczyk, H.: Keying Hash Functions for Message Authentication. In: Koblitz, N. (ed.) CRYPTO 1996. LNCS, vol. 1109, pp. 1–15. Springer, Heidelberg (1996)
6. Bellare, M., Kilian, J., Rogaway, P.: The Security of Cipher Block Chaining. In: Desmedt, Y.G. (ed.) CRYPTO 1994. LNCS, vol. 839, pp. 341–355. Springer, Heidelberg (1994)
7. Benadjila, R., Billet, O., Gilbert, H., Macario-Rat, G., Peyrin, T., Robshaw, M., Seurin, Y.: SHA-3 Proposal: ECHO. Submission to NIST (2008), http://crypto.rd.francetelecom.com/echo/
8. Benadjila, R., Billet, O., Gueron, S., Robshaw, M.J.B.: The Intel AES Instructions Set and the SHA-3 Candidates. In: Matsui, M. (ed.) ASIACRYPT 2009. LNCS, vol. 5912, pp. 162–178. Springer, Heidelberg (2009)
9. Bernstein, D.J.: CubeHash specification (2.B.1). Submission to NIST, Round 2 (2009)
10. Bertoni, G., Daemen, J., Peeters, M., Van Assche, G.: Radiogatun, a belt-and-mill hash function. Presented at Second Cryptographic Hash Workshop, Santa Barbara, August 24-25 (2006), http://radiogatun.noekeon.org/
11. Biham, E., Dunkelman, O.: The SHAvite-3 Hash Function. Submission to NIST, Round 2 (2009), http://www.cs.technion.ac.il/~orrd/SHAvite-3/Spec.15.09.09.pdf

12. Brassard, G. (ed.): CRYPTO 1989. LNCS, vol. 435. Springer, Heidelberg (1990)
13. Brier, E., Clavier, C., Olivier, F.: Correlation Power Analysis with a Leakage Model. In: Joye, Quisquater (eds.) [21], pp. 16–29
14. Damgård, I.: A Design Principle for Hash Functions. In: Brassard (ed.) [12], pp. 416–427
15. FIPS 197. Advanced Encryption Standard. Federal Information Processing Standards Publication 197, U.S. Department of Commerce/N.I.S.T. (2001)
16. FIPS 46-3. Data Encryption Standard. Federal Information Processing Standards Publication, U.S. Department of Commerce/N.I.S.T. (1999)
17. Fouque, P.-A., Leurent, G., Réal, D., Valette, F.: Practical Electromagnetic Template Attack on HMAC. In: Clavier, C., Gaj, K. (eds.) CHES 2009. LNCS, vol. 5747, pp. 66–80. Springer, Heidelberg (2009)
18. Gauravaram, P., Knudsen, L.R., Matusiewicz, K., Mendel, F., Rechberger, C., Schläffer, M., Thomsen, S.S.: Grøstl – a SHA-3 candidate. Submission to NIST (2008), http://www.groestl.info
19. Gauravaram, P., Okeya, K.: An Update on the Side Channel Cryptanalysis of MACs Based on Cryptographic Hash Functions. In: Srinathan, K., Pandu Rangan, C., Yung, M. (eds.) INDOCRYPT 2007. LNCS, vol. 4859, pp. 393–403. Springer, Heidelberg (2007)
20. Golic, J.D., Tymen, C.: Multiplicative Masking and Power Analysis of AES. In: Jr, et al. (eds.) [22], pp. 198–212
21. Joye, M., Quisquater, J.-J. (eds.): CHES 2004, MA, USA, August 11-13. LNCS, vol. 3156. Springer, Heidelberg (2004)
22. Kaliski Jr., B.S., Koç, Ç.K., Paar, C. (eds.): CHES 2002. LNCS, vol. 2523. Springer, Heidelberg (2003)
23. Kelsey, J.: How to Choose SHA-3, http://www.lorentzcenter.nl/lc/web/2008/309/presentations/Kelsey.pdf
24. Knudsen, L.R., Rechberger, C., Thomsen, S.S.: The Grindahl Hash Functions. In: Biryukov, A. (ed.) FSE 2007. LNCS, vol. 4593, pp. 39–57. Springer, Heidelberg (2007)
25. Kocher, P.C., Jaffe, J., Jun, B.: Differential Power Analysis. In: Wiener, M. (ed.) CRYPTO 1999. LNCS, vol. 1666, pp. 388–397. Springer, Heidelberg (1999)
26. Könighofer, R.: A Fast and Cache-Timing Resistant Implementation of the AES. In: Malkin, T.G. (ed.) CT-RSA 2008. LNCS, vol. 4964, pp. 187–202. Springer, Heidelberg (2008)
27. Lemke, K., Schramm, K., Paar, C.: DPA on n-Bit Sized Boolean and Arithmetic Operations and Its Application to IDEA, RC6, and the HMAC-Construction. In: Joye, Quisquater (eds.) [21], pp. 205–219
28. McEvoy, R.P., Tunstall, M., Murphy, C.C., Marnane, W.P.: Differential Power Analysis of HMAC Based on SHA-2, and Countermeasures. In: Kim, S., Yung, M., Lee, H.-W. (eds.) WISA 2007. LNCS, vol. 4867, pp. 317–332. Springer, Heidelberg (2008)
29. Merkle, R.C.: One Way Hash Functions and DES. In: Brassard (ed.) [12], pp. 428–446
30. National Institute of Standards and Technology. FIPS 180-1: Secure Hash Standard (April 1995), http://csrc.nist.gov
31. National Institute of Standards and Technology. FIPS PUB 197, Advanced Encryption Standard (AES). Federal Information Processing Standards Publication 197, U.S. Department of Commerce (2001)

32. National Institute of Standards and Technology. Announcing Request for Candidate Algorithm Nominations for a NewCryptographic Hash Algorithm (SHA-3) Family. Federal Register, 27(212):62212–62220 (November 2007), `http://csrc.nist.gov/groups/ST/hash/documents/FR_Notice_Nov07.pdf` (2008/10/17)
33. NIST. FIPS 198 – The Keyed-Hash Message Authentication Code (HMAC) (2002)
34. Okeya, K.: Side Channel Attacks Against HMACs Based on Block-Cipher Based Hash Functions. In: Batten, L.M., Safavi-Naini, R. (eds.) ACISP 2006. LNCS, vol. 4058, pp. 432–443. Springer, Heidelberg (2006)
35. Kücük, Ö.: The Hash Function Hamsi. Submission to NIST (updated) (2009)
36. Rivest, R.L.: RFC 1321: The MD5 Message-Digest Algorithm (April 1992), `http://www.ietf.org/rfc/rfc1321.txt`
37. Wang, X., Yin, Y.L., Yu, H.: Finding Collisions in the Full SHA-1. In: Shoup, V. (ed.) CRYPTO 2005. LNCS, vol. 3621, pp. 17–36. Springer, Heidelberg (2005)
38. Wang, X., Yu, H.: How to Break MD5 and Other Hash Functions. In: Cramer, R. (ed.) EUROCRYPT 2005. LNCS, vol. 3494, pp. 19–35. Springer, Heidelberg (2005)

Flash Memory 'Bumping' Attacks

Sergei Skorobogatov

University of Cambridge, Computer Laboratory,
15 JJ Thomson Avenue, Cambridge CB3 0FD, United Kingdom
sps32@cam.ac.uk

Abstract. This paper introduces a new class of optical fault injection attacks called bumping attacks. These attacks are aimed at data extraction from secure embedded memory, which usually stores critical parts of algorithms, sensitive data and cryptographic keys. As a security measure, read-back access to the memory is not implemented leaving only authentication and verification options for integrity check. Verification is usually performed on relatively large blocks of data, making brute force searching infeasible. This paper evaluates memory verification and AES authentication schemes used in secure microcontrollers and a highly secure FPGA. By attacking the security in three steps, the search space can be reduced from infeasible $> 2^{100}$ to affordable $\approx 2^{15}$ guesses per block of data. This progress was achieved by finding a way to preset certain bits in the data path to a known state using optical bumping. Research into positioning and timing dependency showed that Flash memory bumping attacks are relatively easy to carry out.

Keywords: semi-invasive attacks, fault injection, optical probing.

1 Introduction

Confidentiality and integrity of sensitive information stored in smartcards, secure microcontrollers and secure FPGAs is a major concern to both security engineers and chip manufacturers. Therefore, sensitive data like passwords, encryption keys and confidential information is often stored in secure memory, especially in Flash memory. This is mainly because data extraction from an embedded Flash memory is believed to be extremely tedious and expensive [1,2]. In addition, Flash memory offers re-programmability and partial updating, useful for keeping firmware, keys and passwords up to date, and replacing compromised ones. However, in some cases, the Flash memory is vulnerable to several types of attacks. Sometimes the memory access path is the weakest link. In order to prevent unauthorised access to the memory, chip manufacturers widely use security fuses and passwords in microcontrollers, FPGAs, smartcards and other secure chips. This approach did not prove to be very effective against semi-invasive attacks [3]. Furthermore, to make the protection more robust and to prevent some known attacks on the security fuses, some chip manufacturers decided not to implement direct access to internal data from the external programming and

S. Mangard and F.-X. Standaert (Eds.): CHES 2010, LNCS 6225, pp. 158–172, 2010.

debugging interfaces. That way, read access to the embedded memory was unavailable to the external interface and only the verify operation in a form of comparison with uploaded data was left for data integrity check. Usually, such verification is carried over large chunks of data in order to prevent brute force attacks.

Optical fault injection attacks proved to be very effective against many protection schemes [4]. As these attacks require visibility of the chip surface without the need of any mechanical contact, they should be classified as semi-invasive. Backside approach can be used on modern sub-micron chips with multiple metal layers that cover the surface and prevent direct access. Such an approach is simpler than the front-side as it does not require special chemicals for opening up the chip package. Moreover, there is no danger of mechanical damage to the die as the thick silicon substrate protects the active area. Mechanical milling is used to open up the package followed by polishing of the silicon surface. Very cheap engraving tools proved to be sufficient for that.

The results presented in this paper show that optical fault injection can be successfully used to circumvent verify-only protection scheme in secure devices. This technique was demonstrated on the Flash program memory of a "secure" microcontroller and on the Flash array of a "highly secure" FPGA. However, it can be applied to any device, that allows verification of the internal data against uploaded one. The attack was carried out in three steps. The first step was aimed at separating of the whole verification packet into blocks of data according to the communication protocol. The second step, later called 'bumping', involved splitting the data in each block into words corresponding to the width of the data bus. The third step, later called 'selective bumping', was used to reduce the number of guesses required to pass the verification within each word of data. Using these techniques the data extraction time can be dramatically reduced to hours and days compared with many years required for brute force attacks. As these attacks do not require expensive equipment they can pose a big problem to the hardware community. Without proper countermeasures in place security in some devices could be easily compromised.

Another interesting direction of the research presented in this paper concerns the possibility of AES key extraction using bumping attacks on the AES-based authentication process. For that, a secure FPGA and a secure microcontroller were tested. Both chips have hardware AES decryption engines for authentication and decryption of firmware updates. Research into positioning and timing dependability of the bumping attacks was carried out. It was found that Flash memory bumping attacks do not require precise positioning on the chip surface and just reasonable timing precision, thus being also suitable for asynchronously clocked chips. For selective bumping attacks some non-linear effects were observed where the sequence of bits set to a known state depended on the point of fault injection.

This paper is organised as follows. Section 2 describes protection mechanisms used in secure chips and background on the bumping attacks. Section 3 introduces experimental setup, while Section 4 shows the results. Section 5 discusses limits and further improvements to these attacks. Some possible countermeasures are presented in the concluding section.

2   Background

Target of my experiments was embedded Flash memory. It uses floating-gate transistors to store the information [5]. Fig. 1 shows the overall structure of a typical Flash memory, the layout of the memory cells and the details of the floating-gate transistor operation modes. Data inside the Flash memory array can only be accessed one row at a time, with the row itself being sent in smaller chunks via read sense amplifiers. The number of the latter usually matches the width of the data bus.

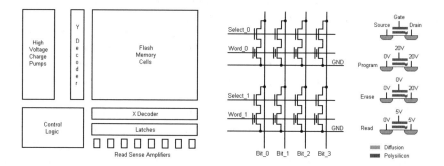

Fig. 1. Structure of Flash memory, layout of memory cells and modes of operation

High voltages are required to operate the Flash memory. Although they are not necessary for reading the memory, the requirement for a special high-voltage control logic and large charge pumps forces chip designers to place the Flash memory separately from the rest of the chip logic. From a security point of view this makes finding and attacking it easier. There are several places where the attack could be targeted. One is the memory cells, but this way is not very practical as it requires precise focusing and positioning not achievable with a backside approach. Another place is the read sense amplifiers, however, in spite of an easy way of locating and attacking them, the number of attack points increases with the width of data bus. Even for a narrow 8-bit data bus, mounting eight lasers will be a non-trivial task. The same difficulties apply to the attack point where the data bus itself is addressed. Contrary to the above mentioned attack points, attacking the control logic that enables the output of the memory array seems to be the most practical way as it requires only a single laser to carry out the attack. Within the control logic itself there might be several vulnerable places ranging from reference current sources to the array and reference voltages for the read sense amplifiers to the data latches and data bus output control. From the implementation point of view, attacking voltage and current sources could be easier both from the locating prospective and laser injection.

There are some references to previous attacks carried out on old smartcards with external programming voltage V_{pp} [6]. A similar effect can be achieved with a laser interfering with the operation of the internal high-voltage supply.

In order to implement optical fault-injection attacks, the chip surface needs to be accessible. This can be done from both front and rear sides. Modern chips have multiple metal layers obstructing the view and preventing optical attacks. Therefore, the only practical way of implementing optical fault injection on chips fabricated with 0.35 µm or smaller technology is from their rear side. Silicon is transparent to infrared light with wavelengths above 1000 nm, thus making it possible to observe the internal structure of the chip with non-filtered CCD cameras. Optical fault injection can be carried out using inexpensive infrared laser diodes. The effective frequency at which lasers can inject signals inside operating chips is limited to several megahertz, as free carriers created by photons require some time for recombination. Therefore, although lasers offer a relatively inexpensive way of controlling internal signals, they are not as effective for direct signal injection as microprobing techniques [7]. One important thing to know is that for Flash memory, optical fault injection causes either '0' to '1' or '1' to '0' state changes, depending on the location of injection, but never both at the same time.

Older microcontrollers had a security protection fuse. Once activated it disabled access to the on-chip memory. This has proved to be not very secure as such fuses are relatively easy to locate and disable using various attacks. Semi-invasive attacks made security-fuse disabling even easier [3]. As a precaution, chip manufacturers started abandoning readback access to the embedded memory by implementing verify-only approach. In this case the content of memory was compared with uploaded data and a single-bit response in the form of pass/fail was sent back. So far this helped a lot, especially when the length of verified blocks was long enough to prevent brute force searching. Moving from 8-bit data bus to 16-bit and later to 32-bit helped in keeping exhaustive searching attacks even further away. All these steps, provided there are no errors in the design, improve security. The verification process can take place both in hardware or in software. It is impossible to verify the whole memory in one go, so the process is split into blocks with their size limited by the available SRAM buffer or hardware register. The result of the verification is either available immediately on the first incorrect block, or it can be checked in a status register, or rarely available only at the end of the whole memory check.

Fig. 2 illustrates an example of a typical secure verify-only implementation. Very often the result of the verify operation is known only at the end. However, due to the limited size of buffers and registers, smaller blocks of data are verified with the result accumulated inside the chip. The division into blocks can usually be observed on the data transfer protocol or via eavesdropping on the communication line. That way the device under test can be put into a mode where it discloses the result of the verification operation for each block. However, for some devices, the intermediate verification result is available as a part of the standard protocol or can be easily requested. It becomes more complicated when the block consists of multiple words of data, for example, if the verification is performed after receiving every packet of 16 bytes. Still, as the verification is done in hardware, memory contents must be read before the values are compared and this is done via a data bus of limited width. This way there will be some inevitable delay between each word of data read from the memory. Hence, with a fast enough fault injection one can influence the value of each word of the data. This will be an example of a bumping attack (Fig. 2b). More interesting results should be expected if the data latching time is slightly changed or if the value of data is influenced on its way to the latches. This is possible because the bits of data cannot reach the registers exactly at the same time. If this process can be influenced with fault injection attacks it may allow certain bits of data to be kept to a known state, thus making it possible to brute force the remaining bits. There are two possible ways of applying selective bumping attacks – on the rising edge of the fault injection or on the falling (Fig. 2b). However, both events should happen close to the time when the data from memory is latched into the data bus drivers.

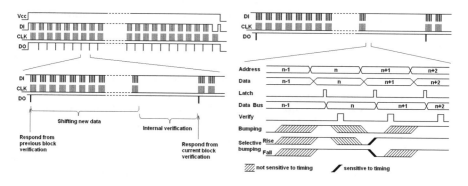

Fig. 2. Timing diagram of a verify-only operation: (a)data blocks, (b)data words level

Term 'bumping' originally comes from a certain type of physical attack on door locks [8]. The idea is to force the key bits into a desired state which will allow access. In the context of the hardware security of semiconductors, 'bumping' shall mean here bypassing the verification of a certain block of data by forcing the data bus into a known state. Alternatively, 'selective bumping' shall mean that certain bits of data are forced into known states allowing the remaining bits to be searched through all possible combinations. Some parallels with lock bumping can be observed. For example, Flash memory bumping attacks allow bypassing the verification for certain words of data without knowing their real value. The more powerful selective bumping attack allows masking of certain bits of data within each word thus substantially reducing the attack time.

3   Experimental Method

There are not many microcontrollers that lack readback access, as this complicates their programming algorithm and many chip manufacturers found this security feature excessive. However, some recent microcontrollers, marketed as "secure", benefit from verify-only feature, as well as AES authentication and SHA-1 integrity check for the firmware. On the other hand, recent non-volatile FPGA chips use a verify-only approach in the high-end security market. Such chips are also marketed as "highly secure".

For my first set of experiments I chose a secure low-end 8-bit microcontroller, the NEC 78K/0S µPD78F9116 with 16 kB of embedded Flash [9]. The chip is fabricated in 0.35 µm and has 3 metal layers. The on-chip firmware bootloader allows the following memory access commands to be executed via SPI, IIC or UART interfaces: Erase, Pre-Write, Write, Verify and Blank check [10]. Only the 'Verify' command seems useful for attempts of data extraction.

My next set of experiments was carried out on a so-called highly secure Flash FPGA, the Actel ProASIC3 A3P250 [11]. Fabricated with a 0.13 µm process with 7 metal layers, this chip incorporates 1,913,600 bits of bitstream configuration data. According to the manufacturer's documentation on this chip: *"Even without any security measures, it is not possible to read back the programming data from a programmed device. Upon programming completion, the programming algorithm will reload the programming data into the device. The device will then use built-in circuitry to determine if it was programmed correctly"*. More information on the programming specification can be obtained indirectly from a programming file which is usually in STAPL format [12]. Alternatively, all the necessary information can be obtained by eavesdropping on the JTAG interface during device programming. The analysis of a simple programming file revealed the following operations on embedded Flash array: Erase, Program and Verify. Again, only the 'Verify' operation can access the internal configuration. Apart from the Flash Array, this FPGA has an AES hardware decryption engine with authentication for firmware updates. As the AES key is stored in Flash memory, it was also evaluated against bumping attacks.

For a detailed demonstration of the bumping attacks I used an engineering sample of a secure microcontroller with AES authentication of the firmware. It was provided by an industrial collaborator under a non-disclosure agreement and is therefore not identified here. This microcontroller has non-volatile programmable memory for an AES key and other security features. As the memory had an independent power supply pin I was able to carry out bumping attacks using non-invasive power glitching. On one hand, it simplified the setup, on the other, it allowed better timing accuracy for characterisation.

Opening up the first two chips was not a difficult task, since only the backside access was a suitable option. No chemicals are required for opening up chips from the rear side. The plastic can be milled away with low-cost hobbyist engraving tools available from DIY shops. The copper heatsink, often attached to the die, can be removed with a knife, followed by removing the glue underneath mechanically and finally cleaning the die with a solvent.

For the experiments both the microcontroller and the FPGA were soldered on test boards (Fig. 3). As the FPGA requires four separate power supply sources, it was plugged into another board with power connectors. The microcontroller board was receiving its power supply from a control board. Both test boards had a power-measurement 10 Ω resistor in the supply line for later power analysis experiments. The microcontroller and the FPGA were programmed with some test data using a universal programmer.

Fig. 3. Test boards with NEC microcontroller and FPGA

A special control board was built for communicating with the tested chips (Fig. 4a). This board was receiving commands from a PC via UART interface and was controlling the test boards. The core of the control board was the Microchip PIC24 microcontroller with 40 MIPS performance thus capable of generating laser fault injection events with 25 ns timing precision. The infrared 1065 nm 100 mW laser diode module was mounted on an optical microscope with 20× NIR objectives and the test board with the chip was placed on a motorised XYZ-stage (Fig. 4b).

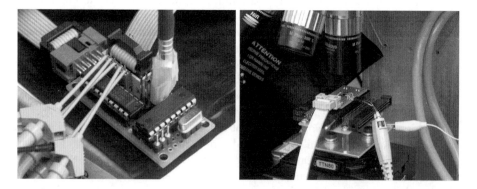

Fig. 4. Test setup: (a)control board, (b)test board under the microscope

The experiments started with finding the minimum length of block verification and possibility of obtaining any intermediate results. Then optical fault injection attacks were used to reduce the verification length down to a single word of the data bus. In this way the width of the data bus can be determined. Alternatively, the width of the data bus can be determined by counting the number of read sense amplifiers under a microscope. As a preliminary step for optical fault injection, laser scanning was performed. This can significantly reduce the area of search for successful fault injection. Also, some power analysis measurements [13] were made using a digital storage oscilloscope to estimate the internal timings of the verification process. This can assist in carrying out selective bumping attacks later, thereby saving the time otherwise required for exhaustive search.

4   Results

Initial experiments were aimed at finding places sensitive to optical fault injection. For that, both chips were exposed to laser scanning imaging, also called OBIC imaging [14]. The same setup as for optical fault injection was used, with the difference being that the photo-current produced by the laser's photons was registered with a digital multimeter. The result of the laser scan is presented in Fig. 5 and allows identification of all major areas. Flash arrays are quite large and easily identifiable structures on a chip die. However, for bumping attacks of more interest is the memory control logic. As that is usually placed next to the array, obvious places for exhaustive search are located nearby. The area for search is designated in Fig. 5 as 'Flash control'. For the NEC microcontroller, many places within the control area were sensitive to optical bumping causing the memory to change its state to all '0' when the laser was switched on for the whole verification process. For the FPGA, an exhaustive search for areas sensitive to bumping was made across the whole chip with a 20 μm grid and the result is presented in Fig. 6a. Each point was verified against the correct initial value and against all '1'. White areas represent successful bumping with all bits set to '1', black areas correspond to faulty verification, while grey area represent no change. Surprisingly, no success was achieved when AES authentication was targeted in the FPGA (Fig. 6b). Only faulty authentications were received for a small area which are of no use since the attacker has no means of observing the result of AES decryption, so he cannot mount differential fault attacks [15]. This could be the result of abandoning read sense amplifiers in the Flash memory that stores the AES key and using direct Flash switches instead, or it could be that some countermeasures against optical fault injection attacks were in place.

Further experiments were aimed at splitting the full verification process into blocks of data. The NEC microcontroller has 16384 bytes of internal memory. According to the Flash memory write application note [10], each verification command receives 128 bytes of data via the serial programming interface before starting the verification process. However, the status register is not updated with the result until all 16384 bytes of data are received and verified. That means there is no simple non-invasive way of splitting up the data sequence. However, the fact that each verification is run on 128-bytes blocks means that it is fairly easy to apply a bumping attack for the duration of the verify operation on all blocks except the chosen one. For the Actel FPGA, splitting 1,913,600

Fig. 5. Laser scanning images: (a)NEC microcontroller, (b)FPGA

bits of the verification data into 832-bit blocks came from the analysis of both the STAPL programming file and the JTAG communication protocol. These 832 bits of data are sent as 32 packets of 26 bits, however, the verification is done on all 832 bits simultaneously in hardware. Moreover, the result of the verification is available after each 832-bit block, making further analysis easier compared to the above microcontroller, as there is no need to do bumping on other blocks.

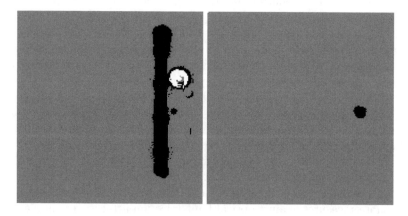

Fig. 6. Fault injection sensitivity mapping: (a)Verify Array operation, (b)AES authentication operation

The next step was aimed at splitting the verification blocks into words down to the width of the memory data bus. Power analysis measurements were carried out on both chips revealing the exact timing for the internal verification process in the form of higher power consumption (Fig. 7). For easier observation, a 10× zoom is presented at the bottom. Block verification takes 16.4 ms for the microcontroller and only 40 μs for the FPGA. Moreover, for the microcontroller 128 peaks in the power consumption are clearly distinguishable making it clear that 8 bits are verified at a time (Fig. 7a). For the FPGA, the granularity of the verification process was invisible in the power trace. However, the number of read sense amplifiers observed under a microscope suggested that the memory data bus is 32 bits wide resulting in 1.5 µs verification time per word. This was later confirmed with bumping experiments.

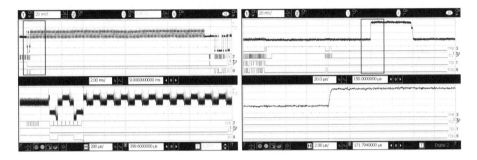

Fig. 7. Power analysis on block verification: (a)NEC microcontroller, (b)FPGA

It was not possible to distinguish any difference between correct and incorrect words of data in the power trace because the signal was too noisy. Averaging might help, but will substantially increase the attack time making it not practical. Therefore, on its own, power analysis was not useful for breaking into the Flash of these chips. However, it gave important information about the timing of the internal verification process which was useful for carrying out bumping and selective bumping attacks later.

Bumping attack to separate words of data on the microcontroller was straightforward as it uses an external clock signal. That way the attack can be easily synchronised to the chips operations. It takes 128 µs to verify each word of data by the internal firmware of the microcontroller. Hence, in order to turn certain bytes into a known zero state the laser should be turned on during that time.

The process was trickier for the FPGA. It runs on its internal clock, which is not very stable and was changing over time. It takes only 1.5 µs to verify each word of data for the on-chip hardware of the FPGA. However, timing control with sub-microsecond precision is not a problem for the laser. The bumping attack aimed at the separation of the words of data within the block worked reliably even for the internally clocked FPGA. As the clock jitter was less than 2%, there was no overlapping between verifications of each word, since there were only 26 words per block.

The last step was aimed at implementing the selective bumping attacks. Triggering the laser pulse at the precise time for data latching in the microcontroller was achieved without much trouble. However, there was no large improvement compared to the standard bumping attack – only reduction of the search field from 2^8 to 2^5 per byte in the block, totalling 2^{12} attempts per whole block. One important thing to know is that for each attempt the full verification process must be run and it takes at least five seconds to get the result of the verification at the end. Hence, selective bumping will improve the full data extraction time from four months to two weeks. Applying the selective bumping attack to the FPGA was much more difficult due to the lack of synchronisation. Therefore, the probability of hitting the data latching at a particular time was reduced by a factor of the data bus width. However, this was not the only obstacle for the selective bumping. It turned out that some bits within the 32-bit word have very close timing characteristics, so it was extremely hard to separate them reliably. Fortunately, as mentioned earlier, there are many points on the chip where the bumping attacks can be applied. By testing different points, I found that some of them offer better separation of bits in time. Selective bumping does not require knowledge about the sequence of bits as this information can be easily obtained during the experiment. Since the all-'1' state always pass the verification, the next attempt involves selective bumping into 32 possible combinations with a single '0' bit. Then followed by 31 combinations of two '0' bits, and so on until the whole word is recovered. The best case has required only 2^{13} attempts per recovery of the word value, totalling 2^{18} searches per 832-bit block. This is slightly more than the theoretically predicted 2^8 per word and 2^{13} per block because of the jitter caused by the internally clocked hardware of the chip. Although each block can be verified independently, it cannot be addressed separately. That means it takes longer to verify higher blocks, on average 10 ms. That way the whole configuration bitstream can be extracted in about one month. Without the use of selective bumping attacks, that is with only bumping attacks, such extraction would take more than fifty thousand years. This is a far more significant improvement than with the microcontroller. However, this was expected for 16-bit and 32-bit chips, which no longer could be brute forced through all possible values within reasonable time.

The final set of experiments was carried out on a secure microcontroller with AES authentication. As the chip under test was not in production at the time of testing, the details of the setup are omitted. Only the statistical results for optical bumping attacks are presented. However, this experiment shows how the selective bumping helps in dramatic reduction of the search field.

Since the chip has a sophisticated secure AES key programming method, it was supplied pre-programmed by the industrial collaborator. Authentication required a special high-speed hardware setup that was also provided. Therefore, my work was limited to finding a way of forcing the chip into bumping and selective bumping and this was achieved by means of just non-invasive power supply glitching attacks. The bumping caused certain bits of key to change to '1'. The result of the authentication was either pass or fail. Further analysis revealed that the key was transferred from the internal secure memory in 8 words, which means the memory bus is 16-bit wide. The probability of bumping depended from the time of glitching with low probability suggesting the time for selective bumping (Fig. 8a).

The result of the selective bumping attack is presented in Fig. 8b showing which bits are changing at a particular time of glitching for one word. It is clear that the bits are well separated in time thus allowing easy key recovery.

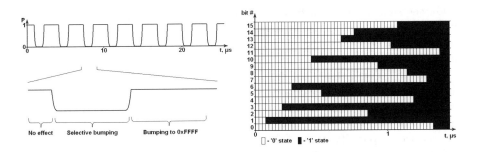

Fig. 8. Attacking AES authentication: (a)bumping, (b)selective bumping

Although the bumping attack was enough to extract the whole key in about one week, since each word was only 16-bit wide, selective bumping brought this figure down to a few minutes.

5   Implications and Further Improvements

The above results were achieved on two microcontrollers and one FPGA. It would be interesting to compare the results with other chips. However, it is not easy to find candidates since most microcontrollers do have readback access through programming or debugging interface. However, when it comes to the AES authentication and hash functions, there might be many chips which store keys in Flash memory. In the FPGA industry, very few manufacturers offer non-volatile chips. The bumping attacks could also be compared with another sample of the same chip, to see if the sequence of bits masked by selective bumping remains the same.

In order to improve the speed of the bumping attacks, a more efficient algorithm could be used. For example, an exhaustive timing analysis could be applied to the first block of data. That way, selective bumping would give faster results for the following blocks.

One can argue that power analysis might give a better result as it was used to break many security protection schemes. However, applying power analysis to the verification process will require too many power traces to be acquired and analysed. This will inevitably lead to a very long extraction time, not practical for a real attack. However, this might work if only a few bytes of data would have to be extracted, for example, a password or an encryption key.

I noticed that FPGA security relies heavily on obscurity. That ranges from the lack of any documentation on the JTAG access interface, and absence of information on the internal operations, down to the data formats. This works well unless an attacker is that determined to find all this information on their own. My experiments showed how some information can be found via systematic testing of device operations. That way, for example, I found the correspondence between bits in the 832-bit verification data and bits in the data bus. Alternatively, for some chips more information can be gained through analysis of development tools and programming files.

On the security of Flash FPGA, one can argue that the demonstrated attack will not entirely compromise its security. This is because on top of the verify-only access control there are other security protection mechanisms, such as FlashLock access control and AES bitstream encryption with proprietary integrity-check algorithm [15]. This is true, but the fact that verify-only schemes do not provide the level of protection anticipated by the manufacturer should cause concern, especially as there might be the possibility of failures in other protection mechanisms. Furthermore, some users were relying solely on the fact that there was no readback capability in the FPGAs.

Bumping attacks demonstrated in this paper used only one parameter – time. It might be possible to find two independent parameters, for example, time and laser power, or time and positioning. That way, more powerful 2D bumping attacks could be implemented.

6   Conclusion

Two types of bumping attacks were introduced – bumping and selective bumping. The first is aimed at bypassing the verification of a block, while the other is aimed at presetting certain bits of data inside the block. Successful attacks on a microcontroller with secure memory verification, on an FPGA with secure firmware verification and on a secure microcontroller with AES authentication were presented. Verify-only memory protection scheme is used in some chips as a more secure alternative to access protection fuses.

The attack was carried out in three steps. The first step was aimed at separating the process into blocks. The second step involved splitting the data in each block into words corresponding to the width of the data bus. The third step was used to reduce the number of guesses required to pass the verification within each word of data. Although for the 8-bit bus selective bumping was not very useful, only a reduction from 2^8 to 2^5 searches, for the 32-bit bus selective bumping attack allowed a tremendous reduction from 2^{32} to just 2^{13} searches, which makes it practical. This is very important as the width of data bus in modern devices is more often 16 or 32 bits rather than 8 bits as in old microcontrollers. The attack was possible because each individual bit within a word has different sensitivity to fault injection thus allowing reliable separation from the others.

My research proved that a verify-only approach does not work on its own. Even from a so-called highly secure FPGA the configuration can be extracted using bumping attacks. Fortunately, this is not the only protection that is available in Actel FPGAs. In addition to the verify-only scheme, the FlashLock can be activated or even more robust AES bitstream encryption with proprietary integrity check algorithm. The latter prevents verification on arbitrary data [16]. Nevertheless, manufacturer's claims that data extraction from these FPGAs is not possible is no longer true. Although the FPGA was developed with some security in mind and has independent clocking and a reasonably wide data bus, it was still possible to successfully apply bumping attacks and get significant improvements over brute force attacks.

My experiments showed that bumping attacks are easy to apply even on chips clocked from an internal asynchronous source. My attempts of applying conventional power analysis to distinguish a single-bit change in the Hamming weight of data were unsuccessful. However, the results from the power analysis were useful for optimising the timing of bumping attacks. Using selective bumping technique the data extraction time can be dramatically reduced to hours and days compared with many years required for brute-force searching. As these attacks do not require expensive equipment they can pose a big problem to the hardware community. As protection against bumping attacks, similar techniques can be used as for other types of optical fault injection attacks. For example, light sensors might prevent optical attacks, while a robust verification process could make bumping attacks very difficult to use. Alternatively, a very long verification process could make finding of each bit not very practical. In addition, clock jitters and dummy cycles would make bumping much harder to synchronise. Secure Flash memory design could also prevent bumping as it was shown on the secure Flash memory storing AES key in the FPGA.

Bumping attacks can find their way in partial reverse engineering of the internal chip structure and its operation. For example, data scrambling in the configuration bitstream of the FPGA could be found using bumping attacks. Flash memory bumping attacks do not require precise positioning on the chip surface and just reasonable timing precision, hence, easily applied. Flash memory bumping attacks complement other semi-invasive methods, such as optical probing [4], laser scanning [14] and optical emission analysis [17]. However, bumping gives the result faster and does not require stopping the clock frequency or placing the device in an idle state which sometimes is not feasible. Once again, semi-invasive attacks such as optical fault injection proved their effectiveness for deep sub-micron chips which should be of concern to secure chip manufacturers. Very likely this will result in the introduction of new countermeasures during the design of semiconductor chips.

References

1. Xilinx CoolRunner-II CPLDs in Secure Applications. White Paper,
   http://www.xilinx.com/support/documentation/white_papers/wp170.pdf
2. Design Security in Nonvolatile Flash and Antifuse FPGAs. Security Backgrounder,
   http://www.actel.com/documents/DesignSecurity_WP.pdf
3. Skorobogatov, S.: Semi-invasive attacks – A new approach to hardware security analysis. Technical Report UCAM-CL-TR-630, University of Cambridge, Computer Laboratory (April 2005),
   http://www.cl.cam.ac.uk/techreports/UCAM-CL-TR-630.pdf
4. Skorobogatov, S., Anderson, R.: Optical Fault Induction Attacks. In: Kaliski Jr., B.S., Koç, Ç.K., Paar, C. (eds.) CHES 2002. LNCS, vol. 2523, pp. 2–12. Springer, Heidelberg (2003)
5. Brown, W.D., Brewer, J.E.: Nonvolatile semiconductor memory technology: a comprehensive guide to understanding and using NVSM devices. IEEE Press, Los Alamitos (1997)

6. Anderson, R.J., Kuhn, M.G.: Tamper resistance – a cautionary note. In: The Second USENIX Workshop on Electronic Commerce, Oakland, California (November 1996)
7. Wagner, L.C.: Failure Analysis of Integrated Circuits: Tools and Techniques. Kluwer Academic Publishers, Dordrecht (1999)
8. Tobias, M.W.: Opening locks by bumping in five seconds or less: is it really a threat to physical security? A technical analysis of the issues, Investigative Law Offices, http://podcasts.aolcdn.com/engadget/videos/lockdown/bumping_040206.pdf
9. NEC PD789104A, 789114A, 789124A, 789134A Subseries User's Manual. 8-Bit Single-Chip Microcontrollers, http://www2.renesas.com/maps_download/pdf/U13037EJ1V0PM00.pdf
10. NEC 78K/0, 78K/0S Series 8-Bit Single-Chip Microcontrollers. Flash Memory Write. Application Note, http://www.necel.com/nesdis/image/U14458EJ1V0AN00.pdf
11. Actel ProASIC3 Handbook. ProASIC3 Flash Family FPGAs, http://www.actel.com/documents/PA3_HB.pdf
12. Actel: ISP and STAPL. Application Note AC171, http://www.actel.com/documents/ISP_STAPL_AN.pdf
13. Kocher, P., Jaffe, J., Jun, B.: Differential Power Analysis. In: Wiener, M. (ed.) CRYPTO 1999. LNCS, vol. 1666, pp. 388–397. Springer, Heidelberg (1999)
14. Ajluni, C.: Two new imaging techniques promise to improve IC defect identification. Electronic Design 43(14), 37–38 (1995)
15. Giraud, C.: DFA on AES. In: Dobbertin, H., Rijmen, V., Sowa, A. (eds.) AES 2004. LNCS, vol. 3373, pp. 27–41. Springer, Heidelberg (2005)
16. Actel ProASIC3/E Production FPGAs. Features and Advantages, http://www.actel.com/documents/PA3_E_Tech_WP.pdf
17. Skorobogatov, S.: Using Optical Emission Analysis for Estimating Contribution to Power Analysis. In: 6th Workshop on Fault Diagnosis and Tolerance in Cryptography (FDTC 2009), Lausanne, Switzerland, September 2009, pp. 111–119. IEEE-CS Press, Los Alamitos (2009) ISBN 978-0-7695-3824-2

Self-referencing: A Scalable Side-Channel Approach for Hardware Trojan Detection

Dongdong Du, Seetharam Narasimhan,
Rajat Subhra Chakraborty, and Swarup Bhunia

Case Western Reserve University, Cleveland OH-44106, USA
sxn124@case.edu

Abstract. Malicious modification of integrated circuits (ICs) in untrusted foundry, referred to as "Hardware Trojan", has emerged as a serious security threat. While side-channel analysis has been reported as an effective approach to detect hardware Trojans, increasing process variations in nanoscale technologies pose a major challenge, since process noise can easily mask the Trojan effect on a measured side-channel parameter, such as supply current. Besides, existing side-channel approaches suffer from reduced Trojan detection sensitivity with increasing design size. In this paper, we propose a novel scalable side-channel approach, named *self-referencing*, along with associated *vector generation algorithm* to improve the Hardware Trojan detection sensitivity under large process variations. It compares transient current signature of one region of an IC with that of another, thereby nullifying the effect of process noise by exploiting spatial correlation across regions in terms of process variations. To amplify the Trojan effect on supply current, we propose a region-based vector generation approach, which divides a circuit-under-test (CUT) into several regions and for each region, finds the test vectors which induce maximum activity in that region, while minimizing the activity in other regions. We show that the proposed side-channel approach is scalable with respect to both amount of process variations and design size. The approach is validated with both simulation and measurement results using an FPGA-based test setup for large designs including a 32-bit DLX processor core ($\sim 10^5$ transistors). Results shows that our approach can find ultra-small (<0.01% area) Trojans under large process variations of up to $\pm$ 20% shift in transistor threshold voltage.

Keywords: hardware Trojan, side-channel analysis, self-referencing.

1 Introduction

Global economics dictates increasing out-sourcing of Integrated Circuit (IC) fabrication to off-shore facilities. Though cost-effective, out-sourcing brings up potential risks for an adversary to maliciously modify a circuit. Such malicious modifications are referred as *Hardware Trojans*. A typical Hardware Trojan would cause an IC to have altered functional behavior during operation in the field, potentially with disastrous consequences in safety-critical applications. The

S. Mangard and F.-X. Standaert (Eds.): CHES 2010, LNCS 6225, pp. 173–187, 2010.

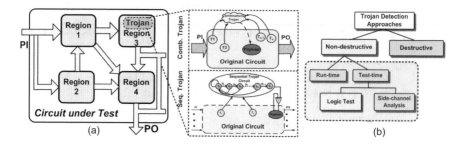

Fig. 1. (a) A circuit with hardware Trojan along with models of two types of Trojans. (b) A taxonomy of existing hardware Trojan detection techniques.

threat of Hardware Trojans has emerged as a major security concern [1], especially since several unexplained military mishaps are attributed to the presence of malicious hardware Trojans [2,3]. Such hardware Trojans can also be inserted in a design house during the design of an IC. Here, we focus on the problem of detecting hardware Trojans inserted during fabrication in an untrusted foundry.

An intelligent adversary can incorporate a hardware Trojan, which is extremely difficult to detect during conventional post-manufacturing test. Due to their stealthiness, Trojans can be triggered only under rare conditions. Upon triggering, they can either cause malfunction by altering internal node values [4] or leak secret information through covert channels [5]. They can also be used to assist software attacks by providing a hardware backdoor [3]. Fig. 1(a) shows an example circuit with Trojan inserted inside one of its constituent blocks. Broadly two types of Trojan can be inserted in a digital circuit: *combinational Trojans*, which are activated by a rare combination of values at internal circuit nodes and *sequential Trojans*, which are activated through a sequence of rare events. Several approaches to detect hardware Trojans have been proposed in recent literature [5]. We show a classification of the Trojan detection techniques in Fig. 1(b). Destructive testing of a chip by de-packaging, de-metallization and micro-photography based reverse-engineering is highly expensive (in time and cost) and not a feasible solution because an attacker may selectively insert Trojan into a small subset of the manufactured ICs [7]. Conventional logic testing, both functional and structural, performs poorly in detecting Trojans, due to their stealthiness, arbitrary nature and size [8]. An alternative approach is to measure a side-channel parameter, such as supply current or path delay, which can be affected due to unintended design modifications. However, the effectiveness of side-channel analysis is limited by large device parameter variations in modern nanometer technologies leading to variations in the measured side-channel parameter, which can mask the effect of a small Trojan.

The issue of process variations on side-channel analysis based Trojan detection has been considered in [9], which explores signal processing techniques to reduce effect of process noise on supply current. Another approach based on power-supply transient [6], measures current signal from multiple power ports and uses a statistical characterization of process noise. Path delays of output

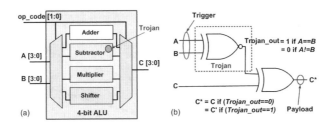

Fig. 2. (a) An simple test circuit: a 4-bit Arithmetic Logic Unit (ALU). (b) A combinational Trojan inserted into the subtractor.

ports have also been used as the fingerprint [11], with extensive characterization for process variations. In this paper, we propose a scalable side-channel approach to hardware Trojan detection based on a concept called "self-referencing". The basic idea is to use supply current signature of one region of a chip as reference to that of another to eliminate the process noise. Such calibration or referencing is possible due to the spatial correlation of process variation effects across regions in a chip. We show that such an approach can be extremely effective in nullifying all forms of process noise, namely inter-die, intra-die random and intra-die systematic variations [13]. Since process noise is eliminated by comparing current signature of regions in an IC, the method is scalable with increasing process noise, unlike existing approaches [9]. To increase the Trojan detection sensitivity, we propose a region-based vector generation approach, which tries to maximize the Trojan effect while minimizing the background current. Current values of n regions are then compared with all other using a slope heuristic and the resultant *region slope matrix* is used to compare a chip with another. We validate the proposed approach using both simulation and measurements for several large open source designs. Simulation results shows high detection sensitivity in presence of large process variations and scalability of the approach with increasing design size. The measurement results with a custom test board validates the effectiveness of the approach.

The rest of the paper is organized as follows. In section 2, we describe the motivation of the proposed self-referencing method. Section 3 presents the methodology along with theoretical analysis. Simulation and experimental results are described in section 4. Section 5 concludes the paper.

2   Motivation of Self-referencing Approach

The idea of self-referencing can be illustrated using an example 4-bit ALU, as shown in Fig. 2(a). The ALU contains four distinct functional units (FUs) - adder, subtractor, multiplier and shifter, which are activated based on the input "opcode" value. There are two 4-bit operands and a 4-bit output. In such a circuit, a single region or FU can be selectively activated by proper choice of opcode, we can easily generate test vectors which target separate activation of the four regions. We consider three different process corners (nominal ±25%) for

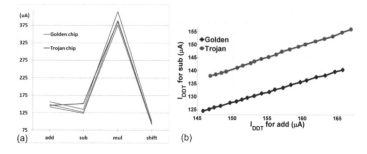

Fig. 3. (a) Comparison of supply current between golden and tampered chip for four regions of a 4-bit ALU. (b) Correlation of region currents at different process points for golden and tampered ICs.

the entire design (modeled as a change in the transistor threshold voltage V_T) and simulate the design in HSPICE for four different vector pairs which activate each of the four regions separately. We also measure the background current. The Trojan circuit, as shown in Fig. 2(b) was designed to invert an output bit of the subtractor if two input bits were equal. We simulated the circuit with the Trojan in the subtractor module (occupying 2.7% area of the ALU) at the nominal process corner for the same set of vectors.

Fig. 3(a) shows the plot of the average I_{DDT} values for the four different vectors activating the four different regions without the background current. We can observe the tampered circuit consumes more current for the vector which activates the subtractor region. We plot the current for one region (adder) with respect that for another (subtractor) for a set of golden and tampered chips at 20 different process points in Fig. 3(b). We expect a correlation between the region currents across process corners. However, since there is a Trojan in the subtractor, it shows uncorrelated behavior in supply current. Hence, the current for the adder can be used to calibrate the process noise and check for the presence of Trojan in other modules. In real life, since we do not know the region which contains the Trojan, we need to compare each region with all others. This also allows us to cancel out the effect of random and systematic intra-die process variations, as explained later.

3   Methodology

For a large design, the golden supply current for a high activity vector can be large compared to the additional current consumed by a small Trojan circuit, and the variation in the current value due to process variation can be very large. This can mask the effect of the Trojan on the measured current, leading to difficulty in detecting a Trojan-infected chip. Most side-channel analysis based approaches perform calibration of the process noise by using golden chips at different process corners. This helps us obtain a limiting threshold value beyond which any chip is classified as Trojan. Since the variation in the measured value can cause a golden chip to be misclassified as a Trojan (we refer to this case as a *false positive - FP*), the limit line has to be close to the nominal golden value. On the other hand, if the Trojan effect does not change the value beyond the limit, the Trojan-containing chip can be misclassified as a golden one (we refer to this case as a *false negative - FN*). To limit the probability of false positives and false negatives, the limiting values need to be chosen carefully.

The Trojan detection sensitivity of this approach reduces with decreasing Trojan or increasing circuit size. In order to detect small sequential/combinational Trojans in large circuits ($> 10^5$ transistors), we need to improve the SNR (Signal-to-Noise Ratio) using appropriate side-channel isolation techniques. At a single V_T point the sensitivity, for an approach where transient current values are compared for different chips, can be expressed as:

$$Sensitivity = \frac{I_{tampered,nominal} - I_{golden,nominal}}{I_{golden,process\ variation} - I_{golden,nominal}} \tag{1}$$

Clearly, the sensitivity can be improved by increasing the current contribution of the Trojan circuit relative to that of the original circuit. We can divide the original circuit into several small regions and measure the supply current (I_{DDT}) for each region. The relationship between region currents also helps to cancel the process variation effects. In Fig. 3(a), if we consider the "slope" or relative difference between the current values of 'add' and 'sub' regions, we can see that there is a larger shift in this value due to Trojan than in the original current value due to process variations. We refer to this approach as the *Self-Referencing* approach, since we can use the relative difference in the region current values to detect a Trojan by reducing the effect of process variations. In the appendix, we present an analysis regarding how the self-referencing approach can help cancel the effect of process variations.

The major steps of the self-referencing approach are as follows. First, we need to perform a *functional decomposition* to divide a large design into several small blocks or regions, so that we can activate them one region at a time. Next, we need a vector generation algorithm which can generate vectors that maximize the activity within one region while producing minimum activity in other regions. Also, the chosen set of test vectors should be capable of triggering most of the feasible Trojans in a given region. Then, we need to perform self-referencing among the measured supply current values. For this we use a *Region Slope Matrix* as described in the appendix. Finally, we reach the decision making process which is to compare the matrix values for the test chip to threshold values derived from golden chips at different process corners, in order to detect the presence or absence of a Trojan. Next we describe each of the steps in detail.

Functional Decomposition: The first step of the proposed self-referencing approach is decomposition of a large design into functional blocks or regions. Sometimes, the circuit under test is designed with clearly-defined functional blocks which can be selectively activated by using control signals, like the 4-bit ALU circuit which we considered for our example in Section 2. Another type of circuit which is amenable to simple functional decomposition is a pipelined processor, where the different pipeline stages correspond to the different regions. However, there can be circuits which are available as a flattened gate-level netlist. For this we could use a hyper-graph based approach to identify partitions which have minimum cut-sets between them. This allows us to isolate the activity in one partition from causing activity in other regions. The region-based partitioning described in [7] can also be used for creating partitions in circuits which do not have well-defined functional blocks or for creating sub-blocks within a functional block. The decomposition should follow a set of properties to maximize the effectiveness of the approach:

1. The blocks should be reasonably large to cancel out the effect of random parameter variations, but small enough to minimize the background current. It should also be kept in mind that if the regions are too small, the number of regions can become unreasonably large for the test vector generation algorithm to handle.
2. The blocks should be functionally as independent of each other as possible so that the test generation process can increase the activity of one block (or few blocks) while minimizing the activity of all others.
3. The decomposition process can be performed hierarchically. For instance, a system-on-a-chip (SoC) can be divided into the constituent blocks which make up the system. But, for a large SoC, one of the blocks could itself be a processor. Hence, we need to further divide this structural block into functional sub-blocks.

Statistical test vector generation: In order to increase the Trojan detection sensitivity, proper test vector generation and application are necessary to reduce the background activity and amplify the activity inside the Trojan circuit. If we partition the circuit into several functional and structurally separate blocks, we can activate them one at a time and observe the switching current for that block with respect to the current values for other blocks. The test vector generation algorithm needs to take into account two factors:

1. Only one region must be activated at a time. If the inputs to different modules are mutually exclusive and the regions have minimal interconnection, it is easy to maximally activate one region while minimizing activity in other regions. If complex interconnections exist between the modules, the inputs need to be ranked in terms of their sensitivity towards activating different modules and the test generation needs to be aware of these sensitivity values.
2. When a particular region is being activated, the test vectors should try to activate possible Trojan trigger conditions and should be aimed at creating activity within most of the innumerable possible Trojans. This motivates us to consider a statistical test generation approach like the one described in [12] for maximizing Trojan trigger coverage. Note that, unlike functional testing approaches, the Trojan payload need not be affected during test time, and the observability of Trojan effect on the side-channel parameter is ensured by the region-based self-referencing approach described earlier.

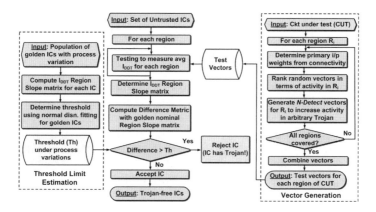

Fig. 4. The major steps of the proposed self-referencing methodology. The steps for test vector generation for increasing sensitivity and threshold limit estimation for calibrating process noise are also shown.

Fig. 4 shows a flow chart of the test vector generation algorithm on the right. For each region, we assign weights to the primary inputs in terms of their tendency to maximize activity in the region under consideration while minimizing activity in other regions. This step can also identify control signals which can direct the activity exclusively to particular regions. Next, we generate weighted random input vectors for activating the region under consideration and perform functional simulation using a graph-based approach, which lets us estimate the activity within each region for each pair of input vectors. We sort the vectors based on a metric C_{ij} which is higher for a vector pair which can maximally activate region R_i while minimizing activity in each of the other regions. Then, we prune the vector set to choose a reduced but highly efficient vector set generated by a statistical approach such as *MERO* [12]. In this approach (motivated by the *N-detect* test generation technique), within a region, we identify internal nodes with rare values, which can be candidate trigger signals for a Trojan. Then we identify the subset of vectors which can take the rare nodes within the region to their rare values at least N times, thus increasing the possibility of triggering the Trojans within the region. Once this process is completed for all the regions, we combine the vectors and generate a test suite which can be applied to each chip for measuring supply current corresponding to each of its regions.

For functional test of a multi-core processor, we can use specially designed small test programs which are likely to trigger and observe rare events in the system such as events on the memory control line or most significant bits of the datapath multiple times. In general a design is composed of several functional blocks and activity in several functional blocks can be turned off using input conditions. For example in a processor, activity in the floating point unit (FPU), branch logic or memory peripheral logic can be turned off by selecting an integer ALU operation. Many functional blocks are pipelined. In these cases, we will focus on one stage at a time and provide initialization to the pipeline such that the activities of all stages other than the one under test are minimized by ensuring that the corresponding stage inputs do not change. Next we describe how the self-referencing approach can be applied to compare the current values for different regions and identify the Trojan-infected region.

Side-Channel Analysis using Self-Referencing: In this step, we measure the current from different blocks which are selectively activated, while the rest of the circuit is kept inactive by appropriate test vector application. Then the average supply current consumed by the different blocks is compared for different chip instances to see whether the relations between the individual block currents are maintained. Any discrepancy in the "slope" of the current values between different blocks indicates the presence of Trojan. This approach can be hierarchically repeated for further increasing sensitivity by decomposing the suspect block into sub-blocks and checking the self-referencing relationships between the current consumed by each sub-block.

The flowchart for this step is shown in Fig. 4. Note that the best Trojan detection capability of region-based comparison will be realized if the circuit is partitioned into regions of similar size. The *Region Slope Matrix* is computed by taking the relative difference between the current values for each region. We estimate the effect of process variations on the "slopes" to determine a threshold for separating the golden chips from the Trojan-infested ones. This can be done by extensive simulations or measurements from several known-golden chips. For a design with n regions, the *Region Slope Matrix* is an $n \times n$ matrix, with entries that can be mathematically expressed as:

$$S_{ij} = \frac{I_i - I_j}{I_i} \ \forall i, j \in [1, n] \tag{2}$$

For each region, we get $2n - 1$ slope values, of which one of them is '0', since the diagonal elements S_{ii} will be zero.

The intra-die systematic variation is eliminated primarily because we use the current from an adjacent block, which is expected to suffer similar variations, to calibrate process noise of the block under test. The intra-die random variations can be eliminated by considering switching of large number of gates. In our simulations we find that even switching of 50 logic gates in a block can effectively cancel out random deviations in supply current.

Decision Making Process: In this step, we make a decision about the existence of Trojan in a chip. The variation in slope values for different regions for a chip from the golden nominal values are combined by taking the L^2 norm (sum of squares of difference of corresponding values) between the two Region Slope matrices. This difference metric for any chip 'k' is defined as

$$D(k) = \sum_{i=1}^{N} \sum_{j=1}^{N} (S_{ij}|_{Chip \ k} - S_{ij}|_{golden,nominal})^2. \tag{3}$$

The limiting "threshold" value for golden chips can be computed by taking the difference $D(golden, process\ variations)$ as defined by

$$Threshold = \sum_{i=1}^{N} \sum_{j=1}^{N} (S_{ij}|_{golden,process\ variation} - S_{ij}|_{golden,nominal})^2. \qquad (4)$$

Any variation beyond the threshold is attributed to the presence of a Trojan. The steps for computing the golden threshold limits are illustrated on the left side of Fig. 4. Since unlike conventional testing, a go/no-go decision is difficult to achieve, we come up with a measure of confidence about the trustworthiness of each region in a chip using an appropriate metric. We compare the average supply current consumed by the different blocks for different chip instances to see whether the expected correlation between the individual block currents is maintained. The Trojan detection sensitivity of the self-referencing approach can be defined as

$$Sensitivity = \frac{D(tampered, nominal)}{Threshold} \qquad (5)$$

Since, the slope values are less affected by process variations compared to the current values alone, we expect to get better sensitivity compared to eqn. (1). Note that since we perform region-based comparison, we can localize a Trojan and repeat the analysis within a block to further isolate the Trojan. This approach can be hierarchically repeated to increase the detection sensitivity by decomposing a suspect block further into sub-blocks and applying the self-referencing approach for those smaller blocks. We can also see that the region-based self-referencing approach is scalable with respect to design size and Trojan size. For the same Trojan size, if the design size is increased two-fold, we can achieve same sensitivity by dividing the circuit into twice as many regions. Similarly we can divide the circuit into smaller regions to increase sensitivity towards detection of smaller Trojan circuits.

4   Results

4.1   Simulation Results

We used two test cases to validate the proposed Trojan detection approach: 1) a 32-bit integer Arithmetic Logic Unit (ALU), and 2) a Finite Impulse Response (FIR) digital filter. The size of the ALU circuit can be scaled by changing the word size parameter. We considered 4 structurally different blocks - adder (*add*), subtracter (*sub*), multiplier (*mul*) and shifter (*shift*) which can be selectively activated by the *opcode* input bits. However, the FIR filter had a flattened netlist and was manually partitioned into four regions with the minimum interconnections, and the test vector generation tool (written in MATLAB) was used to generate test vectors to selectively activate each block. We inserted a small (<0.01% of total area) Trojan in the subtracter of the ALU and the 4th

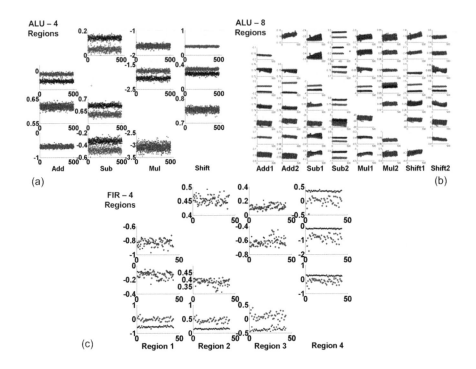

Fig. 5. Self-referencing methodology for detecting Trojan in the 32-bit ALU and FIR circuits. Blue and red lines (or points) denote golden and Trojan chips, respectively.

region of the FIR filter. Both designs were synthesized using Synopsys *Design Compiler* and mapped to a LEDA standard cell library. Circuit simulations were carried out for the 70nm *Predictive Technology Model* (PTM) [15] using Synopsys *HSPICE*. To estimate the effect of process variations, we used Monte Carlo simulations for a maximum of ±20% variation in the nominal V_T value, inter-die variations with $\sigma = 10\%$ and random intra-die variations with $\sigma = 6\%$. We simulated the circuits and separately measured the supply current for different regions for 500 golden chips and 500 infected chips.

The simulated *Region Slope Matrix* values are plotted in Fig. 5(a). The Trojan-infected chip instances can be easily distinguished from the golden ones, even in the presence of process noise. The row and column corresponding to the subtracter (2nd region) show visibly different values for the golden (blue) and Trojan (red) values. Next, we performed simulations with multiple vector pairs activating the same module to show that the Trojan in the subtracter is only selectively activated on the application of one of the two vector pairs activating the subtracter module. The *Region Slope Matrix* for this case is shown in Fig. 5(b). This matrix contains 8 regions since each of the four structurally separate regions of the ALU are further divided into two sub-blocks, corresponding

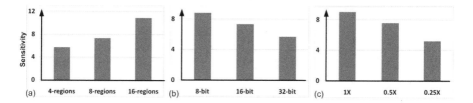

Fig. 6. Sensitivity analysis with (a) different number of regions, (b) different circuit sizes, and (c) different Trojan sizes

Table 1. Probability of Detection and probability of False Alarm (False Positives)

Circuit Name	TN(%)	FP(%)	FN(%)	TP(%)
32-bit ALU	99.10	0.90	5.90	94.10
FIR	97.72	2.28	6.60	93.40

to the two different vector pairs which share the same opcode values. It can be readily observed that increasing the number of regions increases the sensitivity of Trojan detection.

Fig. 5(c) shows the simulation results for the FIR design. The test vectors are chosen by the MATLAB tool and used to dominantly activate different regions of the design. The *Region Slope Matrix* is computed for 50 golden chips and 50 Trojan-infected chips and we can successfully detect the Trojan-infected region (region 4). Fig. 6 shows the variation in sensitivity of the self-referencing approach by varying different parameters of the ALU. For a 16-bit ALU, we see that increasing the number of regions helps increase the sensitivity in Fig. 6(a). In Fig. 6(b), we plot the sensitivity of the approach for increasing circuit sizes. Finally in Fig. 6(c), we show that increasing the number of regions also helps to keep the sensitivity nearly constant as we scale down the Trojan size. The percentage of true positives, true negatives, false positives and false negatives as obtained from the Monte Carlo simulations are presented in Table 1. We used a process point with 20% V_T variation to compute the threshold. For smaller circuits and larger Trojans the sensitivity is higher and hence, the accuracy of classification is also better.

4.2 Experimental Results

We used a custom test board with socketed Xilinx Virtex-II XC2V500 FPGAs to measure current from eight individual supply pins as well as the total current, using 0.5Ω precision current sense resistors to sense the I_{DDT} and an Agilent mixed-signal oscilloscope (100MHz, 2 Gsa/sec) to record the data. The test circuit was a 32-bit DLX processor with a 5-stage pipeline which contains the previously-described 32-bit ALU as part of its execution unit, occupying over 80% of the FPGA slices. The Trojan circuit was a 16-bit serial-in parallel-out

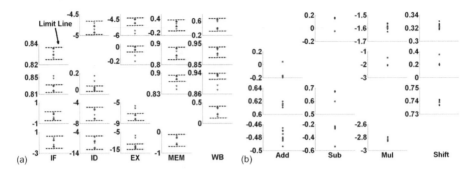

Fig. 7. Experimental results for 8 golden and 2 tampered FPGA chips. *Region slope matrix* for (a) 32-bit DLX processor; (b) 32-bit ALU. The limit lines are obtained by analyzing the 8 golden chips. The red points denote the values for the Trojan-containing test chips while the blue points denote the values for the golden chips.

shift register (sequential Trojan) occupying 0.08% of total area. We performed experiments with 10 FPGA chips from the same lot. We insert a Trojan in two of the ten chips inside the subtracter sub-region of the ALU. The *Region Slope Matrix* is constructed using the measured current values for the five pipeline stages of the DLX processor in the 10 FPGA chips. We use the 8 golden chips to determine the threshold limit and use our self-referencing approach to test 4 test chips (2 golden and 2 Trojan). As can be clearly seen from Fig. 7, the Trojan containing chips are easily identified as well as the region which contains the Trojan in both cases. Next, we repeat the procedure using test vectors which only activate the four sub-regions inside the 32-bit ALU and identify that the Trojan is located within the subtracter.

5   Conclusion

We have presented a side-channel hardware Trojan detection approach that exploits the intrinsic relationship between active-mode current among the different regions of a chip to achieve high signal-to-noise ratio in presence of process variations. We have shown that the self-referencing approach coupled with efficient vector generation provides scalability in terms of increasing process variations (thus being amenable to future scaled technologies) and increasing design size. As a by-product, such an approach also helps to localize the Trojan, which can be helpful for diagnosis. Simulation results for different circuits are supported by the experimental validation for a 32-bit DLX processor core. The approach can be easily extended to multi-core SoC, where the cores can be hierarchically partitioned into multiple regions or functional units. Another possible application involves detecting instances of re-marked chips in a lot of manufactured ICs, which pass functional testing but can cause in-field failure.

References

1. DARPA: TRUST in Integrated Circuits, TIC (2007),
   http://www.darpa.mil/MTO/solicitations/baa07-24
2. Adee, S.: The hunt for the kill switch. IEEE Spectrum 45(5), 34–39 (2008)
3. King, S., et al: Designing and implementing malicious hardware. In: LEET (2008)
4. Wolff, F., et al.: Towards Trojan-free trusted ICs: Problem analysis and detection scheme. In: DATE, pp. 1362–1365 (2008)
5. Chakraborty, R.S., Narasimhan, S., Bhunia, S.: Hardware Trojan: threats and emerging solutions. In: HLDVT (2009)
6. Rad, R., Plusquellic, J., Tehranipoor, M.: A sensitivity analysis of power signal methods for detecting hardware Trojans under real process and environmental conditions. IEEE Tran. VLSI (2010)
7. Banga, M., Hsiao, M.: A region based approach for the identification of hardware Trojans. In: HOST, pp. 40–47 (2008)
8. Adamov, A., Saprykin, A., Melnik, D., Lukashenko, O.: The problem of hardware Trojans detection in system-on-chip. In: CADSM, pp. 178–179 (2009)
9. Agrawal, D., Baktir, S., Karakoyunlu, D., Rohatgi, P., Sunar, B.: Trojan detection using IC fingerprinting. In: Symposium on Security and Privacy, pp. 296–310 (2007)
10. Rad, R., Wang, X., Tehranipoor, M., Plusquellic, J.: Taxonomy of Trojans and methods of detection for IC trust. In: ICCAD (2008)
11. Jin, Y., Makris, Y.: Hardware Trojan detection using path delay fingerprint. In: HOST (2008)
12. Chakraborty, R.S., Wolff, F., Paul, S., Papachristou, C., Bhunia, S.: MERO: A statistical approach for Hardware Trojan detection. In: CHES (2009)
13. Borkar, S., et al.: Parameter variations and impact on circuits and micro-architecture. In: DAC, pp. 338–342 (2003)
14. Papoulis, A., Pillai, S.U.: Probability, Random Variables and Stochastic Processes, 4th edn. McGraw-Hill, New York (2002)
15. Predictive Technology Model, http://www.eas.asu.edu/~ptm/

Appendix

Analysis of the Effect of Process Variations. In order to increase the hardware Trojan detection sensitivity for a large design with ultra-small Trojan, we need to amplify the Trojan effect while nullifying the impact of process variations in the side-channel parameter. There are two types of process variations [13] – *inter-die* variations and *intra-die* variations, with the latter having a systematic component and a random component. Inter-die variations are the parameter variations from one die to another on a wafer and can be modeled by a variation in the transistor threshold voltage (V_T) for the entire design. Intra-die variations are the variations within the same die which can cause different parametric variations than that predicted by inter-die variations. They have a random component which causes random variation in V_T of each transistor about the V_T of the die. There can also be a systematic component to these variations since there are spatial correlations among the V_T variations of the transistors. Fig. 8 shows the effect of the different components of process variation on the V_T of devices in

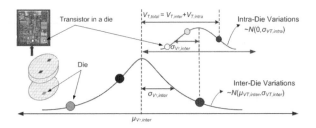

Fig. 8. The effect of process variation on device threshold voltage in an IC

an IC, where each of the "inter-die" and "intra-die" components are modeled as *normal distribution* with certain mean (μ) and standard deviation (σ).

Consider an IC that has been partitioned into N different regions, such that each region can be preferentially activated while the activity of the other partitions are minimized. Consider that the region R_i has been preferentially activated, and consider a gate $g \in R_i$. Then, the switching current of g is approximately given by $I_g = k(V_{DD} - V_{Tg})^2$, where k is a constant depending on the process and the nature of the gate, V_{DD} is the supply voltage and V_{Tg} is the threshold voltage of the i-th gate. Now, V_{Tg} can be expressed as $V_{Tg} = V_T + \Delta V_{Ti} + \Delta v_{Tg1} + \Delta v_{Tg2}$, Here, ΔV_{Ti} represents the effect of the "systematic intra-die" component of variation, and has the same value for all gates in the region R_i; Δv_{Tg1} represents the effect of the "inter-die" component of process variation, and has the same value for all gates in the IC, and Δv_{Tg2} is the effect of the "random intra-die" component of process variation, and has random values for different gates of the IC. Hence,

$$I_g = k \left[V_{DD} - (V_T + \Delta V_{Ti} + \Delta v_{Tg1} + \Delta v_{Tg2})\right]^2$$
$$= k \left[(V_{ov} - \Delta v_{Tg1})^2 + (\Delta V_{Ti} + \Delta v_{Tg2})^2 - 2(V_{ov} - \Delta v_{Tg1})(\Delta V_{Ti} + \Delta v_{Tg2})\right] \tag{6}$$

where $V_{ov} = V_{DD} - V_T$ is the *gate overdrive*. Ignoring all second order terms involving both random and systematic shifts of the threshold voltage, the above equation can be approximated by:

$$I_g \approx \underbrace{k \left[V_{ov}^2 - 2V_{ov}(\Delta v_{Tg1} + \Delta V_{Ti})\right]}_{\text{constant for each gate } g \in R_i} - \underbrace{2V_{ov}\Delta v_{Tg2}}_{\text{random for each gate } g \in R_i} \tag{7}$$

Summing the currents for all the switching gates of the region R_i, the total switching current for region R_i is:

$$I_i = \sum_{g \in R_i} I_g = kn_i \left[V_{ov}^2 - 2V_{ov}(\Delta v_{Tg1} + \Delta V_{Ti})\right] - 2V_{ov} \sum_{g \in R_i} \Delta v_{Tg2} \tag{8}$$

where n_i is the number of switching gates in region R_i. Now, the term $\sum_{g \in R_i} \Delta v_{Tg2}$ represents the sum of n_i (normally distributed) random variables, each with mean $\mu = 0$ and standard deviation σ_T (let). Hence, by the *Central Limit Theorem* [14], the term $\sum_{g \in R_i} \Delta v_{Tg2}$ is approximately normally distributed with mean $\mu = 0$ and a reduced standard deviation $\frac{\sigma_T}{\sqrt{n_i}}$. Hence, for reasonably large value of n_i, this term is approximately equal to zero, and the expression for I_i can be approximated by:

$$I_i \approx \sum_{g \in R_i} I_g = k n_i \left[V_{ov}^2 - 2V_{ov}(\Delta v_{Tg1} + \Delta V_{Ti}) \right] \tag{9}$$

Similarly, for a region R_j. the switching current is given by:

$$I_j \approx \sum_{g \in R_j} I_g = k n_j \left[V_{ov}^2 - 2V_{ov}(\Delta v_{Tg1} + \Delta V_{Tj}) \right] \tag{10}$$

Hence, the difference between the currents of regions R_i and R_j can be expressed as:

$$\begin{aligned} I_i - I_j|_{observed} &= k \left[V_{ov}^2 - 2V_{ov}\Delta v_{Tg1} \right] (n_i - n_j) - 2kV_{ov}(n_i \Delta V_{Ti} - n_j \Delta V_{Tj}) \\ &= c_1(n_i - n_j) + \underbrace{c_2(n_i \Delta V_{Ti} - n_j \Delta V_{Tj})}_{\text{due to } \textit{systematic intra-die} \text{ variation}} \end{aligned} \tag{11}$$

where c_1, c_2 are constants. If the contribution due to the intra-die systematic component is negligible, the above expression can be re-written as:

$$I_i - I_j|_{observed} \approx c_1(n_i - n_j) \qquad \text{and} \qquad I_i|_{observed} \approx c_1 n_i \tag{12}$$

Hence, the mutual *Region Slope* metric for regions R_i and R_j is

$$S_{ij,observed} = \frac{I_i - I_j}{I_i} = \frac{n_i - n_j}{n_i} \tag{13}$$

In the *nominal* case, in the absence of any process variation effects, $\Delta V_{Ti} = \Delta V_{Tj} = \Delta v_{Tg1} = \Delta v_{Tg2} = 0$; hence , $I_i - I_j|_{golden} = c_3(n_i - n_j)$, $I_i = c_3 n_i$ and

$$\boxed{S_{ij,golden} = \frac{n_i - n_j}{n_i} = S_{ij,observed}} \tag{14}$$

Similarly, it can be shown that $S_{ji,golden} = S_{ji,observed}$. This shows that under negligible *systematic intra-die* variations, the ratio of the difference in the switching currents of two regions and the current of each region should remain approximately unchanged. This equality fails to be satisfied in case one of the regions is modified by the insertion of a Trojan, because then the switching current of the gates constituting the Trojan circuit disturbs the balance. This observation is the main motivation behind using the Region Slope values for reducing the process noise. For a circuit with N regions, if we compute the Region Slope values for all pairs of regions, we obtain an $N \times N$ matrix, with zero diagonal elements. It is observed that systematic variations still cause some variations in the Region Slope values, but the effect of process variation has been reduced greatly compared to the variations in individual current values, thus giving us improved sensitivity for Trojan detection.

When Failure Analysis Meets Side-Channel Attacks

Jerome Di-Battista[1,2], Jean-Christophe Courrege[1],
Bruno Rouzeyre[2], Lionel Torres[2], and Philippe Perdu[3]

[1] Thales ITSEF,
18 Avenue Edouard Belin, 31400 Toulouse, France
jerome.dibattista@cnes.fr
[2] Université de Montpellier, Laboratoire du LIRMM,
161 rue Ada, 34095 Montpellier Cedex 5, France
[3] Centre National d'Etudes Spatiales CNES,
18 Avenue Edouard Belin, 31400 Toulouse, France

Abstract. The purpose of failure analysis is to locate the source of a defect in order to characterize it, using different techniques (laser stimulation, light emission, electromagnetic emission...). Moreover, the aim of vulnerability analysis, and particularly side-channel analysis, is to observe and collect various leakages information of an integrated circuit (power consumption, electromagnetic emission ...) in order to extract sensitive data. Although these two activities appear to be distincted, they have in common the observation and extraction of information about a circuit behavior. The purpose of this paper is to explain how and why these activities should be combined. Firstly it is shown that the leakage due to the light emitted during normal operation of a CMOS circuit can be used to set up an attack based on the DPA/DEMA technique. Then a second method based on laser stimulation is presented, improving the "traditional" attacks by injecting a photocurrent, which results in a punctual increase of the power consumption of a circuit. These techniques are demonstrated on an FPGA device.

Keywords: Side-channel, Failure analysis, Light emission, Laser stimulation, FPGA.

1 Introduction

During the last 20 years, failure analysis has become a serious concern for the electronics industry. Its purpose is to locate the source of a defect in order to characterize it, the defect being a problem linked to the environmental conditions, an intrinsic problem in the circuit, or both. More generally, failure analysis should ensure that the detected problem does not occur again [1]. The strongest constraints are the size reduction for CMOS technology components and the increasing complexity of integrated chips (several millions of gates). Currently the most used analysis tools are based on laser stimulation and light emission techniques. Concurrently, during the last 10 years, non-invasive and semi-invasive

S. Mangard and F.-X. Standaert (Eds.): CHES 2010, LNCS 6225, pp. 188–202, 2010.

techniques have received a lot of attention from the hardware security community. Among them, so-called side-channel attacks are the most popular. Different leakage sources [2][3] such as power consumption, electromagnetic field, or time response of the circuit, are correlated to the processed data. Thus, by inspecting this information, and with the help of appropriate software tools, it is possible to retrieve the secret data used in the embedded cryptography circuits, typically the cipher key. From an attacker point of view, side-channel attacks present many advantages, as most of them require only low-cost instrumentations, and they are non-destructive.

These two activities apparently different can be combined. Indeed, the failure analysis techniques can be used to extract another kind of side-channel signal or to improve existing side-channel attacks. Inversely, the vulnerability analysis can be used to extract complementary information about the circuit behavior. In this paper two examples of application, light emission and laser stimulation, are presented.

The light emission phenomenon has been mainly studied for failure analysis. Many techniques have been developed to extract and process the light emitted by the electronic components in order to localize different kinds of defects [4] (junction avalanche, oxide breakdown...). In this paper we mainly focus on the light emitted by NMOS transistors during commutation. Indeed, in [5], the author demonstrates the possibility to set up an attack based on light emission, by implementing part of an AES algorithm on a PIC16F84A microcontroller previously opened from the backside. The purpose of this attack was to recover the secret key stored in the microcontroller RAM. Using this work as a starting point, two approaches have been developed; in [6] the author demonstrates the possibility of using a low-cost system to perform the same kind of experiments on a PIC16F628 and provides some interesting results for a FPGA circuit. In parallel, we had chosen to study the Time Resolved Emission (TRE) technique which allows us to count the number of photons emitted by a transistor or by a group of transistors as a function of time, implemented on a more expensive failure analysis equipment [7]. Our purpose is to show that the extracted TRE signal can be used to gather sensitive data, such as a side-channel signal, exploitable by a statistical post-processing method (e.g. DPA or CPA).

In the same way as light emission, the techniques based on laser stimulation have been mainly developed in failure analysis [8][9]. On the one hand, the laser stimulation at a 1064 nm wavelength allows to induce a local photocurrent [10], either to detect a latch-up mechanism and inter-level shorts or to locate open circuits and direct semiconductor damage (LIVA, OBIC). On the other hand, the laser stimulation at a 1340 nm wavelength can also induce a thermal variation to detect a resistance variation by measuring the power consumption across a circuit (TIVA, OBIRCH). In [11], the author demonstrates the possibility to increase the consumption of a SRAM cell transistors in a microcontroller by applying a photocurrent (639 nm laser). Starting from these experiments, we studied the possibility to use this method to improve a side-channel attack, by reducing the number of power consumption curves necessary to perform the attack.

We experimented both methods on an FPGA Actel Proasic3 A3PE600 in flash technology (0.13μm, 7 metal layers, 600k system gates, single chip). This type of circuit offers a very high flexibility, as it is completely customizable, reconfigurable and non-volatile. These particularities make the FPGA a good test sample to be analyzed on different testing or failure analysis equipments. However FPGAs make the analysis more difficult than ASICs, as the regular structure of FPGA logic elements does not permit to localize sensitive components such as SRAMs or EEPROMs (by using for instance an optical microscope). Furthermore, the attacked microchip is in 0.13μm technology, which may complicate the measurements due to lower power supply and light emission. To overcome this problem, the acquisitions have to be performed from the backside of the chip, even though this requires a more sophisticated sample preparation [12].

2   Light Emission as a Side-Channel Signal

2.1   Background

Currently, most digital circuits are based on CMOS technology. One of the particularities of CMOS transistors is that photons are emitted during their commutation. Indeed when a current flows between the source and the drain, the electrons gain energy and accelerate due to the electrical field. The radiative "de-excitation" of the charge carriers in the pinch-off zone generates photons which are visible in the near-infrared spectral range [13]. This emission is predominant for a transition from 0 to 1. For a 1 to 0 transition the emission is usually too low to be acquired. This phenomenon produces an asymmetric light emission profile for the two transition types (0 to 1 and 1 to 0). This asymmetry can then be used to extract relevant information from the circuit.

To observe the light emitted, the chip needs to be opened either from its backside or frontside, depending on its package type. Furthermore, the light emission quality depends on the quality of the package opening process [14]. For the backside package opening, the silicon substrate is mechanically thinned down and polished. Indeed the thinning is necessary to decrease the absorption rate of the silicon substrate, and also to maximize the generation of photocarriers in the silicon [15]. On the other hand, when working on the frontside, a chemical process is used, which is easier to perform. Nevertheless, because of the increasing number of metal layers in the circuits that act as a light screen, this technique is less and less used.

The photons emitted can be collected by a specific device equipped with a high sensitivity photon sensor mounted on the optical axis of a conventional microscope. Many types of optical sensors, working with different wavelength efficiencies, can be used (CCD, InGaAs, InSb...). However, due to small transistor size and high silicon doping in the most recent technologies, at normal power supply voltage, the photon emission is maximum in the 900 nm - 1100 nm range. In this spectral range InGaAs detectors have the best quantum efficiency, as shown in Fig.1.

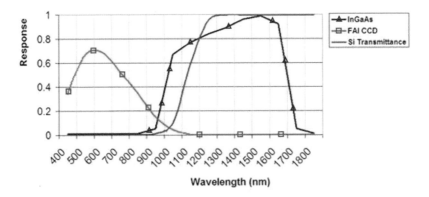

Fig. 1. Comparison of sensor technologies in relation to the silicon transmittance: response as a function of wavelength

In order to perform our experiments we identified two main complementary techniques able to produce time and spatial information: the Picosecond Imaging Circuit Analysis (PICA) and TRE techniques.

The PICA system acquires the light emitted, conserving time and space information. More precisely, the PICA sensor delivers the time and position of each photon emitted by the targeted circuit zone [16]. This technique has been initially developed to identify any functionality problem using temporal information during backside inspections [17].

The PICA system can be coupled with the TRE technique to target a single transistor or a specific zone in the circuit under inspection. The TRE can produce an histogram of the number of photons emitted as a function of time [18]. These histograms are called "TRE curves" and are shown in Fig.4.

2.2 Experimental Method

Since the light emitted depends on the operation executed, there is a straight correlation between TRE waveforms and the cryptographic calculations. This correlation can be exploited through a DPA process. The aim of the DPA is to reveal the secret keys of cryptographic devices based on a large number of power consumption traces that have been recorded during the data encryption of a cipher algorithm. The main advantage of this process is that it only requires knowledge of the cryptographic algorithm that is executed [2]. After extraction of part of a sub-key, the missing parts can be gathered by iterating the process. For our purpose, we replaced the power consumption acquisitions by light emission traces (TRE).

The acquisition system used is a Hamamatsu Tri-PHEMOS equipment [19]. This equipment is composed of an InGaAs camera coupled with a photon counting system. Thanks to this apparatus, we were able to carry out a successful measurement campaign using a Hamamatsu high performance InGaAs camera (high infrared sensitivity in the 950 nm to 1400 nm range). The optical sensor of the InGaAs camera (resolution of 640x480 with a pixel size of 20μm x 20μm) associated with a Solid Immersion Lens (SIL) allows to obtain a resolution of 300 nm and to observe a structure on a 65 nm chip. Moreover, the Tri-PHEMOS equipment is able to perform both static and dynamic light emission measurements with very high precision.

For the experiment we choose to implement part of a cipher algorithm on a FPGA device as shown in Fig.2. A specific test board was built, as shown in Fig.3. It is composed of a FPGA mechanically opened from the backside (silicon s down to 70μm), and laid upside down. In addition, a built-in potentiometer can be used to increase the FPGA core voltage (1.5V to 3V) in order to increase the light emission activity. In this experiment we performed the measurements at the typical voltage level (1.5V).

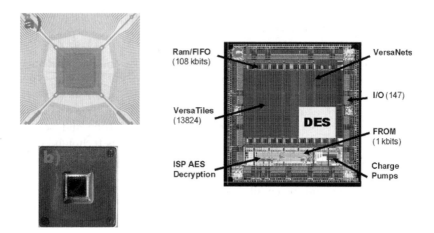

Fig. 2. Different view and informations about the FPGA Actel Proasic3e a) x-ray image b) Picture of the FPGA open from the backside c) Layout informations [20] with location of the DES implementation

The target of our attack was a fragment of a Data Encryption Standard (DES) cipher algorithm. Indeed, in order to simplify our experiment and lighten the data processing, the chosen target was the first round of the DES algorithm, and more specifically the first SBOX. Our goal was to validate the theory and the method efficiency on a small part of the DES algorithm.

Prior to these acquisitions, the light emission activity induced by the 'cryptoprocessor' needs to be localized in order to start the acquisition. This is done by a static scan, consisting in acquiring the light emitted for a few minutes in order to obtain a photon cartography of the whole FPGA. During this time, the cryptoprocessor encrypts the same message. Then the acquisition window is placed on the emissive area of the cryptoprocessor. Indeed, one asset of this method is that, if the cryptoprocessor can be turn on/off it can be easy to locate the area where it is implemented. The most relevant point is that it is usually sufficient to know the location of the cipher block in order to position a TRE acquisition

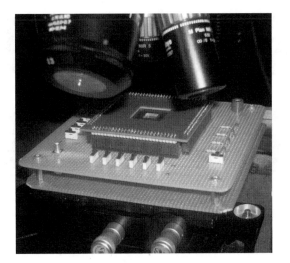

Fig. 3. FPGA test board

window on it. Furthermore, it is not necessary to know either the architecture of the algorithm, or its implementation, as the overall light emission of the cipher block is collected instead of a specific area (SBOX output, XOR operation...). It is then the data post-treatment on the TRE curves which will give us the expected results.

2.3   Results

In Fig.4 the light emission activity of the area where the cipher algorithm is implemented and the corresponding TRE curves are shown. A first message M1 (Fig.4a) is sent to the algorithm, followed by a second M2 (Fig.4b). We can notice that the variation of the input vector sent to the cipher algorithm generates a time and space variation of the emitted light, producing some sensitive differences between the TRE curves. In this way we obtain a TRE curve for each cryptographic calculation, which can then be used as a side-channel signal.

The full set of message vectors ($2^6 = 64$) is sent to the device. In order to obtain the TRE curves, each of the vectors is sent at a frequency of $10MHz$ during 20 seconds, also it has been verified that 5 seconds are sufficient. We used a longer acquisition time to ensure that the camera acquired a number of photons high enough to generate meaningful TRE curves. Each vector is sent to the FPGA in alternance with a zero message. This alternation is needed to force the transisitors to reset. When reseted transistors switch to 1, light emission happens; therefore reseting the transistors force them to emit light. This process generates a set of 64 TRE curves.

Once the TRE curves are acquired, it becomes possible to process them in order to try to extract the key. In our case, the chosen discriminant to classify the curves in the transition groups (0 to 1 or 0 to 0) is based on the chosen

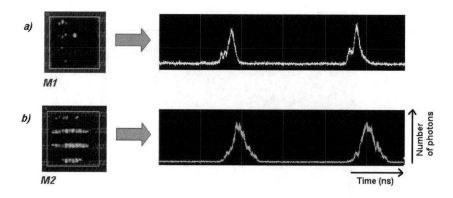

Fig. 4. Variation of TRE curves in function of the light emission activity

bit at the SBOX output, since during the acquisition we forced a reset between each message by sending a zero value. The differential curves resulting from the statistical processing on each output bit are shown in Fig.5.

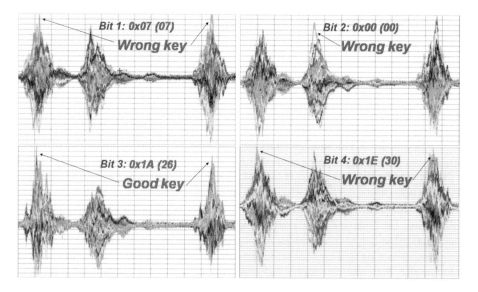

Fig. 5. Differential curves of the 64 key hypothesis for each output bit (number of photons emitted as a function of time(μs))

The attack performed on the third bit reveals the right key; however, the attacks on the first, second and fourth bits are inconclusive. On the other hand, if we sum up the four output bits to enhance the differences between the differential curves [21] we obtain the results shown in Fig.6. These results show that the curve for the right sub-key stands out. This result demonstrate, by targeting the whole cipher block with a TRE acquisition window (without a real precision), the possibility to extract a sub-key by using light emission leakage. In the next section, we propose to demonstrate that laser stimulation coupled with the DPA method could be an innovative technique for side-channel analysis.

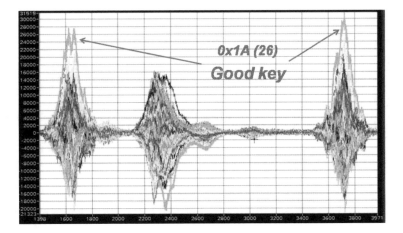

Fig. 6. Sum of differential curves for each output bit (number of photons emitted as a function of time(μs))

3   Laser Stimulation to Improve Side-Channel Attacks

3.1   Background

The photoelectric laser stimulation is a failure analysis technique that uses a scanning laser beam to induce a current flow. This one can be collected and analyzed to generate images that represent the semiconductor sample properties [10]. Indeed, when the laser beam scans the surface of the sample, some electrons into the conduction band are excited thanks to the 'single-photon absorption' phenomenon. In the single-photon absorption process, a single photon excites one conduction band electron. This can only occur if that single photon carries enough energy to overcome the band gap of the semiconductor (1.2 eV for Silicon) and provide the electron with enough energy to make it jump into the conduction band. The creation of charge carriers by excitation of the semiconductor with an optical beam results in a current flow that can be collected and used for imaging. The IC current variations induced by the laser beam is converted into a contrast variation to form an image [8].

One limitation of this technique is that for modern integrated circuits, it is hard to transmit light uniformly to the semiconductor itself. This non-uniform transmission of light is caused by the presence of several metal layers and other materials above the semiconductor. In such instances, one solution is to perform the imaging from the backside through the substrate. However the spatial resolution is limited due to a compromise between being able to transmit the beam through the substrate, and allowing the beam to be absorbed by the semiconductor for the generation of electron-hole pairs that are measurable as a current, as shown in Fig.7.

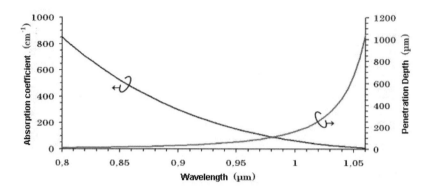

Fig. 7. Absorption coefficient and penetration depth as a function of wavelength [22]

The laser stimulation can be performed by a specific device equipped with a laser beam mounted on the optical axis of a conventional microscope. Two types of laser beams, working with different wavelength, can be used: 1064 nm to induce a photocurrent effect and 1360 nm to induce a thermal effect (and a small photocurrent effect as well). However, the present experiment involves the use of the photocurrent effect, thus the 1064 nm (or less) wavelength is chosen.

3.2   Experimental Method

The aim of this experiment is to extend the method described by Skorobogatov [11] to a DPA attack on a DES cipher algorithm, implemented on a FPGA. With the help of a scanning laser equipment used in failure analysis activities, it becomes possible to scan a chosen area into the FPGA corresponding to the location where a critical function (SBOX, end of round, XOR) of the DES is implemented. Theoretically, the laser induces a current on the chosen scanning area. This additional current should increase the consumption of the circuit during the algorithm encryption, and thus improve the attack by reducing the number of power consumption acquisitions.

The light source used is the Meridian I acquisition system from DCG Systems [23], equipped with a laser scanning microscope system (LSM) with two different lasers for induced current and thermal stimulation (1064 nm and 1340 nm). For the experiment, we implemented a full DES cipher algorithm on the same FPGA device (Actel Proasic3) as for the light emission experiment. This FPGA is opened from the backside and mounted on the same specific test board. The main interest of the FPGA implementation is that it is possible to choose the area where the different DES sub-blocks are implemented on the FPGA programmation grid, as shown in Fig.8.

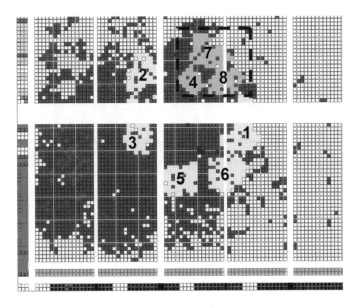

Fig. 8. Location of the DES SBOX on the FPGA programmation grid

The first challenge is the choice of the power laser source, since it is necessary to ensure that the power is neither too low to generate enough photocurrent nor, too strong to avoid a fault injection. For the considerate wavelength, several tests revealed that the maximum power at which the algorithm generates errors is 15-18 mW, therefore a laser power of 10-11 mW is chosen. The second challenge is to select an area for the laser scan, for instance SBOX area. After several trials we selected the area including the 4th, 7th and 8th SBOX (dotted in Fig.8), offering the possibility to scan three SBOX at the same time with a $1\mu m$ laser spot size at 20x zoom lens. Once these steps are complete, a first DPA attack is performed without any laser scan in order to have a reference, followed by a second DPA attack with the laser scan on the area previously identified.

3.3 Results

During the first acquisition process, without any scanning laser, 16000 random messages are sent. The differential process results, considerate as a reference, are shown in the table in Fig.9. The table details the attack results on each bit of each SBOX. The discriminant used is on the one hand a DPA chosen bit at the End of Round (first four rows), and on the other hand a CPA [24] Hamming Weight at the End of round (last row).

A second acquisition process with a scanning laser (with a scan frequency of 200 khz) is then performed, with the same random messages. In Fig.10 the table shows the comparison between the numbers of power traces necessary to perform the attack with and without the laser scan. The discriminant used

	Sbox 1	Sbox 2	Sbox 3	Sbox 4	Sbox 5	Sbox 6	Sbox 7	Sbox 8
Bit 0	YES	NO	YES	YES	NO	YES	NO	NO
Bit 1	NO	YES	YES	YES	YES	YES	NO	NO
Bit 2	YES	YES	YES	NO	NO	NO	YES	YES
Bit 3	YES	YES	YES	NO	NO	NO	YES	YES
CPA	YES	YES	YES	NO	NO	YES	YES	NO

Fig. 9. DPA result laser stimulation at the end of the round without (16000 curves)

is again a DPA chosen bit at the End of Round (first eight columns), or a CPA Hamming Weight at the End of round (last two columns). In each case the number of curves necessary to obtain the right sub-key is shown.

	Bit 0		Bit 1		Bit 2		Bit 3		CPA	
Laser state	OFF	ON	OFF	ON	OFF	ON	OFF	ON	OFF	ON
SBOX 4	~11000	~6500	~11500	~6500	NO	~9000	NO	~9500	NO	YES
SBOX 5	NO	~14500	~10000	~9500	NO	NO	NO	NO	NO	NO
SBOX 6	~11500	~9500	~10000	~7500	NO	NO	NO	~12500	YES	YES
SBOX 7	NO	~9000	NO	~8500	~10500	~6500	~11500	~6500	YES	YES
SBOX 8	NO	NO	NO	NO	~12000	~9500	~13500	~10000	NO	NO

Fig. 10. Comparison between both DPA results with and without laser stimulation and numbers of curves necessary to perform the attack - (laser 1064 nm / power 11 mW)

These results highlight several interesting facts. First, the number of curves required to perform a successful attack are decreased by approximately half on bits (0,1) of SBOX 4 and bits (2,3) of SBOX 7. Moreover, the attack on bits (2,3) of SBOX 4 and bits (0,1) of SBOX 7 with the laser scan allows to recover the good sub-key, whereas the attack on the same bits without laser stimulation are unsuccessful. Likewise, the CPA on SBOX 4 is successful. These results suggest that the laser has a real influence on the power consumption of the circuits and in particular on the targeted SBOX. Fig.11 highlights the amplitude differences between differential curves with and without a scanning laser (16000 power consumption curves were used). This comparison underlines the influence of laser scans on the efficiency of the attack. Nevertheless the 8th SBOX, although scanned by the laser, is not so impacted (this fact is not yet explained). On the opposite, bits (0,1,3) of SBOX 6 and bits (0,1) of SBOX 5 seems to react to the laser stimulation in a lesser extent. This could be explained

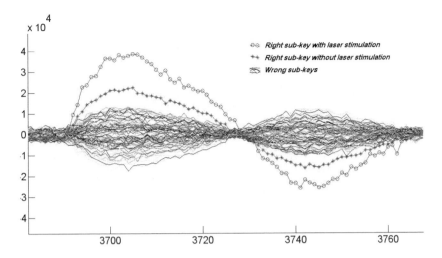

Fig. 11. Amplitude comparison between differential curves on the right key with and without laser stimulation (DPA in 16000 curves on bit 0 of SBOX 4)

by a spreading of photocurrent neighboring SBOX, or by an indirect influence of the scanning laser on the interconnection lines between the two SBOX (e.g. 6 and 7).

4   Conclusion

These different experiments show how failure analysis tools could be effectively applied to perform or enhance side-channel attacks. The light emission techniques allow to localize the different functions implemented on a circuit. With only partial knowledge of the circuit design and by using the TRE technique, light emission enables determination of the internal behavior of the circuit functions. Using the DPA method, we have shown that a differential light emission analysis allows retrieval of the cipher sub-key from a fraction of the DES algorithm implemented on an FPGA that uses a 130 nm technology. Many developments can be carried out based on this method, and multiple perspectives can be considered. First, the efficiency of this technique should be compared to other side-channel methods to further highlight the specific contributions of this method. Moreover, in the experiments reported here, only time information is used. Space information, which is also available, offers the opportunity to refine the process and to improve the method. On the other hand, some countermeasures for this type of attacks have to be developed. A "natural" one is certainly the lack of resolution of the sensors versus the latest CMOS technologies (45 nm or less). The light emission profiles are yet to be investigated for these technologies. A countermeasure, for FPGA devices, could be a dynamic reconfiguration to change the light emission profile, or the insertion of sensors inside the package to detect its opening. We can also notice that the number of TRE curves that need to be acquired to break this type of unprotected implementation is much higher than those for EMA and DPA.

The second experiment based on a laser stimulation technique allows to partially increase the power consumption of a circuit, by scanning a specific area during the encryption of the cipher algorithm. This way it is possible to significantly reduce the number of curves necessary to perform a DPA attack. For this technique also many developments and perspective can be considered. These recent results require further investigation in order to specify how the laser method could be used. For example it could be interesting to repeat several DPA attacks by scanning the SBOX individually (or all the SBOX chained) to establish a comparative statement detailing how the laser method improves the attack. It would be also interesting to apply this method on a secure cipher algorithm, for instance that uses dual rail implementation [25]. Laser scans could be used to induce an unbalance power consumption and thus extract the sub-key. Concerning the method itself, it would also be interesting to reproduce the attack using a laser with a wavelength below 1064 nm to increase the photogeneration of free carriers, or to attempt to use other light source (such as halogen light) instead of monowave light laser source. The using of a static laser (without scan mode) to continuously illuminate the area of interest would also be interesting.

The main constraint for these methods is the backside opening of the component and more particularly the silicon thinning step, a process quite hard to control. In any case, at the present time, due to the price of the equipments used in these experiments (beyond 2M euros for the Tri-PHEMOS and 500K euros for the Meridian), the cost of these attack-enhancing method appears to be very high. An interesting point for future research will be to re-do these experiments on a low-cost sytems to validate the real benefit of them, for example in [6] [11] the author shows the possibility to design a low-cost system based on PMT detector coupled with a CCD camera to perform light emission experiments and a 639 nm laser coupled with a CCD camera to perform laser experiments. However, proof that failure analysis and side-channel attacks are compatible has been provided, and further studies are currently under way based on these promising results.

Acknowledgment

We would like to thank the Hamamatsu Photonics company that allowed us to perform a successful measurement campaign on their Tri-PHEMOS system, and also the Hamamatsu team for their technical support. Thanks also to Fabien Battistela and Kevin Sanchez for their precious advice and discussions.

References

1. Perdu, P.: Contribution a l'Etude et au Developpement de Techniques de Localisation de Defauts dans les Circuits Intgrs VLSI, Ph.D. diss., Bordeaux University (2001)
2. Kocher, P., Jaffe, J., Jun, B.: Differential Power Analysis. In: Wiener, M. (ed.) CRYPTO 1999. LNCS, vol. 1666, pp. 388–397. Springer, Heidelberg (1999)

3. Quisquater, J.-J., Samyde, D.: ElectroMagnetic Analysis (EMA): Measures and Counter-Measures for Smart Cards. In: Attali, S., Jensen, T. (eds.) E-smart 2001. LNCS, vol. 2140, pp. 200–210. Springer, Heidelberg (2001)

4. Barton, D.L., Tangyunyong, P., Soden, J.M., Liang, A.Y., Low, F.J., Zaplatin, A.N., Shivanandan, K., Donohoe, G.: Infrared Light Emission from Semiconductor Devices. In: 22th International Symposium for Testing and Failure Analysis, pp. 9–17 (1996)

5. Ferrigno, J., Hlavac, M.: When AES Blinks: Introducing Optical side-channel. IET Information Security 2(3), 94–98 (2008)

6. Skorobogatov, S.: Using Optical Emission Analysis for Estimating Contribution to Power Analysis. In: 6th Workshop on Fault Diagnosis and Tolerance in Cryptography (FDTC), pp. 111–119. IEEE-CS Press, Los Alamitos (2009)

7. Di-Battista, J., Perdu, P., Courrege, J.C., Rouzeyre, B., Torres, L., Lionel: Light emission analysis on FPGA: a new side channel possibility. In: 7th Workshop on Cryptographic Architectures Embedded in Reconfigurable Devices, CryptArchi 2009 (2009)

8. Stevens, K.C., Wilson, T.J.: Locating IC Defects in Process Monitors and Test Structures Using Optical Beam Induced Current. Microelectronic Engineering 12, 397–404 (1990)

9. Soelkner, G.: Optical beam testing and its potential for electronic device characterization. Microelectronic Engineering 24, 341–353 (1994)

10. Fritz, J., Lackman, R.: Optical Beam Induced Currents in MOS Transistors. Microelectronic Engineering 12, 381–388 (1990)

11. Skorobogatov, S.: Optically Enhanced Position-Locked Power Analysis. In: Goubin, L., Matsui, M. (eds.) CHES 2006. LNCS, vol. 4249, pp. 61–75. Springer, Heidelberg (2006)

12. Desplats, R., Beaudoin, F., Perdu, P.: Chip Unzip for Backside Sample Preparation. In: 27th International Symposium for Testing and Failure Analysis, pp. 179–187 (2001)

13. Wallinger, T.: Characterization of Device Structure by Spectral Analysis of Photoemission. In: 17th International Symposium for Testing and Failure Analysis, pp. 325–334 (1991)

14. Barton, D.L., Bernhard-Hofer, K., Cole Jr., E.I.: Flip-Chip and Backside techniques. Microelectronics Reliability 39, 721–730 (1999)

15. Baudouin, F.: Localisation de defaut par la face arriere des circuits integres. Ph.D. diss., Bordeaux University, 38–40 (2002)

16. Tsang, J.C., Kash, J.A., Vallett, D.P.: Picosecond imaging circuit analysis. IBM Journal of Research and Development 44, 583–603 (2000)

17. McManus, M.K., Kash, J.A., Steen, S.E., Polansky, S., Tsang, J.C., Knebel, D.R., Huott, W.: Huott: PICA: Backside Failure Analysis of CMOS Circuits Using Picosecond Imaging Circuit Analysis. Microelectronic Reliability 40, 1353–1358 (2000)

18. Kolzer, J., Boit, C., Dallmann, A., Deboy, G., Otto, J., Weinmann, D.: Quantitative Emission Microscopy. Journal of Applied Physics 71(11), R23–R41 (1992)

19. Hamamatsu Photonics, http://www.hamamatsu.com/

20. Actel Proasic3 Handbook: 144, http://www.actel.com/products/pa3/docs.aspx

21. Bevan, R., Knudsen, E.: Ways to Enhance Differential Power Analysis. In: Lee, P.J., Lim, C.H. (eds.) ICISC 2002. LNCS, vol. 2587, pp. 327–342. Springer, Heidelberg (2003)

22. Sanchez, K.: Développement et applications de techniques d'analyse par stimulation dynamique laser pour la localisation de défauts et le diagnostic de circuits intégrés. Ph.D. diss., Bordeaux University (2007)
23. DCG systems, http://www.dcgsystems.com/
24. Brier, E., Clavier, C., Oliver, F.: Correlation Power Analysis with a Leakage Model. In: Joye, M., Quisquater, J.-J. (eds.) CHES 2004. LNCS, vol. 3156, pp. 16–29. Springer, Heidelberg (2004)
25. Bystrov, A., Yakovlev, A., Sokolov, D., Murphy, J.: Design and Analysis of Dual-Rail Circuits for Security Applications. IEEE Transactions on Computers 54(4), 449–460 (2005)

Fast Exhaustive Search for Polynomial Systems in $\mathbb{F}_2$

Charles Bouillaguet[1], Hsieh-Chung Chen[2], Chen-Mou Cheng[3],
Tung Chou[3], Ruben Niederhagen[3,4], Adi Shamir[1,5], and Bo-Yin Yang[2]

[1] Ecole Normale Supérieure, Paris, France
charles.bouillaguet@ens.fr
[2] Institute of Information Science, Academia Sinica, Taipei, Taiwan
{kc,by}@crypto.tw
[3] National Taiwan University, Taipei, Taiwan
{doug,blueprint}@crypto.tw
[4] Technische Universiteit Eindhoven, The Netherlands
ruben@polycephaly.org
[5] Weizmann Institute of Science, Israel
adi.shamir@weizmann.ac.il

Abstract. We analyze how fast we can solve general systems of multivariate equations of various low degrees over $\mathbb{F}_2$; this is a well known hard problem which is important both in itself and as part of many types of algebraic cryptanalysis. Compared to the standard exhaustive search technique, our improved approach is more efficient both asymptotically and practically. We implemented several optimized versions of our techniques on CPUs and GPUs. Our technique runs more than 10 times faster on modern graphic cards than on the most powerful CPU available. Today, we can solve 48+ quadratic equations in 48 binary variables on a 500-dollar NVIDIA GTX 295 graphics card in 21 minutes. With this level of performance, solving systems of equations supposed to ensure a security level of 64 bits turns out to be feasible in practice with a modest budget. This is a clear demonstration of the computational power of GPUs in solving many types of combinatorial and cryptanalytic problems.

Keywords: multivariate polynomials, solving systems of equations, exhaustive search, parallelization, Graphic Processing Units (GPUs).

Extended Version of this paper is at eprint.iacr.org/2010/313.

1 Introduction

Solving a system of m nonlinear polynomial equations in n variables over $\mathbb{F}_q$ is a natural mathematical problem that has been investigated by various research communities. The cryptographers are among the interested parties since an NP-complete problem whose random instances seem hard could be used to design cryptographic primitives, as witness the development of multivariate cryptography in the last few decades, using one-way trapdoor functions such as HFE, SFLASH, and QUARTZ [12,20,21], as well as stream ciphers such as QUAD [4]. Conversely, in "algebraic cryptanalysis" one distills from a cryptographic primitive a system of multivariate polynomial equations

S. Mangard and F.-X. Standaert (Eds.): CHES 2010, LNCS 6225, pp. 203–218, 2010.

with the secret among the variables. This does not break AES as first advertised, but does break KeeLoq [11], for a recent example, and find a faster collision on 58-round SHA-1 [24].

Since the pioneering work by Buchberger [9], Gröbner-basis techniques have been the most prominent tool for this problem, especially after the emergence of faster algorithms such as $\mathbf{F_4}$ or $\mathbf{F_5}$ [15,16], which broke the first HFE challenge [17]. The cryptographic community independently rediscovered some of the ideas underlying efficient Gröbner-basis algorithms as of the XL algorithm [13] and its variants. They also introduced techniques to deal with special cases, particularly that of sparse systems [1,23].

In this paper we take a different path, namely improving the standard and seemingly well-understood exhaustive search algorithm. When the system consists of n randomly chosen quadratic equations in n variables, all the known solution techniques have exponential complexity. In particular, Gröbner-basis methods have an advantage on very overdetermined systems (with many more equations than unknowns) and systems with certain algebraic "weaknesses", but were shown to be exponential on "generic" enough systems in [2,3]. In addition, the computation of a Gröbner basis is often a memory-bound process; since memory is more expensive than time at the scale of interest, such sophisticated techniques can be inferior in practice when compared to simple testing of all the possible solutions, which uses almost no memory.

For "generic" quadratic systems, experts believe [2,25] that Gröbner basis methods will go up to degree D_0, which is the minimum possible D where the coefficient of t^D in $(1+t)^n(1+t^2)^{-m}$ goes negative, and then require the solution of a system of linear equations with $T \gtrsim \binom{n}{D_0-1}$ variables. This will take at least $\text{poly}(n) \cdot T^2$ bit-operations, assuming we can afford a sufficiently large amount of memory and that we can solve such a linear system of equations with non-negligible probability in $O(N^{2+o(1)})$ time for N variables. For example, if we assume we can operate a Wiedemann solver on a $T \times T$ submatrix of the extended Macaulay matrix of the original system, then the polynomial is $3n(n-1)/2$. When $m = n = 200$, $D_0 = 25$, making the value of T exceeds 2^{102}; even taking into consideration guessing before solving [6,26], we can still easily conclude that Gröbner-basis methods would not outperform exhaustive search in the practically interesting range of $m = n \le 200$.

The questions we address are therefore: how far can we go, on both theoretical and practical sides, by pushing exhaustive search further? Is it possible to design more efficient exhaustive search algorithms? Can we get better performance using different hardware such as GPUs? Is it possible to solve *in practice*, with a modest budget, a system of 64 equations in 64 unknowns over $\mathbb{F}_2$? Less than 15 years ago, this was considered so difficult that it even underlied the security of a particular signature scheme [19]. Intuitively, some people may consider an algebraic attack that reduces a cryptosystem to 64 equations of degree 4 in 64 $\mathbb{F}_2$-variables to be a successful practical attack. However, the matter is not that easily settled because the complexity of a naïve exhaustive search algorithm would actually be *much higher* than 2^{64}: simply testing all the solutions in a naïve way results in $2 \cdot \binom{64}{4} \cdot 2^{64} \approx 2^{84}$ logical operations, which would make the attack hardly feasible even on today's best available hardware.

Our Contribution. Our contribution is twofold. On the theoretical side, we present a new type of exhaustive search algorithm which is both asymptotically and practically faster than existing techniques. In particular, we show that finding *all* zeroes of a single degree-d polynomial in n variables requires just $d \cdot 2^n$ bit operations. We then extend this technique and show how to find all the common zeroes of m random quadratic polynomials in $\log_2 n \cdot 2^{n+2}$ bit operations, which is only slightly higher. Surprisingly, this complexity is *independent of the number of equations* m.

On the practical side, we have implemented our new algorithms on x86 CPUs and on NVIDIA GPUs. While our CPU implementation is fairly optimized using vector instructions, our GPU implementation running on one single NVIDIA GTX 295 graphics card runs up to 13 times faster than the CPU implementation using all four cores of an Intel quad-code Core i7 at 3 GHz, one of the fastest CPUs currently available. Today, we can solve 48+ quadratic equations in 48 binary variables using just an NVIDIA GTX 295 graphics card in 21 minutes. This device is available for about $500. It would be 36 minutes for cubic equations and two hours for quartics. The 64-bit signature challenge [19] can thus be broken with 10 such cards in 3 months, using a budget of $5000. Even taking into account Moore's law, this is still quite an achievement.

Table 1. Performance results for $n = 48$ and projected budgets for solving $n = 64$ in one month

Time (minutes)			Testing platform				#cores	est. cost
$d = 2$	$d = 3$	$d = 4$	GHz	Arch.	Name	USD	(#used)	(USD)
1217	2686	3191	2.2	K10	Phenom 9550	120	4(1)	54,000
1157	1992	2685	2.3	K10+	Opteron 2376	184	4(1)	113,316
142	240	336	2.3	K10+	Opteron 2376×2	368	8(8)	
780	1364	1819	2.4	C2	Xeon X3220	210	4(1)	60,720
671	1176	1560	2.83	C2+	Core2 Q9550	225	4(1)	55,575
179	294	390	2.83	C2+	Core2 Q9550	225	4(4)	
761	1279	1856	2.26	Ci7	Xeon E5520	385	4(1)	78,720
95	154	225	2.26	Ci7	Xeon E5520×2	770	8(8)	
41	73	271	1.3	G200	GTX 280	n/a	240	n/a
21	36	126	1.25	G200	GTX 295	500	480	15,500

In contrast, the implementation of F_4 in MAGMA-2.16, often cited as the best Gröbner-basis solver *commercially* available today, will completely use up 64 GB of RAM in solving just 25 cubic equations in as many $\mathbb{F}_2$-variables. We have also tested it with overdefined systems, for which Gröbner-basis algorithms are known to work better. While it does not run out of memory, the results are not satisfying: 2.5 hours to solve 20 cubic equations in 20 variables, half an hour for 45 quadratic equations in 30 variables, and 7 minutes for 60 quadratic equations in 30 variables on one 2.2-GHz Opteron core. Some very recent improvements on Gröbner-basis solvers have reported speed-up over MAGMA F_4 of two- to five-fold [10]. However, even with such significant improvements, Gröbner-basis solvers do not seem to be able to compete with exhaustive search algorithms in this range, as each of the above is solved in a split second using negligible amount of memory on the same CPU by the latter.

Implications. The new exhaustive search algorithm can be used as a black box in cryptanalysis that needs to solve quadratic equations. This includes, for instance, several algorithms for the Isomorphism of Polynomials problem [7,22], as well as attacks that rely on such algorithms, e.g., [8].

We also show with a concrete example that (relatively simple) computations requiring 2^{64} operations can be easily carried out in practice with readily available hardware and a modest budget. Lastly, we highlight the fact that GPUs have been used successfully by the cryptographic community to obtain very efficient implementations of combinatorial algorithms or cryptanalytic attacks, in addition to the more numeric-flavored cryptanalysis algorithm demonstrated by the implementation of the ECM factorization algorithm on GPUs [5].

Organization of the Paper. Section 2 establishes a formal framework of exhaustive search algorithms including useful results on Gray Codes and derivatives of multivariate polynomials. Known exhaustive search algorithms are reviewed in Section 3. Our algorithm to find the zeroes of a single polynomial of any degree is given in Section 4, and it is extended to find the common zeroes of a collection of polynomials in Section 5. Section 6 describes the two platforms on which we implemented the algorithm, and Section 7 describes the implementation and performance evaluation results.

2   Generalities

In this paper, we will mostly be working over the finite vector space $(\mathbb{F}_2)^n$. The canonical basis is denoted by $(e_0, \ldots, e_{n-1})$. We use $\oplus$ to denote addition in $(\mathbb{F}_2)^n$, and $+$ to denote integer addition. We use $i \ll k$ (resp. $i \gg k$) to denote binary left-shift (resp. right shift) of the integer i by k bits.

Gray Code. Gray Codes play a crucial role in this paper. Let us denote by $b_k(i)$ the index of the k-th lowest-significant bit set to 1, or -1 if the hamming weight of i is less than k. For example, $b_k(0) = -1, b_1(1) = 0, b_1(2) = 1$ and $b_2(3) = 1$.

Definition 1. $\text{GRAYCODE}(i) = i \oplus (i \gg 1)$.

Lemma 1. *For* $i \in \mathbb{N}$*:* $\text{GRAYCODE}(i + 1) = \text{GRAYCODE}(i) \oplus e_{b_1(i+1)}$.

Derivatives. Define the $\mathbb{F}_2$ *derivative* $\frac{\partial f}{\partial i}$ of a polynomial with respect to its i-th variable as $\frac{\partial f}{\partial i} : \mathbf{x} \mapsto f(\mathbf{x} + e_i) + f(\mathbf{x})$. Then for any vector $\mathbf{x}$, we have:

$$f(\mathbf{x} + e_i) = f(\mathbf{x}) + \frac{\partial f}{\partial i}(\mathbf{x}) \qquad (1)$$

If f is of total degree d, then $\frac{\partial f}{\partial i}$ is a polynomial of degree $d - 1$. In particular, if f is quadratic, then $\frac{\partial f}{\partial i}$ is an affine function. In this case, it is easy to isolate the constant part (which is a constant in $\mathbb{F}_2$) : $c_i = \frac{\partial f}{\partial i}(0) = f(e_i) + f(0)$. Then, the function $\mathbf{x} \mapsto \frac{\partial f}{\partial i}(\mathbf{x}) + c_i$ is by definition a linear form and can be represented by a vector $D_i \in (\mathbb{F}_2)^n$. More precisely, we have $D_i[j] = f(e_i + e_j) + f(e_i) + f(e_j) + f(0)$. Then equation (1) becomes:

$$f(\mathbf{x} + e_i) = f(\mathbf{x}) + D_i \cdot \mathbf{x} + c_i \qquad (2)$$

Enumeration Algorithms. We are interested in *enumeration algorithms, i.e.,* algorithms that evaluate a polynomial f over all the points of $(\mathbb{F}_2)^n$ to find its zeroes. Such an enumeration algorithm is composed of two functions: INIT and NEXT. INIT(f, x_0, k_0) returns a *State* containing all the information the enumeration algorithm needs for the remaining operations. The resulting State is configured for the evaluation of f over $x_0 \oplus (\text{GRAYCODE}(i) \ll k_0)$, for increasing values of i. NEXT$(State)$ advance to the next value and updates *State*. Three values can be directly read from the state: *State*.x, *State*.y and *State*.i. These are linked at all times by $State.\text{y} = f(State.\text{x})$, $State.\text{x} = x_0 \oplus (\text{GRAYCODE}(State.i) \ll k_0)$, NEXT$(State).i = State.i+1$. Finding all the zeroes of f is then achieved with the loop shown below.

```
1: procedure ZEROES(f)
2:     State ← INIT(f, 0, 0)
3:     for i from 0 to 2^n − 1
4:         if State.y = 0 then State.x is a zero of f
5:         NEXT(State)
6:     end for
7: end procedure
```

3 Known Techniques for Quadratic Polynomials

We briefly discuss the enumeration techniques known to the authors.

Naïve Evaluation. The simplest way to implement an enumeration algorithm is to evaluate the polynomial f from scratch at each point of $(\mathbb{F}_2)^n$. This requires two AND per quadratic monomial, and (almost) as many XORs. Since the evaluation takes place many times for the same f with different values of the variables, we will usually assume that the polynomial can be *hard-coded, i.e.,* that it is not necessary to include the terms for which $a_{ijk} = 0$. Each call to NEXT would then require at most $n(n + 1)$ bit operations, half-AND and half-XOR (not counting the cost of enumerating $(\mathbb{F}_2)^n$, *i.e.,* incrementing a counter). This can be improved a bit, by factoring out the monomials:

$$f(\mathbf{x}) = \sum_{i=0}^{n-1} x_i \cdot \left(\sum_{j=i}^{n-1} a_{ij} \cdot x_j \right) + c \tag{3}$$

The bit-operation count falls down to $n(n + 3)/2$, and in general for degree-d polynomials to a sum dominated by $\binom{n}{d}$. This method is simple but not without its advantages, chiefly (a) insensitivity to the order in which the points of $(\mathbb{F}_2)^n$ are enumerated, and (b) we can bit-slice and get a speed up of nearly ω, where ω is the maximum width of the CPU logical instructions.

The Folklore Differential Technique. It was pointed out in Sec. 2 that once $f(\mathbf{x})$ is known, computing $f(\mathbf{x} + e_i)$ amounts to compute $\frac{\partial f}{\partial i}(\mathbf{x})$, and this derivative happens to be a linear function which can be efficiently evaluated by computing a vector-vector product and a few scalar additions. This strongly suggests to evaluate f on $(\mathbb{F}_2)^n$

using a *Gray Code*, *i.e.*, an ordering of the elements of $(\mathbb{F}_2)^n$ such that two consecutive elements differ in only one bit. This leads to the algorithm shown below.

1: **function** INIT(f, _, _)
2:    $i \leftarrow 0$
3:    $\mathbf{x} \leftarrow 0$
4:    $\mathbf{y} \leftarrow f(0)$
5:    **For all** $0 \leq k \leq n-1$,
initialize c_k and D_k
6: **end function**

(a) Initialize

1: **function** NEXT($State$)
2:    $i \leftarrow i+1$
3:    $k = b_1(i)$
4:    $\mathbf{z} \leftarrow \text{VECTORVECTORPROD}(D_k, \mathbf{x}) \oplus c_k$
5:    $\mathbf{y} \leftarrow \mathbf{y} \oplus \mathbf{z}$
6:    $\mathbf{x} \leftarrow \mathbf{x} \oplus e_k$
7: **end function**

(b) Update

We believe this technique to be folklore, and in any case it appears more or less explicitly in the existing literature. Each call to NEXT requires n ANDs, as well as $n + 2$ XORs, which makes a total bit operation count of $2(n+1)$. This is about $n/4$ times less than the naive method. Note that when we describe an enumeration algorithm, the variables that appear inside NEXT are in fact implicit functions of $State$. The dependency has been removed to lighten the notational burden.

4   A Faster Recursive Algorithm for Any Degree

We now describe one of the main contributions of this paper, a new algorithm which is both asymptotically and practically faster than standard exhaustive search in enumerating the solutions of one polynomial equation, as summarized by Theorem 1 below.

Theorem 1. *All the zeroes of a single multivariate polynomial f in n variables of degree d can be found in essentially $d \cdot 2^n$ bit operations (plus a negligible overhead), using n^{d-1} bits of read-write memory, and accessing n^d bits of constants, after an initialization phase of negligible complexity $\mathcal{O}\left(n^{2d}\right)$.*

The proof is given in the full version.

Construction of the Recursive Enumeration Algorithm. We will construct an enumeration algorithm in two stages. First, if f is of degree 0, then we only need to "enumerate" through all vectors by updating with $\mathbf{x} \leftarrow \mathbf{x} \oplus e_{b_1(i)}$ at the i-th step.

When f is of higher degree, we need a little more effort. The main idea is that in the folklore differential algorithm of Sec. 3, the computation of $\mathbf{z}$ essentially amounts to evaluate $\frac{\partial f}{\partial k}$ on something that looks like a Gray Code. We may then use the enumeration algorithm recursively on $\frac{\partial f}{\partial k}$, since it is a polynomial of strictly smaller degree. The resulting algorithm is shown below.

It is not difficult to see that the complexity of NEXT is $\mathcal{O}(d)$, where d is the degree of f. The temporal complexity of INIT is n^d times the time of evaluating f, which is itself upper-bounded by n^d and its spatial complexity is also of order $\mathcal{O}\left(n^d\right)$. This means that the complexity of the algorithm is $\mathcal{O}\left(d \cdot 2^n + n^{2d}\right)$. When $d = 2$, this is about n times faster than the algorithm described in Sec. 3.

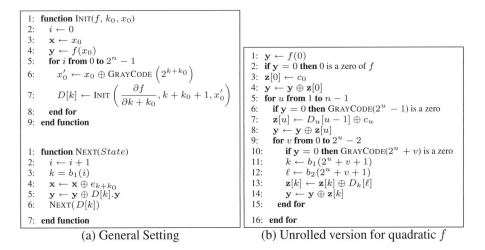

(a) General Setting   (b) Unrolled version for quadratic f

5 Common Zeroes of Several Multivariate Polynomials

We will use several time the following simple idea: all the techniques we discussed above perform a sequence of operations that is independent of the coefficients of the polynomials. Therefore, m instances of (say) algorithm in Sec. 4 could be run in parallel on $f_1, \ldots, f_m$. All the parallel runs would execute the same instruction on different data, making it efficient to implement on vector or SIMD architectures. In each iteration of the main loop, it is easy to check if *all* the polynomials vanished on the current point of $(\mathbb{F}_2)^n$. Evaluating all the m (quadratic) polynomials in parallel using algorithm in Sec. 4 would take $2m2^n$ bit operations. The point of this section is that it is possible to do much better than this.

Note that for the sake of simplicity, we limit our discussion to the case of quadratic polynomials (this case being the most relevant in practice). Our objective is now to show the following result.

Theorem 2. *The common zeroes of m (random) quadratic polynomials in n variables can be found after having performed in expectation $\log_2 n \cdot 2^{n+2}$ bit operations.*

We sketch a proof (a complete one given in the extended version) to the theorem. Let us introduce a useful notation. Given an ordered set U, we denote the common zeroes of $f_1, \ldots, f_m$ belonging to U by $Z([f_1, \ldots, f_m], U)$. Let us also denote $Z_0 = (\mathbb{F}_2)^n$, and $Z_i = Z([f_i], Z_{i-1})$. It should be clear that $Z = Z_m$ is the set of common zeroes of the polynomials, and therefore the object we wish to obtain.

The key idea is to compute Z_k using k parallel runs of algorithm in Sec. 4, and then computing $Z_{k+1}, \ldots, Z_m$ one by one. Computing Z_k requires $2k2^n$ bit operations. It then remains to compute Z_m from Z_k, and to find the best possible value of k. If we use the naïve evaluation strategy with early abort to compute Z_m from Z_k, then it can be shown that the best value of k is $k = 2\log_2 n - \log_2 \log_2 n + o(\log \log n)$, yielding a total complexity of about $8 \log_2 n \cdot 2^n$. In general, for degree-d systems, the same reasoning would lead to a total complexity of about $4d \cdot \log_2 n \cdot 2^n$. In practice, it makes more sense to choose k to be the word width on a microprocessor in order to use the hardware in the most efficient way.

6 A Brief Description of the Hardware Platforms

6.1 Vector Units on x86-64

The most prevalent SIMD (single instruction, multiple data) instruction set today is SSE2, available on all current Intel-compatible CPUs. SSE2 instructions operate on 16 architectural xmm registers, each of which is 128-bit wide. We use integer operations, which treat xmm registers as vectors of 8-, 16-, 32- or 64-bit operands.

The highly non-orthogonal SSE instruction set includes Loads and Stores (to/from xmm registers, memory — both aligned and unaligned, and traditional registers), Packing/Unpacking/Shuffling, Logical Operations (AND, OR, NOT, XOR, Shifts Left, Right Logical and Arithmetic — bit-wise on units and byte-wise on the entire xmm register), and Arithmetic (add, substract, multiply, max-min) with some or all of the arithmetic widths. The interested reader is referred to Intel and AMD's manuals for details on these instructions, and to references such as [18] for throughput and latencies.

6.2 G2xx-Series Graphics Processing Units from NVIDIA

We choose NVIDIA's G2xx GPUs as they have the least hostile GPU parallel programming environment called CUDA (Compute Unified Device Architecture). In CUDA, we program in the familiar C/C++ programming language plus a small set of GPU extensions.

An NVIDIA GPU contains anywhere from 2–30 streaming multiprocessors (MPs). There are 8 ALUs (streaming processors or SPs in market-speak) and two super function units (SFUs) on each MP. A top-end "GTX 295" card has two GPUs, each with 30 MPs, hence the claimed "480 cores". The theoretical throughput of each SP per cycle is one 32-bit integer or floating-point instruction (including add/subtract, multiply, bitwise AND/OR/XOR, and fused multiply-add), and that of an SFU 2 floating-point multiplications or 1 special operation. The arithmetic units have 20+-stage pipelines.

Main memory is slow and forms a major bottleneck in many applications. The read bandwidth from memory on the card to the GPU is only one 32-bit read per cycle per MP and has a latency of > 200 cycles. To ease this problem, the MP has a register file of 64 KB (16,384 registers, max. of 128 per thread), a 16-bank shared memory of 16 KB, and an 8-KB cache for read-only access to a declared "constant region" of at most 64 KB. Every cycle, each MP can achieve one read from the constant cache, *which can broadcast to many thread at once.*

Each MP contains a scheduling and dispatching unit that can handle a large number of lightweight threads. However, the decoding unit can only decode once every 4 cycles. *This is typically 1 instruction, but certain common instructions are "half-sized", so two such instructions can be issued together if independent.* Since there are 8 SPs in an MP, CUDA programming is always on a Single Program Multiple Data basis, where a "warp" of threads (32) should be executing the same instruction. If there is a branch which is taken by some thread in a warp but not others, we are said to have a "divergent" warp; from then on only part of the threads will execute until all threads in that warp are executing the same instruction again. Further, as the latency of a typical instruction is about 24 cycles, NVIDIA recommends a minimum of 6 warps on each MP, although it is sometimes possible to get acceptable performance with 4 warps.

7 Implementations

We describe the structure of our code, the approximate cost structure, our design choices and justify what we did. Our implementation code always consists of three stages:

Partial Evaluation: We substitute all possible values for s variables $(x_{n-s}, \dots, x_{n-1})$ out of n, thus splitting the original system into 2^s smaller systems, of w equations each in the remaining $(n - s)$ variables $(x_0, \dots, x_{n-s-1})$, and output them in a form that is suitable for ...

Enumeration Kernel: Run the algorithm of Sec. 4 to find all candidate vectors $\mathbf{x}$ satisfying w equations out of m ($\approx 2^{n-w}$ of them), which are handed over to ...

Candidate Checking: Checking possible solutions $\mathbf{x}$ in remaining $m - w$ equations.

7.1 CPU Enumeration Kernel

Typical code fragments from the unrolled inner loops can be seen below:

```
(a) quadratics, C++ x86 instrinsics            (b) quadratics, x86 assembly
...                                            .L746:
diff0 ^= deg2_block[ 1 ];                        movq      976(%rsp), %rax   //
res ^= diff0;                                     pxor      (%rax), %xmm2     // d_y ^= C_yz
Mask = _mm_cmpeq_epi16(res, zero);                pxor      %xmm2, %xmm1      // res ^= d_y
mask = _mm_movemask_epi8(Mask);                   pxor      %xmm0, %xmm0      //
if(mask) check(mask, idx, x^155);                 pcmpeqw   %xmm1, %xmm0      // cmp words for eq
...                                               pmovmskb  %xmm0, %eax       // movemask
                                                  testw     %ax, %ax         // set flag for branch
                                                  jne       .L1266           // if needed, check and
                                               .L747:                        // comes back here

.L1624:
  movq      2616(%rsp), %rax   // load C_yza
  movdqa    2976(%rsp), %xmm0  // load d_yz
  pxor      (%rax), %xmm0      // d_yz ^= C_yza
  movdqa    %xmm0, 2976(%rsp)  // save d_yz
  pxor      8176(%rsp), %xmm0  // d_y ^= d_yz        ...
  pxor      %xmm0, %xmm1       // res ^= d_y         diff[0]  ^= deg3_ptr1[0];
  movdqa    %xmm0, 8176(%rsp)  // save d_y           diff[325] ^= diff[0];
  pxor      %xmm0, %xmm0       //                    res ^= diff[325];
  pcmpeqw   %xmm1, %xmm0       // cmp words for eq    Mask = _mm_cmpeq_epi16(res, zero);
  pmovmskb  %xmm0, %eax                              mask = _mm_movemask_epi8(Mask);
  testw     %ax, %ax          // ...                 if(mask) check(mask, idx, x^2);
  jne       .L2246            // branch to check     ...
.L1625:                       // and comes back
(c) cubics, x86 assembly                        (d) cubics, C++ x86 instrinsics
```

testing. All zeroes in one byte, word, or dword in a XMM register can be tested cheaply on x86-64. We hence wrote code to test 16 or 32 equations at a time. Strangely enough, even though the code above is for 16 bits, the code for checking 32/8 bits at the same time is nearly identical, the only difference being that we would subtitute the intrinsics `_mm_cmpeq_epi32/8` for `_mm_cmpeq_epi16` (leading to the SSE2 instruction `pcmpeqd/b` instead of `pcmpeqw`). Whenever one of the words (or double words or bytes, if using another testing width) is non-zero, the program branches away and queues the candidate solution for checking.

unrolling. One common aspect of our CPU and GPU code is deep unrolling by upwards of $1024\times$ to avoid the expensive bit-position indexing. To illustrate with quartics as an example, instead of having to compute the positions of the four rightmost non-zero bits in every integer, we only need to compute the first four rightmost non-zero bits in bit 10 or above, then fill in a few blanks. This avoids most of the indexing calculations and all the calculations involving the most commonly used differentials.

We wrote similar Python scripts to generate unrolled loops in C and CUDA code. Unrolling is even more critical with GPU, since divergent branching and memory accesses are prohibitively expensive.

7.2   GPU Enumeration Kernel

register usage.   Fast memory is precious on GPU and register usage critical for CUDA programmers. Our algorithms' memory complexity grows exponentially with the degree d, which is a serious problem when implementing the algorithm for cubic and quartic systems, compounded by the immaturity of NVIDIA's nvcc compiler which tends to allocate more registers than we expected.

Take quartic systems as an example. Recall that each thread needs to maintain third derivatives, which we may call d_{ijk} for $0 \leq i < j < k < K$, where K is the number of variables in each small system. For $K = 10$, there are 120 d_{ijk}'s and we cannot waste all our registers on them, especially as all differentials are not equal — d_{ijk} is accessed with probability $2^{-(k+1)}$.

Our strategy for register use is simple: Pick a suitable bound u, and among third differentials d_{ijk} (and first and second differentials d_i and d_{ij}), put the most frequently used — i.e., all indices less than u — in registers, and the rest in device memory (which can be read every 8 instructions without choking). We can then control the number of registers used and find the best u empirically.

fast conditional move.   We discovered during implementation an undocumented feature of CUDA for G2xx series GPUs, namely that nvcc reliably generates conditional (predicated) move instructions, dispatched with exceptional adeptness.

```
...
xor.b32 $r19, $r19, c0[0x000c]     // d_y^=d_yz
xor.b32 $p1|$r20, $r17, $r20
mov.b32 $r3, $r1
mov.b32 $r1, s[$ofs1+0x0038]
xor.b32 $r4, $r4, c0[0x0010]
xor.b32 $p0|$r20, $r19, $r20       // res^=d_y
@$p1.eq mov.b32 $r3, $r1
@$p1.eq mov.b32 $r1, s[$ofs1+0x003c]
xor.b32 $r19, $r19, c0[0x0000]
xor.b32 $p1|$r20, $r4, $r20
@$p0.eq mov.b32 $r3, $r1           // cmov
@$p0.eq mov.b32 $r1, s[$ofs1+0x0040] // cmov
...
```

```
...
diff0 ^= deg2_block[ 3 ];   // d_y^=d_yz
res ^= diff0;               // res^=d_y
if( res == 0 ) y = z;       // cmov
if( res == 0 ) z = code233; // cmov
diff1 ^= deg2_block[ 4 ];
res ^= diff1;
if( res == 0 ) y = z;
if( res == 0 ) z = code234;
diff0 ^= deg2_block[ 0 ];
res ^= diff0;
if( res == 0 ) y = z;
if( res == 0 ) z = code235;
...
```

(a) decuda result from cubin   (b) CUDA code for a inner loop fragment

Consider, for example, the code displayed above right. According to our experimental results, the repetitive 4-line code segments average less than three SP (stream-processor) cycles. However, decuda output of our program shows that each such code segment corresponds to at least 4 instructions including 2 XORs and 2 conditional moves [as marked in above left]. The only explanation is that conditional moves can be dispatched by the SFUs (Special Function Units) so that the total throughput can exceed 1 instruction per SP cycle. Further note that the annotated segment on the right corresponds to actual instructions far apart because *an NVIDIA GPU does opportunistic dispatching but is nevertheless a purely in-order architecture,* so proper scheduling must interleave instructions from different parts of the code.

testing. The inner loop for GPUs differs from CPUs due to the fast conditional moves. Here we evaluate 32 equations at a time using Gray code. The result is used to set a flag if it happens to be all zeroes. We can now conditional move of the index based on the flag to a register variable z, and at the end of the loop write z out to global memory.

However, how can we tell if there are too many (here, *two*) candidate solutions in one small subsystem? Our answer to that is to use a buffer register variable y and a second conditional move using the same flag. At the end of the thread, (y, z) is written out to a specific location in device memory and sent back to the CPU.

Now subsystems which have all zero constant terms are automatically satisfied by the vector of zeroes. Hence we note them down during the partial evaluation phase include the zeros with the list of candidate solutions to be checked, and never have to worry about for all-zero candidate solution. The CPU reads the two doublewords corresponding to y and z for each thread, and:

1. z==0 means no candidate solutions,
2. z!=0 but y==0 means exactly one candidate solution, and
3. y!=0 means 2+ candidate solutions (necessitating a re-check).

7.3 Checking Candidates

Checking candidate solutions is always done on CPU because the programming involves branching and hence is difficult on a GPU even with that available. However, the checking code for CPU enumeration and GPU enumeration is different.

CPU. With the CPU, the check code receives a list of candidate solutions. Today the maximum machine operation is 128-bit wide. Therefore we should collect solutions into groups of 128 possible solutions. We would rearrange 128 inputs of n bits such that they appear as n __int128's, then evaluate one polynomial for 128 results in parallel using 128-bit wide ANDs and XORs. After we finish all candidates for one equation, go through the results and discard candidates that are no longer possible. Repeat the result for the second and any further equations (cf. Sec. 3).

We need to transpose a bit-matrix to achieve the effect of a block of w inputs n-bit long each, to n machine-words of w-bit long. This looks costly, however, there is an SSE2 instruction PMOVMSKB (packed-move-mask-bytes) that packs the top bit of each byte in an XMM register into a 16-bit general-purpose register *with 1 cycle throughput.* We combine this with simultaneous shifts of bytes in an XMM and can, for example, on a K10+ transpose a 128-batch of 32-bit vectors (0.5kB total) into 32 __int128's in about 800 cycles, or an overhead of 6.25 cycles per 32-bit vector. In general the transposition cost is at most a few cycles per byte of data, negligible for large systems.

GPU. As explained above, for the GPU we receive a list with 3 kinds of entries:

1. The knowledge that there are two or more candidate solutions within that same small system, with only the position of the last one in the Gray code order recorded.
2. A candidate solution (and no other within the same small system).
3. Marks to subsystems that have all zero constant terms.

For Case 1, we take the same small system that was passed into the GPU and run the Enumerative Kernel subroutine in the CPU code and find all possible small systems.

Since most of the time, there are exactly two candidate solutions, we expected the Gray code enumeration to go two-thirds of the way through the subsystem. Merge remaining candidate solutions with those of Case 2+3, collate for checking in a larger subsystem if needed, and pass off to the same routine as used in the CPU above. Not unexpectedly, the runtime is dominated by the thread-check case, since those does millions of cycles for two candidate solutions (most of the time).

7.4   Partial Evaluation

The algorithm for Partial Evaluation is for the most part the same Gray Code algorithm as used in the Enumeration Kernel. Also the highest degree coefficients remain constant, need no evaluation and and can be shared across the entire Enumeration Kernel stage. As has been mentioned in the GPU description, these will be stored in the *constant memory*, which is reasonably cached on NVIDIA CUDA cards. The other coefficients can be computed by Gray code enumeration, so for example for quadratics we have $(n - s) + 2$ XOR per w bit-operations and per substitution. In all, the cost of the Partial Evaluation stage for w' equations is $\sim 2^s \frac{w'}{8} \left(\binom{n-s}{d-1} + (\text{smaller terms}) \right)$ byte memory writes. The only difference in the code to the Enumerative Kernel is we write out the result (smaller systems) to a buffer, and *check for a zero constant term only* (to find all-zero candidate solutions).

Peculiarities of GPUS. Many warps of threads are required for GPUs to run at full speed, hence we must split a kernel into many threads, the initial state of each small system being provided by Partial Evaluation. In fact, for larger systems on GPUs, we do two stages of partial evaluation because

1. there is a limit to how many threads can be spawned, and how many small systems the device memory can hold, which bounds how small we can split; *but*
2. increasing s decreases the fast memory pressure; and
3. a small systems reporting two or more candidate solutions is costly, yet we can't run a batch check on a small system with only one candidate solution — hence, an intermediate partial evaluation so we can batch check with fewer variables.

7.5   More Test Data and Discussion

Some minor points which the reader might find useful in understanding the test data, a full set of which will appear in the extended version.

Candidate Checking. **The check code is now 6–10% of the runtime.** In theory (cf. Sec. 3) evaluation should start with a script which hard-wires a system of equations into C and compiling to an excutable, eliminating half of the terms, and leading to $\binom{n-s}{d}$ SSE2 (half XORs and half ANDs) operations to check one equation for $w = 128$ inputs. The check code can potentially become more than an order of magnitude faster. We do not (yet) do so presently, because compiling may take more time than the checking code. However, we may want to go this route for even larger systems, as the overhead from testing for zero bits, re-collating the results, and wasting due to the number of candidate solutions is not divisible by w would all go down proportionally.

Without hard-wiring, the running time of the candidate check is dominated by loading coefficients. E.g., for quartics with 44 variables, 14 pre-evaluated, K10+ and Ci7 averages 4300 and 3300 cycles respectively per candidate. With each candidate averaging 2 equations of $\binom{44-14}{4}$ terms each, the 128-wide inner loop averages about 10 and 7.7 cycles respectively per term to accomplish 1 `PXOR` and 1 `PAND`.

Partial Evaluation. We point out that Partial Evaluation also reduces the complexity of the Checking phase. The simplified description in Sec. 5 implies the cost of checking each candidate solution to be $\approx \frac{1}{w}\binom{n}{d}$ instructions. But we can get down to $\approx \frac{1}{w}\binom{n-s}{d}$ instructions by partially evaluating $w' > w$ equations and storing the result for checking. For example, when solving a quartic system with $n = 48$, $m = 64$, the best CPU results are $s = 18$, and we cut the complexity of the checking phase by factor of at least $4\times$ even if it was not the theoretical $7\times$ (i.e., $\binom{n}{d}/\binom{n-s}{d}$) due to overheads.

The Probability of Thread-Checking for GPUs. If we have n variables, pre-evaluate s, and check w equations via Gray Code, then the probability of a subsystem with 2^{n-s} vectors including at least two candidates $\approx \binom{2^{n-s}}{2}(1 - 2^{-w})^{2^{n-s}}(2^{-w})^2 \approx 1/2^{2(s+w-n)+1}$, provided that $n < s+w$. As an example, for $n = 48$, $s = 22$, $w = 32$, the thread-recheck probability is about 1 in 2^{13}, and we must re-check about 2^9 threads using Gray Code. This pushes up the optimal s for GPUs.

Architecture and Differences. All our tests with a huge variety of machines and video cards show that the kernel time in cycles per attempt is almost a constant of the architecture, and the speed-up in multi-cores is almost completely linear for almost all modern hardware. So we can compute the time complexity given the architecture, the frequency, the number of cores, and n. The marked cycle count difference between Intel and AMD cores is explained by Intel dispatching *three* XMM (SSE2) logical instructions to AMD's *two* per cycle and handling branch prediction and caching better.

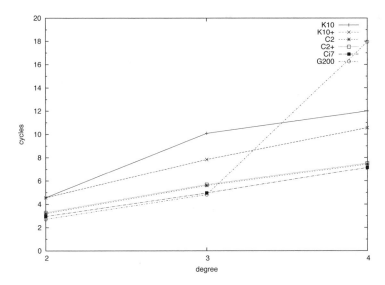

Fig. 1. Cycles per candidate tested for degree 2,3 and 4 polynomials

As the Degree d increases. We plot how many cycles is taken by the inner loop (which is 8 vectors per core for CPUs and 1 vector per SP for GPUs) on different architectures in Fig. 1. As we can see, all except two architectures have inner loop cycle counts that are increasing roughly linearly with the degree. The exceptions are the AMD K10 and NVIDIA G200 architectures, which is in line with fast memory pressure on the NVIDIA GPUs and fact that K10 has the least cache among the CPU architectures.

More Tuning. We can conduct a Gaussian elimination among the m equations and such that $m/2$ selected terms in $m/2$ of the equations are all zero. We can of course make this the most commonly used coefficients (i.e., $c_{01}, c_{02}, c_{12}, \ldots$ for the quadratic case). The corresponding XOR instructions can be removed from the code by our code generator. This is not yet automated and we have to test everything by hand. However, this clearly leads to significant savings. On GPUs, we have a speed up of 21% on quadratic cases, 18% for cubics, and 4% for quadratics. [The last is again due to the memory problems.]

Table 2. Efficiency comparison: cycles per candidate tested on one core

$n = 32$			$n = 40$			$n = 48$			Testing platform			
$d=2$	$d=3$	$d=4$	$d=2$	$d=3$	$d=4$	$d=2$	$d=3$	$d=4$	GHz	Arch.	Name	USD
0.58	1.21	1.41	0.57	1.27	1.43	0.57	1.26	1.50	2.2	K10	Phenom9550	120
0.57	0.91	1.32	0.57	0.98	1.31	0.57	0.98	1.32	2.3	K10+	Opteron2376	184
0.40	0.65	0.95	0.40	0.70	0.94	0.40	0.70	0.93	2.4	C2	Xeon X3220	210
0.40	0.66	0.96	0.41	0.71	0.94	0.41	0.71	0.94	2.83	C2+	Core2 Q9550	225
0.50	0.66	1.00	0.38	0.65	0.91	0.37	0.62	0.89	2.26	Ci7	Xeon E5520	385
2.87	4.66	15.01	2.69	4.62	17.94	2.72	4.82	17.95	1.296	G200	GTX280	n/a
2.93	4.90	14.76	2.70	4.62	15.54	2.69	4.57	15.97	1.242	G200	GTX295	500

Notes and Acknowledgements

C. Bouillaguet thanks Jean Vuillemin for helpful discussions. The Taiwanese authors thank Ming-Shing Chen for assistance with programming and fruitful discussion, Taiwan's National Science Council for partial sponsorship under grants NSC96-2221-E-001-031-MY3, 98-2915-I-001-041, and 98-2219-E-011-001 (Taiwan Information Security Center), and Academia Sinica for the Career Development Award. Questions and esp. corrections about the extended version should be addressed to by@crypto.tw.

References

1. Bard, G.V., Courtois, N.T., Jefferson, C.: Efficient methods for conversion and solution of sparse systems of low-degree multivariate polynomials over GF(2) via SAT-solvers, http://eprint.iacr.org/2007/024
2. Bardet, M., Faugère, J.-C., Salvy, B.: On the complexity of Gröbner basis computation of semi-regular overdetermined algebraic equations. In: Proc. Int'l Conference on Polynomial System Solving, pp. 71–74 (2004) INRIA report RR-5049

3. Bardet, M., Faugère, J.-C., Salvy, B., Yang, B.-Y.: Asymptotic expansion of the degree of regularity for semi-regular systems of equations. In: Proc. MEGA 2005 (2005)

4. Berbain, C., Gilbert, H., Patarin, J.: QUAD: A practical stream cipher with provable security. In: Vaudenay, S. (ed.) EUROCRYPT 2006. LNCS, vol. 4004, pp. 109–128. Springer, Heidelberg (2006)

5. Bernstein, D.J., Chen, T.-R., Cheng, C.-M., Lange, T., Yang, B.-Y.: ECM on graphics cards. In: Joux, A. (ed.) EUROCRYPT 2009. LNCS, vol. 5479, pp. 483–501. Springer, Heidelberg (2010)

6. Bettale, L., Faugére, J.-C., Perret, L.: Hybrid approach for solving multivariate systems over finite fields. J. Math. Crypto. 3(3), 177–197 (2009)

7. Bouillaguet, C., Faugére, J.-C., Fouque, P.-A., Perret, L.: Differential-algebraic algorithms for the isomorphism of polynomials problem, http://eprint.iacr.org/2009/583

8. Bouillaguet, C., Fouque, P.-A., Joux, A., Treger, J.: A family of weak keys in HFE (and the corresponding practical key-recovery), http://eprint.iacr.org/2009/619

9. Buchberger, B.: Ein Algorithmus zum Auffinden der Basiselemente des Restklassenringes nach einem nulldimensionalen Polynomideal. PhD thesis, Innsbruck (1965)

10. Buchmann, J., Cabarcas, D., Ding, J., Mohamed, M.S.E.: Flexible Partial Enlargement to Accelerate Gröbner Basis Computation over $\mathbb{F}_{\not\leq}$. In: Bernstein, D.J., Lange, T. (eds.) AFRICACRYPT 2010. LNCS, vol. 6055, pp. 69–81. Springer, Heidelberg (2010)

11. Courtois, N., Bard, G.V., Wagner, D.: Algebraic and slide attacks on Keeloq. In: Nyberg, K. (ed.) FSE 2008. LNCS, vol. 5086, pp. 97–115. Springer, Heidelberg (2008)

12. Courtois, N., Goubin, L., Patarin, J.: SFLASH: Primitive specification (second revised version) (2002), https://www.cosic.esat.kuleuven.be/nessie

13. Courtois, N.T., Klimov, A., Patarin, J., Shamir, A.: Efficient algorithms for solving overdefined systems of multivariate polynomial equations. In: Preneel, B. (ed.) EUROCRYPT 2000. LNCS, vol. 1807, pp. 392–407. Springer, Heidelberg (2000), Extended ver., http://www.minrank.org/xlfull.pdf

14. de Bruijn, N.: Asymptotic methods in analysis. 2nd edition. Bibliotheca Mathematica. Vol. 4., 200 p. P. Noordhoff Ltd. XII, Groningen (1961)

15. Faugère, J.-C.: A new efficient algorithm for computing Gröbner bases (F_4). J. of Pure and Applied Algebra 139, 61–88 (1999)

16. Faugère, J.-C.: A new efficient algorithm for computing Gröbner bases without reduction to zero (F_5). In: ACM ISSAC 2002, pp. 75–83 (2002)

17. Faugère, J.-C., Joux, A.: Algebraic cryptanalysis of Hidden Field Equations (HFE) using Gröbner bases. In: Boneh, D. (ed.) CRYPTO 2003. LNCS, vol. 2729, pp. 44–60. Springer, Heidelberg (2003)

18. Fog, A.: Instruction Tables. Copenhagen University, College of Engineering, Lists of Instruction Latencies, Throughputs and micro-operation breakdowns for Intel, AMD, and VIA CPUs (February 2010),
http://www.agner.org/optimize/instruction_tables.pdf

19. Patarin, J.: Asymmetric cryptography with a hidden monomial. In: Koblitz, N. (ed.) CRYPTO 1996. LNCS, vol. 1109, pp. 45–60. Springer, Heidelberg (1996)

20. Patarin, J.: Hidden Field Equations (HFE) and Isomorphisms of Polynomials (IP): two new families of asymmetric algorithms. In: Maurer, U.M. (ed.) EUROCRYPT 1996. LNCS, vol. 1070, pp. 33–48. Springer, Heidelberg (1996), Extended ver.:
http://www.minrank.org/hfe.pdf

21. Patarin, J., Courtois, N., Goubin, L.: QUARTZ, 128-bit long digital signatures. In: Naccache, D. (ed.) CT-RSA 2001. LNCS, vol. 2020, pp. 282–297. Springer, Heidelberg (2001), http://www.minrank.org/quartz/

22. Patarin, J., Goubin, L., Courtois, N.: Improved algorithms for Isomorphisms of Polynomials. In: Nyberg, K. (ed.) EUROCRYPT 1998. LNCS, vol. 1403, pp. 184–200. Springer, Heidelberg (1998); Extended ver.: http://www.minrank.org/ip6long.ps
23. Raddum, H.: MRHS equation systems. In: Adams, C., Miri, A., Wiener, M. (eds.) SAC 2007. LNCS, vol. 4876, pp. 232–245. Springer, Heidelberg (2007)
24. Sugita, M., Kawazoe, M., Perret, L., Imai, H.: Algebraic cryptanalysis of 58-round SHA-1. In: Biryukov, A. (ed.) FSE 2007. LNCS, vol. 4593, pp. 349–365. Springer, Heidelberg (2007)
25. Yang, B.-Y., Chen, J.-M.: Theoretical analysis of XL over small fields. In: Wang, H., Pieprzyk, J., Varadharajan, V. (eds.) ACISP 2004. LNCS, vol. 3108, pp. 277–288. Springer, Heidelberg (2004)
26. Yang, B.-Y., Chen, J.-M., Courtois, N.: On Asymptotic Security Estimates in XL and Gröbner Bases-Related Algebraic Cryptanalysis. In: López, J., Qing, S., Okamoto, E. (eds.) ICICS 2004. LNCS, vol. 3269, pp. 401–413. Springer, Heidelberg (2004)

256 Bit Standardized Crypto for 650 GE
– GOST Revisited*

Axel Poschmann, San Ling, and Huaxiong Wang

Division of Mathematical Sciences
School of Physical and Mathematical Sciences
Nanyang Technological University, Singapore
{aposchmann,lingsan,hxwang}@ntu.edu.sg

Abstract. The former Soviet encryption algorithm GOST 28147-89 has
been standardized by the Russian standardization agency in 1989 and extensive security analysis has been done since. So far no weaknesses have
been found and GOST is currently under discussion for ISO standardization. Contrary to the cryptographic properties, there has not been much
interest in the implementation properties of GOST, though its Feistel
structure and the operations of its round function are well-suited for
hardware implementations. Our post-synthesis figures for an ASIC implementation of GOST with a key-length of 256 bits require only 800
GE, which makes this implementation well suitable for low-cost passive RFID-tags. As a further optimization, using one carefully selected
S-box instead of 8 different ones -which is still fully compliant with the
standard specifications!- the area requirement can be reduced to 651 GE.

Keywords: lightweight cryptography, ASIC, GOST.

1 Introduction

Increasingly, everyday items are enhanced to pervasive devices by embedding
computing power and their interconnection leads to Mark Weiser's famous vi-
sion of *ubiquitous computing* (ubicomp) [27], which is widely believed to be the
next paradigm in information technology. Pervasiveness requires mass deployment which in turn implies harsh cost constraints on the used technology. The
cost constraints imply in particular for Application Specific Integrated Circuits
(ASICs) that power, energy, and area requirements must be kept to a minimum.
One counter-argument might be that *Moore's Law* will provide abundant computing power in the near future. However, Moore's Law needs to be interpreted
contrary here: rather than doubling of performance, the price for constant computing power halves each 18 months. This interpretation leads to interesting
conclusions, because many foreseen applications require a minimum amount of
computing power, but at the same time have extremely tight cost constraints

---

* The research was supported in part by the Singapore National Research Foundation
under Research Grant NRF-CRP2-2007-03.

S. Mangard and F.-X. Standaert (Eds.): CHES 2010, LNCS 6225, pp. 219–233, 2010.

(e.g. RFID on consumer items). As a consequence these applications are not realized yet, simply because they do not pay off. Moore's law however halves the price for a constant amount of computing power every 18 months, and consequently enables such applications after a certain period of time. Therefore, a constant or even increasing demand for the cheapest (read lightweight) solutions can be foreseen.

There are physical limits for the chip area: the smaller the area of a chip, the higher the relative costs for handling, packing and cutting of each of the chips, and thus there exists an optimal minimal chip area. Sometimes in the literature it is concluded that there is no need to develop lightweight cryptographic algorithms, because Moore's Law will provide an ever growing amount of computing power for this minimal area. However, we disagree with this viewpoint due to the following points. Firstly, there are plenty of engineering optimisation efforts ongoing to ever minimize the cutting losses and improve other manufacturing steps. Thus the optimal minimal chip area is constantly shrinking. Secondly, and more importantly, there are many envisioned functionalities for pervasive devices, security being only one amongst them. Thus decreasing the area demand for cryptographic primitives, on which security solutions are based, will increase the available space for other functionalities, that are maybe more valued by the users. In fact for RFID-tags used in supply chains there is a strong demand for storage in order to store status information directly on the tag. Thirdly, the smaller the overhead for cryptographic primitives is, the more likely security functionalities will be deployed.

1.1   Previous Work

The demand for small hardware area implementations of cryptographic algorithms has been widely recognised and so different approaches have been published. A lightweight AES core requiring 3400 GE and more than 1000 clock cycles has been first published in [7]. A different implementation of the AES requires only 3100 GE and is more than six times faster than the previous one [9]. In [14] a different approach is followed: DES and DESX have been slightly modified to DESL/DESXL and yield a more compact implementation without scrutinizing the security. There are also block cipher designs from scratch that aim at lightweight hardware implementations. Most notably are here PRESENT [2],[20], which requires only 1000 GE and is currently under ISO standardization and the recently proposed KTANTAN family [5], that can be scaled down to 462 GE, with hardwired key and block size of only 32 bits though.

In this paper we follow the first approach of optimizing the implementation of an existing standardized block ciphers, but we shift our focus far back in time, even before the demand of lightweight cryptography has been recognised. Back in 1989 the Soviet Union has standardized the block cipher GOST 28147-89 as a counterpart to DES (sometimes it is also called the "Russian DES"). For the sake of simplicity in the following we use the term GOST as a synonym for the encryption algorithm GOST 28147-89, though GOST is an abbreviation for *Gosudarstvennyi Standard*, the meaning of *Government Standard* in russian language. Over the past 20 years quite some security analysis on GOST has been published, which we will briefly summarize here.

In [13] it has been shown that using a related key differential characteristic, GOST can be distinguished from a random permutation with probability $1 - 2^{-64}$. The authors propose further attacks on reduced-round versions of GOST and on full GOST, but can only obtain 12 bits of the master key, which leaves another 244 bits. Kara has performed a reflection attack on the full-round GOST in [12]. The attack assumes that the S-boxes are bijective and works only on a subset of approximately 2^{224} keys. Furthermore it requires 2^{32} chosen plaintexts and has a time complexity of 2^{192} steps, which is impractical.

A plain differential cryptanalysis can break up to 13 rounds of GOST using 2^{51} chosen plaintexts [23]. Combining this result with a related-key attack 21 rounds of GOST can be broken using 2^{56} chosen plaintexts [23].

The iterated structure of GOST can be exploited by slide attacks and in [1] a method is presented to break reduced-round versions of GOST with time complexity of 2^{63} encryptions. In case the S-boxes are known an adversary can break up to 30 rounds and 24 rounds if they are not known. In summary we can conclude that despite a considerable cryptanalytic effort has been spent over the past 20 years, GOST is still not broken and provides a security level of 256 bits.

Contrary to the cryptographic properties, there has not been much interest in the implementation properties of GOST. Software implementations of GOST have been presented in [18], but to the best of our knowledge there have been no hardware implementations of GOST reported so far. Therefore one contribution of this article is to provide lightweight hardware implementation details of GOST. Furthermore, we exploit the fact that the S-boxes in GOST can be chosen freely and propose to use the same S-box that has been used in PRESENT [2] to further decrease the area footprint. Our results reveal that GOST is well suited as an encryption algorithm for passive RFID-tags.

1.2   Outline

The remainder of this article is organized as follows: In the next Section we are going to briefly introduce the GOST encryption algorithm. Since the S-boxes are not specified in the original standard, we discuss the selection of an appropriate approach for the selection of S-boxes in the subsequent Section 3. There we also compare the linear and differential properties of the S-boxes as used by the Central Bank of Russian Federation and the PRESENT S-box. We propose a standard conform variant of GOST that uses the cryptographically strong PRESENT S-box. The hardware implementation results of both variants are presented in Section 4. Finally, this paper is concluded in Section 5.

2   Description of the GOST Encryption Algorithm

The former Soviet encryption standard GOST 28147-89 was published in [17]. There is an IETF draft [6] on GOST and GOST is currently discussed for inclusion into the ISO/IEC Standard 18033-3 on Block Ciphers [11]. GOST has a

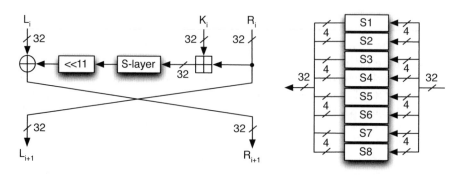

Fig. 1. One round of the Feistel network of GOST (left) and a detailed view of the S-layer (right)

block size of 64 bits and a key size of 256 bits. Its overall structure is a two branch Feistel network with 32 rounds and its inner roundfunction F consists of an integer addition, a non-linear substitution layer and an 11 bits left rotation. Let us denote the 64-bit data state of GOST by $STATE_i = L_i||R_i$, where $||$ denotes concatenation, then $STATE_0$ is the plaintext and $STATE_{32}$ the ciphertext. The integer addition adds the 32 bits roundkey K_i to the right half of the state R_i, the result is then substituted by eight 4-bit to 4-bit S-box look-up tables and finally rotated by 11 bits to the left. The result of the roundfunction $F(K_i, R_i)$ is XORed to the left half of the state L_i and is stored as the right half of the subsequent round R_{i+1}, while R_i is stored without any modifications as L_{i+1}. In a formal notation we have

$$L_{i+1} = R_i,$$
$$R_{i+1} = L_i \oplus (S(K_i + R_i \bmod 2^{32}) \lll 11),$$

where $\oplus$ denotes a bitwise exclusive OR and $\lll a$ a rotation to the left by a bits. In the final round, the halves are not swapped, i.e. $R_{32} = R_{31}$ and $L_{32} = L_{31} \oplus (S(K_{31} + R_{31} \bmod 2^{32}) \lll 11)$. This enables to use the same hardware with the reverse round-key order for decryption. Figure 1 depicts one round of GOST graphically.

There is no real key schedule for GOST, instead the 256-bit key K is considered as eight 32-bit keys, $K = K_0||K_1||K_2||K_3||K_4||K_5||K_6||K_7$. For rounds $0 \leq r \leq 23$ the roundkey K_i is derived as $K_i = K_{(r \bmod 8)}$ and for the last eight rounds $24 \leq r \leq 31$ the order is reversed, i.e. $K_i = K_{7-(r \bmod 8)}$. Table 1 provides an overview of the roundkeys used in every round.

Table 1. Key scheduling of GOST

Round	0	1	2	3	4	5	6	7	8	9	10	11	12	13	14	15
Key	K_0	K_1	K_2	K_3	K_4	K_5	K_6	K_7	K_0	K_1	K_2	K_3	K_4	K_5	K_6	K_7
Round	16	17	18	19	20	21	22	23	24	25	26	27	28	29	30	31
Key	K_0	K_1	K_2	K_3	K_4	K_5	K_6	K_7	K_7	K_6	K_5	K_4	K_3	K_2	K_1	K_0

3   The Choice of a Set of S-Boxes

The GOST standard does not specify a set of S-boxes. In fact, one aim of the designers was to have an encryption algorithm with a flexible security level [4], and the selection of the S-boxes was part of the secret key. There are $2^8 \cdot 16!$ possible sets of such S-boxes [21] and thus the theoretical security level of GOST would be $256 + log_2(2^8 \cdot 16!) = 256 + 354 = 610$ bits. However, Saarinen has shown in [21] that a chosen key attack can reveal the set of S-boxes with a complexity of only 2^{32} encryptions.

Schneier states that the Central Bank of Russian Federation has been using the S-boxes described in Table 2. This set of S-boxes serves as *one* example of GOST, but according to the standard the appropriate choice of S-boxes is a design decision. It is clear that the selection of the S-boxes has a significant influence on the cryptographic strength of the cipher, thus a careful selection is crucial. Please note that the standard does neither specify if the S-boxes used shall be different. Thus, with a small area footprint in mind, we opt for selecting one S-box that is used eight times in parallel – a similar approach as used in DESL/DESXL [14]. While DESL/DESXL lead to a slightly modified standard algorithm, in the case of GOST we end up with a solution that is even conform to the standard. As Biham *et al.* have pointed out in [1], the S-boxes do not even have to be permutations. Hence theoretically, it would be even possible to use simple wiring, which further decreases the area footprint. Of course the differential and linear properties of such an implementation will be very weak and thus this is not an interesting option.

Table 2. Set of GOST S-boxes as used by the Central Bank of Russian Federation [22]

x	0	1	2	3	4	5	6	7	8	9	A	B	C	D	E	F
$S_1(x)$	4	A	9	2	D	8	0	E	6	B	1	C	7	F	5	3
$S_2(x)$	E	B	4	C	6	D	F	A	2	3	8	1	0	7	5	9
$S_3(x)$	5	8	1	D	A	3	4	2	E	F	C	7	6	0	9	B
$S_4(x)$	7	D	A	1	0	8	9	F	E	4	6	C	B	2	5	3
$S_5(x)$	6	C	7	1	5	F	D	8	4	A	9	E	0	3	B	2
$S_6(x)$	4	B	A	0	7	2	1	D	3	6	8	5	9	C	F	E
$S_7(x)$	D	B	4	1	3	F	5	9	0	A	E	7	6	8	2	C
$S_8(x)$	1	F	D	0	5	7	A	4	9	2	3	E	6	B	8	C

Instead we focus on the linear and the differential properties and use the classification of 4-bit S-boxes published in [15] as a guideline for the selection of an appropriate S-box. In fact we have chosen to use the PRESENT S-box, since it is strong with regard to linear and differential properties and has the lowest area footprint of 4-bit S-boxes [2]. Let us denote the *Fourier* coefficient of S by

$$S_b^W(a) = \sum_{x \in F_2^4} (-1)^{<b,S(x)>+<a,x>}.$$

Further, we denote a fixed non-zero input difference with $\Delta_I \in \mathbb{F}_2^4$ and a fixed non-zero output difference with $\Delta_O \in \mathbb{F}_2^4$. The design criteria of the PRESENT S-box are [2]:

1. For any fixed non-zero input difference $\Delta_I \in \mathbb{F}_2^4$ and any fixed non-zero output difference $\Delta_O \in \mathbb{F}_2^4$ we require

$$\#\{x \in \mathbb{F}_2^4 \,|\, S(x) + S(x + \Delta_I) = \Delta_O\} \leq 4.$$

2. For any fixed non-zero input difference $\Delta_I \in \mathbb{F}_2^4$ and any fixed output difference $\Delta_O \in \mathbb{F}_2^4$ such that $\mathrm{wt}(\Delta_I) = \mathrm{wt}(\Delta_O) = 1$ we have

$$\{x \in \mathbb{F}_2^4 \,|\, S(x) + S(x + \Delta_I) = \Delta_O\} = \emptyset.$$

3. For all non-zero $a \in \mathbb{F}_2^4$ and all non-zero $b \in \mathbb{F}_2^4$ it holds that $|S_b^W(a)| \leq 8$.
4. For all $a \in \mathbb{F}_2^4$ and all non-zero $b \in \mathbb{F}_2^4$ such that $\mathrm{wt}(a) = \mathrm{wt}(b) = 1$ it holds that $|S_b^W(a)| \leq 4$.

We have calculated the differential and linear distribution tables of the S-boxes used by the Central Bank of Russian Federation (S_1 to S_8) and PRESENT (S_{PS}) and list them in the appendix. From Table 3, where we summarize the linear and differential characteristics of these S-boxes, it becomes clear that the PRESENT S-box is stronger both with regard to linear and differential cryptanalysis due to the strict design criteria. Therefore in the following we will also consider a GOST implementation that uses eight times the PRESENT S-box. We will refer to this variant with the term GOST-PS while GOST-FB denotes the GOST variant that uses the S-boxes as used by the Central Bank of Russian Federation.

Table 3. Linear and Differential characteristics of GOST and PRESENT S-boxes

Sbox	S_1	S_2	S_3	S_4	S_5	S_6	S_7	S_8	S_{PS}
max S_b^W	8	12	12	12	12	12	12	12	8
max DC	6	6	6	6	4	6	8	8	4

4   Hardware Implementations

A round-based implementation of GOST can be done straight forwardly, while a serialized implementation poses some challenges for a hardware designer. Thus we spare the details of the former architecture and focus on the latter with a data path width of 4 bits. Most challenging is the permutation step, since it rotates by 11 bit positions. Thus it is not possible to operate on 4 bit chunks, but instead we have to operate on the whole state. In our architecture (see Figure 2) it takes 8 clock cycles to process all chunks of the state and to perform one round of GOST. Then we swap the content of the registers as it is required by the Feistel structure within one clock cycle, i.e. we operate on the whole state. We use this clock cycle to perform the 11 bit rotation, but in our architecture we have XORed both halves of the state already. Thus we have to shift the right halve in the previous clock cycles by 11 bit positions to the right, before storing it as the new left halve. Then when the XOR sum of both halves is rotated by 11 bit positions to the left, the final step of one round of GOST is performed. In short, the following operations are carried out when the content of the registers is swapped:

$$L_{i+1} = R_i \gg 11,$$
$$R_{i+1} = (L_i \oplus S(K_i + R_i \bmod 2^{32})) \ll 11.$$

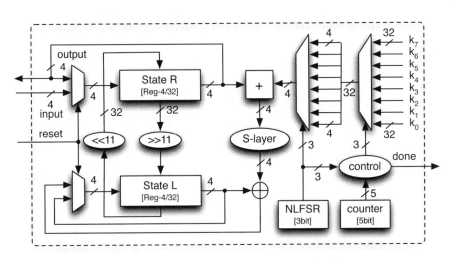

Fig. 2. Architecture of a lightweight hardware architecture with a 4-bit data path for GOST

Recall that the key schedule of GOST is very simple: in every round a 32-bit chunk of the 256-bit key is used as the round key and for the last eight rounds the order is swapped (see Table 1). Thus apart from a 32-bit wide 8-to-1 MUX to select the right round key there is only very little logic required for a round-based implementation. For a serialized implementation we need an additional 4-bit wide 8-to-1 MUX to select the right chunk of the round key. In total the key schedule sums up to only 99 GE for a serialized implementation and 50 GE for a round-based. If the application requires the key to be updated, 256 additional flip-flops are required for storage. However, especially in passive RFID-tag scenarios it is very unlikely that the key needs to be changed – as [2,19] have pointed out before. Therefore the main target for our implementations are applications with a fixed key. Then only a small amount of (cheap) tie cells are required to hard-wire the key.

For functional simulation we used *Mentor Graphics ModelSimXE 6.4b* and *Synopsys DesignCompiler* version *A-2007.12-SP1* [24] was used to synthesize the designs to the *Virtual Silicon* (VST) standard cell library *UMCL18G212T3*, which is based on the *UMC L180 0.18μm 1P6M* logic process and has a typical voltage of 1.8 Volt [26]. Table 4 summarizes the synthesis results and compares them to a selection of other lightweight hardware implementations.

When the synthesis compiler is advised to use the *compile ultra* option, the smallest area can be achieved. In this case the compiler optimizes the whole design without taking care of the single entities, which makes it very difficult (if not impossible) to assign the area requirements to a single component. Thus we also advised the compiler to use the *compile simple* option, in which case the area is larger, but a more detailed breakdown is possible. Below we give such a breakdown of both GOST variants. Recall that GOST-FB refers to the variant that uses the set of S-boxes as used by the Central Bank of Russian Federation while GOST-PS refers to the variant that uses eight times the PRESENT S-box. It is noteworthy to highlight again that both variants are fully standard conform.

GOST		Federal Bank		PRESENT	
		serial	round	serial	round
cycles		264	32	264	32
t'put @100 KHz (Kbps.)		24.24	200	24.24	200
compile ultra	sum	800	1000	651	1017
compile simple	sum	876	1028	664	1055
sequential:	State	384	384	384	384
	Round counter	50	41	50	41
	serial counter	23	0	23	0
combinational:	MUX	15	0	15	0
	KeyAdd	41	271	41	271
	rotation	0	0	0	0
	sBoxLayer	239	187	27	214
	XOR	11	85	11	85
	key scheduling	99	50	99	50
	control	14	10	14	10

As one can see, a serialized implementation leads to smaller area requirements, mostly due to a scaled down key addition module (saves 230 GE) and XOR gates (saves 70 GE). However, the serialized architecture also introduces some area overhead, because additional MUXes are required: one for the state register (15 GE) and one to select the right chunk of the round key (49 GE). If the GOST variant uses different S-boxes such as GOST-FB, but not GOST-PS, another MUX is required to select the correct S-box (52 GE). Furthermore, an NLFSR (23 GE) is required as a serial counter and the key addition requires a flip-flop to store the carry bit. The round counter is smaller in the round-based architectures, which we believe is because in the serialized architecture gated registers are required, but not in the round-based.

The sBoxLayer module in the serialized GOST-PS is by far the smallest of all our variants, because we only need to implement one PRESENT S-box, while GOST-FB needs all 8 S-boxes and an additional MUX. It is also interesting to see that the S-boxes of GOST-FB require less area than the PRESENT S-box. We believe that this is closely related to the fact that these S-boxes are weaker with regard to differential and linear cryptanalysis (see Table 3 and the tables in the appendix).

Table 4. Hardware implementation results of selected symmetric encryption algorithms. GOST-FB uses eight different S-boxes as used by the Central Bank of Russian Federation and GOST-PS uses eight times the PRESENT-S-box. Note that both variants are fully standard compliant.

Algorithm		key size	block size	cycles/ block	Throughput (@100 KHz)	Tech. [μm]	Area [GE]
Stream Ciphers							
Trivium	[8]	80	1	1	100	0.13	2,599
Grain	[8]	80	1	1	100	0.13	1,294
Block Ciphers							
KATAN32	[5]	80	32	255	12.5	0.13	802
KATAN48	[5]	80	48	255	18.8	0.13	927
KATAN64	[5]	80	64	255	25.1	0.13	1054
KTANTAN32	[5]	80	32	255	12.5	0.13	462
KTANTAN48	[5]	80	48	255	18.8	0.13	588
KTANTAN64	[5]	80	64	255	25.1	0.13	688
PRESENT	[20]	80	64	547	11.7	0.18	1,075
SEA	[16]	96	96	93	103	0.13	3,758
mCrypton	[3]	96	64	13	492.3	0.13	2,681
HIGHT	[10]	128	64	34	188	0.25	3,048
AES	[7]	128	128	1,032	12.4	0.35	3,400
AES	[9]	128	128	160	80	0.13	3,100
DESXL	[14]	184	64	144	44.4	0.18	2,168
GOST-FB		256	64	264	24.24	0.18	800
GOST-FB		256	64	32	200	0.18	1000
GOST-PS		256	64	264	24.24	0.18	651
GOST-PS		256	64	32	200	0.18	1017

We used *Synopsys PowerCompiler* version *A-2007.12-SP1* [25] to estimate the power consumption of our implementations. All power estimates for the smallest wire-load model (10K GE) at a supply voltage of 1.8 Volt and a frequency of 100 KHz are below 2.6 μW, which indicates that all GOST variants are well suited for demanding applications, including passive RFID tags. However, the accuracy level of simulated power figures greatly depends on the simulation tools and parameters used. Furthermore, the power consumption also strongly depends on the target library used. Thus to have a fair comparison, we do not include any power figures in Table 4.

5   Conclusions

In this paper we have revisited the former Soviet encryption standard GOST 28147-89, that has been standardized since 1989 already. Despite considerable cryptanalytic efforts spent in the past 20 years, GOST is still not broken. Since to the best of our knowledge no hardware implementations of GOST have been published so far, we have implemented GOST in hardware with a focus on a low area footprint to close this gap. As a further optimization, we exploit the fact that the standard does not specify a set of S-boxes and use a single S-box repeated eight times. We have selected the PRESENT S-box, which has the best linear and differential properties among all 4-bit S-boxes while at the same time requiring the least amount of area. Our synthesis results show that a standard conform GOST variant that uses the PRESENT S-box requires only 651 GE.

Many of the recently proposed lightweight block ciphers are not yet mature for standardization, while others, i.e. PRESENT and HIGHT, are currently undergoing standardization by ISO. At the same time, GOST is already standardized since 20 years. Furthermore, GOST offers 256 bits of security, while many lightweight proposal are limited to 128 or 80 bits. It is therefore an interesting candidate for low-cost applications that also require a very strong security level.

References

1. Biham, E., Dunkelman, O., Keller, N.: Improved slide attacks. In: Biryukov, A. (ed.) FSE 2007. LNCS, vol. 4593, pp. 153–166. Springer, Heidelberg (2007)
2. Bogdanov, A., Leander, G., Knudsen, L., Paar, C., Poschmann, A., Robshaw, M., Seurin, Y., Vikkelsoe, C.: PRESENT - An Ultra-Lightweight Block Cipher. In: Paillier, P., Verbauwhede, I. (eds.) CHES 2007. LNCS, vol. 4727, pp. 450–466. Springer, Heidelberg (2007)
3. Lim, C., Korkishko, T.: mCrypton - A Lightweight Block Cipher for Security of Low-cost RFID Tags and Sensors. In: Song, J., Kwon, T., Yung, M. (eds.) WISA 2005. LNCS, vol. 3786, pp. 243–258. Springer, Heidelberg (2006)
4. Charnes, C., O'Connor, L., Pieprzyk, J., Safavi-Naini, R., Zheng, Y.: Further comments on the soviet encryption algorithm. In: De Santis, A. (ed.) EUROCRYPT 1994. LNCS, vol. 950, pp. 433–438. Springer, Heidelberg (1995)
5. de Cannière, C., Dunkelman, O., Knezević, M.: Katan and ktantan–a family of small and efficient hardware-oriented block ciphers. In: Clavier, C., Gaj, K. (eds.) CHES 2009. LNCS, vol. 5747, pp. 272–288. Springer, Heidelberg (2009)
6. Dolmatov, V.: Gost 28147-89 encryption, decryption and mac algorithms (December 3, 2009), http://tools.ietf.org/html/draft-dolmatov-cryptocom-gost2814789
7. Feldhofer, M., Wolkerstorfer, J., Rijmen, V.: AES Implementation on a Grain of Sand. IEE Proceedings of Information Security 152(1), 13–20 (2005)
8. Good, T., Benaissa, M.: Hardware Results for Selected Stream Cipher Candidates. In: State of the Art of Stream Ciphers 2007 (SASC 2007), Workshop Record (February 2007), http://www.ecrypt.eu.org/stream
9. Hämäläinen, P., Alho, T., Hännikäinen, M., Hämäläinen, T.D.: Design and Implementation of Low-Area and Low-Power AES Encryption Hardware Core. In: DSD, pp. 577–583 (2006)

10. Hong, D., Sung, J., Hong, S., Lim, J., Lee, S., Koo, B.S., Lee, C., Chang, D., Lee, J., Jeong, K., Kim, H., Kim, J., Chee, S.: HIGHT: A New Block Cipher Suitable for Low-Resource Device. In: Goubin, L., Matsui, M. (eds.) CHES 2006. LNCS, vol. 4249, pp. 46–59. Springer, Heidelberg (2006)

11. ISO/IEC. International Standard ISO/IEC 18033 Information technology – Security techniques – Encryption algorithms – Part 3: Block ciphers

12. Kara, O.: Reflection cryptanalysis of some ciphers. In: Chowdhury, D.R., Rijmen, V., Das, A. (eds.) INDOCRYPT 2008. LNCS, vol. 5365, pp. 294–307. Springer, Heidelberg (2008)

13. Ko, Y., Hong, S., Lee, W.L.S., Kang, J.-S.: Related Key Differential Attacks on 27 Rounds of XTEA and Full-Round GOST. In: Roy, B., Meier, W. (eds.) FSE 2004. LNCS, vol. 3017, pp. 299–316. Springer, Heidelberg (2004)

14. Leander, G., Paar, C., Poschmann, A., Schramm, K.: New Lightweight DES Variants. In: Biryukov, A. (ed.) FSE 2007. LNCS, vol. 4593, pp. 196–210. Springer, Heidelberg (2007)

15. Leander, G., Poschmann, A.: On the classification of 4-Bit s-boxes. In: Carlet, C., Sunar, B. (eds.) WAIFI 2007. LNCS, vol. 4547, pp. 159–176. Springer, Heidelberg (2007)

16. Mace, F., Standaert, F.-X., Quisquater, J.-J.: ASIC Implementations of the Block Cipher SEA for Constrained Applications. In: RFID Security — RFIDsec 2007, Workshop Record, Malaga, Spain, pp. 103–114 (2007)

17. National Soviet Bureau of Standards. Informtation Processing System - Cryptographic Protection - Cryptographic Algorithm GOST 28147-89 (1989)

18. Oreku, G.S., Li, J., Pazynyuk, T., Mtenzi, F.J.: Modified s-box to archive accelerated gost. IJCSNS International Journal of Computer Science and Network Security 7(6), 88–98 (2007)

19. Robshaw, M.: Searching for compact algorithms: CGEN. In: Nguyen, P. (ed.) VIETCRYPT 2006. LNCS, vol. 4341, pp. 37–49. Springer, Heidelberg (2006)

20. Rolfes, C., Poschmann, A., Leander, G., Paar, C.: Ultra-Lightweight Implementations for Smart Devices - Security for 1000 Gate Equivalents. In: Grimaud, G., Standaert, F.-X. (eds.) CARDIS 2008. LNCS, vol. 5189, pp. 89–103. Springer, Heidelberg (2008)

21. Saarinen, M.-J.: A chosen Key attack against the secret S-boxes of GOST (unpublished manuscript) (1998)

22. Schneier, B.: Applied Cryptography, 2nd edn. John Wiley & Sons, Chichester (1996)

23. Seki, H., Kaneko, T.: Differential Cryptanalysis of Reduced Rounds of GOST. In: Stinson, D.R., Tavares, S. (eds.) SAC 2000. LNCS, vol. 2012, pp. 315–323. Springer, Heidelberg (2001)

24. Synopsys. Design Compiler User Guide - Version A-2007.12 (December 2007), http://tinyurl.com/pon88o

25. Synopsys. Power Compiler User Guide - Version A-2007.12 (March 2007), http://tinyurl.com/lfqhy5

26. Virtual Silicon Inc. 0.18 µm VIP Standard Cell Library Tape Out Ready, Part Number: UMCL18G212T3, Process: UMC Logic 0.18 µm Generic II Technology: 0.18µm (July 2004)

27. Weiser, M.: The computer for the 21st century. ACM SIGMOBILE Mobile Computing and Communications Review 3(3), 3–11 (1999)

Appendix

The following tables display the differential and linear properties of the PRESENT S-box (S_{PS}) and the GOST S-boxes as used by the Central Bank of Russian Federation (S_1 to S_8).

S_{PS} — Differential (DC), Δ_O rows indexed by Δ_I

Δ_I	0	1	2	3	4	5	6	7	8	9	A	B	C	D	E	F
0	16	0	0	0	0	0	0	0	0	0	0	0	0	0	0	0
1	0	0	0	4	0	0	0	4	0	4	0	0	0	4	0	0
2	0	0	0	2	0	4	2	0	0	0	2	0	2	2	2	0
3	0	2	0	2	2	0	4	2	0	0	2	2	0	0	0	0
4	0	0	0	0	0	4	2	2	0	2	2	0	2	0	2	0
5	0	2	0	0	2	0	0	0	0	2	2	2	4	2	0	0
6	0	0	2	0	0	0	2	0	2	0	0	4	2	0	0	4
7	0	4	2	0	0	0	2	0	2	0	0	0	2	0	0	4
8	0	0	0	2	0	0	0	2	0	2	0	4	0	2	0	4
9	0	0	2	0	4	0	2	0	2	0	0	0	2	0	4	0
A	0	0	2	2	0	4	0	0	2	0	2	0	0	2	2	0
B	0	2	0	0	2	0	0	0	4	2	2	2	0	2	0	0
C	0	0	2	0	0	4	0	2	2	2	2	0	0	0	2	0
D	0	2	4	2	2	0	0	2	0	0	2	2	0	0	0	0
E	0	0	2	2	0	0	2	2	2	0	0	2	2	0	0	0
F	0	4	0	0	4	0	0	0	0	0	0	0	0	0	4	4

S_{PS} — Linear (LC), b rows indexed by a

a	0	1	2	3	4	5	6	7	8	9	A	B	C	D	E	F
0	16	0	0	0	0	0	0	0	0	0	0	0	0	0	0	0
1	0	0	0	0	0	-8	0	-8	0	0	0	0	0	-8	0	8
2	0	0	4	4	-4	-4	0	0	4	-4	0	8	0	8	-4	4
3	0	0	4	4	4	-4	-8	0	-4	4	-8	0	0	0	-4	-4
4	0	0	-4	4	-4	-4	0	8	-4	-4	0	-8	0	0	-4	4
5	0	0	-4	4	-4	4	0	0	4	4	-8	0	8	0	4	4
6	0	0	0	-8	0	0	-8	0	0	-8	0	0	8	0	0	0
7	0	0	0	8	8	0	0	0	0	-8	0	0	0	0	8	0
8	0	0	4	-4	0	0	-4	4	-4	4	0	0	-4	4	8	8
9	0	8	-4	-4	0	0	4	-4	-4	-4	-8	0	-4	4	0	0
A	0	0	8	0	4	4	4	-4	0	0	0	-8	4	4	-4	4
B	0	-8	0	0	-4	-4	4	-4	-8	0	0	0	4	4	4	-4
C	0	0	0	0	-4	-4	-4	-4	8	0	0	-8	-4	4	4	-4
D	0	8	8	0	-4	-4	4	4	0	0	0	0	4	-4	4	-4
E	0	0	4	4	-8	8	-4	-4	-4	-4	0	0	-4	-4	0	0
F	0	8	-4	4	0	0	-4	-4	-4	4	8	0	4	4	0	0

S_1 — Differential (DC), Δ_O rows indexed by Δ_I

Δ_I	0	1	2	3	4	5	6	7	8	9	A	B	C	D	E	F
0	16	0	0	0	0	0	0	0	0	0	0	0	0	0	0	0
1	0	0	0	0	0	2	2	0	2	0	0	2	0	4	4	0
2	0	0	2	0	0	0	2	4	2	0	0	0	2	4	0	0
3	0	0	0	4	2	0	2	0	2	0	6	0	0	0	0	0
4	0	2	2	0	4	0	0	0	0	4	0	0	2	0	0	2
5	0	0	4	0	0	0	0	4	0	4	0	0	4	0	0	0
6	0	0	0	4	6	0	2	0	2	0	2	0	0	0	0	0
7	0	2	0	0	0	2	0	0	0	0	4	2	0	0	4	2
8	0	2	2	0	0	2	0	2	2	0	2	0	0	2	2	0
9	0	0	2	4	0	2	0	0	0	0	0	2	2	0	0	4
A	0	2	0	0	0	2	2	2	2	0	2	0	0	0	0	2
B	0	0	2	0	2	0	0	0	2	2	0	2	0	2	2	2
C	0	4	2	4	0	2	0	0	0	0	0	2	2	0	0	0
D	0	0	0	0	0	0	2	2	0	2	2	2	2	2	2	2
E	0	2	0	0	2	2	2	0	0	2	0	0	2	2	2	0
F	0	2	0	0	0	2	2	2	2	2	0	2	0	0	0	2

S_1 — Linear (LC), b rows indexed by a

a	0	1	2	3	4	5	6	7	8	9	A	B	C	D	E	F
0	16	0	0	0	0	0	0	0	0	0	0	0	0	0	0	0
1	0	-4	8	-4	-4	0	4	0	8	4	0	4	-4	0	4	0
2	0	0	-4	4	-4	4	0	0	-4	4	0	0	-8	-8	4	-4
3	0	4	-4	0	-8	-4	-4	0	4	0	-8	4	4	0	0	-4
4	0	4	0	-4	4	-8	-4	-8	0	4	0	-4	-4	0	4	0
5	0	0	0	0	0	-8	8	0	0	0	0	0	0	-8	-8	0
6	0	4	-4	0	0	-4	-4	8	4	0	8	4	-4	0	0	4
7	0	8	4	4	-4	-4	0	0	4	4	0	-8	0	0	-4	4
8	0	8	4	-4	4	4	0	0	-4	4	0	8	0	0	-4	-4
9	0	-4	4	8	0	-4	-4	0	-4	8	0	4	4	0	0	4
A	0	0	8	0	0	0	-8	0	0	-8	0	0	0	-8	0	0
B	0	-4	0	-4	-4	0	-4	0	0	4	8	-4	4	0	-4	-8
C	0	-4	-4	0	0	4	-4	-8	4	0	0	4	-4	0	-8	4
D	0	0	4	4	-4	-4	0	0	-4	-4	0	0	-8	8	-4	-4
E	0	4	0	4	-4	0	4	-8	0	-4	8	4	4	0	4	0
F	0	0	0	8	8	0	0	0	8	0	0	0	0	0	0	-8

S_2

DC	Δ_O 0	1	2	3	4	5	6	7	8	9	A	B	C	D	E	F
0	16	0	0	0	0	0	0	0	0	0	0	0	0	0	0	0
1	0	2	0	0	0	4	0	2	2	2	0	2	2	0	0	0
2	0	0	2	0	0	2	0	4	0	2	4	0	0	0	2	0
3	0	0	6	2	0	0	0	0	0	2	0	2	2	0	0	2
4	0	0	2	0	2	0	4	0	4	0	0	2	0	2	0	0
5	0	2	0	6	2	2	0	0	0	0	0	0	0	2	2	0
6	0	6	2	0	0	0	2	2	2	0	2	0	0	0	0	0
Δ_I 7	0	2	0	0	4	0	2	0	0	2	2	2	0	0	0	2
8	0	0	0	2	0	0	2	0	2	0	4	0	4	2	0	0
9	0	2	0	0	2	2	2	0	0	2	0	0	0	4	0	2
A	0	0	0	2	2	0	4	0	0	0	2	0	0	2	0	4
B	0	0	0	2	0	0	0	2	4	0	2	0	0	0	2	4
C	0	2	0	0	2	2	0	2	0	0	0	2	2	0	4	0
D	0	0	2	0	0	2	0	0	0	4	0	2	0	2	2	2
E	0	0	2	0	2	0	0	0	0	2	0	4	2	2	2	0
F	0	0	0	2	0	2	0	4	2	0	0	0	4	0	2	0

LC	b 0	1	2	3	4	5	6	7	8	9	A	B	C	D	E	F
0	16	0	0	0	0	0	0	0	0	0	0	0	0	0	0	0
1	0	8	0	0	-4	4	4	4	4	4	4	-4	0	0	-8	0
2	0	0	-8	0	0	0	0	8	4	4	-4	4	4	4	4	-4
3	0	8	0	-8	4	-4	4	-4	0	0	0	0	4	4	4	4
4	0	4	0	4	4	-8	-4	0	0	-4	0	-4	4	0	-4	-8
5	0	4	0	-4	0	4	0	-4	-4	0	-4	8	-4	0	-4	-8
6	0	-4	-8	-4	4	0	-4	0	-4	0	4	0	0	4	-8	4
a 7	0	-4	0	-4	-8	4	0	-4	0	-4	0	-4	8	4	0	-4
8	0	4	-4	0	-8	-4	-4	0	-8	4	4	0	0	-4	4	0
9	0	-4	-4	0	-4	-8	8	-4	4	0	0	4	0	-4	-4	0
A	0	-4	4	-8	0	-4	4	8	-4	0	0	-4	-4	0	0	-4
B	0	4	-4	0	-4	0	0	4	0	-12	-4	0	-4	0	0	4
C	0	0	4	4	-4	-4	0	0	0	0	4	4	-4	12	0	0
D	0	0	4	-4	0	0	-4	4	4	-4	8	8	4	-4	0	0
E	0	0	-4	4	4	4	8	0	-4	-4	8	0	0	0	4	-4
F	0	0	4	4	0	0	4	4	-8	0	-4	4	8	0	-4	4

S_3

DC	Δ_O 0	1	2	3	4	5	6	7	8	9	A	B	C	D	E	F
0	16	0	0	0	0	0	0	0	0	0	0	0	0	0	0	0
1	0	2	2	0	0	0	4	0	0	2	0	2	2	2	0	0
2	0	2	2	0	2	2	0	0	2	0	0	2	0	0	2	2
3	0	0	0	2	0	0	0	2	4	6	0	0	0	2	0	0
4	0	0	0	0	0	4	0	0	2	0	0	2	2	0	0	6
5	0	0	2	2	0	0	2	2	0	4	0	0	0	0	4	0
6	0	2	0	0	2	0	0	4	0	0	4	2	0	0	2	0
Δ_I 7	0	2	2	0	0	2	2	4	0	0	0	0	4	0	0	0
8	0	0	0	2	0	0	0	2	0	2	2	2	2	4	0	0
9	0	2	0	0	0	2	4	0	0	0	4	2	0	0	0	2
A	0	0	6	2	0	0	0	0	2	2	0	0	0	0	0	4
B	0	2	2	2	6	0	0	0	0	0	2	0	0	0	2	0
C	0	0	0	2	2	2	2	0	6	0	0	0	2	0	0	0
D	0	0	0	2	2	4	0	0	0	0	2	0	0	2	4	0
E	0	0	0	2	2	0	2	2	0	0	2	0	2	4	0	0
F	0	4	0	0	0	0	0	0	0	0	0	4	2	2	2	2

LC	b 0	1	2	3	4	5	6	7	8	9	A	B	C	D	E	F
0	16	0	0	0	0	0	0	0	0	0	0	0	0	0	0	0
1	0	4	4	-8	-4	-8	0	-4	0	-4	4	0	-4	0	0	4
2	0	4	-4	0	0	4	4	8	0	-4	4	0	0	-4	-4	8
3	0	0	-8	-8	-4	-4	-4	4	0	0	0	0	-4	4	4	-4
4	0	-4	4	0	-8	4	4	0	-4	-8	0	-4	4	0	0	-4
5	0	0	0	0	4	-4	-4	4	4	-4	4	-4	8	8	0	0
6	0	0	0	0	0	0	0	0	-4	4	4	-4	-4	4	-12	-4
a 7	0	-4	4	0	-4	0	0	4	4	0	0	12	0	4	-4	0
8	0	0	4	-4	4	4	-8	0	4	-4	0	0	0	-8	-4	-4
9	0	-4	0	4	0	4	0	-4	4	0	12	0	-4	0	4	0
A	0	-4	0	4	4	0	-4	0	-4	-8	-4	0	-8	4	0	4
B	0	0	4	-4	0	8	-4	-4	-4	4	0	0	4	4	0	8
C	0	-4	8	-4	-4	0	4	8	0	4	0	-4	-4	0	4	0
D	0	-8	-4	-4	0	0	4	-4	8	0	-4	-4	0	0	-4	4
E	0	8	4	4	-4	-4	0	0	8	0	-4	-4	-4	4	0	0
F	0	-4	0	4	-8	-4	-8	4	0	4	0	-4	0	-4	0	4

S_4

Δ_O (DC), rows indexed by Δ_I

Δ_I	0	1	2	3	4	5	6	7	8	9	A	B	C	D	E	F
0	16	0	0	0	0	0	0	0	0	0	0	0	0	0	0	0
1	0	0	0	0	0	4	0	2	2	6	2	0	0	0	0	0
2	0	2	0	0	0	0	2	4	2	0	2	0	2	2	2	0
3	0	2	4	0	0	0	2	4	2	0	0	0	0	0	0	2
4	0	0	0	4	0	4	2	2	0	0	0	0	0	0	2	2
5	0	0	0	0	0	4	0	0	2	2	0	0	2	2	0	4
6	0	0	2	0	0	0	0	2	0	2	2	2	0	2	4	0
7	0	4	2	0	4	0	0	2	2	0	0	0	0	2	0	0
8	0	0	0	0	0	0	0	0	0	4	2	2	6	2	0	0
9	0	0	2	6	0	0	2	2	0	0	4	0	0	0	0	0
A	0	4	2	0	2	4	0	0	0	0	0	2	0	2	0	0
B	0	0	0	2	2	0	0	0	0	0	0	6	0	2	2	2
C	0	0	2	2	0	0	0	0	0	0	0	0	4	0	2	6
D	0	0	0	0	4	4	4	0	0	4	0	4	0	0	0	0
E	0	2	2	2	2	0	2	2	0	0	0	2	0	0	2	0
F	0	2	0	0	2	0	0	0	4	0	2	0	2	2	2	0

b (LC), rows indexed by a

a	0	1	2	3	4	5	6	7	8	9	A	B	C	D	E	F
0	16	0	0	0	0	0	0	0	0	0	0	0	0	0	0	0
1	0	0	-4	-4	0	0	4	4	0	0	4	4	-8	8	4	4
2	0	4	0	4	0	4	0	4	0	4	0	4	0	4	0	-12
3	0	-4	-4	8	0	-4	4	0	0	-4	4	0	8	4	4	0
4	0	4	0	-4	-8	-4	0	-4	0	-4	0	4	0	-4	8	-4
5	0	4	-12	0	0	4	4	0	0	-4	-4	0	0	-4	-4	0
6	0	-8	0	0	-8	0	0	0	0	0	0	8	0	0	-8	0
7	0	0	4	-4	0	0	4	-4	0	-8	-4	-4	0	8	-4	-4
8	0	-4	4	0	4	8	8	-4	-4	0	0	4	0	-4	4	0
9	0	4	0	4	4	0	-4	-8	4	0	-4	8	0	4	0	4
A	0	0	4	4	-4	4	0	8	4	-4	-8	0	0	0	4	4
B	0	0	0	0	-4	4	-4	12	4	4	-4	0	0	0	0	0
C	0	-8	-4	-4	4	-8	0	4	-4	0	0	0	0	4	0	-4
D	0	0	0	8	-4	4	-4	-4	-4	-4	4	-4	-8	0	0	0
E	0	-4	-4	0	-4	0	0	-4	-4	8	-8	-4	0	4	4	0
F	0	-4	0	4	4	-8	4	0	4	0	-4	0	-8	-4	0	-4

S_5

Δ_O (DC), rows indexed by Δ_I

Δ_I	0	1	2	3	4	5	6	7	8	9	A	B	C	D	E	F
0	16	0	0	0	0	0	0	0	0	0	0	0	0	0	0	0
1	0	0	0	2	0	2	2	2	2	0	2	4	0	0	2	0
2	0	4	0	0	2	0	0	2	2	0	0	2	0	4	0	0
3	0	0	4	2	0	0	0	2	2	0	2	2	0	2	0	0
4	0	0	2	4	2	0	0	0	0	4	2	0	2	0	0	0
5	0	0	0	0	0	2	0	2	0	4	2	2	2	0	0	2
6	0	0	2	0	2	0	0	2	2	0	2	0	2	2	2	0
7	0	4	0	0	2	0	2	0	2	0	2	0	0	0	4	0
8	0	0	2	0	0	2	4	0	0	0	2	0	2	0	2	2
9	0	0	0	2	0	0	2	0	4	2	0	0	2	0	0	4
A	0	0	2	2	0	0	0	0	0	0	0	4	0	4	2	2
B	0	0	0	0	2	4	0	2	4	0	0	0	0	2	2	0
C	0	2	0	2	2	2	4	0	0	0	0	0	2	0	0	2
D	0	2	0	2	0	4	0	0	0	0	2	2	2	0	0	2
E	0	2	4	0	0	0	0	2	0	2	0	0	0	2	2	2
F	0	2	0	0	4	0	2	4	0	0	0	2	2	0	0	0

b (LC), rows indexed by a

a	0	1	2	3	4	5	6	7	8	9	A	B	C	D	E	F
0	16	0	0	0	0	0	0	0	0	0	0	0	0	0	0	0
1	0	-4	4	0	-4	0	0	4	4	8	0	4	-8	4	4	0
2	0	4	0	4	-4	0	-4	0	4	0	4	0	0	-4	0	12
3	0	8	4	-4	0	0	4	4	8	0	4	-4	0	0	-4	-4
4	0	4	0	-4	-4	-8	4	-8	0	-4	0	4	-4	0	4	0
5	0	0	-4	4	0	0	4	-4	4	4	-8	0	-4	-4	-8	0
6	0	8	0	0	0	0	-8	0	-4	-4	-4	-4	-4	-4	4	-4
7	0	-4	-4	0	-4	-8	0	4	0	4	4	4	0	4	-8	0
8	0	-4	4	0	-8	4	4	0	0	-4	-4	-8	0	-4	4	0
9	0	0	-8	-8	-4	4	-4	4	4	-4	-4	4	0	0	0	0
A	0	0	-4	-4	-4	-4	0	0	-4	4	0	-8	0	8	-4	4
B	0	-4	0	-4	8	-4	0	4	0	-4	0	-4	-8	-4	0	4
C	0	0	-4	4	4	-4	0	0	8	0	-4	-4	4	4	8	0
D	0	4	-8	4	0	4	8	4	-4	0	4	0	-4	0	4	0
E	0	4	4	0	0	-4	4	8	-4	0	-8	4	4	0	0	4
F	0	0	0	-8	4	4	4	-4	0	8	0	0	4	-4	4	4

S_6

Δ_O (DC), rows indexed by Δ_I

Δ_I	0	1	2	3	4	5	6	7	8	9	A	B	C	D	E	F
0	16	0	0	0	0	0	0	0	0	0	0	0	0	0	0	0
1	0	2	0	0	0	6	0	0	0	0	2	0	2	2	0	2
2	0	0	2	2	0	0	4	0	0	0	0	4	0	0	2	2
3	0	2	0	4	2	0	2	2	0	0	2	0	0	0	2	0
4	0	0	0	2	0	0	0	2	0	2	4	4	0	2	0	0
5	0	2	0	0	0	0	4	2	0	0	2	0	2	0	0	4
6	0	2	2	0	0	2	2	0	2	2	0	0	2	2	0	0
7	0	0	0	0	2	0	0	2	2	4	2	0	2	2	0	0
8	0	0	2	2	0	2	0	2	0	0	0	0	0	2	6	0
9	0	0	4	0	0	0	0	0	0	4	0	0	4	0	0	4
A	0	2	0	0	0	0	2	0	4	2	0	0	4	0	2	0
B	0	2	0	4	2	0	0	0	0	2	0	0	2	4	0	0
C	0	0	0	0	4	2	0	2	2	2	0	0	0	2	2	0
D	0	4	2	0	4	2	0	0	2	0	0	0	0	0	0	2
E	0	0	2	2	0	2	0	2	0	0	0	4	2	0	0	2
F	0	0	2	0	2	0	2	2	0	2	4	0	0	0	2	0

b (LC), rows indexed by a

a	0	1	2	3	4	5	6	7	8	9	A	B	C	D	E	F
0	16	0	0	0	0	0	0	0	0	0	0	0	0	0	0	0
1	0	-4	0	4	4	0	4	-8	0	-4	0	4	-4	-8	-4	0
2	0	0	-4	4	4	0	-8	-4	-4	4	4	-8	0	-4	4	0
3	0	-4	4	0	-4	0	0	4	4	0	0	12	0	4	-4	0
4	0	4	0	4	4	0	-4	8	4	0	4	0	-8	-4	0	4
5	0	8	0	0	0	0	-8	0	-4	-4	-4	4	4	-4	-4	-4
6	0	-4	-4	0	-4	-8	0	4	0	4	4	0	4	-8	0	-4
7	0	0	4	4	0	0	-4	-4	8	0	4	-4	8	0	-4	4
8	0	0	0	0	4	4	-4	-4	4	4	4	4	0	0	8	-8
9	0	4	0	-4	-8	-4	0	-4	4	-8	4	0	-4	0	4	0
A	0	0	-4	4	-4	-4	0	0	0	0	4	-4	-4	-8	-8	0
B	0	4	4	8	-8	4	4	0	0	-4	0	-4	0	-4	4	0
C	0	4	0	4	0	-4	0	-4	-8	4	8	4	0	4	0	4
D	0	0	0	-8	-4	4	-4	-4	0	8	0	0	-4	-4	-4	4
E	0	-4	12	0	0	-4	-4	0	-4	0	0	-4	-4	0	0	-4
F	0	8	4	-4	4	-4	8	0	4	4	0	0	0	0	-4	-4

S_7

DC / LC		Δ_O																b														
Δ_I / a	0	1	2	3	4	5	6	7	8	9	A	B	C	D	E	F	0	1	2	3	4	5	6	7	8	9	A	B	C	D	E	F
0	16	0	0	0	0	0	0	0	0	0	0	0	0	0	0	0	16	0	0	0	0	0	0	0	0	0	0	0	0	0	0	0
1	0	0	0	0	0	2	2	0	0	2	2	0	4	0	4	0	0	4	0	-4	-4	0	-4	8	8	4	0	4	4	0	-4	0
2	0	0	0	0	4	0	4	0	0	2	2	0	0	2	2	0	0	0	-4	4	4	4	0	8	-4	4	0	0	0	-8	4	4
3	0	0	0	0	2	0	0	2	0	0	8	2	0	0	0	2	0	-4	4	0	0	-4	4	0	4	0	-8	4	4	0	8	4
4	0	2	2	0	2	0	2	0	2	0	0	2	2	0	2	0	0	0	0	0	0	0	0	0	0	0	8	8	0	0	8	-8
5	0	0	4	0	2	2	0	0	4	0	0	0	2	2	0	0	0	4	8	4	-4	0	4	0	-8	4	0	4	4	0	-4	0
6	0	0	4	0	0	0	2	2	4	0	0	0	0	0	2	2	0	0	4	-4	4	4	-8	0	-4	4	0	0	0	8	4	4
7	0	2	2	0	2	0	2	0	2	0	0	2	2	0	2	0	0	-4	4	0	0	12	4	0	4	0	0	-4	4	0	0	-4
8	0	2	0	0	0	4	2	4	0	0	2	0	0	2	0	0	0	-12	4	0	0	-4	-4	0	0	4	4	0	0	-4	-4	0
9	0	0	0	2	0	0	0	2	0	4	0	6	0	0	0	2	0	0	4	4	-4	4	-8	0	0	-8	-4	4	-4	-4	0	0
A	0	4	0	6	2	0	0	0	0	0	0	2	2	0	0	0	0	-4	-8	4	-4	0	-4	0	-4	0	-4	0	8	4	0	-4
B	0	2	0	0	0	2	0	0	0	0	2	0	0	4	2	4	0	0	0	-8	8	0	0	0	-4	-4	-4	4	4	-4	-4	-4
C	0	0	0	4	0	2	2	0	0	0	0	4	0	2	2	0	0	4	4	0	0	-4	-4	0	0	4	-4	-8	0	-4	4	-8
D	0	0	2	2	0	2	0	2	2	2	0	0	0	2	0	2	0	0	-4	-4	-4	4	0	-8	0	8	-4	4	-4	-4	0	0
E	0	0	2	2	0	2	0	2	2	2	0	0	0	2	0	2	0	-4	0	-4	-4	0	4	8	-4	0	-4	0	-8	4	0	-4
F	0	4	0	0	2	0	0	2	0	4	0	0	2	0	0	2	0	0	0	-8	-8	0	0	0	-4	-4	4	-4	4	-4	4	4

S_8

DC / LC		Δ_O																b														
Δ_I / a	0	1	2	3	4	5	6	7	8	9	A	B	C	D	E	F	0	1	2	3	4	5	6	7	8	9	A	B	C	D	E	F
0	16	0	0	0	0	0	0	0	0	0	0	0	0	0	0	0	16	0	0	0	0	0	0	0	0	0	0	0	0	0	0	0
1	0	0	2	0	2	0	0	0	0	0	0	2	0	6	4	0	0	-4	4	-8	4	0	0	4	0	-4	-4	0	-4	-8	0	4
2	0	0	0	2	0	0	0	2	0	0	2	0	4	0	2	4	0	-8	-4	-4	0	0	4	-4	4	4	0	-8	-4	4	0	0
3	0	6	2	2	0	0	0	2	0	0	2	0	0	2	0	0	0	4	8	-4	-4	-8	4	0	4	0	4	0	0	4	0	4
4	0	0	2	0	4	0	0	2	2	2	0	2	0	0	0	2	0	-4	0	-4	4	0	4	0	0	4	0	4	12	0	-4	0
5	0	0	2	0	2	0	4	0	0	2	4	0	0	0	0	2	0	-8	4	4	0	-8	-4	-4	0	0	-4	4	0	0	4	-4
6	0	2	0	0	0	4	0	2	2	0	0	4	0	0	2	0	0	4	4	0	0	4	0	0	-12	-4	0	0	-4	0	-4	0
7	0	0	0	0	0	8	0	0	4	0	4	0	0	0	0	0	0	0	0	0	8	0	8	0	-4	-4	4	4	-4	4	4	-4
8	0	0	2	2	0	0	0	0	4	0	0	0	2	2	4	0	0	-4	4	0	-4	8	0	-4	4	0	8	4	0	-4	4	0
9	0	2	0	6	0	0	4	0	0	0	0	0	2	0	2	0	0	0	0	-8	0	0	-8	0	-4	-4	4	-4	4	4	4	-4
A	0	2	4	0	2	0	0	0	0	0	0	2	2	2	0	2	0	-4	0	4	-4	0	4	0	0	-12	0	-4	4	0	-4	0
B	0	2	2	0	0	0	0	0	4	0	0	2	0	0	0	6	0	0	4	4	8	0	-4	4	8	0	4	-4	0	0	-4	-4
C	0	0	0	0	2	4	0	2	0	2	2	0	4	0	0	0	0	0	4	4	0	0	4	4	-4	4	0	-8	4	-8	4	0
D	0	2	0	0	2	0	0	4	2	2	2	0	0	0	2	0	0	4	0	-4	-4	0	4	0	4	0	-4	0	0	-4	0	-12
E	0	0	4	0	0	4	0	0	4	0	4	0	0	0	0	0	0	0	-8	0	0	-8	0	0	0	0	8	0	0	-8	0	0
F	0	0	0	0	2	0	4	2	2	0	0	2	0	4	0	0	0	4	-4	0	4	0	0	-4	8	-4	-4	0	4	0	8	4

Mixed Bases for Efficient Inversion in $\mathbb{F}_{((2^2)^2)^2}$ and Conversion Matrices of SubBytes of AES

Yasuyuki Nogami, Kenta Nekado, Tetsumi Toyota,
Naoto Hongo, and Yoshitaka Morikawa

Graduate School of Natural Science and Technology, Okayama University
3-1-1, Tsushima–naka, Kita–ku, Okayama, Okayama 700-8530, Japan
nogami@trans.cne.okayama-u.ac.jp

Abstract. A lot of improvements and optimizations for the hardware implementation of SubBytes of Rijndael, in detail *inversion* in $\mathbb{F}_{2^8}$ have been reported. Instead of the Rijndael original $\mathbb{F}_{2^8}$, it is known that its isomorphic tower field $\mathbb{F}_{((2^2)^2)^2}$ has a more efficient inversion. For the towerings, several kinds of bases such as polynomial and normal bases can be used in *mixture*. Different from the meaning of this *mixture* of bases, this paper proposes another *mixture* that contributes to the reduction of the critical path delay of SubBytes. To the $\mathbb{F}_{(2^2)^2}$–inversion architecture, for example, the proposed *mixture* inputs and outputs elements represented with normal and polynomial bases, respectively.

1 Introduction

SubBytes of the Advanced Encryption Standard (AES), that is Rijndael, uses arithmetic operations in $\mathbb{F}_{2^8}$, especially *inversion* [8]. From the viewpoint of *hardware implementation*, it is said that *tower field* technique efficiently works and then a lot of efficient techniques have been reported [4,9]. In detail, instead of the Rijndael original $\mathbb{F}_{2^8}$, its *isomorphic* tower field $\mathbb{F}_{((2^2)^2)^2}$ is efficiently applied for calculating an inversion in SubBytes. According to Canright's work [2], there are 432 possible combinations of the modular polynomials and bases for constructing tower field $\mathbb{F}_{((2^2)^2)^2}$. Morioka et al's work [7] adopted only polynomial bases and Canright's work [2] did only normal bases; however, the difference causes little influence for the critical path delays. For example, another efficient construction that is introduced at **Sec.** 2.4 of this paper has the same *critical path delay*. It uses two normal bases and one polynomial basis for the towering bases in *mixture*. Different from the meaning of this *mixture* of bases, this paper proposes to use normal and polynomial bases in *mixture*.

When the tower field $\mathbb{F}_{((2^2)^2)^2}$ is used in SubBytes, it needs certain conversion matrices between the Rijndael original $\mathbb{F}_{2^8}$ and the tower field $\mathbb{F}_{((2^2)^2)^2}$. A few papers [2,6] have discussed the efficiency of conversion matrices. Most of those previous works just discuss the number of 1's in the conversion matrices; however, this paper focuses on their critical path delays only, in detail, the Hamming weights of the row vectors of the conversion matrices. It has been experimentally shown that there are some rare conversion matrices whose row vectors all have the Hamming weights less than or equal to 4. It is very important for

S. Mangard and F.-X. Standaert (Eds.): CHES 2010, LNCS 6225, pp. 234–247, 2010.

the hardware implementation. For such efficient conversion matrices, Canright's approach [2] such as *greedy* algorithm and *tree structure analysis* will be also applied to decrease the number of 1's in the matrices.

The *mixture* of bases proposed in this paper, in brief *mixed bases*, means the following usage of two different bases such as polynomial and normal bases. Let $A = a_0\beta + a_1\beta^4, a_0, a_1 \in \mathbb{F}_{2^2}$ be a non–zero element represented with normal basis $\{\beta, \beta^4\}$ in $\mathbb{F}_{(2^2)^2}$, where β is a zero of $g(x) = x^2 + x + \alpha$ and α is a zero of $e(x) = x^2 + x + 1$. Then, calculate its inverse $D = A^{-1}$ in $\mathbb{F}_{(2^2)^2}$ as

$$D = A^{-1} = (a_0\beta + a_1\beta^4)^{-1} = d_0 + d_1\beta, \ d_0, d_1 \in \mathbb{F}_{2^2}. \tag{1}$$

The most important point is that the input A is represented with normal basis $\{\beta, \beta^4\}$ but the output D is represented with polynomial basis $\{1, \beta\}$. This paper especially applies the *mixed bases* to the inversions in $\mathbb{F}_{(2^2)^2}$ and $\mathbb{F}_{((2^2)^2)^2}$. It is shown that the former contributes to the reduction of the critical path delay of $\mathbb{F}_{((2^2)^2)^2}$–inversion and the latter connects the $\mathbb{F}_{((2^2)^2)^2}$–inversion to some efficient conversion matrices. As previously introduced, the conversion matrices have smaller critical path delays and they are quite rare cases. In addition, it is shown that the use of the *mixed bases* has little influence to the number of gates needed for the logical architectures but reduces the critical path delays.

2   Preliminaries

This section briefly introduces the conventional construction of an inversion in tower field $\mathbb{F}_{((2^2)^2)^2}$ for the use in S-Box (SubBytes) of AES. In detail, let us review the adopted bases, modular polynomials, calculation procedure of an inversion in $\mathbb{F}_{((2^2)^2)^2}$, and then Morioka's work [7], Canrigt's work [2], and another efficient construction. Since the tower field is used with two conversion matrices for the isomorphism between $\mathbb{F}_{2^8}$ and $\mathbb{F}_{((2^2)^2)^2}$, the conventional viewpoints of the efficiency of conversion matrices are also introduced.

2.1   Extension Field $\mathbb{F}_{2^8}$ and Its Tower Construction $\mathbb{F}_{((2^2)^2)^2}$

8-bit inputs and outputs of the S-Box are dealt as elements in binary field of extension degree 8, that is $\mathbb{F}_{2^8}$. Among the arithmetic operations in the binary field, *inversion* plays an important role in SubBytes. In detail, the SubBytes calculates the multiplicative inverse A^{-1} of a non–zero input element $A \in \mathbb{F}_{2^8}^*$ and then carries out a certain Affine transformation. For the preparation of $\mathbb{F}_{2^8}$, AES [8] originally adopts an irreducible polynomial $x^8 + x^4 + x^3 + x + 1$ as the modular polynomial; however, it is well known that its isomorphic *tower* construction $\mathbb{F}_{((2^2)^2)^2}$ achieves a more efficient inversion together with Itoh–Tsujii inversion algorithm (ITA) [5]. In detail, first construct $\mathbb{F}_{2^2}$ by using the irreducible polynomial $e(x) = x^2 + x + 1$ over $\mathbb{F}_2{}^1$, then construct $\mathbb{F}_{(2^2)^2}$ by using a certain irreducible polynomial $f(x)$ of degree 2 over $\mathbb{F}_{2^2}$, and then construct $\mathbb{F}_{((2^2)^2)^2}$ by using a certain irreducible polynomial $g(x)$ of degree 2 over $\mathbb{F}_{(2^2)^2}$. Thus, the efficiencies of the arithmetic operations in $\mathbb{F}_{((2^2)^2)^2}$ are closely related to the

---

[1] Any other irreducible polynomials of degree 2 over $\mathbb{F}_2$ does not exist.

selection of the modular polynomials and the bases for the towerings. For example, polynomial and normal bases are efficient for multiplication and Frobenius mapping, respectively. In the case that the characteristic is equal to 2 such as AES, Frobenius mapping is equivalent to squaring.

2.2 Morioka's Construction [7]

Conventional works such as [4] have often referred to Morioka et al.'s construction [7] for achieving efficient inversion in $\mathbb{F}_{((2^2)^2)^2}$. Morioka's work [7] adopts $e(x) = x^2 + x + 1$ with its polynomial basis $\{1, \alpha\}$ for $\mathbb{F}_{2^2}$, $f(x) = x^2 + x + \alpha$ with its polynomial basis $\{1, \beta\}$ for $\mathbb{F}_{(2^2)^2}$, and $g(x) = x^2 + x + \lambda$, $\lambda = \alpha^2 \beta$ with its polynomial basis $\{1, \gamma\}$ for $\mathbb{F}_{((2^2)^2)^2}$, where $\alpha \in \mathbb{F}_{2^2}$, $\beta \in \mathbb{F}_{(2^2)^2}$, and $\gamma \in \mathbb{F}_{((2^2)^2)^2}$ are zeros of $e(x)$, $f(x)$, and $g(x)$, respectively. Note that it adopts polynomial bases for all towerings. Its *critical path delays* are summarized in **Table 1**.

2.3 Canright's Construction [2]

Different from Morioka's work, Canright's work [2] adopts $e(x) = x^2 + x + 1$ with its normal basis $\{\alpha, \alpha^2\}$ for $\mathbb{F}_{2^2}$, $f(x) = x^2 + x + \alpha$ with its normal basis $\{\beta, \beta^4\}$ for $\mathbb{F}_{(2^2)^2}$, and $g(x) = x^2 + x + \lambda$, $\lambda = \alpha^2 \beta$ with its normal basis $\{\gamma, \gamma^{16}\}$ for $\mathbb{F}_{((2^2)^2)^2}$, where $\alpha \in \mathbb{F}_{2^2}$, $\beta \in \mathbb{F}_{(2^2)^2}$, and $\gamma \in \mathbb{F}_{((2^2)^2)^2}$ are zeros of $e(x)$, $f(x)$, and $g(x)$, respectively. It is noted that it adopts normal bases for all towerings. Its *critical path delays* are summarized in **Table 1**.

2.4 Another Efficient Construction

This section introduces another efficient construction. Different from Morioka's and Canright's works, it adopts $e(x) = x^2 + x + 1$ with its normal basis $\{\alpha, \alpha^2\}$ for $\mathbb{F}_{2^2}$, $f(x) = x^2 + x + \alpha$ with its polynomial basis $\{1, \beta\}$ for $\mathbb{F}_{(2^2)^2}$, and $g(x) = x^2 + x + \lambda$, $\lambda = \alpha^2 \beta$ with its normal basis $\{\gamma, \gamma^{16}\}$ for $\mathbb{F}_{((2^2)^2)^2}$, where $\alpha \in \mathbb{F}_{2^2}$, $\beta \in \mathbb{F}_{(2^2)^2}$, and $\gamma \in \mathbb{F}_{((2^2)^2)^2}$ are zeros of $e(x)$, $f(x)$, and $g(x)$, respectively. Its *critical path delays* are summarized in **Table 1**.

The improvements proposed in this paper are started from this construction, thus in what follows let us briefly review its arithmetic operations in $\mathbb{F}_{2^2}$, $\mathbb{F}_{(2^2)^2}$, and $\mathbb{F}_{((2^2)^2)^2}$. Their calculation architectures are summarized in **App. A**.

Arithmetic operations in $\mathbb{F}_{2^2}$. In the same of Canright's work [2], construct $\mathbb{F}_{2^2}$ with the modular polynomial $e(x) = x^2 + x + 1$ and its normal basis $\{\alpha, \alpha^2\}$ as follows. According to the coefficients of $e(x)$ whose zero is α, $\alpha + \alpha^2 = 1$ and $\alpha^3 = 1$. Let $A = a_0 \alpha + a_1 \alpha^2$, $B = b_0 \alpha + b_1 \alpha^2$, $a_0, a_1, b_0, b_1 \in \mathbb{F}_2$, a multiplication $C = AB$ becomes as follows (**Fig. 7**).

$$
\begin{aligned}
AB &= (a_0 \alpha + a_1 \alpha^2)(b_0 \alpha + b_1 \alpha^2) \\
&= a_1 b_1 \alpha + a_0 b_0 \alpha^2 + (a_1 b_0 + a_0 b_1)(\alpha + \alpha^2) \\
&= \{(a_0 + a_1)(b_0 + b_1) + a_0 b_0\} \alpha + \{(a_0 + a_1)(b_0 + b_1) + a_1 b_1\} \alpha^2 \\
&= c_0 \alpha + c_1 \alpha^2 = C.
\end{aligned}
\tag{2}
$$

For a non–zero element A in $\mathbb{F}_{2^2}$, Frobenius mapping with respect to $\mathbb{F}_2$, that is squaring, is equivalent to inversion as follows (**Fig.** 8).

$$A^2 = A^{-1} = (a_0\alpha + a_1\alpha^2)^2 = a_0\alpha^2 + a_1\alpha^4 = a_1\alpha + a_0\alpha^2. \tag{3}$$

Times α and times α^2 are carried out as follows (**Fig.** 9).

$$\alpha A = a_0\alpha^2 + a_1\alpha^3 = a_1\alpha + (a_0 + a_1)\alpha^2, \tag{4a}$$
$$\alpha^2 A = a_0\alpha^3 + a_1\alpha^4 = (a_0 + a_1)\alpha + a_0\alpha^2. \tag{4b}$$

Arithmetic operations in $\mathbb{F}_{(2^2)^2}$. In the same of Morioka et al.'s work [7], construct $\mathbb{F}_{(2^2)^2}$ with the modular polynomial $g(x) = x^2 + x + \alpha$ and its polynomial basis $\{1, \beta\}$. Thus, the arithmetic operations and calculation procedures become as follows. Let $A = a_0 + a_1\beta$, $B = b_0 + b_1\beta, a_0, a_1, b_0, b_1 \in \mathbb{F}_{2^2}$, a multiplication $C = AB$ in $\mathbb{F}_{(2^2)^2}$ is carried out as follows (**Fig.** 10).

$$\begin{aligned} AB &= (a_0 + a_1\beta)(b_0 + b_1\beta) \\ &= (a_0 b_0 + a_1 b_1\alpha) + \{(a_0 + a_1)(b_0 + b_1) + a_0 b_0\}\beta \\ &= c_0 + c_1\beta = C. \end{aligned} \tag{5}$$

Frobenius mapping of A with respect to $\mathbb{F}_{2^2}$, that is 4–th power operation, becomes as follows.

$$A^{2^2} = a_0 + a_1\beta^4 = a_0 + a_1(\beta + 1) = (a_0 + a_1) + a_1\beta. \tag{6}$$

The square of A is calculated as follows (**Fig.** 11).

$$A^2 = a_0^2 + a_1^2\beta^2 = a_0^2 + a_1^2(\beta + \alpha) = (a_0^2 + a_1^2\alpha) + a_1^2\beta. \tag{7}$$

Let A be a non–zero element in $\mathbb{F}_{(2^2)^2}$, its inverse $D = A^{-1}$ is calculated by ITA as follows (**Fig.** 12).

$$\begin{aligned} A^{-1} &= (AA^4)^{-1}A^4 \\ &= \left\{(a_0 + a_1\beta)(a_0 + a_1\beta^4)\right\}^{-1}((a_0 + a_1) + a_1\beta) \\ &= \left\{a_0(a_0 + a_1) + a_1^2\alpha\right\}^{-1}((a_0 + a_1) + a_1\beta) \\ &= d_0 + d_1\beta = D. \end{aligned} \tag{8}$$

Times $\lambda = (\alpha + 1)\beta = \alpha^2\beta$, that is the constant term of the modular polynomial $g(x)$, is carried out as follows (**Fig.** 13).

$$\alpha^2\beta A = a_0\alpha^2\beta + a_1\alpha^2\beta^2 = a_0\alpha^2\beta + a_1\alpha^2(\beta + \alpha) = a_1 + (a_0 + a_1)\alpha^2\beta. \tag{9}$$

Inversion in $\mathbb{F}_{((2^2)^2)^2}$. In the same of Canright's construction [2], construct $\mathbb{F}_{((2^2)^2)^2}$ with the modular polynomial $g(x) = x^2 + x + \lambda$, $\lambda = \alpha^2\beta$ with its normal basis $\{\gamma, \gamma^{16}\}$. Let $A = a_0\gamma + a_1\gamma^{16}$, $a_0, a_1 \in \mathbb{F}_{(2^2)^2}$ be a non–zero element in $\mathbb{F}_{((2^2)^2)^2}$, ITA calculates its inverse $D = A^{-1}$ as follows (**Fig. 14**).

$$
\begin{aligned}
A^{-1} &= (AA^{16})^{-1}A^{16} \\
&= \left\{ a_0 a_1 (\gamma + \gamma^{16})^2 + (a_0^2 + a_1^2)\gamma\gamma^{16} \right\}^{-1} \left(a_1\gamma + a_0\gamma^{16} \right) \\
&= \left\{ a_0 a_1 + (a_0 + a_1)^2\lambda \right\}^{-1} \left(a_1\gamma + a_0\gamma^{16} \right) \\
&= d_0\gamma + d_1\gamma^{16} = D.
\end{aligned}
\tag{10}
$$

Efficiencies of various tower fields. One of typical features of Morioka's work [7] is that all of the towering bases are polynomial bases such as $\{1, \alpha\}$ for $\mathbb{F}_{2^2}$. As introduced in Canright's work [2], not only polynomial bases but also normal bases are available for the towering bases and it is said that there are 432 possible combinations. Canright's work [2] has introduced an efficient construction of tower field $\mathbb{F}_{((2^2)^2)^2}$ that uses normal bases for all towerings. As introduced in [2], it will be one of the best combinations for tower field $\mathbb{F}_{((2^2)^2)^2}$; however, such *good* constructions of *inversion* in tower field $\mathbb{F}_{((2^2)^2)^2}$ have a comparable compactness. According to his detail report [3], the best *inversion* introduced in [2] and that shown in **Sec.** 2.4 have almost the same compactness. In addition, the improvements and optimizations introduced in Morioka et al.'s and Canright's works [7], [2] will be also efficiently applied to the inversions shown in this paper. Thus, this paper focuses on the inversion in $\mathbb{F}_{((2^2)^2)^2}$ and the conversion matrices with the viewpoint of *critical path delay* and without discussing the compactness.

2.5   Conversion Matrices with the Viewpoint of Conjugates

As shown in **Fig.** 1 and the following **Eqs.** (12), when the inversion in the isomorphic *tower* field $\mathbb{F}_{((2^2)^2)^2}$ is applied to SubBytes instead of that of the Rijndael original $\mathbb{F}_{2^8}$, the input 8–bit vector needs to be converted to the corresponding element in $\mathbb{F}_{((2^2)^2)^2}$. Then, after calculating its inverse in $\mathbb{F}_{((2^2)^2)^2}$, the result needs to be returned to the Rijndael–original vector representation. Thus, two conversion matrices together with a certain Affine transformation are required before and after the inversion in $\mathbb{F}_{((2^2)^2)^2}$ (**Fig.** 1).

(a) encryption phase

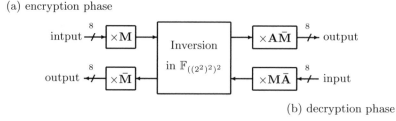

(b) decryption phase

Fig. 1. Sharing the inversion for encryption/decryption with conversion matrices

In detail, let $\{1, \omega, \cdots, \omega^6, \omega^7\}$ be the polynomial basis of $\mathbb{F}_{2^8}$, where ω is a zero of the modular polynomial $x^8 + x^4 + x^3 + x + 1$, the Rijndael originally represents 8–bit vector as an element $\tilde{X}$ in $\mathbb{F}_{2^8}$ as follows.

$$\tilde{X} = \tilde{x}_0 + \tilde{x}_1 \omega + \cdots + \tilde{x}_6 \omega^6 + \tilde{x}_7 \omega^7 = (\tilde{x}_0, \tilde{x}_1, \cdots, \tilde{x}_6, \tilde{x}_7). \tag{11}$$

Then, SubBytes for *encryption* phase calculates

$$\tilde{Z}^T = \mathbf{A}\left(\bar{\mathbf{M}}\left(\left(\mathbf{M}\tilde{X}^T\right)^{-1}\right)\right) + (0, 1, 1, 0, 0, 0, 1, 1)^T, FS \tag{12a}$$

where $\mathbf{M}$, $\bar{\mathbf{M}} = \mathbf{M}^{-1}$, and $\mathbf{A}$ denote the conversion, inverse conversion, and Affine transformation matrices, respectively. Thus, $X = \mathbf{M}\tilde{X}$ becomes an element in the tower field $\mathbb{F}_{((2^2)^2)^2}$ and then its inverse X^{-1} is efficiently calculated in $\mathbb{F}_{((2^2)^2)^2}$. As understood from **Eq.** (12a), $\mathbf{A}\bar{\mathbf{M}}$ is precomputed.

Inversely, SubBytes for *decryption* phase calculates

$$\tilde{X}^T = \bar{\mathbf{M}}\left(\left(\mathbf{M}\left(\bar{\mathbf{A}}\tilde{Z}^T + (0, 0, 0, 0, 0, 1, 0, 1)^T\right)\right)^{-1}\right).BS \tag{12b}$$

In this case, $Z = \mathbf{M}\left(\bar{\mathbf{A}}\tilde{Z}^T + (0, 0, 0, 0, 0, 1, 0, 1)^T\right)$ becomes an element in the tower field $\mathbb{F}_{((2^2)^2)^2}$ and then its inverse Z^{-1} is efficiently calculated in $\mathbb{F}_{((2^2)^2)^2}$. In the same of the encryption phase, $\mathbf{M}\bar{\mathbf{A}}$ and $\mathbf{M}(0, 0, 0, 0, 0, 1, 0, 1)^T$ are precomputed. Note here that the inversions in *encryption* phase **Eq.** (12a) and *decryption* phase **Eq.** (12b) can be carried out in the same procedure such as **Fig. 14**. Thus, previous works such as Canright's [2] works have mostly focused on the compact construction of inversion in tower field $\mathbb{F}_{((2^2)^2)^2}$ but not together with the efficiency of the conversion matrices in detail.

In the case of the efficient construction shown in **Sec. 2.4**, for example, the conversion matrices are given as follows (**Table 1**).

$$\mathbf{M} = \begin{bmatrix} 1 & 0 & 1 & 0 & 0 & 1 & 0 & 0 \\ 1 & 0 & 0 & 0 & 1 & 0 & 1 & 0 \\ 0 & 0 & 1 & 1 & 1 & 1 & 0 & 0 \\ 0 & 1 & 0 & 1 & 0 & 1 & 0 & 0 \\ 1 & 1 & 0 & 1 & 0 & 1 & 0 & 0 \\ 1 & 0 & 1 & 1 & 0 & 0 & 0 & 1 \\ 0 & 1 & 0 & 0 & 0 & 1 & 1 & 1 \\ 0 & 0 & 1 & 0 & 1 & 0 & 1 & 0 \end{bmatrix} \cdot \mathbf{A}\bar{\mathbf{M}} = \begin{bmatrix} 1 & 1 & 0 & 1 & 1 & 1 & 0 & 0 \\ 0 & 1 & 1 & 1 & 0 & 1 & 1 & 1 \\ 0 & 1 & 0 & 0 & 0 & 1 & 1 & 0 \\ 1 & 0 & 0 & 1 & 1 & 1 & 1 & 1 \\ 0 & 1 & 1 & 0 & 0 & 0 & 1 & 0 \\ 1 & 0 & 0 & 0 & 0 & 0 & 1 & 0 \\ 0 & 0 & 0 & 1 & 1 & 0 & 1 & 0 \\ 0 & 1 & 1 & 1 & 0 & 0 & 1 & 1 \end{bmatrix} \cdot \tag{13a}$$

$$\bar{\mathbf{M}} = \begin{bmatrix} 0 & 1 & 0 & 0 & 0 & 0 & 0 & 0 \\ 1 & 0 & 1 & 0 & 1 & 0 & 1 & 0 \\ 1 & 0 & 1 & 0 & 0 & 1 & 1 & 1 \\ 0 & 1 & 1 & 0 & 1 & 0 & 1 & 1 \\ 1 & 0 & 1 & 0 & 0 & 1 & 0 & 0 \\ 1 & 0 & 0 & 1 & 0 & 1 & 1 & 0 \\ 0 & 1 & 1 & 1 & 0 & 1 & 1 & 0 \\ 1 & 0 & 1 & 0 & 0 & 1 & 0 & 1 \end{bmatrix} \cdot \mathbf{M}\bar{\mathbf{A}} = \begin{bmatrix} 1 & 1 & 0 & 1 & 0 & 0 & 0 & 0 \\ 1 & 0 & 1 & 0 & 0 & 1 & 0 & 0 \\ 1 & 1 & 0 & 0 & 1 & 0 & 0 & 1 \\ 1 & 1 & 0 & 1 & 0 & 1 & 1 & 1 \\ 1 & 0 & 0 & 1 & 0 & 0 & 0 & 0 \\ 0 & 1 & 0 & 1 & 0 & 0 & 1 & 1 \\ 1 & 1 & 0 & 0 & 0 & 0 & 0 & 0 \\ 0 & 0 & 0 & 1 & 1 & 0 & 0 & 0 \end{bmatrix} \cdot \tag{13b}$$

Efficiency of conversion matrices. These conversion matrices are easily determined but they are not uniquely determined because the modular polynomials such as $e(x) = x^2 + x + 1$ have conjugate elements as zeros. In detail, in the case of **Sec**. 2.2, since α has its conjugate α^2 with respect to $\mathbb{F}_2$, $\{1, \alpha^2\}$ can be the basis of $\mathbb{F}_{2^2}$. In the same, $\{1, \beta^4\}$ and $\{1, \gamma^{16}\}$ can be the *towering* bases of $\mathbb{F}_{(2^2)^2}$ and $\mathbb{F}_{((2^2)^2)^2}$, respectively. Thus, there are 8 variants for each matrix and they play the same role on the connection to Rijndael original $\mathbb{F}_{2^8}$. Most of previous works such as Mentens's work [6] have basically focused on the number of 1's in the conversions matrices to evaluate their efficiencies.

This paper focuses on that every Hamming weight of row vectors of $\mathbf{M}$ shown in **Eq**. (13a) is smaller than or equal to 4. It is very important for the hardware implementation. For example, let us consider the following vector multiplications (inner products). Its hardware calculation will be implemented as **Fig**. 2.

$$(1,1,1,1,0,0,0,0)(x_0, x_1, x_2, x_3, x_4, x_5, x_6, x_7)^T, \tag{14a}$$

$$(1,1,1,1,1,0,0,0)(x_0, x_1, x_2, x_3, x_4, x_5, x_6, x_7)^T. \tag{14b}$$

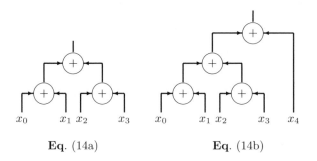

Eq. (14a)   **Eq**. (14b)

Fig. 2. Implementations of **Eq**. (14a) and **Eq**. (14b)

Thus, in the case of **Eq**. (13a), since every Hamming weight of row vectors of $\mathbf{M}$ is smaller than or equal to 4, it is efficiently implemented as shown in **Fig**. 2 and then its critical path delay becomes $2\,T_X$, where in what follows T_X and T_A denote the delays of XOR and AND, respectively. Such an efficient conversion matrix is a quite rare case, therefore, as shown in **Eqs**. (13), $\mathbf{M}$ has the efficiency but the other matrices such as $\mathbf{A\bar{M}}$ do not (**Table** 1).

Since it has been introduced that the Hamming weights of the matrices are reduced by some techniques such as *tree structure* [2], this paper does not discuss the weights of matrices into detail. Then, from the viewpoint of *critical path delay*, this paper proposes an efficient inversion in $\mathbb{F}_{((2^2)^2)^2}$ and conversion matrices to which *polynomial* and *normal* bases are used in *mixture*.

3   Main Proposal

This paper proposes an efficient architecture for *inversion* in tower field $\mathbb{F}_{((2^2)^2)^2}$ to which, different from Morioka et al.'s proposal [7] and Canright's approaches [2], polynomial and normal bases are used in *mixture*, in brief *mixed bases*.

Table 1. Comparison of the efficiencies of three constructions

construction		# of 1's	critical path delay[‡]
	M	32	$3\ T_X$
	A$\bar{\text{M}}$	29	$3\ T_X$
Morioka et al. [7]	inv. in $\mathbb{F}_{((2^2)^2)^2}$	–	$17\ T_X + 4\ T_A$
	$\bar{\text{M}}$	27	**$2\ T_X$**
	M$\bar{\text{A}}$	29	$3\ T_X$
	M	32	$3\ T_X$
	A$\bar{\text{M}}$	25	$3\ T_X$
Canright [2]	inv. in $\mathbb{F}_{((2^2)^2)^2}$	–	**$15\ T_X + 4\ T_A$**
	$\bar{\text{M}}$	29	$3\ T_X$
	M$\bar{\text{A}}$	26	$3\ T_X$
	M	28	**$2\ T_X$**
another efficient	**A$\bar{\text{M}}$**	33	$3\ T_X$
construction	inv. in $\mathbb{F}_{((2^2)^2)^2}$	–	**$15\ T_X + 4\ T_A$**
	$\bar{\text{M}}$	31	$3\ T_X$
	M$\bar{\text{A}}$	26	$3\ T_X$

Especially based on the *inversion* in $\mathbb{F}_{((2^2)^2)^2}$ constructed as **Fig. 14**, the *mixed bases* are mainly applied to two calculation parts: I_4 and I_8. In detail, denote their new versions by $\hat{I}_4$ and $\hat{I}_8$, respectively,

- $\hat{I}_4$ has the input and output for $\mathbb{F}_{(2^2)^2}$–elements represented with normal basis $\{\beta, \beta^4\}$ and polynomial basis $\{1, \beta\}$, respectively,
- $\hat{I}_8$ has the input and output for $\mathbb{F}_{((2^2)^2)^2}$–elements represented with normal basis $\{\gamma, \gamma^{16}\}$ and polynomial basis $\{1, \gamma\}$, respectively.

Then, the critical path delay for *encryption* phase of SubBytes of AES becomes

$$2\ T_X + (\ 14\ T_X + 4\ T_A\ ) + 2\ T_X. \tag{15}$$

Together with the meaning of the *mixed* bases, in what follows, several improvements using *mixed bases* especially at I_4 and I_8 are shown in detail. Note here that the modular polynomials and bases are as introduced in **Sec. 2.4**.

3.1   Mixed Bases for I_4 of Fig. 14

As also introduced in [2], it is often said that *inversion* with normal basis is more efficient than that with polynomial basis because several Frobenius mappings are needed in ITA–based inversion. Inversely, it is often said that *multiplication* with polynomial basis is more efficient than that with normal basis because Karatsuba–based multiplication needs polynomial multiplications [1].

First, let us consider an *inversion* in $\mathbb{F}_{(2^2)^2}$ with the normal basis $\{\beta, \beta^4\}$, where β is a zero of $g(x) = x^2 + x + \alpha$. Let $A = a_0\beta + a_1\beta^4$ be a non–zero element in $\mathbb{F}_{(2^2)^2}$, its inverse $D = A^{-1}$ is calculated by ITA as follows.

$$
\begin{aligned}
A^{-1} &= (AA^4)^{-1}A^4 \\
&= \left\{(a_0\beta + a_1\beta^4)(a_1\beta + a_0\beta^4)\right\}^{-1}\left(a_1\beta + a_0\beta^4\right) \\
&= \left\{a_0a_1 + (a_0 + a_1)^2\alpha\right\}^{-1}\left(a_1\beta + a_0\beta^4\right) \\
&= d_0\beta + d_1\beta^4 = D.
\end{aligned}
\tag{16a}
$$

However, the following multiplications in $\mathbb{F}_{(2^2)^2}$ denoted by M_4 in **Fig**. 14 cannot accept $\mathbb{F}_{(2^2)^2}$–elements represented with the normal basis. Because, they accept ones represented with the polynomial basis $\{1, \beta\}$. Thus, consider the following inversion in $\mathbb{F}_{(2^2)^2}$ with a non–zero element $A = a_0\beta + a_1\beta^4$.

$$
\begin{aligned}
A^{-1} &= (AA^4)^{-1}A^4 \\
&= \left\{(a_0\beta + a_1\beta^4)(a_1\beta + a_0\beta^4)\right\}^{-1}\left(a_1\beta + a_0\beta^4\right) \\
&= \left\{a_0a_1 + (a_0 + a_1)^2\alpha\right\}^{-1}\left((a_0 + a_1) + a_0\beta\right) \\
&= d_0 + d_1\beta = D.
\end{aligned}
\tag{16b}
$$

Based on **Eq**. (16b), the calculation architecture of the new version $\hat{I}_4$ is constructed as **Fig**. 3. It is the meaning of *mixed bases*. If I_4 in **Fig**. 14 that is constructed with the polynomial basis $\{1, \beta\}$ is replaced to the *inversion* with normal basis $\{\beta, \beta^4\}$, that is denoted by $\hat{I}_4$ (**Fig**. 3), the critical path delay of I_8 constructed as **Fig**. 14 is reduced to $14\,T_X + 4\,T_A$.

input (Normal basis)   output (Polynomial basis)

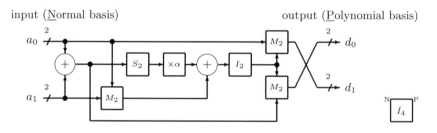

Fig. 3. Inversion in $\mathbb{F}_{(2^2)^2}$ with normal and polynomial bases ($\hat{I}_4$)

On the other hand, $\hat{I}_4$ (**Fig**. 3) needs a non–zero input represented with the normal basis $\{\beta, \beta^4\}$ in $\mathbb{F}_{(2^2)^2}$. Without increasing the critical path delay, it needs two changes at $\times\lambda$ and M_4 in **Fig**. 14 before the inversion in $\mathbb{F}_{(2^2)^2}$. Their output elements are originally represented with the polynomial basis $\{1, \beta\}$. Thus, change them so as to output $\mathbb{F}_{(2^2)^2}$–elements represented with the *normal basis* $\{\beta, \beta^4\}$. In detail, let $A = a_0 + a_1\beta$, $B = b_0 + b_1\beta, a_0, a_1, b_0, b_1 \in \mathbb{F}_{2^2}$ and based on the following calculations, their new versions denoted by $\times\hat{\lambda}$ and $\hat{M}_4$ are constructed as **Fig**. 4 and **Fig**. 5, respectively.

$$
\begin{aligned}
\lambda A &= a_0\alpha^2\beta + a_1\alpha^2\beta^2 \\
&= \{a_1 + (a_0 + a_1)\alpha^2\}\beta + a_1\beta^4 \\
&= (a_1\alpha + a_0\alpha^2)\beta + a_1\beta^4.
\end{aligned}
\tag{17}
$$

$$
\begin{aligned}
AB &= (a_0 + a_1\beta)(b_0 + b_1\beta) \\
&= \{(a_0 + a_1)(b_0 + b_1) + a_1b_1\alpha\}\beta + (a_0b_0 + a_1b_1\alpha)\beta^4 \\
&= c_0\beta + c_1\beta^4 = C.
\end{aligned}
\tag{18}
$$

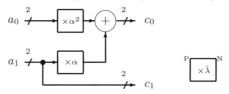

Fig. 4. Times λ in $\mathbb{F}_{(2^2)^2}$ with polynomial and normal bases ($\times\hat{\lambda}$)

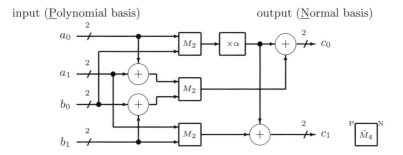

Fig. 5. Multiplication in $\mathbb{F}_{(2^2)^2}$ with polynomial and normal bases ($\hat{M}_4$)

3.2 Mixed Bases for the Inversion in $\mathbb{F}_{((2^2)^2)^2}$

The input and output elements for the inversion architecture constructed as **Fig**. 14 both need to be represented with the normal basis $\{\gamma, \gamma^{16}\}$. However, this paper changes only the representation of the output element to that with the polynomial basis $\{1, \gamma\}$. In detail, let $A = a_0\gamma + a_1\gamma^{16}$, $a_0, a_1 \in \mathbb{F}_{(2^2)^2}$ be a non–zero element in $\mathbb{F}_{((2^2)^2)^2}$, based on ITA, calculate its inverse $D = A^{-1}$ as

$$
\begin{aligned}
A^{-1} &= (AA^{16})^{-1}A^{16} \\
&= \{a_0a_1(\gamma + \gamma^{16})^2 + (a_0^2 + a_1^2)\gamma\gamma^{16}\}^{-1}(a_1\gamma + a_0\gamma^{16}) \\
&= \{a_0a_1 + (a_0 + a_1)^2\lambda\}^{-1}\{a_0 + (a_0 + a_1)\gamma\} \\
&= d_0 + d_1\gamma = D.
\end{aligned}
\tag{19}
$$

Note that, for a non–zero input represented with the normal basis $\{\gamma, \gamma^{16}\}$, it calculates its inverse represented with the polynomial basis $\{1, \gamma\}$. **Fig.** 6 shows its calculation architecture to which $\hat{I}_4$, $\hat{M}_4$, and $\times\hat{\lambda}$ are also applied.

input (<u>N</u>ormal basis)   output (<u>P</u>olynomial basis)

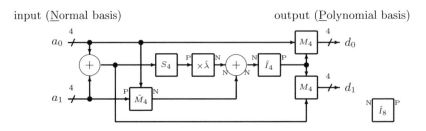

Fig. 6. Inversion in $\mathbb{F}_{((2^2)^2)^2}$ with normal and polynomial bases

As previously introduced, this inversion achieves $14\ T_X + 4\ T_A$; however, the last *mixed bases* used in **Eq.** (19) is not related to this efficiency. It is related to the efficiency of the conversion matrices. When the output is represented with the normal basis $\{\gamma, \gamma^{16}\}$, the calculated inverse A^{-1} is multiplied by the conversion matrix $\mathbf{A\bar{M}}$ shown in **Eqs.** (13a). On the other hand, in the case of the inversion constructed as **Fig.** 6, since the output is represented with the polynomial basis $\{1, \gamma\}$, it needs to be multiplied by the following conversion matrix $\mathbf{A\bar{M}M'}$,

$$\mathbf{A\bar{M} \times M'} = \begin{bmatrix} 1\,1\,0\,1\,1\,1\,0\,0 \\ 0\,1\,1\,1\,0\,1\,1\,1 \\ 0\,1\,0\,0\,0\,1\,1\,0 \\ 1\,0\,0\,1\,1\,1\,1\,1 \\ 0\,1\,1\,0\,0\,0\,1\,0 \\ 1\,0\,0\,0\,0\,0\,1\,0 \\ 0\,0\,0\,1\,1\,0\,1\,0 \\ 0\,1\,1\,1\,0\,0\,1\,1 \end{bmatrix} \times \begin{bmatrix} 1\,0\,0\,0\,1\,0\,0\,0 \\ 0\,1\,0\,0\,0\,1\,0\,0 \\ 0\,0\,1\,0\,0\,0\,1\,0 \\ 0\,0\,0\,1\,0\,0\,0\,1 \\ 1\,0\,0\,0\,0\,0\,0\,0 \\ 0\,1\,0\,0\,0\,0\,0\,0 \\ 0\,0\,1\,0\,0\,0\,0\,0 \\ 0\,0\,0\,1\,0\,0\,0\,0 \end{bmatrix} = \begin{bmatrix} 0\,0\,0\,1\,1\,1\,0\,1 \\ 0\,0\,0\,0\,0\,1\,1\,1 \\ 0\,0\,1\,0\,0\,1\,0\,0 \\ 0\,1\,1\,0\,1\,0\,0\,1 \\ 0\,1\,0\,0\,0\,1\,1\,0 \\ 1\,0\,1\,0\,1\,0\,0\,0 \\ 1\,0\,1\,1\,0\,0\,0\,1 \\ 0\,1\,0\,0\,0\,1\,1\,1 \end{bmatrix}, \quad (20a)$$

where $\mathbf{M'}$ is given by

$$\mathbf{M'} = \begin{bmatrix} 1\,0\,0\,0\,1\,0\,0\,0 \\ 0\,1\,0\,0\,0\,1\,0\,0 \\ 0\,0\,1\,0\,0\,0\,1\,0 \\ 0\,0\,0\,1\,0\,0\,0\,1 \\ 1\,0\,0\,0\,0\,0\,0\,0 \\ 0\,1\,0\,0\,0\,0\,0\,0 \\ 0\,0\,1\,0\,0\,0\,0\,0 \\ 0\,0\,0\,1\,0\,0\,0\,0 \end{bmatrix} \qquad (20b)$$

and it converts the vector representation with the polynomial basis $\{1, \gamma\}$ to that with the normal basis $\{\gamma, \gamma^{16}\}$. According to **Eq.** (20a), the conversion matrix $\mathbf{A\bar{M}M'}$ after the inversion in $\mathbb{F}_{((2^2)^2)^2}$ shown in **Fig.** 6 fortunately has the efficiency introduced in **Sec.** 2.5. Such an efficient conversion matrix is a quite rare case and it is experimentally found. Thus, the last *mixed bases* shown in **Fig.** 6 is just for obtaining this efficient conversion matrix $\mathbf{A\bar{M}M'}$.

3.3 Evaluation

Finally, the proposed architecture with conversion matrices, especially its *encryption* phase, has the critical path delays shown in **Table** 2.

Table 2. Critical path delays of the proposed architecture

construction			# of 1's	critical path delay[‡]
proposal	**M**	**Eq.** (13a)	28	**2** T_{X}
	AM̄M′	**Eq.** (20a)	27	**2** T_{X}
	inv. in $\mathbb{F}_{((2^2)^2)^2}$	**Fig.** 6	–	**14** T_{X} + **4** T_{A}

According to the result, this paper could show that the *mixed bases* contributes to some improvements of SubBytes of AES with tower field technique.

4 Conclusion and Future Work

This paper has proposed an efficient architecture for *inversion* in tower field $\mathbb{F}_{((2^2)^2)^2}$ to which, different from the conventional works, polynomial and normal bases are used in *mixture*, in brief *mixed bases*. Then, this paper has especially shown some improvements of the inversion architecture in $\mathbb{F}_{((2^2)^2)^2}$ and the conversion matrices in the *encryption* phase. As a future work, using *mixed bases*, those in the *decryption* phase should be also improved. Then, the detailed comparison with some other efficient implementations is needed. After that, a consideration for *side channel attacks* will be also required.

Acknowledgments

The authors would like to thank the anonymous referees for detailed review. We adequately appreciate their observations and helpful suggestions.

References

1. Bailey, D., Paar, C.: Optimal Extension Fields for Fast Arithmetic in Public–Key Algorithms. In: Krawczyk, H. (ed.) CRYPTO 1998. LNCS, vol. 1462, pp. 472–485. Springer, Heidelberg (1998)
2. Canright, D.: A Very Compact S-Box for AES. In: Rao, J.R., Sunar, B. (eds.) CHES 2005. LNCS, vol. 3659, pp. 441–455. Springer, Heidelberg (2005)
3. Canright, D.: Naval Postgraduate School Technical Report: NPS–MA–05–001 (2005), http://web.nps.navy.mil/~dcanrig/pub/NPS-MA-05-001.pdf
4. Canright, D., Batina, L.: A Very Compact "Perfectly Masked" S–Box for AES. In: Bellovin, S.M., Gennaro, R., Keromytis, A.D., Yung, M. (eds.) ACNS 2008. LNCS, vol. 5037, pp. 446–459. Springer, Heidelberg (2008)
5. Itoh, T., Tsujii, S.: A Fast Algorithm for Computing Multiplicative Inverse in GF(2^m) using Normal Basis. Inf. Comput. 78, 171–177 (1988)

6. Mentens, N.: Secure and Efficient Coprocessor Design for Cryptographic Applications on FPGAs, Doctor thesis, Katholieke Universiteit Leuven (2007)
7. Morioka, S., Satoh, A.: An optimized S–box circuit arthitecture for low power AES design. In: Kaliski Jr., B.S., Koç, Ç.K., Paar, C. (eds.) CHES 2002. LNCS, vol. 2523, pp. 172–186. Springer, Heidelberg (2003)
8. National Institute of Standards and Technology (NIST), Advanced Encryption Standard (AES), FIPS publication 197 (2001),
http://csrc.nist.gov/encryption/aes/index.html
9. Satoh, A., Morioka, S., Takano, K., Munetoh, S.: A compact Rijndael hardware architecture with S-box optimization. In: Boyd, C. (ed.) ASIACRYPT 2001. LNCS, vol. 2248, pp. 239–254. Springer, Heidelberg (2001)

A   Architectures of the Construction Shown in Sec. 2.4

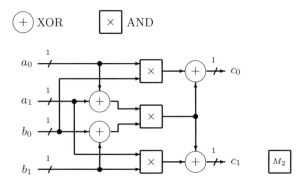

Fig. 7. Multiplication in $\mathbb{F}_{2^2}$ (M_2)

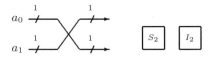

Fig. 8. Squaring (Frobenius mapping) in $\mathbb{F}_{2^2}$ (S_2, I_2)

Fig. 9. Times α and times α^2 in $\mathbb{F}_{2^2}$ ($\times\alpha$, $\times\alpha^2$)

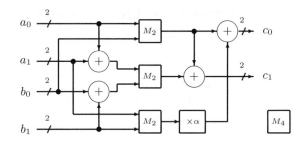

Fig. 10. Multiplication in $\mathbb{F}_{(2^2)^2}$ (M_4)

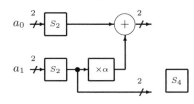

Fig. 11. Squaring in $\mathbb{F}_{(2^2)^2}$ (S_4)

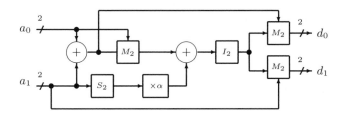

Fig. 12. Inversion in $\mathbb{F}_{(2^2)^2}$ (I_4)

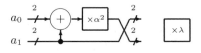

Fig. 13. Times λ in $\mathbb{F}_{(2^2)^2}$ $(\times \lambda)$

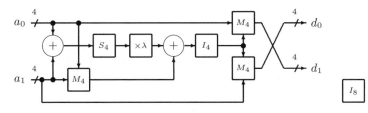

Fig. 14. Inversion in $\mathbb{F}_{((2^2)^2)^2}$

Developing a Hardware Evaluation Method for SHA-3 Candidates

Luca Henzen[1], Pietro Gendotti[2], Patrice Guillet[2], Enrico Pargaetzi[2],
Martin Zoller[2], and Frank K. Gürkaynak[3]

[1] Integrated Systems Laboratory, ETH Zurich
henzen@iis.ee.ethz.ch
[2] Department of Information Technology and Electrical Enginnering, ETH Zurich
{gpietro,pguillet,penrico,mzoller}@ee.ethz.ch
[3] Microelectronics Designs Center, ETH Zurich
kgf@ee.ethz.ch

Abstract. The U.S. National Institute of Standards and Technology encouraged the publication of works that investigate and evaluate the performances of the second round SHA-3 candidates. Besides the hardware characterization of the 14 candidate algorithms, the main goal of this paper is the description of a reliable methodology to efficiently characterize and compare VLSI circuits of cryptographic primitives. We took the opportunity to apply it on the ongoing SHA-3 competition. To this end, we implemented several architectures in a 90 nm CMOS technology, targeting high- and moderate-speed constraints separately. Thanks to this analysis, we were able to present a complete benchmark of the achieved post-layout results of the circuits.

1 Introduction

In 2007, the U.S. National Institute of Standards and Technology (NIST) started a public competition aiming at the selection of a new standard for cryptographic hashing [9]. Hash functions are cryptographic primitives that generate a sort of digital fingerprint of an arbitrary-length file, following some fundamental principles. Due to their flexibility, hash functions are used in a wide range of communication protocols where they provide data integrity, user authentication and many other security features. The motivation behind the NIST competition has been the growing concern of the security of two widely deployed hash functions MD5 and SHA-1 following a series of successful attacks [12,1,2]. The structural similarity of MD5 and SHA-1 with the current standard SHA-2 encouraged the NIST to start a new evaluation and selection process similar to the competition which promoted the Rijndael block cipher as new Advanced Encryption Standard (AES) in 2001. The cryptographic community was asked to propose new hash functions and to evaluate the security level of other candidates. In 2008, a total of 51 functions were accepted to the first round, while in July 2009 this number has been reduced to 14 second round candidates. The final decision, i.e., the proclamation of the winner algorithm, has been scheduled for 2012. To this end, the organizers are not only interested in the cryptographic strength of the candidates but also in the evaluation of the performance of the algorithm implemented in different platforms. The new SHA-3

S. Mangard and F.-X. Standaert (Eds.): CHES 2010, LNCS 6225, pp. 248–263, 2010.

standard is indeed expected to provide at least the security of SHA-2 with significantly improved efficiency. Several applications, from multi-gigabit mass storage devices to radio-frequency identification (RFID) tags, are expected to utilize SHA-3. It is therefore crucial that the final SHA-3 function should be flexible enough to be used in both high-performance and resource constrained environments. From a pure hardware point of view, the SHA-3 algorithm should provide good performance in terms of speed, area, and power.

Our interest in the SHA-3 selection process started with our involvement with the development of the candidate algorithm BLAKE. We participated in the algorithm specification, providing relevant information on the hardware performance and possible optimizations in this direction. When the SHA-3 competition entered the second phase, we started a VLSI characterization of several candidates within three separate student projects at our institute. The resulting designs were manufactured in three different ASICs, each containing a dedicated interface for I/O communication and the selected algorithms. At this time, we had implemented twelve out of fourteen candidate algorithms (all apart from ECHO and SIMD). We then decided to extend the analysis to all candidate algorithms.

In this paper we develop and present one methodology to evaluate the ASIC implementation of all SHA-3 second round algorithms. Rather than going for extremes of performance (fastest or smallest implementation) we propose to optimize all algorithms for multiple clearly defined specifications. We have applied our methodology and have evaluated several architectural variations of all candidate algorithms and presented the results.

The organization of the paper is as follows: A discussion of our methodology is the focus of Sect. 2. We present our approach to have a fair comparison, provide details and reasoning for key design decisions. Implementation details are given in Sect. 3. Due to limited space we were unable to provide implementation details for the architectures, an abbreviated summary of all architectures is provided in the Appendix. The results of our evaluation are presented in Sect. 4 together with a subsection that explains the errors in our methodology. We hope that this "open" approach will allow independent researchers to validate our findings. Finally in Sect. 5 we have concluding remarks.

2   Evaluation Methodology

In this work we will attempt to make a fair comparison between VLSI implementations of a set of algorithms all of which realize a similar function, but have very different structures. The main difficulty in this particular evaluation is the lack of concrete hardware specifications for the secure hash function candidates.

In practice, the specifications of the hardware are determined by the application. The hardware designers can then make several well-known trade-offs to come up with a design that offers the best compromise between, the required silicon area, the amount of energy required for the operation and the throughput/latency of the operation. For this study the requirements state *efficient hardware implementation* without being specific[1].

---

[1] This should not necessarily be understood as criticism for the NIST specifications. However, lack of concrete specifications make a fair comparison more difficult.

In some cases, such as telecommunication algorithms which have to fulfill requirements of certain well-defined standards, the application field alone sets sufficient constraints on the system. However cryptographic functions, like the SHA-3 hash function candidates that is the topic of this paper, are used for a very wide range of applications with different requirements. This makes it difficult to determine which of the performance parameters is more important. A hash function that is part of a battery operated wireless transmitter would probably be optimized for energy consumption, while the same algorithm when implemented in a telecommunication base station would most likely favor a high-throughput realization.

For comparative studies, if concrete specifications are not present, the authors will usually determine one parameter to be more important (i.e. throughput) [11,8,7], or will come up with aggregate performance metrics such as throughput per mm^2 [10, 3, 5]. Both approaches have their problems. Focusing on one parameter will favor algorithms which are strong on one parameter (i.e. throughput), but will not merit algorithms which perform better in other scenarios. Aggregate performance metrics on the other hand, may end up hiding the absolute performance of an implementation, impractical design corners (i.e. very large area, very low throughput) may perturb the results.

In the following subsection we will first define the performance metrics that we will consider in this evaluation. The next step will be to define specifications that will set limits on these performance metrics.

2.1   Performance Metrics

The most common metrics for hardware include the operation speed, the circuit area and the power consumption. For this analysis we have decided to use the following three main metrics for performance:

– **Circuit Area**
   Generally speaking the *cost* of an ASIC implementation of a function for a particular technology directly depends on the area required to realize the function[2]. In this evaluation we will use the net circuit area of a placed and routed design, including the overhead for power routing, clock trees. The area will be reported in kilo gate equivalents (kGE), where a gate equivalent corresponds to the area of a nominal drive strength 2-input NAND (or NOR) gate in the standard cell library used for the design realization. This metric covers the evaluation criteria *4.B.ii Memory requirements* in the NIST specification [9].
– **Throughput**
   We need a measure to determine how *fast* the implementation is. To this end we define the throughput of a hash function as the amount of message (input information) in bits for which a message digest can be computed per second. Furthermore, we assume that the hash function has been properly initialized, and the message sizes are matched to individual candidate functions for best case performance. The

---

[2] This is only true if the area is within a certain range. Extremely large circuits will have yield penalties, while very small circuits will not be able to justify the overhead associated with manufacturing.

throughput numbers are given in Gigabits per second (Gbps). This metric covers the evaluation criteria *4.B.i Computational Efficiency* in the NIST specification [9].

– **Energy Consumption**

Power and energy metrics have gained more importance in recent years. On one hand there are power density limits the circuits have to comply for sub 100 nm technologies, and on the other hand for systems with scarce energy resources (handheld devices, smartcards, RFID devices etc.) reduced energy consumption equals to increased functionality or longer operating time. In this evaluation we will consider the energy consumption as our metric and will calculate the energy per bit of input information processed by the hash function. This will be obtained by dividing the total power consumption (in Watts) by the throughput (Gigabits/s) described above. The energy consumption will be given in milli Joules per Gigabit (mJ/Gbit). This metric partly covers the evaluation criteria *4.C.i.b Flexibility* in the NIST specification [9] as the energy efficiency is a deciding factor for implementation in constrained environments.

2.2 SHA-3 Parameters

The SHA-3 Minimum Acceptability Requirements state that all candidates should support message digest sizes of 224, 256, 384, and 512 bits, and support a maximum message length of at least $2^{64} - 1$ bits. All algorithms process the message in blocks. The so-called *message block size* differs from algorithm to algorithm. In addition several submissions have included a *salt* input that can be used as a parameter in the hash function.

In our evaluation we have chosen:

– **Message Digest Size of 256**

Several algorithms use (slightly) different architectures for different output lengths. Additional circuitry is then required to support all possible digest sizes. By selecting a single length, we aim to focus on the core algorithm which also simplifies certain architectural decisions. Out of the four required sizes, we have eliminated 224 and 384 as they are not a power of two (always an advantage in hardware design). We have settled on 256 as it will usually result in smaller hardware and faster implementations.

– **Use the largest message block size available**

For each algorithm we have used the largest message block size and we have assumed that the message has already been padded (i.e. the length of the padded message is an exact multiple of the message block size). For throughput computation we always give the maximum achievable values, e.g., very long message for algorithms that have an initialization procedure.

– **No *salt* inputs**

Since not all algorithms provide such an input, we have not included any *salt* inputs. For algorithms that provide a salt, the inputs are set to their default values according to the specification, and these constants have been propagated during synthesis to allow further optimizations whenever possible.

2.3 Defining Specifications

As mentioned earlier, the main difficulty in this evaluation is the lack of precise specifications that the candidate algorithms have to fulfill. Hardware design is based on finding a compromise between competing parameters that determine circuit performance. For example, there are several architectural transformations that allow to increase the throughput at the expense of the circuit area (see [6]). Without guiding specifications, it is difficult to determine which of the circuit metrics is more important for a design.

In summary, the NIST specifications in [9] require that the candidate algorithms to be computationally efficient (4.B.i), have limited memory requirements (4.B.ii), to be flexible (4.C.i) and simple (4.C.ii) [3].

The classical way to perform this analysis would be to concentrate on only the throughput metric and try to find out which algorithms are the fastest. In the last year, several groups presented comparative works and, almost certainly, others will be publishing new results to this effect. However, if only the maximum throughput requirement is investigated the *flexibility* of candidate algorithms may not be visible. Therefore we suggest to use two separate specifications: an aggressive **high-throughput** target and a **moderate-throughput** target.

The high throughput target has been chosen to be beyond the expected performance of most algorithms, and would therefore still be able to rank the algorithms in their maximum throughput capability. Our observation has been that even with older fabrication technologies, such as 180 nm CMOS, several candidate algorithms are able to reach throughputs of multiple Gigabits/s.

There are certainly applications which could make use of such throughputs, however such data rates are way beyond the requirements for many applications. For the moderate throughput requirement we have decided to determine a throughput which is at least two orders of magnitude lower than that used in the first case.

Fixing one of the performance metrics, allows us to make a fairer comparison between the remaining performance metrics (area and energy), and by considering two distinct throughput targets, we hope to uncover the *flexibility* of the candidate algorithms for different operational requirements. In particular, we will be interested in the circuit area for our high-throughput target, while we will be more interested in the energy consumption for our moderate-throughput target.

The maximum achievable throughput by a circuit implementing a cryptographic algorithm depends on the specific technology into which the circuit will be mapped. A throughput value that is easily achieved in 65 nm process, may not be feasible at all when using a 180 nm process. Therefore the specifications for our two scenarios have to be chosen while considering the capabilities of our target process.

We have decided to use the 90 nm CMOS process by UMC with the free libraries from Faraday Technology Corporation, mainly because we already had experience in designing ASICs with this technology and it was readily available within our design environment at the time of this study.

---

[3] Note that, computational efficiency could be interpreted in different ways, however, in the NIST specification it is stated that the "*computational efficiency essentially refers to the speed of the algorithm*". Similarly the memory requirements refer to the circuit area in hardware implementations.

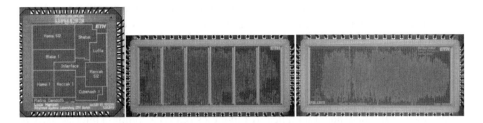

Fig. 1. From left to right: Photograph of the fabricated 90 nm chip implementing BLAKE, Cubehash, Hamsi, Keccak, Luffa and Shabal. Photograph of the 180 nm chips implementing BMW, Fugue, Grøstl, JH, SHAvite and Skein.

Table 1. Post-Layout results of the implemented algorithms

Algorithm	Area [kGE]	Throughput [Gbps]	Energy [mJ/Gbit]	Technology [nm]
BLAKE-32	33.55	7.314	15.291	UMC 90
BMW-256	95.00	3.527	31.407	UMC 180
CubeHash16/32-256	39.69	8.000	20.700	UMC 90
Fugue-256	26.00	2.806	122.506	UMC 180
Grøstl-256	65.00	4.064	73.075	UMC 180
Hamsi-256	32.25	7.467	23.624	UMC 90
Hamsi-512	68.66	7.467	46.605	UMC 90
JH-256	44.00	2.371	72.885	UMC 180
Keccak-256[†]	27.85	39.822	5.726	UMC 90
Keccak-512[†]	26.94	19.911	11.933	UMC 90
Luffa-256	29.70	22.400	9.482	UMC 90
Shabal-256	35.99	4.923	30.713	UMC 90
SHAvite-3$_{256}$	48.00	2.452	93.764	UMC 180
Skein-256-256	27.00	1.917	44.329	UMC 180

[†] First round specification.

Our experiences from designing the three ASICs (one of which was manufactured using this target technology) have given us a good estimation for the expected performance of all algorithms in the 90 nm process. We have decided to use 20 Gigabits/s for our high throughput target and 0.2 Gigabits/s for our moderate performance specifications. In the high-speed mode, almost all designs should be pushed to their speed limit, while with the latter we could evaluate the scalability and therefore the flexibility of each candidate algorithm.

2.4   ASIC Realizations

During this work twelve out of the fourteen second round SHA-3 candidates (some with several architectural variations) were fabricated in three different ASICs as shown in Fig. 1. Table 1 shows a list of algorithms that were implemented and their performances measured on the manufactured chips.

Actually implementing the designs in real silicon is certainly the best way to validate a design and determine its true potential. However, during this work we have realized that several practical factors have affected these results. The maximum available silicon area (that can be afforded for this project), the total number of I/O pins, the capabilities of the test infrastructure that is available for the test of the ASIC have all set limits on the implementations.

Since none of the designs was large enough to merit its own ASIC, each ASIC comprised of several independent modules. All modules shared a common interface which provided the inputs and collected the outputs from individual hash function realizing cores. For practical reasons, cores with similar clock frequencies were grouped together and were optimized using common constraints. In many cases compromises had to be made to allow two or more cores to be optimized at the same time. All of these had non-negligible influence on the outcome.

Practical considerations for testing of the systems has brought even more constraints. The necessity to include test structures (scan chains) adds some overhead, but more importantly, the maximum achievable clock rate greatly depends on the capabilities of the ASIC test infrastructure available. Designs with a high clock frequency (more than 500 MHz for 90 nm designs) put yet other constraints. When compared to designs running at lower frequencies, these designs suffer more from clock and power distribution problems, and are difficult to test at speed.

When designing these three ASICs we were forced to make many design decisions (i.e. blocks running faster than 700 MHz were deemed to be impractical within our environment) based on practical constraints which had its influence on the results. Scheduling constraints have also played a role in the choice of technology used to implement the designs. For the last two ASICs, there were no feasible 90 nm MPW (Multi Project Wafer) runs available. Consequently we had to submit these designs to a 180 nm run, which in turn made direct comparisons more difficult.

For this reason we have taken the design experience from the actual implementation of the individual cores, and have decided to re-implement all cores without considering these practical limitations. In particular we have decided:

- **No limits on the clock frequency**
  In this study we will not set any artificial limits on the clock rate. Obviously designs with high clock rates will still face the penalties for clock distribution, but we will not deal with practical considerations such as test, crosstalk and I/O limitations.
- **No test structures**
  Testing is an essential part of IC design. The exact overhead for testing depends on many factors, such as the desired test quality, and a one-size fits all solution is difficult to find[4]. Since the designs in this study will not be manufactured directly we chose not to include any test specific structures into the designs to have a fair comparison.
- **Assumed an ideal interface**
  The candidate algorithms differ in the number of I/Os they require. We have assumed that these core will eventually be part of a larger system which has an

---

[4] Simply using a full-scan methodology for example would not ensure that all designs have the same test coverage. Furthermore certain designs could be partially tested using functional vectors, or would be more amenable to BIST structures.

adequate I/O interface matching the requirements of each core. In this way, every function could express its maximum potentiality without suffering from any external limiation. However, we made no assumptions about how long the inputs stayed valid, all required inputs were sampled by the cores at the beginning of the operation. In other words, we implemented an internal message block memory for designs that require the input to be stable for more than one clock cycle.

- **No macro blocks**

  We have not used any macro blocks to realize look-up tables or register files for portability reasons. All look-up tables and memory blocks were realized by standard cells.

3 Implementation

3.1 Design Flow

The same design procedure was used for all candidate algorithms. We have first developed a *golden* model based on the *Known Answer Tests* provided by the submission package. This golden model was then used to generate the stimuli vectors and expected responses that we have used to verify the RTL description of the algorithm written in VHDL.

We have then used Synopsys Design Vision-2009.06 to map the RTL description to the UMC 90 nm technology using the fsd0a_a_2009Q2v2.0 RVT standard cell library from Faraday Technology Corporation. All outputs are assumed to have a capacitive loading of 50 fF (equivalent to the input capacitance of about 9 medium strength buffers), and the input drive strength is assumed to be that of a medium strength buffer (BUFX8).

We use the worst case condition (1.08 V, 125 °C) characterization of the standard cell libraries. We have decided to use worst case characterized libraries in order to guarantee that we can meet the specifications. Table 2 is given as a reference to be able to compare the three characterizations that are commonly available (worst, typical, best) for one of the candidate algorithms.

Table 2. Comparison of different characterizations, synthesis results for the ECHO algorithm

	Worst Case	Typical Case	Best Case
Supply Voltage	1.08 V	1.2 V	1.32 V
Temperature	125 °C	27 °C	-40 °C
Critical Path	3.49 ns	2.24 ns	1.59 ns
Throughput	13.75 Gbps	21.42 Gbps	30.19 Gbps
Relative Performance	64.2 %	100 %	140.9 %

Depending on the throughput requirements, we try different architectural transformations such as parallelization, pipelining to come up with an architecture that meets (or comes closest to meeting) the requirements. We then use the Cadence Design Systems Velocity-9.1 tool for the back-end design. The technology used in this evaluation uses 8 metal layers (metallization option 8m026), out of which the top-most two are double pitch (wider and thicker). A square floorplan is generated, leaving 30 µm space around the core for the power connections. For all designs we have used a 85 % utilization of the core area, in other words we have left 15 % of the area for post-layout optimization and power and ground distribution overhead. For power routing we have used a power grid utilizing Metal-7 and Metal-8.

Then the design is placed, a clock tree is synthesized and subsequently the design is routed. After every step the timing is checked, and if necessary a timing optimization is performed. At the end, if a valid layout without any Design Rule Check (DRC) violations are found, the *total core area* is reported as the area of the system. The total core area excludes the 30 µm space reserved for power rings, but includes all the available area that the placement and routing tool can use for the design. By default, all designs start with a 15 % overhead for post-layout optimizations. Depending on the design some amount of this overhead is used during various optimization phases during the back-end design. However it is difficult to quantify the *minimum required overhead* for every design reliably. We have decided to start all designs with the same initial placement density, and verified that the final design was not overly-congested. In a congested design, the routing solution includes many detours which adversely affect timing. For these designs the initial row utilization would have been reduced by 5 %, increasing the overhead. This was not necessary for any designs in this study[5]. In some designs, the routing resources are sparsely utilized. Such designs could have benefited from a higher initial row utilization, which could have resulted in a slightly smaller circuit without noticeable timing penalties. As mentioned earlier, it is not trivial to make sure that two designs have exactly the same amount of overhead. Therefore, we have not considered changing the default row utilization, unless there was a noticeable problem.

The timing results are taken from the finalized design. First, the Velocity tool is used to extract the post-layout parasitics and an SDF file containing the delays of all interconnections and instances is generated. The final netlist and the SDF files is read by the Mentor Graphics Modelsim-6.5a simulator and the functionality of the design is verified. At the same time, a Value Change Dump (VCD) file that records the switching activity of all the nodes during the simulation is produced. To have more realistic results, the start of the VCD file is chosen after the circuit has been properly initialized. This VCD file is then read back into the Velocity tool and a statistical power analysis is performed. The *Total Power* number is used to determine the energy consumption of the system.

3.2 Algorithms

For a given candidate algorithm, there are several well-known architectural transformations such as parallelization, pipelining, loop-unrolling etc. that will allow different trade-offs between circuit size and throughput. In addition, within the submission document, the authors often suggest different computational methods to perform a specific transformation of their candidate function. A good example is the frequently used

---

[5] Note that the initial density strongly depends on the technology options such as used metal layers. We have used 85 % as a result of our previous experience with this particular technology.

substitution boxes. They can be implemented as look-up tables, or can be realized as a circuit that computes the underlying function mathematically. To make matters worse, the exact trade-off between alternative realizations may only be visible after placement and routing. All these aspects broaden the spectrum of the possible hardware architectures. For a single candidate, there is often a large set of circuits with different trade-offs between size and speed. To identify the *best* design among many possibilities is not a trivial task. Despite all attempts to formalize architectural exploration, our experience has been that optimizing the circuit still remains a manual task, that relies on the skill and experience of the designer.

In this work, for each candidate algorithm we have selected what we believe was the most appropriate architecture that was able to reach the target throughput (20 and 0.2 Gbps) with minimal resources. For every candidate we designed and implemented two different architectures. The specifications of the single designs used within this work, is given in App. A. We make no claims that any of the architectures we have reported in this paper is the *best* possible architecture for a given candidate algorithm. In our opinion, it is not possible to make such a claim, and the exact implementations should be open to public scrutiny and review. For this purpose we have made all the source code that was used for this evaluation public on our www site [4].

4 Results

In this section we present the performance of the circuits implemented for high and moderate speed environments. The comparison between these two scenarios gives a further overview of the efficiency and flexibility of the candidate algorithms. We will refrain from concluding remarks about the performance of the algorithms, as we do not consider the results complete without public scrutiny.

For each architecture we report two operating frequencies/throughputs. The *Maximum Clock Frequency* is the maximum achievable clock frequency of the given architecture. When operating with this clock frequency the circuit can achieve the given *Maximum Achievable Throughput*. In most cases, this throughput is not exactly the same as the required throughput (either 20 or 0.2 Gbps). The second clock frequency states the clock frequency required to reach the target throughput. The final value in the tables is a relative indicator of how close the architecture is in achieving the target clock frequency. A number lower than one means that the architecture failed to achieve the target throughput. One can take this as a ratio of how closely we were able to optimize the circuit to the given target performance.

4.1 High Throughput Scenario

As expected, not all the circuits optimized for high-speed were able to reach the target throughput. Only two algorithms, Keccak and Luffa, were able to achieve the constraint. Table 3 lists the main performance figures for all architectures. In this scenario both area and energy were sacrificed to achieve high-throughput. The corresponding layouts can be seen in Fig. 2. The scale is given in the lower right corner of the figure. Circuits with a higher congestion rate (i.e. BMW or SIMD) require indeed the entire core for routing,

Blake   BMW   CubeHash ECHO   Fugue   Grøstl   Hamsi

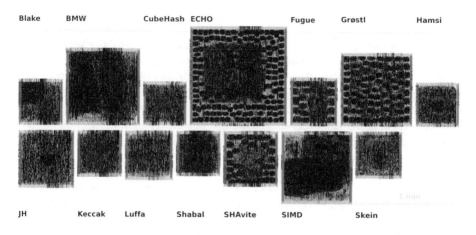

JH   Keccak   Luffa   Shabal   SHAvite   SIMD   Skein

Fig. 2. The final layouts of all candidate algorithms for a target throughput of 20 Gbps

and would probably reach a faster throughput with more core area, i.e., a lower row utilization. Particularly interesting is also the local congestion for the 8-bit LUT-based S-boxes which makes them easily identifiable within ECHO, Grøstl, Fugue, and partly SHAvite.

4.2 Medium Throughput Scenario

The moderate-throughput circuits match the target throughput of 0.2 Gbps without difficulty. As can be seen in Table 4 the maximum achievable clock rate always exceeds the clock frequency required for 0.2 Gbps operation. To some extent, the additional speed can be traded to reduce the overall energy consumption, by lowering the supply voltage. It must be noted that there is a lower limit for the supply voltage (around 0.5 V for this process). Such voltage scaling techniques were not considered in this comparison, all results are listed for 1.2 V supply voltage.

Since in this scenario, timing was quite relaxed, the main figure of merit becomes the area and the energy dissipation. The layouts of all fourteen architectures are compared in Fig. 3, with the scale indicated on the bottom left.

The most interesting result is that a smaller area (or indeed throughput) does not always equal lower energy consumption (see Hamsi or Skein compared to BMW or SIMD). It must be noted that, no special precautions were taken for a low-power design (i.e. proper clock-gating, input-silencing). In addition some architectural decisions resulted in increased number of operations and/or increased circuit activity which affected the energy consumption differently for separate algorithms. We believe that there is much room for improvement in terms of low-power performance of the architectures. We must conclude that the present specifications do not necessarily result in low-power realizations in the medium-throughput corner. In a next step, the design methodology could be extended to provide a low-power scenario.

Table 3. Post-layout performances of all candicate algorithms for a target throughput of 20 Gbps in the UMC 90 nm process

Algorithm	Area [kGE]	Energy [mJ/Gbit]	Maximum Achievable Throughput [Gbps]	Clock Frequency [MHz]	Clock Freq. for 20 Gbps Throughput [MHz]	Max. / Target Frequency Ratio
BLAKE-32	47.5	11.00	9.752	400	820	0.49
BMW-256	150.0	16.86	8.486	298	703	0.42
CubeHash16/32-256	42.5	13.71	10.667	667	1250	0.53
ECHO-256	260.0	43.41	13.966	291	417	0.70
Fugue-256	55.0	15.60	8.815	551	1250	0.44
Grøstl-256	135.0	14.13	16.254	667	820	0.81
Hamsi-256	45.0	15.90	8.686	814	1876	0.43
JH-256	80.0	17.54	10.807	760	1406	0.54
Keccak-256	50.0	2.42	43.011	949	441	2.15
Luffa-256	55.0	6.92	23.256	727	625	1.16
Shabal-256	45.0	14.83	6.819	693	2033	0.34
SHAvite-3_{256}	75.0	19.21	7.999	562	1406	0.40
SIMD-256	135.0	35.66	5.177	364	1406	0.26
Skein-256-256	50.0	30.47	3.558	264	1484	0.18

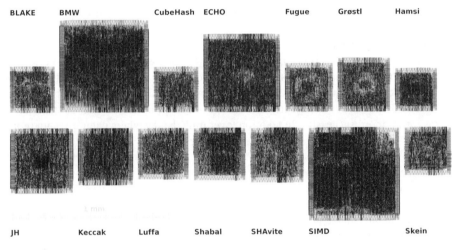

Fig. 3. The final layouts of all candidate algorithms for a target throughput of 0.2 Gbps

4.3 Sources of Error

Although we have tried our best to ensure a fair comparison, there are many factors that could have influenced the results. In this section we try to outline the possible sources of error in our results, and outline what we have done to address them.

– **Conflict of interest**

   One of the authors of this paper, Luca Henzen, is involved with the SHA-3 candidate algorithm BLAKE. Our interest in implementing the SHA-3 candidate algorithms has started by investigating optimal hardware implementations of BLAKE. We have tried to be as impartial as possible when implementing other candidate

Table 4. Post-layout performances of all candidate algorithms for a target throughput of 0.2 Gbps in the UMC 90 nm process

Algorithm	Area [kGE]	Energy [mJ/Gbit]	Maximum Achievable Throughput [Gbps]	Clock Frequency [MHz]	Clock Freq. for 0.2 Gbps Throughput [MHz]	Max. / Target Frequency Ratio
BLAKE-32	16.0	13.00	0.463	73.282	31.646	2.32
BMW-256	85.0	14.04	1.845	64.876	7.031	9.23
CubeHash16/32-256	16.0	10.50	1.741	217.581	25.000	8.70
ECHO-256	60.0	59.44	0.204	137.061	134.771	1.02
Fugue-256	19.0	9.02	1.828	114.260	12.500	9.14
Grøstl-256	25.0	22.28	0.412	128.750	62.500	2.06
Hamsi-256	15.0	35.12	0.200	150.083	149.925	1.00
JH-256	37.5	13.03	1.909	134.228	14.063	9.54
Keccak-256	27.5	5.50	6.767	149.276	4.412	33.83
Luffa-256	22.0	21.79	1.265	118.624	18.751	6.33
Shabal-256	25.0	26.57	0.399	128.634	64.475	2.00
SHAvite-3$_{256}$	25.0	11.43	1.871	131.527	14.063	9.35
SIMD-256	90.0	32.49	0.943	66.295	14.063	4.71
Skein-256-256	19.0	32.67	0.200	118.765	118.765	1.00

algorithms. However, it is true that we are more familiar with this algorithm than any other algorithm.

- **Designer experience**
  The algorithms have been implemented by a group of students over a period of several months. Different designers may have more or less success in optimizing a given design. We have confidence in our team, but it is possible that for some algorithms we have inadvertently missed a possible optimization while for the others we were more successful. In addition, over time the designers naturally gain more experience and are more successful with the designs.

  We believe that the most important aspect of a fair comparison is openness. For this reason we have made the source code and run scripts for the EDA tools used to implement all designs presented in this paper available on our website [4]. In this way, other groups can replicate our results, and can find and correct any mistakes we might have made in the process.

- **Accuracy of numbers**
  The numbers delivered by synthesis and analysis tools rely on the library files provided by the manufacturer. The values in the libraries are essentially statistical entities and sometimes have large uncertainties associated with it. In addition most of the design process involves heuristic algorithms which depending on a vast number of parameters can return different results. Our experience with synthesis tools suggest that the results have around ± 5% variation. We therefore consider results that are within 10% of each other to be comparable.

  In an effort to be more accurate we have chosen to report post-layout area numbers that include clock and power distribution overhead. We have designed all circuits with the same overhead. For some circuits this overhead is adequate, for others it is too much, and for others is insufficient. We made sure that there is an acceptable solution for all cases.

– **Bias through specification**

We have chosen two design corners in our applications, these specifications have helped us to have a common base for comparing all 14 algorithms. Regardless of how these specifications are chosen, it is possible that they benefit some algorithms more than the others. We hope that, similar studies by other groups which use different specifications will help to give a clearer picture.

– **Simplification due to assumptions**

All our assumptions, the specific choices we made for SHA-3 parameters and the practical choices we made in the design flow will have some effect on the results. For example, we have decided not take IR-drop or crosstalk effects into account. As a result, the cores that achieve their reported performance by using very high clock frequencies will be more difficult to realize in practice. The assumptions in the design flow are a practical necessity and were designed to create a methodology in which the same solution could be used for all designs.

5   Conclusions

In this paper we have presented a methodology to compare the SHA-3 candidate algorithms. Our previous experiences in designing ASIC implementations of candidate algorithms (Table 1) has been instrumental in developing what we believe is a fair set of specifications. Rather than targeting outright performance, we have set limits for one performance metric (throughput) and re-implemented all algorithms to meet two distinct throughput requirements. This enabled us to compare the *flexibility* of the algorithms (Tables 3 and 4).

A public selection process, such as the SHA-3 invariably attracts a large number of submissions with many different algorithms. In early stages of the selection process, the sheer number of algorithms (51 in the first round) makes it impractical to employ a detailed analysis for hardware suitability. Our experience has shown that even with the 14 second round candidates, it is difficult to present an authoritative and fair evaluation of all candidates. We believe that for the final round of evaluations, a similar approach to what we have demonstrated in this paper should be utilized: Clear constraints should be set for the implementations, preferably more than one performance corner should be targeted, the evaluation process should be well documented and the errors in the evaluation process should be openly discussed. We would also suggest the addition of a low-power corner that also considers voltage scaling for low-power operation to our methodology.

In many parts of this paper, we have extensively commented on limitations of our methodology, and have included a whole subsection on sources of error. We strongly believe that any such comparison must be thorough with its analysis of error sources and clear with its performance metrics.

References

1. De Cannière, C., Rechberger, C.: Finding SHA-1 characteristics: General results and applications. In: Lai, X., Chen, K. (eds.) ASIACRYPT 2006. LNCS, vol. 4284, pp. 1–20. Springer, Heidelberg (2006)

2. De Cannière, C., Rechberger, C.: Preimages for reduced SHA-0 and SHA-1. In: Wagner, D. (ed.) CRYPTO 2008. LNCS, vol. 5157, pp. 179–202. Springer, Heidelberg (2008)

3. El-Hadedy, M., Gligoroski, D., Knapskog, S.J., Aas, E.J.: Low area FPGA and ASIC implementations of the hash function "Blue Midnight Wish-256". In: International Conference on Computer Engineering & Systems, ICCES 2009, Cairo, pp. 10–14 (2009)

4. Gürkaynak, F.K., Henzen, L., Gendotti, P., Guillet, P., Pargaetzi, E., Zoller, M.: Hardware evaluation of the second-round SHA-3 candidate algorithms (2010),
http://www.iis.ee.ethz.ch/~sha3/

5. Gürkaynak, F.K., Luethi, P., Bernold, N., Blattmann, R., Goode, V., Marghitola, M., Kaeslin, H., Felber, N., Fichtner, W.: Hardware evaluation of eSTREAM candidates: Achterbahn, grain, mickey, mosquito, sfinks, trivium, vest, zk-crypt. eSTREAM, ECRYPT Stream Cipher Project, Report 2006/015 (2006), http://www.ecrypt.eu.org/stream

6. Kaeslin, H.: Digital Integrated Circuit Design, from VLSI Architectures to CMOS Fabrication. Cambridge University Press, Cambridge (2008)

7. Kobayashi, K., Ikegami, J., Matsuo, S., Sakiyama, K., Ohta, K.: Evaluation of hardware performance for the SHA-3 candidates using SASEBO-GII. Cryptology ePrint Archive, Report 2010/010 (2010), http://eprint.iacr.org/

8. Namin, A.H., Hasan, M.A.: Hardware implementation of the compression function for selected SHA-3 candidates. CACR 2009-28 (2009),
http://www.vlsi.uwaterloo.ca/~ahasan/hasan_report.html

9. NIST. Announcing request for candidate algorithm nominations for a new cryptographic hash algorithm (SHA-3) family. Federal Register 72(212) (2007),
http://www.nist.gov/hash-competition

10. Tillich, S., Feldhofer, M., Issovits, W., Kern, T., Kureck, H., Mühlberghuber, M., Neubauer, G., Reiter, A., Köfler, A., Mayrhofer, M.: Compact hardware implementations of the SHA-3 candidates ARIRANG, BLAKE, Grøstl, and Skein. Cryptology ePrint Archive: Report 2009/349 (2009)

11. Tillich, S., Feldhofer, M., Kirschbaum, M., Plos, T., Schmidt, J.-M., Szekely, A.: High-speed hardware implementations of BLAKE, Blue Midnight Wish, CubeHash, ECHO, Fugue, Grøstl, Hamsi, JH, Keccak, Luffa, Shabal, SHAvite-3, SIMD, and Skein. Cryptology ePrint Archive, Report 2009/510 (2009)

12. Wang, X., Yu, H.: How to break MD5 and other hash functions. In: Cramer, R. (ed.) EUROCRYPT 2005. LNCS, vol. 3494, pp. 19–35. Springer, Heidelberg (2005)

A   Hardware Architectures

Table 5 gives an overview of the architectures, used within this work. For some candidates we used the same design for the 20 Gbps (HS) and 0.2 Gbps (MS) analysis. In such cases, different optimization parameters were used. The detailed description of the architectures has been omitted because of the limited article length. Refer to [4] for the complete source code for all the architectures used in this evaluation

Table 5. Design specification of the HS and MS-target architectures. For the latency, the enclosed value refers to the finalization cycles.

Algorithm	Message Block Size [bits]	Arch.	Latency [cycles]	Implementation details
BLAKE	512	HS	21	Four parallel G function modules, anticipation of the first message-constant addition.
		MS	81	One G function module.
BMW	512	HS-MS	18 (+18)	f_0 and f_2 computed in one cycle, while f_1 iteratively decomposed in a single *expand* block.
CubeHash	256	HS	16 (+160)	Single round per cycle, initial state stored.
		MS	32 (+320)	Half round, initial state stored.
ECHO	1536	HS	32	8 AES rounds per clock cycle.
		MS	1034	Single 32-bit AES core, one parallel BigMixColumn unit.
Fugue	32	HS	2 (+37)	S-box as LUT.
		MS	2 (+37)	S-box as composite field logic.
Grøstl	512	HS	21 (+21)	Interleaved P and Q permutation with one pipeline stage, *SubBytes* as LUT.
		MS	160 (+160)	Single-column round (64-bit datapath), *SubBytes* as composite field.
Hamsi	32	HS	3 (+6)	Message expansion in three 256×256 LUTs, single round per cycle, substitution layer as logic.
		MS	24 (+48)	Same as HS, datapath reduced to 128 bits.
JH	512	HS-MS	36	S-boxes S_0 and S_1 stored in LUTs, constants stored.
Keccak	1088	HS-MS	24	Single round per cycle.
Luffa	256	HS	8	Three parallel *Step* function modules, *SubCrumb* function as logic.
		MS	24	One *Step* function modules, *SubCrumb* function as logic.
Shabal	512	HS	52 (+156)	One keyed permutation round per cycle. In total, 30 adders and 16 subtractors.
		MS	165	One adder and one subtractor only.
SHAvite-3	512	HS	36	One AES round for message expansion and one AES round for the F^3 round, *SubBytes* as LUT.
		MS	36	Same as HS, *SubBytes* in composite field.
SIMD	512	HS-MS	36 (+36)[†]	Four parallel Feistel modules, message expansion based on NNT_8 and eight multipliers for tweadle mult.
Skein	256	HS	19 (+19)	Four unrolled Threefish rounds.
		MS	152 (+152)	Half Threefish round.

[†] Further 36 cycles of initialization required for message expansion.

Fair and Comprehensive Methodology for Comparing Hardware Performance of Fourteen Round Two SHA-3 Candidates Using FPGAs*

Kris Gaj, Ekawat Homsirikamol, and Marcin Rogawski

ECE Department, George Mason University, Fairfax, VA 22030, U.S.A.
{kgaj,ehomsiri,mrogawsk}@gmu.edu
http://cryptography.gmu.edu

Abstract. Performance in hardware has been demonstrated to be an important factor in the evaluation of candidates for cryptographic standards. Up to now, no consensus exists on how such an evaluation should be performed in order to make it fair, transparent, practical, and acceptable for the majority of the cryptographic community. In this paper, we formulate a proposal for a fair and comprehensive evaluation methodology, and apply it to the comparison of hardware performance of 14 Round 2 SHA-3 candidates. The most important aspects of our methodology include the definition of clear performance metrics, the development of a uniform and practical interface, generation of multiple sets of results for several representative FPGA families from two major vendors, and the application of a simple procedure to convert multiple sets of results into a single ranking.

Keywords: benchmarking, hash functions, SHA-3, FPGA.

1 Introduction and Motivation

Starting from the Advanced Encryption Standard (AES) contest organized by NIST in 1997-2000 [1], open contests have become a method of choice for selecting cryptographic standards in the U.S. and over the world. The AES contest in the U.S. was followed by the NESSIE competition in Europe [2], CRYPTREC in Japan, and eSTREAM in Europe [3].

Four typical criteria taken into account in the evaluation of candidates are: security, performance in software, performance in hardware, and flexibility. While security is commonly recognized as the most important evaluation criterion, it is also a measure that is most difficult to evaluate and quantify, especially during a relatively short period of time reserved for the majority of contests. A typical outcome is that, after eliminating a fraction of candidates based on security flaws, a significant number of remaining candidates fail to demonstrate any easy

---

* This work has been supported in part by NIST through the Recovery Act Measurement Science and Engineering Research Grant Program, under contract no. 60NANB10D004.

S. Mangard and F.-X. Standaert (Eds.): CHES 2010, LNCS 6225, pp. 264–278, 2010.

to identify security weaknesses, and as a result are judged to have adequate security.

Performance in software and hardware are next in line to clearly differentiate among the candidates for a cryptographic standard. Interestingly, the differences among the cryptographic algorithms in terms of hardware performance seem to be particularly large, and often serve as a tiebreaker when other criteria fail to identify a clear winner. For example, in the AES contest, the difference in hardware speed between the two fastest final candidates (Serpent and Rijndael) and the slowest one (Mars) was by a factor of seven [1][4]; in the eSTREAM competition the spread of results among the eight top candidates qualified to the final round was by a factor of 500 in terms of speed (Trivium x64 vs. Pomaranch), and by a factor of 30 in terms of area (Grain v1 vs. Edon80) [5][6].

At this point, the focus of the attention of the entire cryptographic community is on the SHA-3 contest for a new hash function standard, organized by NIST [7][8]. The contest is now in its second round, with 14 candidates remaining in the competition. The evaluation is scheduled to continue until the second quarter of 2012.

In spite of the progress made during previous competitions, no clear and commonly accepted methodology exists for comparing hardware performance of cryptographic algorithms [9]. The majority of the reported evaluations have been performed on an ad-hoc basis, and focused on one particular technology and one particular family of hardware devices. Other pitfalls included the lack of a uniform interface, performance metrics, and optimization criteria. These pitfalls are compounded by different skills of designers, using two different hardware description languages, and no clear way of compressing multiple results to a single ranking. In this paper, we address all the aforementioned issues, and propose a clear, fair, and comprehensive methodology for comparing hardware performance of SHA-3 candidates and any future algorithms competing to become a new cryptographic standard.

The hardware evaluation of SHA-3 candidates started shortly after announcing the specifications and reference software implementations of 51 algorithms submitted to the contest [7][8][10]. The majority of initial comparisons were limited to less than five candidates, and their results have been published at [10]. The more comprehensive efforts became feasible only after NISTs announcement of 14 candidates qualified to the second round of the competition in July 2009. Since then, two comprehensive studies have been reported in the Cryptology ePrint Archive [11][12]. The first, from the University of Graz, has focused on ASIC technology, the second from two institutions in Japan, has focused on the use of the FPGA-based SASEBO-GII board from AIST, Japan. Although both studies generated quite comprehensive results for their respective technologies, they did not quite address the issues of the uniform methodology, which could be accepted and used by a larger number of research teams. Our study is intended to fill this gap, and put forward the proposal that could be evaluated and commented on by a larger cryptographic community.

2   Choice of a Language, FPGA Devices, and Tools

Out of two major hardware description languages used in industry, VHDL and Verilog HDL, we choose VHDL. We believe that either of the two languages is perfectly suited for the implementation and comparison of SHA-3 candidates, as long as all candidates are described in the same language. Using two different languages to describe different candidates may introduce an undesired bias to the evaluation.

FPGA devices from two major vendors, Xilinx and Altera, dominate the market with about 90% of the market share. We therefore feel that it is appropriate to focus on FPGA devices from these two companies. In this study, we have chosen to use seven families of FPGA devices from Xilinx and Altera. These families include two major groups, those optimized for minimum cost (Spartan 3 from Xilinx, and Cyclone II and III from Altera) and those optimized for high performance (Virtex 4 and 5 from Xilinx, and Stratix II and III from Altera). Within each family, we use devices with the highest speed grade, and the largest number of pins.

As CAD tools, we have selected tools developed by FPGA vendors themselves: Xilinx ISE Design Suite v. 11.1 (including Xilinx XST, used for synthesis) and Altera Quartus II v. 9.1 Subscription Edition Software.

3   Performance Metrics for FPGAs

Speed. In order to characterize the speed of the hardware implementation of a hash function, we suggest using Throughput, understood as a throughput (number of input bits processed per unit of time) for long messages. To be exact, we define Throughput using the following formula:

$$Throughput = \frac{block\_size}{T \cdot (HTime(N+1) - HTime(N))} \tag{1}$$

where $block\_size$ is a message block size, characteristic for each hash function (as defined in the function specification, and shown in Table 3), $HTime(N)$ is a total number of clock cycles necessary to hash an N-block message, T is a clock period, different and characteristic for each hardware implementation of a specific hash function.

In this paper, we provide the exact formulas for $HTime(N)$ for each SHA-3 candidate, and values of $f = 1/T$ for each algorithm–FPGA device pair (see Tables 3 and 6).

For short messages, it is more important to evaluate the total time required to process a message of a given size (rather than throughput). The size of the message can be chosen depending on the requirements of an application. For example, in the eBASH study of software implementations of hash functions, execution times for all sizes of messages, from 0-bytes (empty message) to 4096 bytes, are reported, and five specific sizes 8, 64, 576, 1536, and 4096 are featured in the tables [13]. The generic formulas we include in this paper (see Table 3) allow the calculation of the execution times for any message size.

In order to characterize the capability of a given hash function implementation for processing short messages, we present in this study the comparison of execution times for an empty message (one block of data after padding) and a 100-byte (800-bits) message before padding (which becomes equivalent for majority, but not all, of the investigated functions to 1024 bits after padding).

Resource Utilization/Area. Resource utilization is particularly difficult to compare fairly in FPGAs, and is often a source of various evaluation pitfalls. First, the basic programmable block (such as CLB slice in Xilinx FPGAs) has a different structure and different capabilities for various FPGA families from different vendors. Taking this issue into account, we suggest avoiding any comparisons across family lines. Secondly, all modern FPGAs include multiple dedicated resources, which can be used to implement specific functionality. These resources include Block RAMs (BRAMs), multipliers (MULs), and DSP units in Xilinx FPGAs, and memory blocks, multipliers, and DSP units in Altera FPGAs. In order to implement a specific operation, some of these resources may be interchangable, but there is no clear conversion factor to express one resource in terms of the other.

Therefore, we suggest in the general case, treating resource utilization as a vector, with coordinates specific to a given FPGA family. For example,

$$Resource\_Utilization_{Spartan3} = (\#CLBslices, \#BRAMs, \#MULs) \quad (2)$$

Taking into account that vectors cannot be easily compared to each other, we have decided to opt out of using any dedicated resources in the hash function implementations used for our comparison. Thus, all coordinates of our vectors, other than the first one have been forced (by choosing appropriate options of the synthesis and implementation tools) to be zero. This way, our resource utilization (further referred to as Area) is characterized using a single number, specific to the given family of FPGAs, namely the number of CLB slices ($\#CLBslices$) for Xilinx FPGAs, the number of Logic Elements ($\#LE$) for Cyclone II and Cyclone III, and the number of Adaptive Look-Up Tables ($\#ALUT$) in Stratix II and Stratix III.

4   Uniform Interface

In order to remove any ambiguity in the definition of our hardware cores for SHA-3 candidates, and in order to make our implementations as practical as possible, we have developed an interface shown in Fig. 1a, and described below. In a typical scenario, the SHA core is assumed to be surrounded by two standard FIFO modules: Input FIFO and Output FIFO, as shown in Fig. 1b. In this configuration, SHA core is an active module, while a surrounding logic (FIFOs) is passive. Passive logic is much easier to implement, and in our case is composed of standard logic components, FIFOs, available in any major library of IP cores.

Each FIFO module generates signals empty and full, which indicate that the FIFO is empty and/or full, respectively. Each FIFO accepts control signals write and read, indicating that the FIFO is being written to and/or read from, respectively.

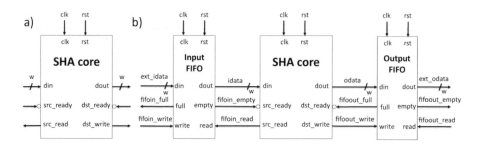

Fig. 1. a) Input/output interface of a SHA core. b) A typical configuration of a SHA core connected to two surrounding FIFOs.

The aforementioned assumptions about the use of FIFOs as surrounding modules are very natural and easy to meet. For example, if a SHA core implemented on an FPGA communicates with an outside world using PCI, PCI-X, or PCIe interface, the implementations of these interfaces most likely already include Input and Output FIFOs, which can be directly connected to the SHA core. If a SHA core communicates with another core implemented on the same FPGA, then FIFOs are often used on the boundary between the two cores in order to accommodate for any differences between the rate of generating data by one core and the rate of accepting data by another core.

Additionally, the inputs and outputs of our proposed SHA core interface do not need to be necessarily generated/consumed by FIFOs. Any circuit that can support control signals src_ready and src_read can be used as a source of data. Any circuit that can support control signals dst_ready and dst_write can be used as a destination for data.

The exact format of an input to the SHA core, for the case of pre-padded messages, is shown in Fig. 2. Two scenarios of operation are supported. In the first scenario, the message bitlength after padding is known in advance and is smaller than 2^w. In this scenario, shown in Fig. 2a, the first word of input represents message length after padding, expressed in bits. This word has the least significant bit, representing a flag called last, set to one. This word is followed by the message length before padding. This value is required by several SHA-3 algorithms using internal counters (such as BLAKE, ECHO, Shavite-3, and Skein), even if padding is done outside of the SHA core. These two control words are followed by all words of the message.

The second format, shown in Fig. 2b, is used when either message length is not known in advance, or it is greater than 2^w. In this case, the message is processed in segments of data denoted as seg_0, seg_1,...,seg_n-1. For the ease of processing data by the hash core, the size of the segments, from seg_0 to seg_n-2 is required to be always an integer multiple of the block size b, and thus also of the word size w. The least significant bit of the segment length expressed in bits is thus naturally zero, and this bit, treated as a flag called last, can be used to differentiate between the last segment and all previous segments of the message.

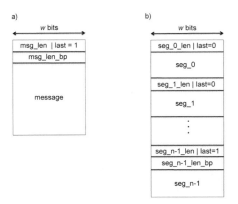

Fig. 2. Format of input data for two different operation scenarios: a) with message bitlength known in advance, and b) with message bitlength unknown in advance. Notation: msg_len – message length after padding, msg_len_bp – message length before padding, seg_i_len – segment i length after padding, seg_i_len_bp – segment i length before padding, last – a one-bit flag denoting the last segment of the message (or one-segment message), "|" – bitwise OR.

The last segment before padding can be of arbitrary length $< 2^w$. Scenario a) is a special case of scenario b). In case the SHA core supports padding, the protocol can be even simpler, as explained in [14].

5   Optimization Target and Design Methodology

Our study is performed using the following assumptions. Only the SHA-3 candidate variants with a 256-bit output are compared in this paper. Padding is assumed to be done outside of the hash cores (e.g., in software). All investigated hash functions have very similar padding schemes, which would lead to similar absolute area overhead if implemented as a part of the hardware core.

Only the primary mode of operation is supported for all functions. Special modes, such as tree hashing or MAC mode are not implemented. The salt values are fixed to all zeros in all SHA-3 candidates supporting this special input (namely BLAKE, ECHO, SHAvite-3, and Skein).

We believe that the choice of the primary optimization target is one of the most important decisions that needs to be made before the start of the comparison. The optimization target should drive the design process of every SHA-3 candidate, and it should also be used as a primary factor in ranking the obtained SHA-3 cores. The most common choices are: Maximum Throughput, Minimum Latency, Minimum Area, Throughput to Area Ratio, etc.

Our choice is the Throughput to Area Ratio, where Throughput is defined as Throughput for long messages, and Area is expressed in terms of the number of basic programmable logic blocks specific to a given FPGA family. This choice has multiple advantages. First, it is practical, as hardware cores are typically applied in situations, where the size of the processed data is significant and the speed of processing is essential. Otherwise, the input/output latency overhead associated with using a hardware accelerator dominates the total processing time, and the cost of using dedicated hardware (FPGA) is not justified. Optimizing for the best ratio provides a good balance between the speed and the cost of the solution.

Secondly, this optimization criterion is a very reliable guide throughout the entire design process. At every junction where the decisions must be made, starting from the choice of a high-level hardware architecture down to the choice of the particular FPGA tool options, this criterion facilitates the decision process, leaving very few possible paths for further investigation.

On the contrary, optimizing for Throughput alone, leads to highly unrolled hash function architectures, in which a relatively minor improvement in speed is associated with a major increase in the circuit area. In hash function cores, latency, defined as a delay between providing an input and obtaining the corresponding output, is a function of the input size. Since various sizes may be most common in specific applications, this parameter is not a well-defined optimization target. Finally, optimizing for area leads to highly sequential designs, resembling small general-purpose microprocessors, and the final product depends highly on the maximum amount of area (e.g., a particular FPGA device) assumed to be available.

Our design of all 14 SHA-3 candidates followed an identical design methodology. Each SHA core is composed of the Datapath and the Controller. The Controller is implemented using three main Finite State Machines, working in parallel, and responsible for the Input, Main Processing, and the Output, respectively. As a result, each circuit can simultaneously perform the following three tasks: output hash value for the previous message, process a current block of data, and read the next block of data. The parameters of the interface are selected in such a way that the time necessary to process one block of data is always larger or equal to the time necessary to read the next block of data. This way, the processing of long streams of data can happen at full speed, without any visible input interface overhead. The finite state machines responsible for input and output are almost identical for all hash function candidates; the third state machine, responsible for main data processing, is based on a similar template. The similarity of all designs and reuse of common building blocks assures a high fairness of the comparison.

The design of the Datapath starts from the high level architecture. At this point, the most complex task that can be executed in an iterative fashion, with the minimum overhead associated with multiplexing inputs specific to a given iteration round, is identified. The proper choice of such a task is very important, as it determines both the number of clock cycles per block of the message and the circuit critical path (minimum clock period).

It should be stressed that the choice of the most complex task that can be executed in an iterative fashion should not follow blindly the specification of a function. In particular, quite often one round (or one step) from the description of the algorithm is not the most suitable component to be iterated in hardware.

Table 1. Main iterative tasks of the hardware architectures of SHA-3 candidates optimized for the maximum Throughput to Area ratio

Function	Main Iterative Task	Function	Main Iterative Task
BLAKE	$G_i..G_{i+3}$	**JH**	Round function R_8
BMW	entire function	**Keccak**	Round R
CubeHash	one round	**Luffa**	The Step Function, Step
ECHO	AES round/AES round/ BIG.SHIFTROWS, BIG.MIXCOLUMNS	**Shabal**	Two iterations of the main loop
Fugue	2 subrounds (ROR3, CMIX, SMIX)	**SHAvite-3**	AES round
Groestl	Modified AES round	**SIMD**	4 steps of the compression function
Hamsi	Truncated Non-Linear Permutation P	**Skein**	8 rounds of Threefish-256

Table 2. Major operations of SHA-3 candidates (other than permutations, fixed shifts and fixed rotations). mADDn denotes a multioperand addition with n operands.

Function	NTT	Linear code	S-box	GF MUL	MUL	mADD	ADD /SUB	Boolean
BLAKE						mADD3	ADD	XOR
BMW						mADD17	ADD,SUB	XOR
CubeHash							ADD	XOR
ECHO			AES 8x8	x02, x03				XOR
Fugue			AES 8x8	x04..x07				XOR
Groestl			AES 8x8	x02..x07				XOR
Hamsi		LC[128, 16,70]	Serpent 4x4					XOR
JH			Serpent 4x4	x2, x5				XOR
Keccak								NOT,AND,XOR
Luffa			4x4	x2				XOR
Shabal					x3, x5		ADD,SUB	NOT,AND,XOR
SHAvite-3			AES 8x8	x02, x03				NOT,XOR
SIMD	NTT$_{128}$				x185, x233	mADD3	ADD	NOT,AND,OR
Skein							ADD	XOR
SHA-256						mADD5		NOT,AND,XOR

Either multiple rounds (steps) or fractions thereof may be more appropriate. In Table 1 we summarize our choices of the main iterative tasks of SHA-3 candidates. Each such task is implemented as combinational logic, surrounded by registers.

The next step is an efficient implementation of each combinational block within the DataPath. In Table 2, we summarize major operations of all SHA-3 candidates that require logic resources in hardware implementations. Fixed shifts, fixed rotations, and other more complex permutations are omitted because they appear in all candidates and require only routing resources (programmable interconnects). The most complex out of logic operations are the Number Theoretic Transform (NTT) [15] in SIMD, linear code (LC) [16] in Hamsi, and basic operations of AES (8x8 AES S-box and multiplication by a constant in the Galois Field $GF(2^8)$) in ECHO, Fugue, Groestl, and SHAvite-3; and multioperand additions in BLAKE, BMW, SIMD, and SHA-256.

For each of these operations we have implemented at least two alternative architectures. NTT was optimized by using a 7-stage Fast Fourier Transform (FFT) [15]. In Hamsi, the linear code was implemented using both logic (matrix by vector multiplications in $GF(4)$), and using look-up tables. AES 8x8 S-boxes (SubBytes) were implemented using both look-up tables (stored in distributed memories), and using logic only (following method described in [17], Section 10.6.1.3). Multi-operand additions were implemented using the following four methods: carry save adders (CSA), tree of two operand adders, parallel counter, and a "+" in VHDL. Finally, integer multiplications by 3 and 5 in Shabal have been replaced by a fixed shift and addition.

All optimized implementations of basic operations have been applied uniformly to all SHA-3 candidates. In case the initial testing did not provide a strong indication of superiority of one of the alternative methods, the entire hash function unit was implemented using two alternative versions of the basic operation code, and the results for a version with the better throughput to area ratio have been listed in the result tables.

All VHDL codes have been thoroughly verified using a universal testbench, capable of testing an arbitrary hash function core that follows interface described in Section 4 [18]. A special padding script was developed in Perl in order to pad messages included in the Known Answer Test (KAT) files distributed as a part of each candidates submission package. An output from the script follows a similar format as its input, but includes apart from padding bits also the lengths of the message segments, defined in Section 4, and shown schematically in Fig. 2b. The generation of a large number of results was facilitated by an open source tool ATHENa (Automated Tool for Hardware EvaluatioN) [18]. This benchmarking environment was also used to optimize requested synthesis and implementation frequencies and other tool options.

6   Results

In Table 3, we summarize the major parameters of our hardware architectures for all 14 SHA-3 candidates, as well as the current standard SHA-256. Block size, b, is a feature of the algorithm, and is described in the specification of each SHA-3 candidate. The I/O Data Bus Width, w, is a feature of our interface described in Section 4. It is the size of the data buses, din and dout, used to connect the SHA core with external logic (such as Input and Output FIFOs). The parameter w has been chosen to be equal to 64, unless there was a compelling reason to make it smaller. The value of 64 was considered to be small enough so that the SHA cores fit in all investigated FPGAs (even the smallest ones) without exceeding the maximum number of user pins. At the same time, setting this value to any smaller power of two (e.g., 32) would increase the time necessary to load input data from the input FIFO and store the hash value to the output FIFO. In some cases, it would also mean that the time necessary for processing a single block

Table 3. Timing characteristics of our hardware architectures of SHA-3 candidates. Notation: T – minimum clock period in ns (specific for each algorithm and each FPGA device, see Table 6), N - Number of blocks of an input message after padding.

Function	Block size, b [bits]	I/O Data Bus Width, w [bits]	Time to hash N message blocks [clock cycles]	Throughput [Mbit/s]
BLAKE	512	64	2+8+20·N+4	512/(20·T)
BMW	512	64	2+⌈8/8⌉+N+⌈4/8⌉	512/T
CubeHash	256	64	2+4+16·N+160+4	256/(16·T)
ECHO	1536	64	3+24+27·N+4	1536/(27·T)
Fugue	32	32	2+N+18+8	32/T
Groestl	512	64	3+8+21·N+4	512/(21·T)
Hamsi	32	32	3+1+3·(N-1)+6+8	32/(3·T)
JH	512	64	3+8+36·N+4	512/(36·T)
Keccak	1088	64	3+17+24·N+4	1088/(24·T)
Luffa	256	64	3+4+9·N+9+4	256/(9·T)
Shabal	512	64	3+8+1+25·N+3·25+4	512/(25·T)
Shavite-3	512	64	3+8+37·N+4	512/(37·T)
SIMD	512	64	3+8+8+9·N+4	512/(9·T)
Skein	256	64	2+4+9·N+4	256/(9·T)
SHA-256	512	32	2+1+65·N+8	512/(65·T)

of data would be smaller than the time of loading the next block of data, which would decrease the overall throughput. The only exceptions are made in case of Fugue and Hamsi, which have a block size b equal to 32 bits. Additionally, in the old standard SHA-256, the input/output data bus is set naturally to 32-bits, as the message scheduling unit accepts only one word of data per clock cycle.

In case of BMW, an additional faster i/o clock was used on top of the main clock shown in Fig. 1a. This faster clock is driving input/output interfaces of the SHA core, as well as surrounding FIFOs. The ratio of the i/o clock frequency to the main clock frequency was selected to be 8, so the entire block of message (512 bits) can be loaded in a single clock cycle of the main clock (8 cycles of the fast i/o clock).

The forth column of Table 3 contains the detailed formulas for the number of clock cycles necessary to hash N blocks of the message after padding. The formulas include the time necessary to load the message length, load input data from the FIFO, perform all necessary initializations, perform main processing, perform all required finalizations, and then send the result to the output FIFO. Finally, the last column contains the formula for the circuit throughput for long messages as defined by equation (1).

In Table 4, we list absolute values of the major parameters describing our implementations for one particular FPGA family, Xilinx Virtex 5. According to this table the highest throughput to area ratio is achieved by Keccak, Luffa, Groestl, and CubeHash. The highest absolute throughput is accomplished by

Table 4. Major performance measures of SHA-3 candidates when implemented in Xilinx Virtex 5 FPGAs. Notation: T_{empty} – Time to hash an empty message (after this message is padded in software), T_{100B} – Time to hash a 100-byte message (after this message is padded in software).

Function	Clk Freq [MHz]	Area [CLB slices]	Throughput [Mbits/s]	Throughput to Area Ratio	T_{empty} [ns]	T_{100B} [ns]
BLAKE	102.0	1851	2610.6	1.4	333.4	529.5
BMW	10.9	4400	5576.7	1.3	459.1	550.9
CubeHash	199.4	730	3189.8	4.4	922.9	1163.7
ECHO	178.1	6453	10133.4	1.6	308.8	308.8
Fugue	98.5	956	3151.2	3.3	304.6	558.5
Groestl	355.9	1884	8676.5	4.6	101.2	160.2
Hamsi	248.1	946	2646.2	2.8	96.7	399.1
JH	282.2	1275	4013.5	3.1	180.7	308.3
Keccak	238.4	1229	10806.5	8.8	201.4	201.4
Luffa	281.5	1154	8008.0	6.9	103.0	198.9
Shabal	128.1	1266	2624.0	2.1	905.4	1100.5
SHAvite-3	208.6	1130	2885.9	2.6	249.3	426.8
SIMD	40.9	9288	2325.9	0.3	635.9	1076.2
Skein	49.8	1312	1416.1	1.1	381.6	924.0
SHA-256	207.0	433	1630.5	3.8	352.7	1294.7

Table 5. Results for the reference design of SHA-256

	Spartan 3	Virtex 4	Virtex 5	Cyclone II	Cyclone III	Stratix II	Stratix III
Max. Clk Freq. [MHz]	90.8	183.0	207.0	111.0	126.9	158.1	212.8
Throughput [Mbit/s]	715.6	1441.6	1630.5	874.7	999.3	1245.2	1676.3
Area	838	838	433	1655	1653	973	963
Throughput to Area Ratio	0.85	1.72	3.77	0.53	0.60	1.28	1.74

Table 6. Clock frequencies of all SHA-3 candidates and SHA-256 expressed in MHz (post placing and routing)

Function	Spartan 3	Virtex 4	Virtex 5	Cyclone II	Cyclone III	Stratix II	Stratix III
BLAKE	41.87	79.82	101.98	52.40	52.37	85.77	109.21
BMW	4.19	12.37	10.89	7.69	8.41	13.45	16.45
CubeHash	84.70	187.58	199.36	115.67	133.83	179.40	237.64
ECHO	52.10	131.90	176.24	N/A	105.70	109.50	164.20
Fugue	39.67	72.86	98.47	53.25	60.71	83.75	123.64
Groestl	105.72	234.74	355.87	132.00	148.46	216.73	270.27
Hamsi	90.37	200.88	248.08	148.83	183.52	193.87	294.81
JH	119.36	221.58	282.20	173.43	215.89	267.45	364.30
Keccak	96.32	202.47	238.38	165.07	174.28	198.65	296.30
Luffa	129.84	260.28	281.53	171.64	173.43	219.88	307.31
Shabal	30.99	114.03	128.12	69.57	68.76	105.40	126.87
SHAvite-3	84.60	152.23	208.55	95.40	114.40	170.00	255.00
SIMD	17.20	29.25	40.89	21.66	23.97	37.07	47.40
Skein	18.22	38.16	49.79	22.30	25.14	38.89	52.29
SHA-512	90.84	183.02	207.00	111.04	126.86	158.08	212.81

Table 7. Throughput of all SHA-3 candidates normalized to the throughput of SHA-256. N/A means that the design did not fit within any device of a given family.

Function	Spartan 3	Virtex 4	Virtex 5	Cyclone II	Cyclone III	Stratix II	Stratix III	Overall
Keccak	6.10	6.37	6.63	8.56	7.91	7.23	8.01	7.21
ECHO	4.14	5.21	6.15	N/A	6.02	5.00	5.57	5.30
Luffa	5.16	5.14	4.91	5.58	4.94	5.02	5.21	5.13
Groestl	3.60	3.97	5.32	3.68	3.62	4.24	3.93	4.02
BMW	3.00	4.39	3.42	4.50	4.31	5.53	5.02	4.48
JH	2.37	2.19	2.46	2.82	3.07	3.05	3.09	2.70
CubeHash	1.89	2.08	1.96	2.12	2.14	2.31	2.27	2.10
Fugue	1.77	1.62	1.93	1.95	1.94	2.15	2.36	1.95
SHAvite-3	1.64	1.46	1.77	1.51	1.58	1.89	2.11	1.70
Hamsi	1.35	1.49	1.62	1.82	1.96	1.66	1.88	1.67
BLAKE	1.50	1.42	1.60	1.53	1.34	1.76	1.67	1.54
Shabal	0.89	1.62	1.61	1.63	1.41	1.73	1.55	1.46
SIMD	1.37	1.15	1.43	1.41	1.36	1.69	1.61	1.38
Skein	0.72	0.75	0.87	0.73	0.72	0.89	0.89	0.79

Table 8. Area (utilization of programmable logic blocks) of all SHA-3 candidates normalized to the area of SHA-256

Function	Spartan 3	Virtex 4	Virtex 5	Cyclone II	Cyclone III	Stratix II	Stratix III	Overall
CubeHash	1.81	1.81	1.69	1.87	1.88	1.99	2.01	1.86
Hamsi	2.17	2.16	2.18	1.92	1.94	2.40	2.41	2.16
BLAKE	4.96	4.87	4.27	2.17	2.16	2.00	2.04	2.96
Luffa	3.28	3.29	2.67	2.74	2.77	3.40	3.43	3.07
Skein	3.41	3.45	3.03	3.28	3.34	3.68	3.74	3.41
Shabal	3.75	3.84	2.92	3.67	3.68	3.90	3.74	3.63
Keccak	3.97	3.99	2.84	3.77	3.62	4.20	4.63	3.82
JH	4.84	4.78	2.94	4.37	4.31	3.18	3.24	3.88
SHAvite-3	4.91	4.91	2.61	5.68	5.64	2.57	2.59	3.89
Fugue	4.26	4.44	2.21	5.85	5.87	3.70	3.73	4.11
Groestl	15.96	16.01	4.35	4.60	4.50	3.21	3.22	5.86
BMW	12.07	13.45	10.16	12.00	12.02	12.99	13.12	12.24
SIMD	20.97	19.99	21.45	18.53	18.57	23.03	23.24	20.39
ECHO	30.87	28.48	14.90	N/A	39.77	22.29	22.52	25.29

Table 9. Throughput to Area Ratio of all SHA-3 candidates normalized to the throughput to area ratio of SHA-256

Function	Spartan 3	Virtex 4	Virtex 5	Cyclone II	Cyclone III	Stratix II	Stratix III	Overall
Keccak	1.54	1.60	2.34	2.27	2.18	1.72	1.73	1.89
Luffa	1.57	1.56	1.84	2.04	1.78	1.48	1.52	1.67
CubeHash	1.04	1.15	1.16	1.13	1.14	1.16	1.13	1.13
Hamsi	0.62	0.69	0.74	0.94	1.01	0.69	0.78	0.77
JH	0.49	0.46	0.84	0.65	0.71	0.96	0.95	0.70
Groestl	0.23	0.25	1.22	0.80	0.81	1.32	1.22	0.69
BLAKE	0.30	0.29	0.37	0.71	0.62	0.88	0.82	0.52
Fugue	0.42	0.36	0.88	0.33	0.33	0.58	0.63	0.47
SHAvite-3	0.33	0.30	0.68	0.27	0.28	0.74	0.81	0.44
Shabal	0.24	0.42	0.55	0.44	0.38	0.44	0.41	0.40
BMW	0.25	0.33	0.34	0.38	0.36	0.43	0.38	0.37
Skein	0.21	0.22	0.29	0.22	0.21	0.24	0.24	0.23
ECHO	0.13	0.18	0.41	N/A	0.15	0.22	0.25	0.21
SIMD	0.07	0.06	0.07	0.08	0.07	0.07	0.07	0.07

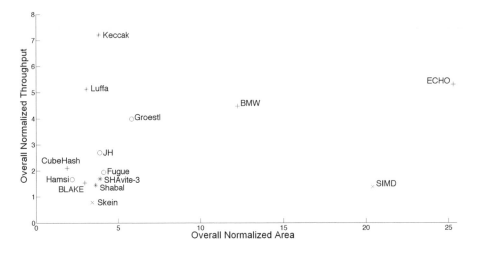

Fig. 3. Relative performance of all Round 2 SHA-3 Candidates in terms of the overall normalized throughput and the overall normalized area (with SHA-256 used as a reference point)

Keccak, ECHO, Groestl, Luffa, and BMW. The smallest hashing time for an empty message is achieved by Hamsi, Groestl, Luffa, and JH. For a 100 byte message, the list of the first four candidates changes to Groestl, Luffa, Keccak, and JH. As one can see, the execution time for small messages is not strongly correlated with the throughput for long messages, and therefore it must be treated as a separate evaluation criterion (as discussed in Section 3).

In Table 5, we summarize the absolute results obtained for our implementation of the current standard SHA-256. The results are repeated for all seven FPGA families used in our study. As hardware architecture, we have selected the architecture by Chaves et al., presented at CHES 2006 [19]. This architecture has been specifically optimized for the maximum throughput to area ratio [19][20], and is considered one of the best known SHA-2 architectures of this type.

In Table 6, the maximum clock frequencies are listed for each pair: hash algorithm–FPGA family. These frequencies can be used together with the formulas provided in Table 3, in order to compute the exact execution times of each algorithm (depending on the number of the message blocks, N) and the values of the throughputs for long messages.

In the following analysis, the absolute values of the three major performance measures: throughput, area, and the throughput to area ratio, for all SHA-3 candidates, have been normalized by dividing them by the corresponding values for the reference implementation of SHA-256. The corresponding ratios, referred to as normalized throughput, normalized area, and normalized throughput to area ratios are summarized in Tables 7, 8, and 9. The Overall column represents the geometric mean of all normalized results available for a given algorithm. The candidate algorithms are ranked based on the value of this Overall metric, representing the performance for a wide range of different FPGA families.

Interestingly, based on Table 9, only three candidates, Keccak, Luffa, and CubeHash outperform SHA-256 in terms of the throughput to area ratio. The additional four candidates, Hamsi, JH, Groestl, and BLAKE, have the overall normalized ratio higher than 0.5.

In Fig. 3, we present a two dimensional diagram, with Normalized Area on the X-axis and Normalized Throughput on the Y-axis. The algorithms seem to fall into several major groups. Group with the high normalized throughput (>5), medium normalized area (<4), and the high normalized throughput to area ratio (>1.5), include Keccak and Luffa. Groestl, BMW, and ECHO, have all high normalized throughput (>4), but their normalized area varies significantly from about 6 in case of Groestl, through 12 for BMW, up to over 25 in case of ECHO. SIMD is both relatively slow (less then 2 times faster than SHA-256) and big (more than 20 times bigger than SHA-256). The last group includes 8 candidates covering the range of the normalized throughputs from 0.8 to 2.7, and the normalized areas from 1.9 to 4.1.

7   Conclusions and Future Work

Our evaluation methodology, applied to 14 Round 2 SHA-3 candidates, has demonstrated large differences among competing candidates. The ratio of the best result to the worst result was equal to about 9 in terms of the throughput (Keccak vs. Skein), over 13 times in terms of area (CubeHash vs. ECHO), and about 27 in terms of our primary optimization target, the throughput to area ratio (Keccak vs. SIMD). Only three candidates, Keccak, Luffa, and CubeHash, have demonstrated the throughput to area ratio better than the current standard SHA-256. Out of these three algorithms, Keccak and Luffa have also demonstrated very high throughputs, while CubeHash outperformed other candidates in terms of minimum area. All candidates except Skein outperform SHA-256 in terms of the throughput, but at the same time none of them matches SHA-256 in terms of the area.

Future work will include the evaluation of the remaining variants of SHA-3 candidates (such as variants with 224, 384, and 512 bit outputs, and an all-in-one architecture). The uniform padding units will be added to each SHA core, and their cost estimated. We will also investigate the influence of synthesis tools from different vendors (e.g., Synplify Pro from Synopsys). The evaluation may be also extended to the cases of hardware architectures optimized for the minimum area (cost), maximum throughput (speed), or minimum power consumption. Each algorithm will be also evaluated in terms of its suitability for implementation using dedicated FPGA resources, such embedded memories, dedicated multipliers, and DSP units. Finally, an extension of our methodology to the standard-cell ASIC technology will be investigated.

Acknowledgments. The authors would like to acknowledge all students from the Fall 2009 edition of the George Mason University course entitled "Digital System Design with VHDL," for conducting initial exploration of the design space of all SHA-3 candidates.

References

1. Nechvatal, J., et al.: Report on the Development of the Advanced Encryption Standard (AES), http://csrc.nist.gov/archive/aes/round2/r2report.pdf
2. NESSIE, https://www.cosic.esat.kuleuven.be/nessie/
3. eSTREAM, http://www.ecrypt.eu.org/stream/
4. Gaj, K., Chodowiec, P.: Fast Implementation and Fair Comparison of the Final Candidates for Advanced Encryption Standard Using Field Programmable Gate Arrays. In: Naccache, D. (ed.) CT-RSA 2001. LNCS, vol. 2020, pp. 84–99. Springer, Heidelberg (2001)
5. Hwang, D., Chaney, M., Karanam, S., Ton, N., Gaj, K.: Comparison of FPGA-targeted Hardware Implementations of eSTREAM Stream Cipher Candidates. In: State of the Art of Stream Ciphers Workshop, SASC 2008, February, pp. 151–162 (2008)
6. Good, T., Benaissa, M.: Hardware Performance of eStream Phase-III Stream Cipher Candidates. In: State of the Art of Stream Ciphers Workshop, SASC 2008, February 2008, pp. 163–173 (2008)
7. SHA-3 Contest, http://csrc.nist.gov/groups/ST/hash/sha-3/index.html
8. SHA-3 Zoo, http://ehash.iaik.tugraz.at/wiki/TheSHA-3Zoo
9. Drimer, S.: Security for Volatile FPGAs. ch. 5: The Meaning and Reproducibility of FPGA Results. Ph.D. Dissertation, University of Cambridge, Computer Laboratory, uCAM-CL-TR-763 (Nov 2009)
10. SHA-3 Hardware Implementations, http://ehash.iaik.tugraz.at/wiki/SHA-3_Hardware_Implementations
11. Tilich, S., et al.: High-speed Hardware Implementations of Blake, Blue Midnight Wish, Cubehash, ECHO, Fugue, Groestl, Hamsi, JH, Keccak, Luffa, Shabal, Shavite-3, SIMD, and Skein. Cryptology, ePrint Archive, Report 2009/510 (2009)
12. Kobayashi, K., et al.: Evaluation of Hardware Performance for the SHA-3 Candidates Using SASEBO-GII. Cryptology, ePrint Archive, Report 2010/010 (2010)
13. ECRYPT Benchmarking of Cryptographic Systems, http://bench.cr.yp.to
14. CERG GMU Group: Hardware Interface of a Secure Hash Algorithm (SHA), http://cryptography.gmu.edu/athena/index.php?id=interfaces
15. Meyer-Baese, U.: Digital Signal Processing with Field Programmable Gate Arrays, ch. 6, 7, 3rd edn., pp. 343–475. Springer, Heidelberg (2007)
16. van Lint, J.H.: Introduction to Coding Theory, 2nd edn. Springer, Heidelberg (1992)
17. Gaj, K., Chodowiec, P.: FPGA and ASIC Implementations of AES. In: Cryptographic Engineering, ch. 10, pp. 235–294. Springer, Heidelberg (2009)
18. ATHENa Project Website, http://cryptography.gmu.edu/athena
19. Chaves, R., Kuzmanov, G., Sousa, L., Vassiliadis, S.: Improving SHA-2 Hardware Implementations. In: Goubin, L., Matsui, M. (eds.) CHES 2006. LNCS, vol. 4249, pp. 298–310. Springer, Heidelberg (2006)
20. Chaves, R., Kuzmanov, G., Sousa, L., Vassiliadis, S.: Cost Efficient SHA Hardware Accelerators. IEEE Trans. Very Large Scale Integration Systems 16, 999–1008 (2008)

Performance Analysis of the SHA-3 Candidates on Exotic Multi-core Architectures

Joppe W. Bos[1] and Deian Stefan[2]

[1] Laboratory for Cryptologic Algorithms, EPFL, CH-1015 Lausanne, Switzerland
[2] Dept. of Electrical Engineering, The Cooper Union, NY 10003, New York, USA

Abstract. The NIST hash function competition to design a new cryptographic hash standard 'SHA-3' is currently one of the hot topics in cryptologic research, its outcome heavily depends on the public evaluation of the remaining 14 candidates. There have been several cryptanalytic efforts to evaluate the security of these hash functions. Concurrently, invaluable benchmarking efforts have been made to measure the performance of the candidates on multiple architectures. In this paper we contribute to the latter; we evaluate the performance of *all* second-round SHA-3 candidates on two *exotic* platforms: the Cell Broadband Engine (Cell) and the NVIDIA Graphics Processing Units (GPUs). Firstly, we give performance estimates for each candidate based on the number of arithmetic instructions, which can be used as a starting point for evaluating the performance of the SHA-3 candidates on various platforms. Secondly, we use these generic estimates and Cell-/GPU-specific optimization techniques to give more precise figures for our target platforms, and finally, we present implementation results of all 10 non-AES based SHA-3 candidates.

Keywords: Cell Broadband Engine, Graphics Processing Unit, Hash function, SHA-3.

1 Introduction

The design and analysis of cryptographic hash functions have come under renewed interest with the public competition[1] commenced by the US National Institute of Standards and Technology (NIST) to develop a new cryptographic hash algorithm intended to replace the current standard Secure Hash Algorithm-2 (SHA-2) [28]. The new hash algorithm will be called 'SHA-3' and will be subject to a Federal Information Processing Standard (FIPS), similar to the Advanced Encryption Standard (AES) [27]. The competition is NIST's response to recent advances in the cryptanalysis of hash functions, particularly those affecting widely deployed algorithms, including MD5 and SHA-1. Although these breakthroughs have no direct consequence on the current cryptographic hash standard SHA-2, a successful attack on SHA-2 would have catastrophic effects on the security of applications relying on hash functions (e.g., digital signatures).

---

[1] See http://csrc.nist.gov/groups/ST/hash/sha-3/index.html

S. Mangard and F.-X. Standaert (Eds.): CHES 2010, LNCS 6225, pp. 279–293, 2010.

Such attacks are believed to be quite probable due to the structural similarities between SHA-2 and its broken ancestors.

Competition History. The NIST competition officially started in late October 2008 with various contributions from academia, industry and government institutions. A total of 64 proposals were submitted worldwide, of which 51 met the minimum submission requirements and were announced as the first-round candidates. Compared to the AES competition, which had 15 candidates, this number was quite large. In late July 2009, NIST narrowed the number of candidates for the second round to a more manageable size of 14. The total number of candidates is expected to be reduced to about 5 (finalists) by the third quarter of 2010. The new hash function standard(s) will be announced in 2012.

Motivation. The candidates are reviewed based on three main evaluation criteria: security, cost, and algorithmic and implementation characteristics [29]. Through the second round, nearly all of the eliminated algorithms were found to suffer from either efficiency or security flaws. Furthermore, despite suffering from minor security issues, some of the high-performing candidates survived the elimination process [35]; this clearly highlights the importance of efficiency in the evaluation procedure.

One of the motivations behind this work is NIST's predisposition for algorithms with greater flexibility [29]; specifically, NIST states that is it preferable if *"the algorithm can be implemented securely and efficiently on a wide variety of platforms."* We endeavor to evaluate the performance of the remaining candidates on two *exotic* platforms: the high-end Cell Broadband Engine architecture (Cell) and the NVIDIA Graphics Processing Units (GPUs). For these platforms, which allow the use of vectorization optimization techniques, multiple input streams of equal length are processed at once using SIMD (single instruction, multiple data) and SIMT (single instruction, multiple threads) techniques for the Cell and GPUs, respectively. Due to the low prices, wide availability, and shift in architecture design towards many-core processors [34], it is of valuable interest to evaluate the performance of the Cell and GPUs as cryptologic accelerators.

There are numerous cryptographic applications in which the computation of a message digest of a fixed-length message is necessary. For instance, the work of Bellare and Rogaway [3], standardized in [36,21,1], proposes a mask generation function used in optimal asymmetric encryption that is based on a hash function which takes a fixed-length input. Further, protocols which use hash-based message authentication codes (HMAC) require the computation of a message digest of fixed-length blocks. Specifically, given hash function H, message m, and key k, HMAC is defined as: $H((k \oplus o_{\mathrm{pad}})||H((k \oplus i_{\mathrm{pad}})||m))$. In this case, $||$ denotes concatenation, and o_{pad} and i_{pad} are fixed-length constants such that the outermost hash is of a fixed-length block (cf. [22] for more details). Thus, computing the message digest of a batch of such fixed-length input messages, e.g., in high-end servers, can be efficiently accomplished with the implementations proposed in this work.

Additionally, in a cryptanalytic setting such implementations may be used to speed up brute-force password cracking, allow for hash function cube attack/tester analysis using high-dimensional cubes, among many other applications.

Our Contribution. We present a new software performance analysis of all second-round SHA-3 candidates on the Cell and GPU. Our results are three-fold:

1. We present an in-depth performance analysis of all SHA-3 candidates by investigating their internal operations. It is worth noting that the aim of this work is not to claim that our techniques are optimal (hence, the provided estimates are indeed subject to change). Rather, our intended goal is to make a fair, reliable, and accurate comparison between all second-round SHA-3 candidates, which might serve as a reference before the final candidates are announced. Due to the significant number of candidates, all using different techniques, this is not a straightforward task. To facilitate the analysis, we separate the AES-inspired candidates from the others. For the former case, we make extensive use of the work by Osvik et al. [33], which introduced the fastest results of AES on our target architectures. For the latter case, however, a more careful analysis, starting from scratch, is required.
2. We propose specific optimization techniques for each of our target platforms; in combination with our estimation framework, more precise estimates per architecture are given for all second-round SHA-3 candidates.
3. We complement this framework by providing real implementations of all non-AES based candidates on the target platforms. We show that our techniques are indeed applicable, and that the base estimates are usually realistic.

Related Work. The PlayStation 3 (PS3) video game console, which contains the Cell architecture, has been previously used to find chosen-prefix collisions for the cryptographic hash function MD5 [37]. Fast multi-stream implementations of MD5, SHA-1 and SHA-256 for the Cell are presented in [10]; from this work, we use the performance numbers for SHA-256 as a comparison to the performance of the SHA-3 candidates, as they outperform the single stream results from [13] by an order of magnitude. Graphics cards have similarly been used for MD5 collision searches [8], password cracking [26], and accelerating cryptographic applications [38,25]. To the best of our knowledge, there is no previous work implementing second-round SHA-3 candidates on the Cell architecture or NVIDIA GT200 GPUs.

Organization. We start with a brief introduction to our target platforms in Section 2. Several optimization techniques are described in Section 3, directly addressing our main target architectures. Then, in Section 4 and 5 we introduce our performance analysis and implementation results on AES-inspired and other second round candidates, respectively. We conclude in Section 6.

2   Target Platforms

Cell Broadband Engine Architecture. The Cell architecture [20], jointly developed by Sony, Toshiba, and IBM, is equipped with one dual-threaded, 64-bit in-order Power Processing Element (PPE) based on the Power 5 architecture and 8 Synergistic Processing Elements (SPEs). Our interest is in the SPEs [39], the main computational cores of the Cell. Each SPE consists of a Synergistic Processing Unit (SPU), 256 KB of private memory called Local Store (LS), and a Memory Flow Controller (MFC). To avoid the complexity of sending explicit direct memory access requests to the MFC, all code and data must fit within the LS.

The SPU is equipped with a large register file containing 128 registers of 128 bits each. Most SPU instructions work on 128-bit operands denoted as *quadwords*. The instruction set is partitioned into two sets: one set consists of (mainly) 4- and 8-way SIMD arithmetic instructions, while the other consists of instructions operating on the whole quadword (including the load and store instructions) in a single instruction, single data (SISD) manner. The SPU is an asymmetric processor; each set of instructions is executed in a separate pipeline, denoted by the *even* and *odd* pipeline for the SIMD and SISD instructions, respectively. For instance, the $\{4, 8\}$-way SIMD left-rotate instruction is an even instruction, while the instruction left-rotating the full quadword is dispatched into the odd pipeline. When dependencies are avoided, a single pair of even and odd instructions can be dispatched every clock cycle.

One of the first applications of the Cell processor was to serve as the heart of Sony's PS3 game console. Although the Cell contains 8 SPEs, in the PS3, one is disabled and a second is reserved by Sony. Thus, with the first generation PS3s the programmer has access to six SPEs, this has been disabled in the current version of the game console. In subsequent applications, serving the supercomputing community, the Cell has been placed in blade servers, with newer variants containing the PowerXCell 8i, a derivative of the Cell that offers enhanced double-precision floating-point capabilities. The SPEs are particularly useful as (cryptographic) accelerators. For this purpose, PCIe cards are available (either equipped with a complete Cell processor or a stripped-down version containing 4 SPEs) so that workstations can benefit from the computational power of the SPEs.

NVIDIA Graphics Processing Units. Unlike the Cell, there are many different GPU architectures, though, most share the primary goal of accelerating 3-dimensional graphics (rendering) applications, such as games. In this work, we focus on programming NVIDIA GPUs using the Compute Unified Device Architecture (CUDA) extension of the C language. With the latest GPUs implementing the Fermi architecture [32], availability and interest in the older G80 series GPUs, which have also been used for cryptologic applications (cf. [25,33,19]), is rapidly decreasing. We therefore restrict our focus to the more-recent GT200 series GPUs.

Each GPU is equipped with several Simultaneous Multiprocessors (SMs), varying from 24 in the GTX 260 to 30 in each of the GPUs of the GTX 295 graphics card. Each SM consists of a large register file (16384 32-bit registers), fast 16-way banked on-chip 16KB shared memory, 8 Scalar Processors (SPs), 2 special function units (used for transcendentals), an instruction scheduler, and

(6-8KB) texture and (8KB) constant memory caches. The SPs are capable of executing many instructions, including 32-bit integer arithmetic and bitwise operations, which can be used to implement most cryptologic algorithms.

Although explicit SIMD access of the SM compute units (the SPs) is desirable for many applications, the programmer is limited to writing parallel code at the thread level [31]. Specifically, using CUDA, the programmer writes code for a *kernel* which is executed by many *threads* (all executing the same instructions of the kernel, though operating on different data) on the SPs. In the SIMT programming model, threads are grouped into a *thread block*, which is executed on a single SM and, consequently, these threads may synchronize execution and use the shared memory to communicate. When launching a kernel, it is common (and highly recommended) to execute multiple thread blocks, grouped in a *grid*, which the hardware then assigns to the available SMs; to hide various latencies, it is recommended that at least 2 blocks be available for scheduling on each SM [31]. Note that although each SM has many resources, the shared memory is divided among the 'co-located' thread blocks, and similarly the registers are divided among the individual threads—careful consideration of an application's use of these resources is critical when trying to achieve high performance. Despite these design 'restrictions', GPUs are very commonly being used as accelerators for workstations, given their wide availability as moderately-priced PCIe cards.

3   Porting the SHA-3 Candidates to the Cell and GPU

Cell Broadband Engine Architecture. On the SPE architecture, all distinct binary operations $f : \{0,1\}^2 \rightarrow \{0,1\}$ are available, making it a suitable platform to implement hash functions. Operations frequently used by the hash candidates, such as rotations, shifts, and additions, are available as 4-way SIMD instructions operating on the 4 32-bit words of a quadword, in parallel. When possible, we use the 32-bit optimized reference code of the SHA-3 candidates as a base and further optimize this code for the SPE architecture.

To make the code more suitable for execution on the Cell, the use of branches is eliminated or reduced to a minimum, since all four input strings need to be processed in an identical way. Most of the instructions used in the various compression functions are arithmetic instructions, which go in the even pipeline. When naively porting the code to the SPE architecture, this results in a highly unbalanced implementation where the odd pipeline is underutilized. In order to improve performance, some even operations, when feasible, are implemented by a sequence of odd instructions (following a similar approach to that described in [33]). This increases the latency of this operation, but if these instructions can be dispatched for free with the surrounding even instructions, the overall number of cycles decreases (while the number of overall instructions increases).

One obvious way to do this is to make use of the `shuffle` instruction that is dispatched in the odd pipeline. The `shuffle` instruction can pick any 16 bytes of the 32-byte (two 128-bit registers) input or select one of the byte-constants $\{$`0x00`, `0xFF`, `0x80`$\}$ and place them in any of the 16-byte positions of the 128-bit output register. For example, when a 4-way SIMD shift or rotate by x (to the left or right) is required this is typically implemented using the even `shift` or `rotate` instruction. When $x \equiv 0 \bmod 8$, this is simply a reordering of bytes, and can be done for 4 32-bit integer values in parallel using the `shuffle` instruction.

Converting a 4-way SIMD left rotation of a quadword V by $x \not\equiv 0 \bmod 8$ bits to odd instructions can be done using two odd `shuffle` and two odd quadword `shift` instructions. When using an odd quadword `rotate` operation, the bits rotated out from each 32-bit boundary are dislocated. To address this, create a quadword W which contains, on byte positions $4i$ and $4i + 1$, the values from the byte location $3 + 4i$ and $4i$ from V respectively, where $0 \leq i \leq 3$ and the most (least) significant byte position is labeled as 0 (15). The other bytes in W, at byte positions $4i + 2$ and $4i + 3$, are set to zero. Next, V and W are shifted left by $x \bmod 8$ using the odd quadword `shift` instruction. Finally, shuffle the three bytes from V and single byte from W per word to the correct positions to complete the 4-way SIMD rotation. This technique allows one to trade 1 even `rotate` instruction for 4 odd instructions. Note that the latency of the operation has increased from 4 cycles for the even `rotate` to $4 \times 4 = 16$ for the odd variant.

One of the NIST submission requirements is to provide an implementation of the SHA-3 candidate suitable to run on a 32- and 64-bit platform [29]. However, some of the candidates, e.g., Skein, provide a 32-bit implementation which requires the use of a 64-bit data type in the compression function. This requires to implement fast 64-bit additions and rotations built from 32-bit instructions, since these operations, on the SPE, are only available in 32-bit flavors. A 2-way SIMD addition can be implemented as follows. First, a 4-way SIMD carry generation (even) instruction is used to provide the carries going from the least to the most significant 32-bit word. An odd `shuffle` instruction is then used to put the two carries in the correct position, while the other two carries corresponding to the most significant 32-bit word of each 64-bit integer are ignored. Finally, the 4-way SIMD extended addition, an addition with carry, is used to add the two quadwords consisting of four 32-bit values considering the carries. Thus, a 64-bit addition can be implemented using a single odd and two even instructions.

To implement an efficient 2-way SIMD rotate, the `select` instruction, which is dispatched in the even pipeline, is used when a rotation by $x \not\equiv 0 \bmod 8$ is required. The `select` instruction acts as a 2-way multiplexer: depending on the input pattern, the corresponding bit from either the first or the second input quadword is selected as output. The approach is to first perform a full quadword rotation to the left by x bit-positions and store this in a quadword V_1. Then, put the incorrectly-positioned rotated bits in the correct positions of a separate quadword V_2 by swapping the 64-bit double-words. Use the `select` instruction to get the correct bits from the two quadwords, using a pattern, defined by concatenating twice the 64-bit unsigned integer value $2^x - 1$, selecting the corresponding bit position from V_1 or V_2 if the bit position in the pattern is set to zero or one respectively. Since the SPE architecture has a quadword rotation instruction up to 7 bits and another instruction rotating by bytes, the 2-way SIMD rotation costs 3 odd rotations and one even selection for rotating by $x > 8$. When $x < 8$, the cost is reduced by one odd rotation.

NVIDIA Graphics Processing Units. Compared to the SPE instruction set architecture (ISA), the GPU parallel thread execution (PTX[2]) ISA [30] is considerably less rich. With respect to integer arithmetic operations, programmers have access to 32-bit bitwise operations (`and`, `or`, `xor`, etc.), left/right shifts, 32-bit additions (with carry-in and carry-out), and 32-bit multiplication (sometimes implemented using several 24-bit multiplication instructions).

Given the simplicity of PTX, to gain the most speedup from the raw computational power, it is imperative that the kernels be very compact (especially with respect to register utilization and shared memory allocation). Compact and non-divergent kernels allow for the execution of more simultaneous threads, and can thus increase the performance of the target hash function. Thus, when implementing common hash function building blocks, a simple approach is also usually the most optimal. For example, a rotation of a 32-bit word is implemented using two shifts (`shl` and `shr`), and an `or` instruction. Furthermore, for many hash functions we can store the full internal state, and sometimes even the input message block, in registers. Although this limits the number of simultaneous threads per SM, it also lowers the copies to and from (shared) memory and thereby contributes to a faster implementation, overall. Additionally, when possible, we manually unroll the compression functions since branching on the SMs can lead to a degradation in performance when threads of a common thread block take divergent paths and execution is serialized. Moreover, conditional statements consisting of a small number of operations in each basic block are implemented using predicate instructions, instead of branches—PTX allows for the predication of almost all instructions. Nevertheless, when branching is necessary (e.g., the compression function of Skein-512), the thread execution is synchronized (at a barrier near the branch) and the branch instruction is executed uniformly by all the threads.

For algorithms with small-to-medium sized chain values (e.g., 256- or 512-bits), we buffer the chain values in registers. To avoid multiple kernel launches, each thread processes multiple message blocks. This, in conjunction with the caching of the chaining values, not only simplifies the multi-block hashing, but also results in a faster implementation (than, for example, executing multiple kernels and having to read/write chain values from/to global memory). For algorithms with larger-sized chain values or internal states, we cache the chain values in shared memory. In implementing algorithms that use shared memory, we require that the thread block size always be a multiple of 16 threads (usually at least 64 threads) and further (implicitly) assert that the n-th thread (counting from 0) loads/stores any shared memory cached values from/to bank n mod 16, as to avoid bank conflicts.

When considering algorithms using 64-bit operations, the number of registers and instructions usually doubles. For example, a 64-bit addition is performed using two additions with carry (`add.cc`). Similarly, rotations by $x \not\equiv 0$ mod 32 is implemented using 4 `shift` and 2 `or` 32-bit instructions. For these algorithms,

---

[2] We note that the PTX is an intermediate description and not the actual GPU ISA. The latter is not publicly available.

Table 1. The number of AES-like operations per b bytes for all AES-inspired candidates and the performance estimation on the SPE and single GTX 295 GPU. (R): One AES encryption round, SB: Substitution operation, MCX: Mix-Column operation over X bytes (i.e., X=4 is identical to the one used in AES). Note that Shift-Row operations are ignored because it can be dispatched through the Mix-Column operation. C/B: Cycles per byte, Gb/sec: 10^9 bits per seconds. The SPE estimates do not use the T-table approach.

Hash function	b	(R)	SB	MC4	MC8	MC16	xor (byte)	SPE C/B	SPE Gb/sec	GPU C/B	GPU Gb/sec
SHA-256 [10]	-	-	-	-	-	-	-	8.2	3.1	-	-
AES-128 [33]	16	10	-	-	-	-	16	11.3	2.3	0.32	30.9
ECHO-256	192	256	-	512	-	-	448	29.6	0.9	0.85	11.7
Fugue-256	4	-	32	-	-	2	60	15.1	1.7	0.62	16.1
Grøstl-256	64	-	1280	-	160	-	1472	41.4	0.6	1.23	8.1
SHAvite-3-256	64	52	-	-	-	-	1280	16.5	1.6	0.42	23.7

rather than using expensive registers to cache chain values or message blocks, we resort to using shared memory for caching. We, again, stress that the restriction on shared memory bank access applies to all our algorithms, and thus a 64-bit cache value requires 2 (non-conflicting) memory accesses per 64-bit word.

4   AES-Inspired SHA-3 Candidates

A popular design choice of the SHA-3 hash function designers was to use AES-like byte oriented operations (and, in some cases the AES round function itself) as building blocks in the compression function of their hash function. The second-round SHA-3 candidates following this paradigm include ECHO [4], Fugue [18], Grøstl [16], and SHAvite-3 [9]. The motivation for using AES-like operations is mainly because AES has successfully withstood much cryptanalytic effort and, moreover, one can exploit the high capabilities of AES-like functions on a wide variety of architectures. Moreover, many of the design teams have pointed out the new Intel AES instruction set and claimed several performance figures outperforming the other candidates (for a more detailed analysis, cf. [5]). Considering the possible widespread use of these processors in the future, these designs will likely have a clear advantage.

Although several optimization methods for these hash functions are possible on particular processors, such as using the Intel AES instruction set, we analyze the performance of AES-inspired candidates in a more generic setting. More precisely, we simply count the number of 'AES-like' operations required for the compression function of each candidate, as this gives an intuition of how these designs behave in architectures without native AES-instructions, such as the PowerPC, SPARC, and most low-power microcontrollers. Table 1 provides these rough estimates. Note that since the operations may differ per candidate, we clearly differentiate all possibilities, particularly the variants of the 'Mix-Column' (MC) operation used in AES.

Table 2. Straight-forward estimates for the different mix-column operations without (left) and with (right) the use of T-tables. Note that the `xor` and `rotate` instruction counts for the T-table approach in MCX operate on $(8 \cdot X)$-bit values.

		XTIME	xor (byte)	size of table(s) in bytes	xor	rotate
MC4	(AES)	4	16	1,024	3	3
				4,096	3	0
MC8	(Grøstl)	16	104	2,048	7	7
				16,384	7	0
MC16	(Fugue)	32	148	4,096	15	15
				65,536	15	0

The estimates given in Table 1 provide a good indication on the performance of the AES-inspired candidates, especially for hashing extremely long messages, where we simply focus on the compression functions. It should, however, be noted that the techniques used to implement the MC operations used by these candidates account for the largest performance loss/gain. Typically, the MC operation is implemented using a number of `xor` operations and the `XTIME` function. The latter treats a byte-value as a polynomial in the finite field $\mathbf{F}_{2^8}$ and performs modular multiplication by a fixed modulus and multiplier. In practice, `XTIME` can be implemented using a `shift` and a conditional `xor`. An upper bound on the required MC-operations, working on single byte-values, is given in Table 2. First, the double and quadruple of the X elements are computed in MCX for $X \in \{8, 16\}$; the octuple for MC16 is not needed since all the constants in Fugue are below 8. We note that these require $2 \cdot X$ `XTIME` operations, and that the number of required `xor` operations depend on the constants. Counting the latter, for MC4 in AES and MC8 in Grøstl, there are at most $4 \times 5 - 4 = 16$ and $14 \times 8 - 8 = 104$ `xor` instructions, since the rows are simply rotations of each other. Similarly, in Fugue there are $4 \times (10+8+14+9-4) = 148$ `xor` instructions, corresponding to its constants. We stress that these (naive) estimates should be treated as an upper bound; as illustrated by the implementation of MC4 in [33], the number of times `XTIME` and `xor` are required is lower: 3 and 15, respectively.

Following the "T-table" approach [14], the MC and substitution steps can be performed by using lookup tables on 32-bit (and larger) processors. The use of T-tables can greatly reduce the number of required operations; estimates of the cost of the different MC steps using a varying number of T-tables (as the different tables are simply rotations of each other) are also stated in Table 2. The MCX T-table variants require $X - 1$ `xor`, and 0 or $X - 1$ `rotate` instructions (depending on the number of tables used) operating on X-byte values. The use of T-tables is, however, not always favorable where, for example, in memory constraint environments, the tables might be too big. This is also the case for certain SIMD environments, such as the SPE, where as indicated in [33], fetching data for multiple streams in parallel is not trivial and may be more expensive than actually computing the MC operation.

Among the four AES-inspired second-round SHA-3 candidates, ECHO and SHAvite-3 make use of the AES round itself and can highly benefit from Intel AES instruction set. Therefore, it is relatively easy to infer the speed estimates for these two hash functions once we have those for AES. We use the recent work by Osvik et al. [33] on AES to obtain estimates for our target platforms. Based on their results, the corresponding workload required to implement the compression function of the AES-inspired candidates is given in Table 1. As an example of how SHAvite-3 performs under this result (given the estimates of Table 1), one requires 52 AES round function evaluations plus 1280 8-bit xors to perform one compression function invocation of SHAvite-3, compressing a 64 byte message block. From [33] we learn that one AES round can be implemented in 300 and 78600 cycles on the SPE and GPU when hashing 16 simultaneous streams and 600 blocks of 256 streams, respectively. Hence, SHAvite-3 is estimated to achieve performance of $\frac{52 \cdot 300 + 1280}{64 \cdot 16} = 16.5$ cycles/byte on a single SPE, and $\frac{52 \cdot 78600 + 1280}{64 \cdot 256 \cdot 600} = 0.42$ cycles/byte on a single GTX 295 GPU.

We note that the performance estimates given in Table 1 for Grøstl and Fugue are conservative. This is because the naive estimates for MC8 and MC16 use the estimate from Table 2, leaving room for significant optimizations. These numbers can be further improved on platforms where a T-table approach is faster than computing the Mix-Column operation. For example, on the GPU, placing the smaller (2KB) table in shared memory, Grøstl would require two 32-bit lookups in addition to the 7 xor and 7 rotate (64-bit) instructions.

5   Other SHA-3 Candidates

The non-AES based SHA-3 candidates use a variety of techniques and ideas in their hash function designs. From a performance perspective, it is interesting to have an indication of the number of required instructions per byte. An approximation of this is given in Table 3. We note that operations ending with a 'c' indicate that one of the input parameters is complemented before use, eqv denotes bitwise equivalence (i.e., xorc) and csub denotes conditional subtraction. These *raw* instruction counts are obtained from the optimized implementations as submitted to NIST and only the number of instructions in the compression function are considered. Since load and store operations are hard to predict (due to possible cache misses), and may be incomparable between platforms, only arithmetic instructions are taken into account (i.e., the required moves, loads/stores, including all the possible table-lookups, are ignored).

We would like to stress that the performance figures presented in Table 3 are estimates for a hypothetical 32-bit architecture, the instruction set of which includes all the operations shown in the columns of Table 3. Moreover, we assume that such a machine can dispatch one instruction clock cycle. Estimating the actual performance number on modern platforms is considerably more difficult because they often have access to a separate SIMD unit, which is ignored by our estimates. However, these estimates can be used as a starting point to create more accurate platform-specific speed estimations, for instance for the Cell and

Table 3. Performance estimates for all non-AES inspired SHA-3 candidates based on the number of 32- and 64-bit arithmetic instructions used in the various compression functions (which process b bytes). The † indicates an alternative implementation approach (on-the-fly interleaving) for Keccak. We assume that all operations stated in the columns are single instruction operations.

Hash function	b	add	sub csub	mul	and	nand andc	eqv	or orc	rotate	shift	xor	Cycles / byte
Hash functions operating on 32-bit words												
BLAKE-32	64	480	-	-	-	-	-	-	320	-	508	20.4
BMW-256	64	296	58	-	-	-	-	-	212	144	277	15.4
CubeHash-16/1	1	512	-	-	-	-	-	-	512	-	512	1536.0
CubeHash-16/32	32	512	-	-	-	-	-	-	512	-	512	48.0
Hamsi-256	4	-	-	-	24	12	-	24	72	24	287	110.8
JH-256	64	-	-	-	1792	*1152*	288	688	-	800	4024	136.6
Keccak-256	136	-	-	-	684	*96*	144	480 144	1248	204	3810	50.1
Keccak-256†	136	-	-	-	756	384	-	624	1248	360	4224	55.9
Luffa-256	32	-	-	-	144	-	96	96	392	-	756	46.4
Shabal-256	64	52	16	96	-	*48*	48	-	112	-	242	9.6
SIMD-256	64	817	901 256	419	852	-	-	256	288	804	176	74.5
Hash functions operating on 64-bit words												
Skein-512	64	497	-	-	1	-	-	-	288	-	305	17.0

GPU architectures. Note that while the multiplications by the candidate SIMD operate on 16-bit operands, the multiplications in Shabal are by one of the constants $\{3, 5\}$. Each of the latter multiplications can be converted into a shift and addition, if cheaper than native multiplication.

Cell Broadband Engine Architecture. Ignoring moves and assuming perfect circumstances, i.e., all even and odd pairs of instructions can be dispatched simultaneously without stalls, an estimate for hashing four messages of equal length in parallel on a single SPE may be obtained by dividing the performance numbers in Table 3 by a factor of four. Note that these are pessimistic estimates, as the balancing techniques from Section 3 are not (implicitly) considered. In Table 4 we present actual implementation results of all non-AES based candidates, with the fine-tuned estimates in parentheses. The performance results are obtained by hashing thousands of long messages (25 KB) and measuring the complete hash function (not only the compression function), in addition to the benchmarking overhead.

The number of shifts and rotations which are replaced by their odd variants is often close to the expected value required to balance the number of odd and even instructions. It might happen that this introduces stalls due to instruction dependencies, the optimal number of operations which are replaced is then decided experimentally. This information is taken into account in the estimates in Table 4. For some candidates, additional optimizations are possible. For instance, the candidate SIMD uses the select operation (bitwise "if X then Y

Table 4. Performance results and estimates (in parentheses) for the non-AES based SHA-3 candidates for the SPE and the GPU architecture. The SPE implementations process four or two (for Skein) messages of equal length. The GPU implementations process 680 blocks of 64 threads on a single NVIDIA GTX 295 GPU. Measurements of only the compression function are shown in [brackets].

Algorithm		SPE		GPU	
	Cycles per byte	Throughput (Gb/sec)	Cycles per byte	Throughput (Gb/sec)	
SHA-256 [10]	8.2	3.1	-	-	
[2] BLAKE-32	5.0 (4.5)	5.1 (5.7)	0.27 [0.13] (0.13)	36.8 (76.4)	
[17] BMW-256	4.2 (3.7)	6.2 (6.9)	0.27 [0.27] (0.10)	36.8 (99.4)	
[6] CubeHash-16/1	326.7 (316.0)	0.1 (0.1)	11.1 [11.0] (10.9)	0.90 (0.91)	
[6] CubeHash-16/32	11.6 (9.9)	2.2 (2.6)	0.36 [0.35] (0.34)	27.6 (29.2)	
[23] Hamsi-256	32.2 (26.9)	0.8 (1.0)	5.19 [0.66] (0.64)	1.91 (15.5)	
[40] JH-256	31.5 (29.8)	0.8 (0.9)	0.76 [0.75] (0.67)	13.1 (14.8)	
[7] Keccak-256	13.0 (11.1)	2.0 (2.3)	0.56 [0.56] (0.31)	17.7 (32.1)	
[12] Luffa-256	11.5 (10.1)	2.2 (2.5)	0.35 [0.34] (0.32)	28.4 (31.1)	
[11] Shabal-256	3.5 (2.8)	7.2 (9.2)	0.69 [0.56] (0.07)	14.4 (141.9)	
[24] SIMD-256	22.6 (19.0)	1.1 (1.4)	3.60 [3.60] (0.43)	2.76 (23.1)	
[15] Skein-512	13.7 (12.1)	1.9 (2.1)	0.46 [0.29] (0.22)	22.1 (45.2)	

else Z"), $(X \wedge Y) \oplus (\bar{X} \wedge Z)$, and the majority operation on three operands, $(X \wedge Y) \oplus (X \wedge Z) \oplus (Y \wedge Z)$, which can be implemented using one and two select instructions, respectively. This optimization is counter-balanced by the fact that the conditional subtraction requires three instructions (a comparison, subtraction and a select) to avoid branching. Another example where instructions on the SPE can be saved is in JH: the swapping of (multiple) bytes requires just a shuffle instruction, and the swapping of bits requires two shift and a single select.

We observe that one of the main reasons the actual performance numbers are slightly higher than the given estimates is that the four input streams of bytes need to be converted to a 4-way SIMD representation. This introduces noticeable overhead, similar to all candidates, which is not accounted for in the estimates. In Hamsi, the overhead is even larger because the message-data is used as an index for a table look-up which further gives rise to extra arithmetic instructions needed to calculate the locations of the loads. Doing this in 4-way SIMD, even when pre-fetching data for subsequent blocks, introduces ample overhead that is not considered in our estimates since all load and store operations are ignored.

NVIDIA Graphics Processing Units. As discussed in Section 3, the PTX ISA is considerably more limited than the Cell's ISA, and therefore some of the instructions in Table 3 will have to be implemented by multiple, simpler, instructions. For example, each rotate is implemented using two shift instruction and an or; each andc is implemented using a not and an and, etc. Taking the implementation of these non-native instructions into account, in addition to the fact that each GPU on the GTX 295 contains 30 SMs (for a total of 240 SPs) we divide the (slightly higher) instruction count of Table 3 by a factor of 240. These estimates are presented in Table 4, along with actual implementation results.

As in the Cell, the GPU estimated performance results of Table 4 do not account for message memory-register copies or moves. Furthermore, they do not account for kernel launch overhead, host-to/from-device copies, or possible table-setup timings (e.g., copying a table to shared memory). For fair comparison, we, however, do account for the chain value copies to/from registers and global memory; this rough figure was measured for the different sizes using a kernel that simply copied the state to registers and back to global memory. Nevertheless, our GPU estimates are certainly optimistic and implementation results, measuring the full hash function, are higher. Additionally, for algorithms with huge internal states or expanded messages, e.g., SIMD, the use of local storage might not be easily avoided and the implementation results are expected to be much worse than the estimates.

Along with considering the techniques of Section 3 when implementing the candidates, we further emphasize the details of Keccak and Hamsi. Since using large tables on the GPU is prohibited, we estimate and implement Keccak with on-the-fly interleaving (Keccak-256† in Table 4) and divide the execution of Hamsi into two kernels. The latter requires the use of a very large 32KB table (which is larger than all the fast memories on the SMs) for the message expansion, and, thus, necessitates a less direct implementation approach. The proposed two-part approach requires: (i) a kernel in which 16 threads expand the 32-bit message to 256-bits (each using 2 $1KB$ tables and an atomic xor), and (ii) a kernel implementing the actual compression function. Because the message expansion requires random access reads and uses atomic instructions (to global memory), estimates without considering the effects of memory operations are expected to diverge.

As expected, we observe that the actual performance numbers in Table 4 are slightly higher than the corresponding estimated figures. In most cases, however, the performance overhead is a result of the memory copies (host-to-device and global memory-to-registers). We confirmed this conjecture by measuring the throughput of the compression functions working on a single message block, the results of which are shown [in brackets] in Table 4. We note that the implementation result of SIMD does not, however, agree with our estimated figure—we attribute the extremely low performance to using local memory for the message expansion (4096 bits) and having a single thread do the full compression; splitting the compression function across multiple threads would likely improve SIMD's performance. Additionally, we highlight the Shabal implementation, for which we heavily used the optimized reference code, required the use of a non-inline function in the permutations as to address a compiler optimization bug; the fully-inlined, but buggy, implementation is twice as fast.

6 Conclusion

Efficiency of hash function algorithms is a very important design criterion, almost parallel with security. This work presents a generic framework for analyzing and evaluating the performance of such algorithms; specifically, we estimate the performance of the second-round candidates in the ongoing competition to establish a new cryptographic hash standard, SHA-3. Using this framework as a base, we then take advantage of platform-specific optimization techniques to provide more precise performance estimates for two *exotic* many-core architectures: the Cell Broadband Engine and NVIDIA Graphics Processing Units. We further support our analysis by presenting multi-stream implementation results of all the non-AES based candidates. Finally, we believe that this work can assist in the decision process of the SHA-3 competition.

Acknowledgements. We gratefully acknowledge useful suggestions by Dag Arne Osvik, Onur Özen and the CHES reviewers.

References

1. American National Standards Institute. ANSI X9.44-2007: Key Establishment Using Integer Factorization Cryptography (2007)
2. Aumasson, J.-P., Henzen, L., Meier, W., Phan, R.C.-W.: SHA-3 proposal BLAKE (2008)
3. Bellare, M., Rogaway, P.: Optimal asymmetric encryption. In: De Santis, A. (ed.) EUROCRYPT 1994. LNCS, vol. 950, pp. 92–111. Springer, Heidelberg (1995)
4. Benadjila, R., Billet, O., Gilbert, H., Macario-Rat, G., Peyrin, T., Robshaw, M., Seurin, Y.: SHA-3 Proposal: ECHO (2009)
5. Benadjila, R., Billet, O., Gueron, S., Robshaw, M.J.B.: The Intel AES instructions set and the SHA-3 candidates. In: Matsui, M. (ed.) ASIACRYPT 2009. LNCS, vol. 5912, pp. 162–178. Springer, Heidelberg (2009)
6. Bernstein, D.J.: CubeHash specification (2.B.1) (2009)
7. Bertoni, G., Daemen, J., Peeters, M., Assche, G.V.: Keccak specifications (2009)
8. Bevand, M.: MD5 Chosen-Prefix Collisions on GPUs. Black Hat, Whitepaper (2009)
9. Biham, E., Dunkelman, O.: The SHAvite-3 Hash Function (2009)
10. Bos, J.W., Casati, N., Osvik, D.A.: Multi-Stream Hashing on the PlayStation 3. In: PARA 2008. LNCS. Springer, Heidelberg (to appear 2008), http://documents.epfl.ch/users/b/bo/bos/public/PARA2008.pdf
11. Bresson, E., Canteaut, A., Chevallier-Mames, B., Clavier, C., Fuhr, T., Gouget, A., Icart, T., Misarsky, J.-F., Naya-Plasencia, M., Paillier, P., Pornin, T., Reinhard, J.-R., Thuillet, C., Videau, M.: The Hash Function Shabal (2008)
12. Canniere, C.D., Sato, H., Watanabe, D.: Hash Function Luffa (2009)
13. Chen, T., Raghavan, R., Dale, J., Iwata, E.: Cell broadband engine architecture and its first implementation: A performance view (November 2005), http://www.ibm.com/developerworks/power/library/pa-cellperf/
14. Daemen, J., Rijmen, V.: The design of Rijndael. Springer, New York (2002)
15. Ferguson, N., Lucks, S., Schneier, B., Whiting, D., Bellare, M., Kohno, T., Callas, J., Walker, J.: The Skein Hash Function Family (2009)
16. Gauravaram, P., Knudsen, L.R., Matusiewicz, K., Mendel, F., Rechberger, C., Schläffer, M., Thomsen, S.S.: Grøstl – a SHA-3 candidate (2008)
17. Gligoroski, D., Klima, V., Knapskog, S.J., El-Hadedy, M., Amundsen, J., Mjo lsnes, S.F.: Cryptographic Hash Function BLUE MIDNIGHT WISH (2009)

18. Halevi, S., Hall, W.E., Jutla, C.S.: The Hash Function Fugue (2009)
19. Harrison, O., Waldron, J.: Practical Symmetric Key Cryptography on Modern Graphics Hardware. In: USENIX Security Symposium, pp. 195–210 (2008)
20. Hofstee, H.P.: Power Efficient Processor Architecture and The Cell Processor. In: HPCA 2005, pp. 258–262. IEEE Computer Society, Los Alamitos (2005)
21. IEEE Std 1363-2000. IEEE Standard Specifications for Public-Key Cryptography. IEEE, New York (2000)
22. Krawczyk, H., Bellare, M., Canetti, R.: HMAC: Keyed-Hashing for Message Authentication. RFC 2104, IETF (1997)
23. Küçük, O.: The Hash Function Hamsi (2009)
24. Leurent, G., Bouillaguet, C., Fouque, P.-A.: SIMD Is a Message Digest (2009)
25. Manavski, S.A.: CUDA Compatible GPU as an Efficient Hardware Accelerator for AES Cryptography. In: ICSPC 2007, November 2007, pp. 65–68. IEEE, Los Alamitos (2007)
26. Marechal, S.: Advances in password cracking. Journal in Computer Virology 4(1), 73–81 (2008)
27. NIST. FIPS-197: Advanced Encryption Standard (AES) (2001),
    http://www.csrc.nist.gov/publications/fips/fips197/fips-197.pdf
28. NIST. Secure hash standard. FIPS 180-2 (August 2002),
    http://www.itl.nist.gov/fipspubs/fip180-2.htm
29. NIST. Announcing request for candidate algorithm nominations for a new cryptographic hash algorithm (SHA-3) family. Technical report, Department of Commerce (November 2007),
    http://csrc.nist.gov/groups/ST/hash/documents/FR_Notice_Nov07.pdf
30. NVIDIA. NVIDIA Compute. PTX: Parallel Thread Execution (March 2008)
31. NVIDIA. NVIDIA CUDA Programming Guide 2.3 (2009)
32. NVIDIA. NVIDIA's Next Generation CUDA Compute Architecture: Fermi. Whitepaper (September 2009)
33. Osvik, D.A., Bos, J.W., Stefan, D., Canright, D.: Fast software AES encryption. In: beyer, i. (ed.) FSE 2010. LNCS, vol. 6147, pp. 75–93. Springer, Heidelberg (2010)
34. Patterson, D., Hennessy, J.: Computer organization and design: the hardware/software interface. Morgan Kaufmann, San Francisco (2008)
35. Regenscheid, A., Perlner, R., jen Chang, S., Kelsey, J., Nandi, M., Paul., S.: Status report on the first round of the SHA-3 cryptographic hash algorithm competition. Technical Report 7620, NIST (September 2009),
    http://csrc.nist.gov/groups/ST/hash/sha-3/Round1/documents/
    sha3_NISTIR7620.pdf
36. RSA Laboratories. PKCS #1 v2.1: RSA Cryptography Standard (2002)
37. Stevens, M., Sotirov, A., Appelbaum, J., Lenstra, A., Molnar, D., Osvik, D.A., de Weger, B.: Short chosen-prefix collisions for MD5 and the creation of a rogue CA certificate. In: Halevi, S. (ed.) CRYPTO 2009. LNCS, vol. 5677, pp. 55–69. Springer, Heidelberg (2009)
38. Szerwinski, R., Güneysu, T.: Exploiting the power of GPUs for asymmetric cryptography. In: Oswald, E., Rohatgi, P. (eds.) CHES 2008. LNCS, vol. 5154, pp. 79–99. Springer, Heidelberg (2008)
39. Takahashi, O., Cook, R., Cottier, S., Dhong, S.H., Flachs, B., Hirairi, K., Kawasumi, A., Murakami, H., Noro, H., Oh, H., Onish, S., Pille, J., Silberman, J.: The circuit design of the synergistic processor element of a Cell processor. In: ICCAD 2005, pp. 111–117. IEEE Computer Society, Los Alamitos (2005)
40. Wu, H.: The Hash Function JH (2009)

XBX:
eXternal Benchmarking eXtension
for the SUPERCOP
Crypto Benchmarking Framework

Christian Wenzel-Benner[1] and Jens Gräf[2]

[1] ITK Engineering AG
Software Center 1, 35037 Marburg, Germany
Christian.Wenzel-Benner@itk-engineering.de
http://www.itk-engineering.de
[2] LiNetCo GmbH
Hauptstrasse 17a, 35684 Dillenburg, Germany
jgraef@linetco.com
http://www.linetco.com

Abstract. SUPERCOP [1] is a benchmarking framework for cryptographic algorithms like ciphers and hash functions. It automatically benchmarks algorithms across several implementations, compilers, compiler options and input data lengths. Since it is freely available for download the results are easily reproducible and benchmark results for virtually every computer that is capable of running SUPERCOP are available. However, since SUPERCOP is a collection of scripts for the Bourne Again Shell and depends on some command line tools from the POSIX standard in it's current form it can not run on any hardware that does not support POSIX. This is a significant limitation since small devices like mobile phones, PDAs and Smart Cards are important target platforms for cryptographic algorithms. The work presented in this paper extends the SUPERCOP concepts to facilitate benchmarking external targets. A combination of hard- and software allows for cross compilation with SUPERCOP and execution/timing of the generated code on virtually any kind of device large enough to hold the object code of the algorithm benchmarked plus some space for communication routines and a bootloader.

Keywords: SUPERCOP, XBX, benchmarking, microcontroller, small device, 8-bit, hash function.

1 Introduction

The design of a cryptographic algorithm is always a trade-off between security and performance. A 'good' algorithm either achieves stronger security than a 'bad' one at the same runtime and memory cost or the same level of security at lesser cost. Yet telling a 'good' algorithm from a 'bad' one is not always trivial.

S. Mangard and F.-X. Standaert (Eds.): CHES 2010, LNCS 6225, pp. 294–305, 2010.

Aside from spotting obvious design flaws, both security and performance are not easily quantified.

1.1   Judging Security

If a newly proposed algorithm were to be broken by applying a well-known mode of attack this would be an obvious design flaw. But if it is not vulnerable to any known attack that does not mean it is flawless. An algorithm that seems secure for a long time may suddenly be affected by a new type of attack that was not anticipated. If and when such a new attack is going to be discovered can not be determined in advance. There is however one universal rule: attacks only get better, they never get worse. The flip-side of this coin is that algorithms only get weaker, never stronger.

1.2   Judging Performance

Judging the performance of an algorithm seems trivial by comparison. Yet it is not uncommon to see different performance numbers claimed by different people for a well known standard algorithm like SHA-256 in the course of a four day conference[2]. Obviously, different implementations of the algorithm, different compilers and different target platforms result in a huge diversity of performance numbers. Which one is the 'true' performance number? A sophisticated benchmarking framework like SUPERCOP can answer this question for a given CPU, a given implementation and a given compiler. Across all these parameters there is a fastest combination of CPU, implementation and compiler that is arguably 'true' because the SUPERCOP framework is freely available and all parameters used to obtain the performance number are clearly stated in the result file. A freely available benchmarking framework that explicitly states all parameters used to obtain a performance number like SUPERCOP provides a universal rule for performance evaluation, too: implementations, CPUs and compilers only get better, they never get worse. Hence a given algorithm's performance only gets better over time.

1.3   Additional Criteria

On desktop computers and servers the size of program code and lookup tables is usually not an issue. When the implementation of a cryptographic algorithm uses table lookups on such machines it is a concern because of timing attacks[3], not because of the amount of memory required. Small devices are different, they impose severe size limitations. There are hard limits, such as a 64k address space on an 8-bit machine, and somewhat softer restrictions, like the price of a smart card that is to be manufactured several million times. In both cases, smaller is better. Implementors compete for the smallest implementation, designers for the smallest algorithm (at comparable security level). The fact that theoretical work[4] concerning memory consumption of cryptographic algorithms is being done indicates that this is an area a growing interest.

1.4 Motivation for the eXternal Benchmarking eXtension

Without a sophisticated framework different performance numbers obtained under different (sometimes not clearly stated) circumstances circulate and make it very hard to judge how well designed a given algorithm really is. The eXternal Benchmarking eXtension presented here brings the advantages of SUPERCOP benchmarking to small devices like 8-bit microcontrollers. The results obtained this way are useful to

- find the fastest algorithm for a given target platform

- find the smallest algorithm for a given target platform

- find the best (cross-)compiler for a given hardware design and algorithm (for either speed or size)

- find the best compiler settings for a given hardware design and algorithm (for either speed or size)

- select a microcontroller for a future hardware design

- compare different implementations, e.g. a proprietary and commercial assembler implementation vs. public domain c code

- design new algorithms to run well on targeted hardware platforms

2 Design Goals

The aim of XBX (as the name suggests) is to extend SUPERCOP to a new domain: benchmarking external devices as opposed to the CPU(s) inside the computer that runs the SUPERCOP framework. To extend means 'to stretch out' and when stretching something it is usually advisable to take care not to rip it in two. In order to keep what the authors of this work perceived as the core of SUPERCOP intact the design goals for XBX were defined as follows:

Goal 1. Automatic testing of algorithms by a simple script invocation

Goal 2. Precise performance numbers for different message lengths that reflect real world user experience

Goal 3. Free source code, for every user to inspect and re-use

Goal 4. Cheap, easily available hardware

Goal 5. Compatibility to standard SUPERCOP algorithm interface

Goal 6. Compatibility to standard SUPERCOP results interface

As in many engineering projects there are also restrictions that are of a more practical nature but nonetheless must be taken into account if any result is to be generated. In this case the main restrictions were limited manpower and no funding at all. The goals derived from those limitations are:

Goal 7. Development using pre-owned or free development tools

Goal 8. Re-use of as many existing components as possible

Goal 9. Focus on SUPERCOP subset of current public interest: eBASH[1]

3   Hardware

3.1   Overview

XBX hardware consists of two main components: the eXternal Benchmarking Harness XBH and the eXternal Benchmarking Device XBD. The three hardware components PC, XBH and XBD are shown in Fig.1, together with their physical connections. The XBH connects to the PC running the eXternal Benchmarking Software (XBS), which is based on SUPERCOP, by means of Ethernet. This provides easy interfacing with any kind of computer that can run the SUPER-COP framework regardless of operating system. An RS232 port is available for low level configuration and debug output during development. Communication between the XBD and the harness is handled by means of a data connection and discrete digital I/O lines. The data connection is implemented using either I^2C or UART, depending on the type of device used as XBD. However, if UART is used for XBH-XBD communication the RS232 port of the XBH is no longer available for debugging purposes. The digital I/O lines are used for special purposes where the data connection would not perform adequately. The first purpose is device reset of the XBD. The XBD's reset pin is connected to a digital output of the XBH (in open collector configuration) and a pull-up resistor to the XBD's supply voltage. This allows the XBH to issue a hardware reset on the XBD, either because of a timeout or due to a command received from the PC. In the event that the XBD crashes, e.g. due to stack overflow, this mechanism provides a fast and reliable way to recover communication to the XBD in a situation where the data connection would be utterly useless.

Timing measurement is the second purpose that uses a dedicated digital I/O line. A digital output on the XBD is hooked up to an event capture pin on the XBH. Edges on that pin are triggered when a piece of code to be benchmarked is called and again when it returns. The event capture pin the XBH captures and timestamps these events and provides a duration from which a clock cycle count can be calculated.

3.2   Microcontroller Family

Due to goals 4, 7 and 8 the Atmel AVR 8-bit microcontroller family was selected to be the project's workhorse.

---

[1] SUPERCOP can benchmark many types of cryptographic algorithms, eBASH is the type for hash functions which are currently in focus due to the NIST SHA-3 competition.

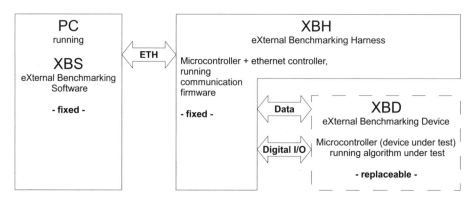

Fig. 1. High level overview of the XBX hardware setup. The PC on the left can be any computer that runs XBS, the XBH is a fixed interface component that does not need any adjustments while the XBD is the device under test and can be replaced at will.

This family has many beneficial features, starting with a low unit price and good availability in a variety of stores that sell to end customers. The performance delivered by the AVR family is high for an 8-bit design and many parts come in dual-in-line packaging, which is one of the few IC packages easily soldered by hand onto a perfboard[2] or similar carrier. On the software side, a port of the GNU Compiler Collection (GCC) exists for years now and is well tested. The GCC is supplemented by a standard C library implementation specifically tailored to the AVR and available for free. These features make the AVR family very popular with hobbyists around the world, which results in the added benefit of a huge user community providing ideas and code snippets for most questions that arise during development of an AVR based application. Also, the fact that one of the authors had a compatible JTAG adapter and some AVR chips in the closet accelerated the decision.

3.3 eXternal Benchmarking Harness XBH

The XBH setup comprises an Atmel ATmega644 microcontroller running at 16 MHz, a Microchip ENC28J60 Ethernet controller and a MAX232 TTL/RS232 voltage level shifter. For simplicity it is based on a commercially available module for home automation[7] which requires only minor modifications to act as XBH. An in system programmer connector is available to update the firmware as well as several signal connectors. The module runs on 9V AC power and can supply XBDs with up to 100mA of operating current.

3.4 eXternal Benchmarking Device XBD

The XBD is the device on which the actual benchmarking takes place. The XBD module consists of a microcontroller and whatever clock source and voltage

---

[2] A perfboard is a prototyping board that component are soldered to, as opposed to a (solderless) breadboard which works with little wires that are just plugged in to form connections.

regulation is necessary for the controller. Due to the design of the XBX setup the XBD can be easily replaced, requiring only the data connection and a digital output pin for timing measurements. Connecting the reset pin is highly recommended, although not strictly required. The timing measurement does not use any timer resources on the XBD, so theoretically even a microcontroller without any timers could be used. However, in order to be able to calibrate the timing measurements before a benchmark run it is advisable to have at least one timer unit available on the XBD.

The reference implementation uses an ATmega644 with an 8 MHz crystal oscillator circuit that runs on the 5V supply of the XBH. The data connection to the XBH is I^2C in this implementation. Running the XBD at half the frequency of the XBH gives the reference implementation the best possible timing accuracy. A different XBD implementation uses a Luminary Micro LM3S811 evaluation board. The LM3S811 is an ARM Cortex-M3 based 32-bit microcontroller and much more powerful that the AVR. It runs on 3.3V and the data connection uses a UART.

4 Software

The software is laid out as a chain of components with different tasks. Most components take the form of either shell or Perl scripts. Keeping several small components makes testing individual functions easier and reduces the likelihood of bugs compared to one big monolithic tool. The components are:

- Object file creation
- Download of binary code and parameters
- Execution framework
- Timing and result collection
- Benchmark control
- Post-processing

4.1 XBS: Benchmark Control

In a top-down view of the software architecture, benchmark control is highest layer. It talks to the user, it controls all actions. The benchmark control functionality is derived from SUPERCOP's control scripts, and called eXternal Benchmarking Software (XBS). SUPERCOP has a very simple user interface script called 'do'. Since one computer running XBS can control many different external targets the XBS scripts have some options that SUPERCOP's 'do' does not require.

XBS needs to be told which target platform to use since every platform potentially has it's own compilers, linkers and compiler options. The platform is selected by a line in a config file.

Once a build process for the selected platform is started the XBS copies the application code that is to be benchmarked into a temporary directory, where it is combined with supporting code called 'application framework' (AF). The AF provides all the communication services that the XBD will provide, except for bootloader functionality. The application code and AF combined for the binary of the application that is later downloaded into the XBD.

The binary just created is subjected to static analysis concerning the size of it's sections. If it is determined that this binary will not even fit into the memory of the XBD, download will not be tried.

If the binary passed the static analysis and looks as if it will fit into the XBD benchmark control calls a helper script that will talk to the XBH, which will talk to the bootloader on the XBD in order to download the newly created and checked application binary into the XBD. After successful download, the helper script is called again to trigger a short benchmark run executed and the performance of the triple [Algorithm, Compiler, Options] is stored for later reference, provided the binary produces the correct result for the known answer test. At this time, stack consumption is measured if the XBD AF for the selected platform supports this feature[3]. This process is repeated until all triples have been built, statically checked and if applicable, downloaded and benchmarked.

From the stored performance numbers for the short benchmark run the fastest tuple [Compiler, Options] per algorithm is selected and the corresponding binary application is downloaded into the XBD and subjected to a detailed benchmark at different message lengths.

The results of the short and detailed benchmark runs are written to a text based output file, with XBS specific information like stack use embedded as comments in SUPERCOPs results format. Thus the result files should be readable by the same tools that process SUPERCOP results[4] although without making use of the additional information like stack usage.

4.2 Algorithms to Benchmark

The major part of the algorithms benchmarked using XBX are taken from the SUPERCOP suite. The XBS scripts are closely modeled on SUPERCOP in that regard, using the same directory structure to hold algorithms and their implementations. Some code that was not submitted to SUPERCOP was adapted by the authors to fit the same interface and subsequently benchmarked. Most of this code came from 'Das Labor'[5], a small device working group related to Ruhr Uni Bochum who wrote a collection of cryptographic primitives implemented specifically for the Atmel AVR family.

4.3 Hardware Abstraction

To separate the benchmarking logic from platform specific code such as communication, execution of binaries and debugging output a simple hardware abstraction layer (HAL) was employed. This HAL hides the device dependent aspects

---

[3] This is currently available for Atmel AVR targets only.

[4] A quick test by Dan Bernstein showed that they are indeed, although no written documentation of that test exists.

of the XBD like special function registers required to use the UART, the exact method used to erase and program flash pages in the bootloader and the fact that some microcontroller families, e.g. the Atmel AVR, are based on Harvard architecture and do not hide this fact from the programmer. The special treatment of constant *data* in what *per definitionem* is *program memory* on such devices is also hidden in the HAL. Without this functionality all constant data would end up in the RAM, rendering most SUPERCOP submitted implementations useless due to the fact that not even the initial stack would fit into RAM anymore.

One function in the HAL is especially important for goal 2: precise performance numbers. Since the actual timing measurement takes place on the XBH but the reported value is the amount of CPU cycles on the XBD it is important that the relation between the time bases on XBH and XBD is known as exactly as possible. To this end, a timing calibration service has been implemented. When the XBS requests a timing calibration, the XBH triggers the timing calibration routine on the XBD. This routine busy loops for a device dependent amount of time, toggles the timing output digital I/O line as it does when benchmarking algorithms and additionally counts the number of CPU cycles it spent between the two digital I/O toggles using an internal timer. This number of cycles is reported to the XBH, which reports it along with it's own timing measurement of the same event to the XBS. A correction factor for clock drift between XBH and XBD and/or a sanity check on the reported values can then be performed on the PC by the XBS. This timing calibration sequence looks as depicted in Fig.2 . Measuring stack usage is another challenge that can only be solved in a device dependent manner, yet should be available to the application via a standard interface. The HAL contains two functions that allow for stack measurement: paintStack and countStack. PaintStack 'paints' the free stack area with a known pattern, called a canary bird. Then the function to be benchmarked is called and after it returns, countStack counts the number of canary birds that did not 'survive'. This gives the maximum amount of stack used by the benchmarked function. Combined with the static RAM requirement obtained from the application binary and the known RAM requirement of the AF, the total RAM consumption of a triple [Algorithm, Compiler, Options] can be measured.

4.4   Application Framework

The application framework provides hardware independent basic management functions like processing requests, parameter handling and so on. It is combined with the algorithm to benchmark and the hardware abstraction layer for the device under test to form the XBX application binary.

4.5   Bootloader

The bootloader is used to download and execute application binaries. It is formed by combining the generic boot loader logic code with the hardware abstraction

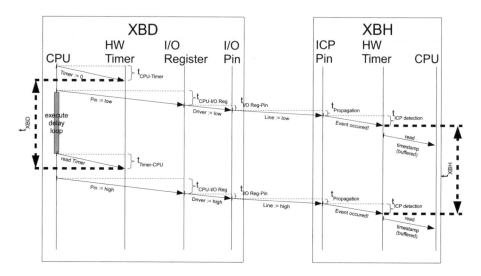

Fig. 2. UML sequence diagram for the XBX timing calibration routine. If it holds that a) reading and writing the timer requires the same time and b) switching the timing indicator pin high takes as long as switching it low then all delays from measurement and indication are symmetrical and cancel each other out. Thus if XBD and XBH have perfectly identical clock sources $t_{XBD} = t_{XBH}$ holds, otherwise $\frac{t_{XBD}}{t_{XBH}}$ can be used for sanity checks and is also logged by the XBS so it can later manually be applied as a correction ratio by the user.

layer. Before benchmarking a target device, the bootloader must once be manually uploaded, e.g. via JTAG. After that, application download and execution is handled over a communication channel established between the bootloader and the benchmarking harness.

4.6 Benchmarking Harness

The firmware for the XBH receives commands from the XBS on the PC by means of UDP packets. The commands form a protocol that is simply called the XBH protocol, which features ASCII based command words with ASCII-encoded hex digits as parameters. Although the upper 4 bit of every byte are wasted, this is no major concern. The bandwidth on the Ethernet is orders of magnitude higher than on the following I^2C or UART link to the XBD. The ASCII encoding allowed for simple testing using the *netcat* command in the early development phases and is easily processed by Perl scripts. Monitoring XBH protocol in a serial terminal also benefits from this choice. The XBH software generates requests to the XBD as necessary. The protocol for these requests is called XBD protocol and uses ASCII commands but binary encoding for the parameters. Answers from the XBD are processed and reported back to the XBS over Ethernet using XBH protocol. Since the XBS never uses the XBD protocol, changes in the XBS do not affect existing XBDs and vice versa. Keeping in line with goal 8: re-use, the XBH firmware consists of an embedded web server software[6] by Ulrich Radig with the XBH functionality added as another UDP service and some modifications to the timer handling code in order to facilitate XBD timing measurements.

5   Benchmarking Results

This sections gives a few examples of benchmarking results obtained using the XBX.

5.1   Different Implementations of Skein512 on Atmel ATmega1281

The ATmega1281 comes from the same family as the ATmega644 in the reference implementation of XBX, it has the same CPU core and performance per MHz. The RAM however is at 8kiB twice as big which helps enormously in running implementations that are not size optimized. The performance numbers for Skein512[8] listed in table 1 are certainly not the best the algorithm can do, they just reflect the best implementations available to the authors at the time this work was written. The C implementations were compiled with the AVR port of GCC, the original SUPERCOP submission by the Skein team had to be modified to fit into the XBD's flash memory. Even though the ATmega1281 comes with 128kiB of flash memory, this implementation was so aggressively loop unrolled that the second author of this work has to re-roll the loops manually to make it fit. This is a typical issue with speed optimized SUPERCOP submissions intended to run on a PC and by no means specific to the Skein team's work. The huge advantage of the assembly implementation in both execution time and

Table 1. Skein512, different implementations on Atmel ATmega1281 in cycles per byte

Property	Skein Team C	Das Labor C	Das Labor ASM
cpb @ 1536 bytes message length	7842.6	8602.1	1571.1
RAM usage(Stack+global/static)	2684	1580	1391

space is evident, yet in this case a size optimized C implementation also gives a 40% advantage in RAM usage at only a small performance loss.

5.2   SHA-3 Candidates on an ARM Cortex-M3 32-Bit CPU Using Two Compilers

One of the most interesting features of SUPERCOP for many users is the ability to benchmark the same implementations using different compilers and compiler options. XBX preserves this property for the target platforms where several compilers are available. One such platform is the ARM Cortex family. Using a free trial license of the ARM C compiler (ARMCC) we obtained performance

Table 2. ARMCC vs. GCC on ARM Cortex-M3 in cycles per byte for a 1536 byte message, fastest algorithms on platform

Compiler	BMW256[9]	Shabal512[10]	BMW512[9]
ARMCC	24.2	33.6	48.2
GCC	26.2	43.5	70.3

Table 3. ARMCC vs. GCC on ARM Cortex-M3 in cycles per byte for a 1536 byte message, unexpected behavior

Compiler	Blake32[11]	Keccak1024c576[12]	CubeHash1632[13]
ARMCC	96.2	125.1	835.8
GCC	72.1	109.7	323.2

Table 4. Cortex-M3 vs. Pentium 3 (683) in cycles per byte for a 1536 byte message

Compiler	BMW256	Shabal512	BMW512
Cortex-M3	24.2	33.6	48.2
Pentium 3	13.82	14.24	30.04

numbers of SHA-3 candidate implementations using both GCC and ARMCC on a Cortex-M3 based Luminary Micro LM3S811 microcontroller. ARMCC is generally considered to be expensive but also the most sophisticated compiler available for ARM CPUs. Expectation was that it would perform better than GCC. The fastest SHA-3 candidates on this platform from the subset available for XBX benchmarking at the time met this expectation, as can be seen in table 2.

Surprisingly, some algorithms suite the optimization strategies of GCC better, resulting in an overall better performance. See table 3 for details.

In general, the Cortex-M3 performs very well when compared to a roughly 50 [14] [15] [16] times larger Intel Pentium 3 desktop processor as benchmarked in [17]. Table 4 shows a performance gap of barely factor 2 between the two CPUs with respect to the three SHA-3 candidates.

6   Conclusion

In this paper we introduced an extension to the SUPERCOP benchmarking suite and described the main design decisions and compromises we made in order to get it running in time for the second round of the SHA-3 competition. We believe that the overall design is sound and that using SUPERCOP-XBX meaningful results both for speed and memory requirements of cryptographic hash functions can be obtained.

Acknowledgments. First of all many thanks to Daniel J. Bernstein and Tanja Lange for their support and encouragement. Their approval really means a lot to us. The Cortex-M3 target was provided by ARM, Ltd. and we have Richard York and Alex Nancekievill to thank for that. Since the XBX project has no funding their support is most appreciated.

References

1. Bernstein, D.J., Lange, T. (eds.): eBACS: ECRYPT Benchmarking of Cryptographic Systems, http://bench.cr.yp.to (accessed November 5, 2009)
2. NIST: First SHA-3 Candidate Conference, http://csrc.nist.gov/groups/ST/hash/sha-3/Round1/Feb2009/program.html (accessed February 27, 2010)
3. Bernstein, D.J.: Cache-timing attacks on AES, http://cr.yp.to/antiforgery/cachetiming-20050414.pdf (accessed February 27, 2010)
4. Ideguchi, K., Owada, T., Yoshida, H.: A Study on RAM Requirements of Various SHA-3 Candidates on Low-cost 8-bit CPUs, http://www.sdl.hitachi.co.jp/crypto/lesamnta/A_Study_on_RAM_Requirements.pdf (accessed February 27, 2010)
5. Otte, D., et al.: AVR Crypto Lib., http://www.das-labor.org/wiki/AVR-Crypto-Lib/en (accessed February 27, 2010)
6. Radig, U.: AVR Webserver Software, http://www.ulrichradig.de/ (accessed February 27, 2010)
7. Pollin: AVR-Net-IO Board, http://www.pollin.de/shop/downloads/D810058B.PDF (accessed February 28, 2010)
8. Ferguson, N., Lucks, S., Schneier, B., Whiting, D., Bellare, M., Kohno, T., Callas, J., Walker, J.: The Skein Hash Function Family Submission to NIST, Round 2 (2009)
9. Gligoroski, D., Klima, V., Knapskog, S.J., El-Hadedy, M., Amundsen, J., Mjølsnes, S.F.: Cryptographic Hash Function BLUE MIDNIGHT WISH Submission to NIST, Round 2 (2009)
10. Bresson, E., Canteaut, A., Chevallier-Mames, B., Clavier, C., Fuhr, T., Gouget, A., Icart, T., Misarsky, J.-F., Naya-Plasencia, M., Paillier, P., Pornin, T., Reinhard, J.-R., Thuillet, C., Videau, M.: - Shabal, a Submission to NIST's Cryptographic Hash Algorithm Competition Submission to NIST (2008)
11. Aumasson, J.-P., Henzen, L., Meier, W., Phan, R.C.-W.: SHA-3 proposal BLAKE Submission to NIST (2008)
12. Bertoni, G., Daemen, J., Peeters, M., Van Assche, G.: Keccak specifications Submission to NIST, Round 2 (2009)
13. Bernstein, D.J.: CubeHash specification (2.B.1) Submission to NIST, Round 2 (2009)
14. ARM: Whitepaper about the Cortex-M3, http://www.arm.com/files/pdf/IntroToCortex-M3.pdf (accessed February 28, 2010)
15. Intel: Presskit on Moore's law, http://www.intel.com/pressroom/kits/events/moores_law_40th/ (accessed February 28, 2010)
16. Intel: Pentium 3 datasheet, http://developer.intel.com/design/pentiumiii/datashts/245264.htm (accessed February 28, 2010)
17. Bernstein, D.J., Lange, T. (eds.): SUPERCOP benchmarking results. See results for computer 'manneke', http://bench.cr.yp.to/results-hash.html (accessed February 27, 2010)

Public Key Perturbation of Randomized RSA Implementations

Alexandre Berzati[1,2], Cécile Canovas-Dumas[1], and Louis Goubin[2]

[1] CEA-LETI/MINATEC, 17 rue des Martyrs, 38054 Grenoble Cedex 9, France
{alexandre.berzati,cecile.canovas}@cea.fr
[2] Versailles Saint-Quentin-en-Yvelines University,
45 Avenue des Etats-Unis, 78035 Versailles Cedex, France
Louis.Goubin@prism.uvsq.fr

Abstract. Among all countermeasures that have been proposed to thwart side-channel attacks against RSA implementations, the exponent randomization method – also known as exponent blinding – has been very early suggested by P. Kocher in 1996, and formalized by J.-S. Coron at CHES 1999. Although it has been used for a long time, some authors pointed out the fact that it does not intrinsically remove all sources of leakage. At CHES 2003, P.-A. Fouque and F. Valette devised the so-called *"Doubling Attack"* that can recover the blinded secret exponent from an SPA analysis. In this paper, we consider the case of fault injections. Although it was conjectured by A. Berzati *et al.* at CT-RSA 2009 that exponent randomization avoids fault attacks, we describe here how to recover the RSA private key under a practical fault model. Our attack belongs to the family of public key perturbations and is the first fault attack against RSA implementations with the exponent randomization countermeasure. In practice, for a 1024-bit RSA signature algorithms, the attack succeeds from about 1000 faulty signatures.

Keywords: RSA, fault attacks, exponent randomization/blinding, public modulus.

1   Introduction

The exponent randomization method – also referred as exponent blinding – has been first suggested by P. Kocher [11]. This method inspired J.-S. Coron [7] that later formalizes it to defeat side channel attacks, such as DPA, that gain the information leaked during the exponentiation. This method is widely used because it is easy to implement and the induced overhead is reasonable. However any implementation may still be a potential source of leakage.

The first attack published against this countermeasure is due to P.-A. Fouque and F. Valette [10]. The so-called *"Doubling Attack"* allows an attacker to recover a blinded secret exponent from an SPA analysis. This attack only works for *"Left-To-Right"*-based implementations of the modular exponentiation. Moreover, the attacker is assumed to be able to send many times the same known

S. Mangard and F.-X. Standaert (Eds.): CHES 2010, LNCS 6225, pp. 306–319, 2010.

message and that no message randomization is performed before the modular exponentiation. At CHES 2006 [8], P.-A. Fouque *et al.* show that if Coron's countermeasure is used with some windowing exponentiation algorithms and a small public key, then a simple SPA combined with a tricky analysis makes it possible to recover both secret key and factorize the public modulus. It is worthwhile to notice that this attack exploits the *non-uniformity* of the exponent randomization countermeasure (see Sect. 3.1). Instead of exploiting the physical leakage due to the execution of a modular exponentiation, like in previous attacks, P.-A. Fouque *et al.* proposed at CHES 2008 [9] to focus on the leakage induced by the computation of the random exponent itself. Since the secret exponent and the blinding part are cut into words, spying on the carries of the adder may reveal information that is used to guess the most significant bits of each word of the secret key. When the number of missing bits is small enough, the attacker can use classical methods, such as Shanks' Baby-Step Giant-Step algorithm, to obtain the whole secret key.

The exponent randomization may also be used to protect implementations against some fault attacks. Namely, this countermeasure is useful to defeat attacks that require multiple faulty signatures to recover the private exponent since each signature is computed with a different exponent. Although the device embedding this countermeasure still remains vulnerable to perturbation, it does not exist, as far as we know, any method for exploiting such faulty outputs. This paper bridges the gap by providing a new fault attack that defeats the exponent randomization. This attack belongs in the recent family of public key perturbations.

Exploiting the perturbation of public elements has been first addressed by I. Biehl *et al.* with several applications to elliptic curves [3]. But it took a half decade before seeing a successful application to RSA [12]. The first exploitation of the RSA public modulus perturbation leading to a full secret key recovery is due to E. Brier *et al.* [5] (see also [6] for further optimizations). In the case of the last attack, the use of the blinded exponent is an efficient countermeasure. A new fault attack based on the public modulus corruption has been proposed lately by A. Berzati *et al.* against both *"Right-To-Left"* and *"Left-To-Right"* implementations of the core RSA modular exponentiation [2,1]. Unlike previous works, the attack takes advantage of a perturbation of the modulus that occurs while the device is performing a signature. Such a fault injection splits the signature into a correct and a faulty part and so, isolates a part of the secret exponent. Then, from a correct/faulty signature pair, the attacker can *guess-and-determine* both faulty modulus and the part of secret exponent. The whole exponent is obtained by cascading the attack on signatures corrupted at different moments of the execution. At CT-RSA'09, authors claimed that the exponent randomization may be used to defeat their fault attack [1].

In this article, we show that even if this countermeasure is used, it is possible to recover the private exponent under a practical fault model. To the best of our knowledge, this is the first fault attack that aims to threaten RSA implementations with the exponent randomization countermeasure. The analysis takes advantage of the *non-uniformity* of the exponent randomization. As a consequence, this work completes the state-of-the-art of the side channel analysis of the exponent randomization.

The remainder of this paper is organized as follows: Section 2 describes classical implementations of RSA and the random exponent countermeasure. Our fault analysis is detailed in Sect. 3 and summarized as an algorithm in Sect. 4. Finally, we conclude in Sect. 5 about the vulnerability of random exponent countermeasure implementations in the context of fault attacks.

2   Background

2.1   Notations

Let N, the public modulus, be the product of two large prime numbers p and q. The length of N is denoted by n. Let e be the public exponent, coprime to $\varphi(N) = (p-1) \cdot (q-1)$, where $\varphi(\cdot)$ denotes Euler's totient function. The public key exponent e is linked to the private exponent d by the equation $e \cdot d \equiv 1 \bmod \varphi(N)$. The private exponent d is used to perform the operations below.

RSA Decryption: Decrypting a ciphertext C boils down to compute $\tilde{m} \equiv C^d \bmod N$. If no error occurs during computation, transmission or decryption of C, then $\tilde{m}$ equals m.

RSA Signature: The signature of a message m is given by $S \equiv \dot{m}^d \bmod N$ where $\dot{m} = \mu(m)$ for some hash and/or deterministic padding function μ. The signature S is validated by checking that $S^e \equiv \dot{m} \bmod N$.

2.2   Modular Exponentiation Algorithms

Algorithm 1. *"Right-To-Left"* modular exponentiation
INPUT: m, d, N
OUTPUT: $A \equiv m^d \bmod N$
1 : $A{:=}1$;
2 : $B{:=}m$;
3 : **for** i **from** 0 **upto** $(n-1)$
4 :   **if** $(d_i == 1)$
5 : $A := (A \cdot B) \bmod N$;
6 :   **end if**
7 : $B := B^2 \bmod N$;
8 : **end for**
9 : **return** A;

Algorithm 2. *"Left-To-Right"* modular exponentiation
INPUT: m, d, N
OUTPUT: $A \equiv m^d \bmod N$
1 : $A{:=}1$;
2 : **for** i **from** $(n-1)$ **downto** 0
3 : $A := A^2 \bmod N$;
4 :   **if** $(d_i == 1)$
5 : $A := (A \cdot m) \bmod N$;
6 :   **end if**
7 : **end for**
8 : **return** A;

Binary exponentiation algorithms are often used to compute the RSA modular exponentiation $\dot{m}^d \bmod N$ where the exponent d is expressed in a binary form as $d = \sum_{i=0}^{n-1} 2^i \cdot d_i$, where d_i stands for the i-th bit of d. Their polynomial complexity with respect to the input length make them very interesting to perform the core RSA operation. Algorithm 1 describes a way to compute modular exponentiations by scanning bits of d from least significant bits (LSB) to most significant bits (MSB). That is why it is usually referred to as the *"Right-To-Left"* modular exponentiation algorithm.

The dual algorithm that implements the binary modular exponentiation is the *"Left-To-Right"* exponentiation described in Algorithm 2. This algorithm scans bits of the exponent from MSB to LSB and is lighter than *"Right-To-Left"* one in terms of memory consumption.

It exists multiple implementations derived from these dual algorithms, such as OpenSSL *fixed/sliding window* implementations or the *Square-and-Multiply-always* variant [7]. For the sake of clarity, we will only focus our presentation on the binary version of the *"Right-To-Left"* method. But the principle of our analysis can be easily adapted to attack its variants.

2.3 Exponent Randomization

The exponent randomization method has been proposed by P. Kocher [11] to defeat side channel attacks, such as DPA, that gain information leaked during the exponentiation. The principle of this countermeasure is based on Fermat's theorem. Indeed, for all $m \in (\mathbb{Z}/N\mathbb{Z})^*$ and $\lambda \in \mathbb{Z}$, $m^{\lambda \cdot \varphi(N)} \equiv 1 \bmod N$. The exponent randomization algorithm derived from this result is detailed below. The complexity of the modular exponentiation algorithm is polynomial with

---

Algorithm 3. RSA exponent randomization algorithm

INPUT: $\dot{m}$, N, $\varphi(N)$, d and l

OUTPUT: $S = \dot{m}^d \bmod N$

1: *//Randomize the private exponent*
2: Pick a random $\lambda \in [\![0; 2^l - 1]\!]$;
3: $\bar{d} = d + \lambda\varphi(N)$;
4: *//Perform the exponentiation*
5: $S = \mathtt{PowMod}(\dot{m}, \bar{d}, N)$;
6: **return** S;

---

respect to the exponent length. Thus, to guarantee a reasonable overhead, the l value as to be small compared to the RSA length n. Typically, for a 1024-bit RSA, $l = 20$ or $l = 32$.

3 Description of Our Attack

3.1 Bit Analysis of a Randomized Exponent

In this section, we aim to analyze the influence of the different variables that are involved in the computation of a randomized exponent. By definition, the blinded exponent $\bar{d}$ is built by adding a random multiple of $\varphi(N)$ to the secret exponent d. Using a different expression of $\varphi(N)$, the expression of $\bar{d}$ can also be written as:

$$\bar{d} = d + \lambda\varphi(N) \tag{1}$$
$$= d + \lambda\,(p-1)\,(q-1)$$
$$= d + \lambda N - \lambda\,(p+q-1)$$

From the previous expression, one can notice that the randomized exponent $\bar{d}$ is built by adding 3 terms of different sizes. As a consequence, the bits of d are not homogeneously masked by this method. Figure 1 illustrates this statement for a n-bit RSA and a l-bit random value λ. This figure highlights that despite

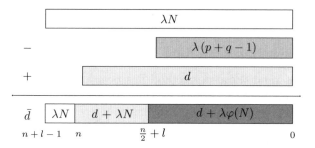

Fig. 1. Bit analysis of a random exponent

being properly masked by a multiple of $\varphi(N)$, the most significant bits of d are masked by a random multiple of N. But the blinding method seems to be more efficient for the least significant bits. In the next section, we will explain how to use the heterogeneity of the exponent randomization method for exploiting faults.

3.2   Fault Model

Description. The model we have chosen to perform the attack is derived from the ones previously used by A. Berzati *et al.* to successfully attack both *"Right-To-Left"* [2] and *"Left-To-Right"* [1] implementations of standard RSA. We suppose that the attacker is able to inject a transient fault that modifies the public modulus N during the execution of a signature with a randomized exponent $\bar{d}$ for a known input message m (see Sect 2.3). The injected fault affects a byte of the modulus by modifying it in a random way, namely:

$$\hat{N} = N \oplus \varepsilon \tag{2}$$

where $\varepsilon = R_8 \cdot 2^{8i}$, $i \in [\![0; \frac{n}{8} - 1]\!]$ and R_8 is a non-zero random byte value. In our assumption, the value of the faulty modulus $\hat{N}$ is not known, *a priori*, by the attacker. In this article, we consider that the exponentiation is implemented with the *"Right-To-Left"* method or a variant. The fault is injected during a square, at the t-th step of the exponentiation and such that the end of the execution is performed with the faulty modulus $\hat{N}$. It is also assumed that the time location of the injection is controlled by the attacker, and so, the parameter t may be set (or known) by the attacker depending on the exponent part he aims to recover.

Discussion. This fault model has been chosen because of its practicability in the smartcard context. Although the effect of a fault injection is highly dependent of the component attacked, it seems that a random modification of the value of a memory word can be easily produced by a laser. This model has been already used in the literature leading to successful applications [13,4,1]. Furthermore, the timing control of the fault injection is not a restrictive assumption since the attacker can trigger the laser shots using a Simple Power Analysis.

3.3   Result of a Faulty Computation

Let $\bar{d} = \sum_{i=0}^{n+l-1} 2^i \cdot \bar{d}_i$ be the binary representation of a randomized exponent $\bar{d}$. According to the fault model described above, the fault occurs during a square at the t-th step of the execution. Hence, if B_{t-1} denotes the internal register value that contains the result of the consecutive squares before the fault injection:

$$\hat{B}_t \equiv B_{t-1}{}^2 \bmod \hat{N}$$
$$\equiv \left(\dot{m}^{2^{t-1}} \bmod N\right)^2 \bmod \hat{N} \tag{3}$$

The subsequent operations of the exponentiation are also performed with the faulty modulus. If we denote by $A_t \equiv \dot{m}^{\sum_{i=0}^{t-1} 2^i \cdot \bar{d}_i}$ the internal state value before the fault injection. The result of the faulty RSA signature $\hat{S}_t$ can be written as:

$$\hat{S}_t \equiv A_t \cdot \hat{B}_t^{\bar{d}_t} \cdot \ldots \cdot \hat{B}_t^{2^{(n+l-1)-t} \cdot \bar{d}_{n+l-1}} \bmod \hat{N} \tag{4}$$
$$\equiv A_t \cdot \hat{B}_t^{\frac{\bar{d}_{[t]}}{2^t}} \bmod \hat{N} \tag{5}$$

where $\bar{d}_{[t]} = \sum_{i=t}^{n+l-1} 2^i \cdot \bar{d}_i$. The previous equation highlights that the fault has isolated the most significant part of the blinded exponent $\bar{d}$. In other words the perturbation gives the opportunity for the attacker to focus on the recovery of a part of the exponent. The next section reminds the general methodology used to exploit faults on the public modulus during the execution of the exponentiation (see also [2,1] for further details).

3.4   Analysis

In the following sections we will detail the effects of faults that have been injected according to the model described above. Then we will propose different ways for exploiting perturbations, depending on their timing location t.

General Methodology. The general principle of the analysis consists in making use of the isolation of a part of exponent by the fault injection. Indeed, if the isolated part of exponent is small enough, it is possible to *guess-and-determine* it from a faulty/correct signature pair $(\hat{S}_t, S_t)$. Therefore, since the faulty modulus is also unknown by the attacker, he chooses a candidate value $\hat{N}'$ and another candidate value $\bar{d}'_{[t]}$ for the most significant part of the randomized exponent he has to determine. Then he computes from the correct signature:

$$A'_t \equiv S_t \cdot \dot{m}^{-\bar{d}'_{[t]}} \bmod N \tag{6}$$

This computation aims to retrieve the value of the internal register A_t when the fault occurred. The next step consists in using the candidate values to simulate a faulty end of exponentiation. To do so, the attacker computes:

$$S'_{(\bar{d}'_{[t]},\hat{N}')} \equiv A'_t \cdot \left(\dot{m}^{2^{t-1}} \bmod N \right)^{2 \cdot \frac{\bar{d}'_{[t]}}{2^t}} \bmod \hat{N}' \tag{7}$$

Finally, he checks if the following equation is satisfied:

$$S'_{(\bar{d}'_{[t]},\hat{N}')} \equiv \hat{S}_t \bmod \hat{N}' \tag{8}$$

In the case of satisfaction, it means that the chosen candidate pair is the correct one with high probability. Otherwise, the attacker has to choose another candidate pair and perform this test again. One can notice that a similar analysis can be performed when the first operation infected by the fault is a multiplication The details of this variant are provided in [2].

Contrary to the attack presented in [2], the subsequent bits of exponent can not be obtained by repeating the analysis on a signature faulted earlier. Indeed, the exponent randomization countermeasure implies that a fresh exponent is used for each execution of the signature. Hence, it is not possible to repeat the attack by using the knowledge of already found bits of blinded exponent $\bar{d}_{[t]}$ as in [2]. As a consequence we have to adapt this general methodology to extract real bits of the private exponent d from bits of $\bar{d}$ recovered by the analysis detailed above. We show in the following parts how to make use of the *non-homogeneity* of the exponent randomization to do so.

Case of unexploitable faults. This case corresponds to fault that have been injected at final steps of the exponentiation, namely for $n \leq t \leq (n + l - 1)$. The few amount of information about the secret key that belongs to this range of data is due to the carry propagation of the addition of d and a random multiple of $\varphi(N)$ (see Fig. 1). So, the analysis of signatures faulted in this timing range is not relevant for extracting information about the secret key d since. As a consequence, it is worthwhile focusing on faults injected earlier in the computation.

Faults on MSB. This section aims to provide a method for analyzing RSA signatures that have been faulted while the t-th bit of the blinded exponent is treated, namely if $\left(\frac{n}{2} + l \right) \leq t < n$. As we said in the previous section, our goal is to extract some bits of the real exponent from the recovered part of randomized exponent. In fact, performing the attack by this way seems to be difficult. First, $\bar{d}$ depends on d but also on a random multiple of $\varphi(N)$ and all these values are unknown by the attacker. This difficulty can be overcome thanks to the non-homogeneity of the exponent randomization (see Fig. 1). Indeed, in the bound $\left(\frac{n}{2} + l\right) \leq t < n$, we have:

$$\bar{d} \approx d + \lambda N \tag{9}$$

As a consequence, the most significant part of $\bar{d}$ only depends on d and λ. Instead of searching $\bar{d}_{[t]}$ and extracting bits of d from it, we have decided to directly guess both d and λ by building "good" candidate values for $\bar{d}_{[t]}$:

$$\bar{d}_{[t]} = \sum_{i=t}^{n+l-1} 2^i \cdot \bar{d}_i \tag{10}$$

$$\approx \sum_{i=t}^{n+l-1} 2^i \cdot (d + \lambda N)_i \tag{11}$$

$$\approx \sum_{i=t}^{n-1} 2^i \cdot d_i + \sum_{i=t}^{n+l-1} 2^i \cdot (\lambda N)_i + carry_t\,(d, \lambda N) \cdot 2^t \tag{12}$$

where $carry_t\,(a, b)$ denotes the carry bit resulting from the bit-wise addition of the t first bits of a and b. Equation (12) shows that guessing $\bar{d}_{[t]}$ boils down to simultaneously guess the random value λ and the $(n - t)$ most significant bits of d. In the general case, this part of d splits into a known (already recovered) part d_{MSB} and w missing bits denoted by d_w. The carry bit is considered as an uncertainty on the parity of d_w. As a consequence, if an attacker builds candidate values for $\bar{d}_{[t]}$ that satisfy (8), then he can directly deduce $w - 1$ new bits of d. Thus, the known part d_{MSB} grows up of $w - 1$ bits, and it can be used again to analyze signatures faulted earlier in the execution and cascade the resolution of almost half of the secret key. That way, even if the isolated part of random exponent grows up, the part of exponent to be determined d_w remains constant. Hence, instead of guessing a pair of candidate values to satisfy (8) as described in the general methodology, we guess the triplet of values $(d_w, \lambda, \hat{N})$ and thus deduce a part of the secret key.

By carefully studying our improvement, one can wonder if using the sole relation (8) is enough for determining with high probability a triplet of values. This remark is all the more relevant since our implementation of the attack showed us that multiple candidate triplets may satisfy (8). The authors of [2,1] previously proved that the order false-acceptance probability is about $\frac{1}{N}$. Thus, it is highly negligible for common RSA length. But, we also noticed that for all false-accepted triplets, the candidate values accepted for λ are always smaller than the correct one. Hence, the triplet that contains the biggest candidate value for λ is always the correct one. This heuristic was successfully adopted to improve our attack algorithm by reducing the number of candidate triplets that satisfy (8) to the correct one only. We also formalized this heuristic in the following theorem. The proof of the theorem is given in Appendix A.

Theorem 1. *Let $\hat{S}_t$ be a faulty signature performed under an exponent randomized by λ, and S the corresponding correct signature. For all candidate pairs $(d'_w, \lambda') \in [\![0; 2^w]\!] \times [\![0; 2^l]\!]$, if $\lambda' > \lambda$, then (8) can not be satisfied.*

As a consequence, by combining the approximation of the randomization (see Eq. (12)) for building candidate values for $\bar{d}_{[t]}$, and the theorem above, an attacker will be able to recover a part of the real private exponent d with high probability from only one correct/faulty signature pair. Moreover, our method enables the attacker to use the already found bits of d and cascade the analysis for signatures faulted earlier in their execution. By this way, it is possible to recover almost all the most significant bits of d.

Faults on LSB. We will focus here in the recovery of the least significant bits of d. From Fig. 1, if $0 \leq t < \left(\frac{n}{2} + l\right)$, then $\varphi(N)$ is, this time, fully involved in the randomization of the secret exponent d. Since this value is private, contrary to the modulus N, the attacker can not run the analysis described for the recovery of the most significant part of d. But, we will describe in this section how we have used the previous analysis on multiple faulty signatures to overcome this difficulty.

As we previously said, we can not approximate the least significant bits of the blinded exponent $\bar{d}$ as the sum of the real exponent d and a random multiple of the public modulus (see Fig. 1). But, let us rewrite the expression of a part of randomized exponent isolated by the fault injection:

$$\bar{d}_{[t]} = \sum_{i=t}^{n+l-1} 2^i \cdot (d + \lambda\varphi(N))_i \tag{13}$$

$$= \sum_{i=t}^{n+l-1} 2^i \cdot (d + \lambda N - \lambda(p + q - 1))_i \tag{14}$$

$$\approx \sum_{i=t}^{n-1} 2^i \cdot \delta_i + \sum_{i=t}^{n+l-1} 2^i \cdot (\lambda N)_i \tag{15}$$

where $\delta_i = (d - \lambda(p + q - 1))_i$. As for the MSB case, for any iteration of the analysis, we assume that the attacker has already determined both most significant parts of d and $(p+q-1)$ respectively denoted by d_{MSB} and $(p+q-1)_{MSB}$. Here the value $\delta = \sum_{i=t}^{n-1} 2^i \cdot \delta_i$ splits into w missing bits denoted by δ_w and a part δ_{MSB} that depends on λ, d_{MSB} and $(p + q - 1)_{MSB}$. This last part is also unknown, but it becomes computable whenever λ is guessed. Equation (15) shows that the attacker can apply the same *guess-and-determine* method to recover the part of randomized exponent $\bar{d}_{[t]}$ that satisfies (8) by building it from the triplet of candidate values for $(\delta_w, \lambda, \hat{N})$. Hence, such an analysis applied on a correct/faulty signature pair only returns the searched triplet with high probability. But, one can notice that contrary to the MSB case, the analysis does not directly returns a part of d, but a more intricate value δ_w. So, the attacker has to perform a complementary analysis on the variable δ_w he has just recovered to extract both expected parts of d and $\lambda(p + q - 1)$. According to Fig. 1, it is relevant to notice that δ_w depends on:

- w unknown bits of the exponent d,
- w unknown bits of the sum of RSA primes $(p + q - 1)$,
- the random value λ that has been just recovered,
- some carry bits.

Thus, from this value, the attacker obtains one equation that involves two unknown variables. In order to recover simultaneously w bits of the exponent and w bits of the sum of RSA primes, it is necessary to get additional equations (at least one more). This can be achieved by repeating this analysis on a signature faulted at the same step of its execution to recover another δ_w value for a different λ.

For the sake of clarity, we have voluntary withdrawn the influence of the carry bits in the system. In practice, as for the MSB case, these carry bits add some uncertainty on the low order bits of d_w. In other words, instead of recovering a unique value for d_w, in practice two or three solutions are returned in the worst case. But, the wrong values obtained for d_w will be discarded when subsequent analysis will be performed. Thus, the number of false-accepted candidates does not grow up exponentially in cascading the resolution, but stay bounded. When the part of d recovered by repeating the described analysis is large enough, the attacker may complete the attack by using classical methods such as Shank's Baby-Step Giant-Step or lattice techniques.

4   Attack Algorithm

4.1   Summary of Our Attack

In this section, we detail the implementation of our Differential Fault Analysis described above. This part completes our previous theoretical approach by providing a more pragmatic description of our attack methodology. This algorithm has been successfully implemented on a standard PC using the GMP Library [1] leading.

Gather faulty signatures. The attacker first chooses a window length w for the recovery of the secret key d. Then he has to gather multiple signatures faulted at different steps of the execution. If we denote by t the time location of the fault injection, the attacker has to gather:
- One faulty signature and the corresponding correct one if $\left(\frac{n}{2} + l\right) \leq t < n$
- Two (or more) faulty signatures and the corresponding correct one if $0 \leq t < \left(\frac{n}{2} + l\right)$

where t is decremented by w each time. The collected signatures are sorted in descending fault locations.

---

[1] The GNU Multiple Precision Library. Available at http://gmplib.org/

Analysis of the MSB. For each correct/faulty signature pairs the attacker *guesses-and-determines* the triplet of values $(d_w, \lambda, \hat{N})$. The part of exponent d_w is composed by the bits obtained from previous analysis and the w bits to guess. Hence, with our method, the analysis of one correct/faulty signature pair reveals each time w bits of d. So, this analysis has to be repeated on all signature pairs $(\hat{S}_t, S_t)$ such that $\left(\frac{n}{2} + l\right) \leq t < n$ to recover almost all the most significant bits of d.

Analysis of the LSB. For all the gathered signatures, the attacker has to perform an analysis split into two parts:

- First, he has to *guess-and-determine* two or more triplets of values $(\delta_w, \lambda, \hat{N})$ from different pairs of faulty/correct signatures
- Then, he extracts both w bits of d and $(p+q-1)$ by solving the obtained system of equations.

As a result, the analysis of the signatures faulted at the same step t of their respective execution allows an attacker to recover both a w-bit part of d and $(p + q - 1)$. As for the MSB case, this step has to be repeated (completed by Baby-Step Giant-Step or lattice techniques if necessary) to recover the missing bits of d.

4.2   Performance

Fault Number. Since our attack is based on fault injection, it seems relevant to evaluate the number of faulty signatures an attacker has to collect to recover the secret key. According to the description of our analysis (see Sect. 3.4), the number of faults depends on the part of d the attacker aims to recover. In the case of the MSB, the attacker will be able to recover w bits of d from one correct/faulty signature pair. For the LSB, the attacker has to collect at least two signatures faulted at the same step, since he has to solve a system of equations to extract w bits of d. As a result, the number of faulty signatures to collect $\mathcal{F}$ is:

$$\mathcal{F} = \mathcal{O}\left(\frac{n}{w}\right) \tag{16}$$

In practice, for a 1024-bit RSA and a resolution window length $w = 2$, our attack succeeded from about 1000 faulty signatures which is a little more than expected. This extra cost is due to the LSB analysis that required an average of 3 faulty signatures to correctly recover both parts of d and $(p+q-1)$. But this number of fault is still reasonable and highlights the practicability of our fault attack.

Complexity. The general principle of our attack is based on extracting a w part of d from each correct/faulty signature pair collected according to the model. This can be achieved by guessing and determining simultaneously the w bits of d isolated by the fault, the l-bit random value λ and the faulty modulus $\hat{N}$. Therefore, according to the fault model and the described algorithm, the computational complexity $\mathcal{C}$ of our attack is:

$$\mathcal{C} = \mathcal{O}\left(\frac{2^{(w+l)} \cdot n^2}{w}\right) \text{ exponentiations} \tag{17}$$

The extra analysis required for the LSB case does not appear in this expression since it is dominated by the search of a triplet that satisfy (8). Moreover one can notice that the complexity exponentially depends on the random length l. So lengthening λ exponentially hardens our analysis. But, some computational optimizations can be done to bypass this problem.

Computational Optimizations. In this part, we propose to speed up the execution of our fault attack by using some optimizations inspired from particular feature of the attack. First, instead of computing candidate values the faulty modulus "on the fly", the attacker can precompute a dictionary of possible values for the faulty modulus according to the fault model chosen. Moreover, by using the Theorem 1, the attacker may advantageously compute the *guess-and-determine* step by decrementing the candidate values for λ and stopping when a triplet satisfies (8). This optimization is all the more interesting if the exponent is blinded with a λ close to 2^l. The last optimization also concerns the *guess-and-determine* step. Indeed, one can notice that for a given faulty signature, all candidate values can be tested independently. As a consequence, this step can be easily computed in parallel. So, if an attacker can get a cluster of k machines, then he can distribute the *guess-and-determine* step and reduce the global complexity of the attack $\mathcal{C}$ to $\dfrac{\mathcal{C}}{k}$.

5   Conclusion

This paper presents the first fault attack against implementations of an RSA signature scheme that embedding the exponent randomization countermeasure. Through their *"Doubling Attack"*, P.-A. Fouque and F. Valette first alerted the community that using this countermeasure may introduce a physical leakage if it is combined with a *"Left-To-Right"*-based modular exponentiation. We complete this work by showing in this paper that implementations of RSA based on the dual exponentiation may be vulnerable to fault. Indeed, we demonstrate that the exploitation of a reasonable number of faulty signatures may lead to a full secret key recovery. Moreover the GMP implementation of our method as well as the use of a practicable fault model provide evidences that the perturbation of public elements represents a real threat for RSA implementation, even randomized. Thus, it might be worthwhile to check the effective robustness of the exponent blinding against other fault attacks.

References

1. Berzati, A., Canovas, C., Dumas, J.-G., Goubin, L.: Fault Attacks on RSA Public Keys: Left-To-Right Implementations are also Vulnerable. In: Fischlin, M. (ed.) CT-RSA 2009. LNCS, vol. 5473, pp. 414–428. Springer, Heidelberg (2009)
2. Berzati, A., Canovas, C., Goubin, L.: Perturbating RSA Public Keys: an Improved Attack. In: Oswald, E., Rohatgi, P. (eds.) CHES 2008. LNCS, vol. 5154, pp. 380–395. Springer, Heidelberg (2008)

3. Biehl, I., Meyer, B., Müller, V.: Differential Fault Attacks on Ellitic Curve Cryptosystems. In: Bellare, M. (ed.) CRYPTO 2000. LNCS, vol. 1880, pp. 131–146. Springer, Heidelberg (2000)
4. Blömer, J., Otto, M.: Wagner's Attack on a secure CRT-RSA Algorithm Reconsidered. In: Breveglieri, L., Koren, I., Naccache, D., Seifert, J.-P. (eds.) FDTC 2006. LNCS, vol. 4236, pp. 13–23. Springer, Heidelberg (2006)
5. Brier, E., Chevallier-Mames, B., Ciet, M., Clavier, C.: Why One Should Also Secure RSA Public Key Elements. In: Goubin, L., Matsui, M. (eds.) CHES 2006. LNCS, vol. 4249, pp. 324–338. Springer, Heidelberg (2006)
6. Clavier, C.: De la sécurité physique des crypto-systèmes embarqués. PhD thesis, Université de Versailles Saint-Quentin (2007)
7. Coron, J.-S.: Resistance Against Differential Power Analysis for Elliptic Curve Cryptosystems. In: Koç, Ç.K., Paar, C. (eds.) CHES 1999. LNCS, vol. 1717, pp. 292–302. Springer, Heidelberg (1999)
8. Fouque, P.-A., Kunz-Jacques, S., Martinet, G., Muller, F., Valette, F.: Power Attack on Small RSA Public Exponent. In: Goubin, L., Matsui, M. (eds.) CHES 2006. LNCS, vol. 4249, pp. 339–353. Springer, Heidelberg (2006)
9. Fouque, P.-A., Réal, D., Valette, F., Drissi, M.: The Carry Leakage on the Randomized Exponent Countermeasure. In: Oswald, E., Rohatgi, P.P. (eds.) CHES 2008. LNCS, vol. 5154, pp. 198–213. Springer, Heidelberg (2008)
10. Fouque, P.-A., Valette, F.: The Doubling Attack – why Upwards Is Better than Downwards. In: Walter, C.D., Koç, Ç.K., Paar, C. (eds.) CHES 2003. LNCS, vol. 2779, pp. 269–280. Springer, Heidelberg (2003)
11. Kocher, P.: Timing attacks on Implementations of Diffie-Hellman, RSA, DSS, and Other Systems. In: Koblitz, N. (ed.) CRYPTO 1996. LNCS, vol. 1109, pp. 104–113. Springer, Heidelberg (1996)
12. Seifert, J.-P.: On Authenticated Computing and RSA-Based Authentication. In: ACM Conference on Computer and Communications Security (CCS 2005), pp. 122–127. ACM Press, New York (2005)
13. Wagner, D.: Cryptanalysis of a provably secure CRT-RSA algorithm. In: Proceedings of the 11th ACM Conference on Computer Security (CCS 2004), pp. 92–97. ACM, New York (2004)

A   Proof of the Theorem 1

Let us consider a candidate value λ' for the random value λ used in a faulty RSA signature $\hat{S}_t$ such that $\lambda' > \lambda$. Then we can also write $\lambda' \geq \lambda + 1$. Now, let us use this relationship to build a candidate value for $d_{[t]}$:

$$\lambda' \cdot N \geq (\lambda + 1)\, N \tag{18}$$

$$\Leftrightarrow \lambda' \cdot N \geq \lambda N + N \tag{19}$$

But, since the secret key d is computed as the invert of e modulo $\varphi(N)$, we also know that:

$$N > \varphi(N) > d \tag{20}$$

As a consequence, we can deduce from the previous relations that:

$$\lambda' N > \lambda N + d \tag{21}$$

$$\Rightarrow \lfloor \frac{\lambda' N}{2^t} \rfloor > \lfloor \frac{\lambda N + d}{2^t} \rfloor \tag{22}$$

Now let us rewrite the previous equation using the binary representation of the operands:

$$\sum_{i=t}^{n+l-1} 2^{i-t} \cdot (\lambda'N)_i > \sum_{i=t}^{n+l-1} 2^{i-t} \cdot (\lambda N + d)_i \qquad (23)$$

$$\Leftrightarrow \sum_{i=t}^{n+l-1} 2^{i} \cdot (\lambda'N)_i > \sum_{i=t}^{n+l-1} 2^{i} \cdot (\lambda N + d)_i \qquad (24)$$

$$\Leftrightarrow \sum_{i=t}^{n+l-1} 2^{i} \cdot (\lambda'N)_i > \bar{d}_{[t]} \qquad (25)$$

From this inequality, we can conclude that any candidate value computed with λ' will be strictly greater than the searched $\bar{d}_{[t]}$. So, (8) can not be satisfied for such a λ'. $\qquad \square$

Fault Sensitivity Analysis

Yang Li[1], Kazuo Sakiyama[1], Shigeto Gomisawa[1], Toshinori Fukunaga[2],
Junko Takahashi[1,2], and Kazuo Ohta[1]

[1] Department of Informatics, The University of Electro-Communications
1-5-1 Chofugaoka, Chofu, Tokyo 182-8585, Japan
{liyang,saki,g-shigeto-1fat,junko,ota}@ice.uec.ac.jp
[2] NTT Information Sharing Platform Laboratories, NTT Corporation
3-9-1 Midori-cho, Musashino-shi, Tokyo 180-8585, Japan
{fukunaga.toshinori,takahashi.junko}@lab.ntt.co.jp

Abstract. This paper proposes a new fault-based attack called the
Fault Sensitivity Analysis (FSA) attack, which unlike most existing fault-
based analyses including Differential Fault Analysis (DFA) does not use
values of faulty ciphertexts. Fault sensitivity means the critical condition
when a faulty output begins to exhibit some detectable characteristics,
e.g., the clock frequency when fault operation begins to occur. We explain that the fault sensitivity exhibits sensitive-data dependency and
can be used to retrieve the secret key. This paper presents two practical
FSA attacks against two AES hardware implementations on SASEBO-R,
PPRM1-AES and WDDL-AES. Different from previous work, we show
that WDDL-AES is not perfectly secure against setup-time violation attacks. We also discuss a masking technique as a potential countermeasure
against the proposed fault-based attack.

Keywords: Side-channel attacks, Fault Sensitivity Analysis, AES, WDDL.

1 Introduction

Nowadays, the security of cryptographic devices such as smart cards is threatened by side-channel attacks that retrieve secret information from side-channel
leakages such as power consumption and electromagnetic radiation. The most
studied fault-based attack is Differential Fault Analysis (DFA) proposed by Biham and Shamir in 1997 [1]. The DFA attacks have been actively studied in
[2, 4, 6–8, 10, 12–14] and [16].

Generally, DFA attacks retrieve the key based on information of the characteristics of the injected fault and the values of the faulty ciphertexts. In this paper,
a faulty ciphertext represents the output after a transient fault is injected, *i.e.*,
the output of the original cryptographic algorithm using a faulty intermediate
value as an intermediate input. On the other hand, when an action of fault injection is performed, we generally call the output a *faulty output*. A faulty output
could be a faulty ciphertext, a fault-free ciphertext when the fault injection fails,
or a nonsense value for implementations with fault attack countermeasures.

S. Mangard and F.-X. Standaert (Eds.): CHES 2010, LNCS 6225, pp. 320–334, 2010.

This paper proposes a new fault-based attack called Fault Sensitivity Analysis (FSA) attack. We notice that in the process of fault injection, there are other types of information that are available to attackers, which we call fault sensitivity. The fault sensitivity is a condition where the faulty output begins to exhibit some detectable characteristics. For example, when gradually increasing the intensity of the fault injection, attackers can discern the critical condition where a fault begins to occur or the fault becomes stable. Similar to most side-channel attacks, if the relationship between the fault sensitivity and the processed sensitive data is known, the FSA attacks can retrieve the secret information from a cryptographic device.

This paper explains the general attack procedures and attack requirements for the FSA attacks. To prove the validity of the FSA attacks, this paper first presents a detailed FSA attack example against PPRM1-AES [9] (1 stage Positive Polarity Reed-Muller) implemented in ASIC mounted on the Side-channel Attack Standard Evaluation Board (SASEBO-R) [5]. For the PPRM1 S-box, based on its structure and a simulation, we explain that there is a correlation between the faulty sensitivity and the Hamming weight of the input signals for the S-box. In the FSA attack against PPRM1-AES herein, the 128-bit key can be retrieved with less than 50 plaintexts.

We note that the FSA attack has the potential to threaten many DFA-resistant implementations, since it does not require the value of the faulty ciphertext. For example, FSA can be applied to Wave Dynamic Differential Logic (WDDL) [15], which was said to be naturally immune to the DFA attacks based on setup-time violation[11]. This paper also describes an FSA attack against WDDL-AES on SASESO-R. Based on experimentation, we find that the fault sensitivity for a WDDL combinational logic tree can be correlated with the values of one bit of the output signal. By retrieving 3 out of 16 key bytes, we show that a practical ASIC implementation of WDDL-AES is not perfectly secure against the FSA attack based on setup-time violations.

Compared to DFA, FSA does not restrict the injected fault to a small subspace by assuming that only a few bits or bytes are disturbed. On the other hand, the masking technique, which is shown not to be effective against the DFA attacks [3], is a potential countermeasure against FSA attacks.

This paper is organized as follows. In Section 2, we briefly review the previous work on fault-based attacks. Section 3 describes the general principle and attack procedures of FSA. We also describe a detailed FSA attack against PPRM1-AES. Finally, we discuss the attack requirements and countermeasures against FSA. In Section 4, we explain the FSA attack against WDDL-AES. Section 5 concludes this paper.

2   Preliminaries

This section reviews several common fault injection techniques and presents the attack assumptions and requirements for DFA.

2.1 Common Fault Injection Techniques

In [2], the common fault injection techniques are listed as spike attacks, glitch attacks, optical attacks, and electromagnetic perturbations attacks. The spike and glitch attacks are likely to be simpler to implement than others since they disturb the external power supply or the external clock, respectively. An illegal power supply or illegal clock will cause a setup-time violation since flip-flops are triggered before the output signals are fixed to a correct value. Compared to spike attacks, it is easier to control the exact time of a fault injection for the glitch attack. Therefore, we use the glitch attack to perform fault injections in this paper.

2.2 DFA and Attack Requirements

In 1997, Biham and Sharmir first proposed the concept of the DFA attack and applied it to DES [1]. Since then, the DFA attack has been the most discussed fault-based attack. DFA assumes that attackers are in physical possession of the cryptographic device, and can use it to obtain two ciphertexts for the same plaintext and secret key[1]. One of the ciphertexts is a fault-free ciphertext denoted by C, and the other denoted by C' is the result after some computational fault is injected. DFA further assumes that the attackers know some characteristics of the injected fault, $e.g.$, only several bits or bytes are disturbed in a specific round operation.

In the DFA attacks, the attackers first make a key guess, K_g. Then fault-free intermediate value I and faulty intermediate value I' are calculated based on (C, K_g) and (C', K_g), respectively. Subsequently, the attackers check whether $I \oplus I'$ satisfies the characteristics of the injected fault. Repeating these procedures for multiple pairs of (C, C'), the attackers can finally identify the secret key.

Generally, there are two major requirements for the DFA attacks.

- First, the DFA attack requires the value of faulty ciphertext C'. A faulty ciphertext is the output of the original cryptographic algorithm using the faulty intermediate value as the intermediate input. In the case of WDDL circuits under the fault injections caused by setup-time violations, this requirement cannot be satisfied [11].
- Second, attackers need to know some characteristics of the injected fault; however, the characteristics of the injected fault cannot be judged from the values of C and C'. Only when the actual injected fault is the expected one, can the DFA attackers identify the secret key.

3 FSA Proposal

In this section, we explain the general principle of the proposed FSA, and discuss the attack scenarios, attack requirements, and countermeasures to it. In a general discussion concerning the FSA attack, we present a detailed FSA procedure using PPRM1-AES as a case study.

---

[1] For simplicity, we consider only the encryption process.

3.1   General Principle of FSA

In the same way as in the DFA case, we assume that the attackers are in physical possession of the device. Starting from a condition where a correct ciphertext is obtained, the attacker gradually increases the intensity to which he disturbs the power supply or external clock. While doing so, there must be a moment where the success rate of the fault injection is non zero and a moment where the success rate is 1.

We call these critical conditions where the faulty output exhibits some detectable characteristics fault sensitivity. The fault sensitivity information can be observed and recorded by attackers and can be utilized as new side-channel information if it exhibits sensitive-data dependency. Consequently, we propose a new side-channel analysis FSA that utilizes the leakage of the fault sensitivity to retrieve secret information.

3.2   Data-Dependency of Fault Sensitivity

Since the transitions of signals in a device are data-dependent, it is natural to believe that the fault sensitivity is data-dependent. For faults caused by the setup-time violation, we explain the data-dependency for the timing delays of signals.

General Mechanism. We use AND, OR and XOR gates as examples to explain the general mechanism of the data-dependency of the signal timing delay. In the following analysis, T_X denotes the timing delay for a signal X.

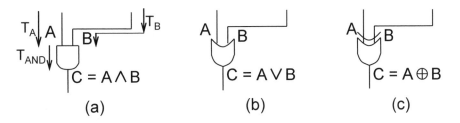

Fig. 1. Examples for data-dependency for fault sensitivity

For a two-input AND gate as shown in Fig. 1 (a), we assume $T_A < T_B$. If signal A is logic 0, signal C is determined after signal A arrives at the AND gate. As a result, $T_C = T_A + T_{AND}$, where T_{AND} is the timing delay caused by the AND gate. On the other hand, if signal A is logic 1, signal C will be determined after signal B arrives, so that $T_C = T_B + T_{AND}$.

Similarly, for a two-input OR gate as shown in Fig. 1 (b), we still assume $T_A < T_B$. If signal A is logic 1, $T_C = T_A + T_{OR}$, otherwise $T_C = T_B + T_{OR}$. In a word, for a two-input AND/OR gate, the input signal with a shorter timing delay is the selector for the timing delay of the output signal. However, for an XOR gate as shown in Fig. 1 (c), the timing delay of the output signal is decided by the maximum timing delay of its input signals without data-dependency.

We call the maximum timing delay among all the output signals for a combinational logic tree the critical timing delay. The fault sensitivity for setup-time violation is dependent on the critical timing delay. Since the timing delays of intermediate signals are data-dependent, the critical timing delay is also data-dependent. Once a circuit is physically decided, the data-dependency of the fault sensitivity is also physically fixed. Attackers can analyze the data-dependency based on the structure of the circuit, software simulation, or implementation of the circuit.

Data-dependency of Critical Timing Delay for PPRM1 S-box. As a case study, we analyze the data-dependency of the critical timing delay for the PPRM1 S-box based on its structure and a simulation.

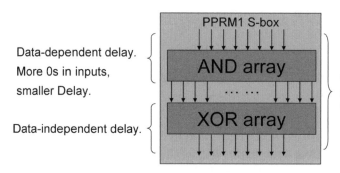

Fig. 2. Structure of PPRM1 S-box and data-dependency for the critical timing delay

PPRM1 (AND-XOR logic) was proposed by Morioka and Satoh at CHES 2002 for a low power AES design [9]. Although PPRM1 is not likely to be used in a practical implementation, its straight-forward structure makes it a perfect attack target for the case study. As shown in Fig. 2, for the S-box of PPRM1, the input signals go through an AND gate array and an XOR gate array to become the output signals. For the AND gate array, the timing delays of the output signals are dependent on the values of the input signals. For the XOR gate array, the timing delays of the output signals are not data-dependent. In general, the structure of the PPRM1 S-box indicates the dependency between the critical timing delays and the values of the input signals.

For the AND gate array, each logic 0 input signal has a probability for decreasing the critical timing delay. Consequently, the more 0s in the input signals, the bigger the possibility that the critical timing delay of the S-box becomes shorter. In conclusion, statistically the critical timing delay of the PPRM1 S-box should be correlated with the Hamming weight of the input signals. Specifically, the input signals with a higher Hamming weight make the PPRM1 S-box more sensitive to a fast clock.

To confirm this, we simulate the transitions and timing delays of signals using the Verilog-HDL codes for the PPRM1 S-box [5] by Xilinx. For each possible input of the S-box, we obtain the corresponding critical timing delay where the

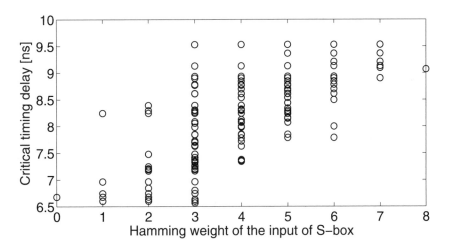

Fig. 3. Relationship between the critical timing delay and Hamming height of the input of PPRM1 S-box

initial values of all wires are reset to logic 0. As shown in Fig. 3, the correlation coefficient between the critical timing delay and the Hamming weight of the input signals is approximately 0.71. Later we show a similar correlation existing in the ASIC implementation of PPRM1 that can lead us to a practical FSA attack.

3.3   General FSA Attacks Scenarios

Without loss of generality, this paper only shows the case in which the rounds near the output are attacked.

Algorithm 1. Collection of Fault Sensitivity Information

Inputs: The number of different plaintexts: N
Outputs: Ciphertexts: $CT[i]$, Critical fault injection intensity: $F^C[i]$
for $i = 1$ to N **do**
   Generate a random plaintext $PT[i]$
   Reset fault injection intensity $F \leftarrow 0$
   $CT[i] \leftarrow Enc(PT[i], F)$ (No fault injection)
   repeat
     Increase F by a little
   until $Enc(PT[i], F) \neq CT[i]$
   $F^C[i] \leftarrow F$
end for

Collection of Fault Sensitivity Information. In a practical FSA attack, attackers first need to collect the fault sensitivity information. For simplicity, we use an unspecific parameter called the fault injection intensity denoted by F. When $F = 0$, no fault injection is performed. An increase in F represents an increase in the intensity for the fault injection, *e.g.*, a decrease in the power supply and a shortening of the clock period. Then we denote the output of the encryption for plaintext PT with fault injection intensity F as $Enc(PT, F)$. We use the intensity where a fault begins to occur as critical fault injection intensity F^C, then the fault sensitivity information is collected according to the procedures in Alg. 1. In Alg. 1, the critical fault injection intensity information is collected by gradually increasing the intensity and checking whether the output is still the same as the fault-free ciphertext.

The Key Retrieval Procedure. We assume that the attackers can use ciphertexts $CT[i]$ and a key guess, K_g, to predict critical fault injection intensity F^C using a function denoted by $f_{F_g^C}$. Algorithm 2 shows the basic procedures for the key retrieval calculation where $\rho(A, B)$ denotes the absolute value of the Pearson correlation coefficient between A and B. The correlation peak among all possible key guesses is expected to be the same as the correct one.

Algorithm 2. Key Retrieval Procedure

Inputs: Bit length of (sub-)key: t, Ciphertexts: $CT[i]$, Critical fault injection intensity: $F^C[i]$
Outputs: Key
for $K_g = 0$ to $2^t - 1$ **do**
    for $i = 1$ to N **do**
        $F_g^C[i] \leftarrow f_{F_g^C}(CT[i], K_g)$
    end for
    $Cor[K_g] \leftarrow \rho(F^C, F_g^C)$
end for
$Key \leftarrow K_g$ where $Cor[K_g]$ is the maximum

3.4   FSA Attack Scenarios against PPRM1-AES

In this section, we propose an FSA attack scenario against 128-bit PPRM1-AES. We denote the calculation results of the i-th round of AES by H^i and $i \in [1, 10]$. In the last round of AES, the MixColumns operation is omitted. Each byte of H^9 is substituted by an S-box, and the 10th round key, K^{10}, is added to become the corresponding byte for ciphertext H^{10}. Since the bytewise calculations in the last round of AES are independent from each other, there are 16 independent combinational logic trees. As a result, each byte of the ciphertext is an independent indicator of whether a fault is injected in its combinational logic tree. Therefore, even through all the S-boxes are calculated in parallel for the PPRM1-AES on SASEBO-R, the 16 bytes of K^{10} can be attacked independently in the FSA attack.

Based on the analysis in Section 3.2, when we inject the fault during the last round of PPRM1-AES by shortening the corresponding clock period, the fault sensitivity is correlated with H^9. Attackers can set the basic attack target as a byte of K^{10}, then the FSA attacks can be applied as shown in Algs. 1 and 2, where $f_{F_g^C}$ is the Hamming Weight of $InvSbox(CT[i] \oplus K_g)$.

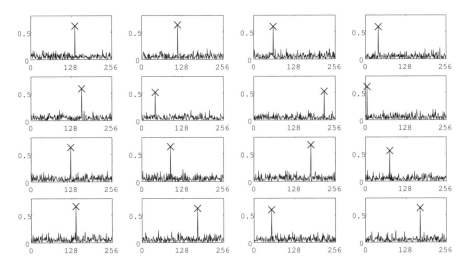

Fig. 4. Attack results for PPRM1-AES using 360 plaintexts. Each sub-figure corresponds to a key byte. The key guess is represented on the horizontal axis. The correlation coefficient between the critical fault injection intensities and Hamming weight of the input is represented on the vertical axis. Each correct key byte is marked by an ×.

FSA Attack Results against PPRM1-AES. This section shows the FSA attack results against PPRM1-AES implemented in ASIC mounted on the SASEBO-R [5]. The detailed experimental setup and parameter settings are shown in Appendix A.

In Fig. 4, we show 16 sub-figures for the correlation coefficients against key guesses for 16 bytes of K^{10} when 360 plaintexts are used. We note that in the practical attack, all 16 key bytes of K^{10} are attacked in parallel, *i.e.*, for each plaintext we collect fault sensitivity data for every combinational logic tree. Figure 5 shows the number of correct key bytes against the number of used plaintexts. We found that full key recovery for the FSA attack against PPRM1-AES requires less than 50 plaintexts.

3.5   Attack Requirements and Countermeasures for FSA

There are two requirements for a practical FSA attack. First, attackers must understand the data-dependency for the fault sensitivity. Even through the sensitive data and the fault sensitivity may not have a clear correlation, as long as attackers have a template for the data-dependency of the fault sensitivity, FSA can retrieve the secret key. Second, the secret key must be able to be divided and attacked independently so that Alg. 2 can be finished in a practical amount of time. All the software implementations where S-boxes are sequentially calculated satisfy the second requirement. Furthermore, most of the parallel implementations without countermeasures satisfy the second requirement as well.

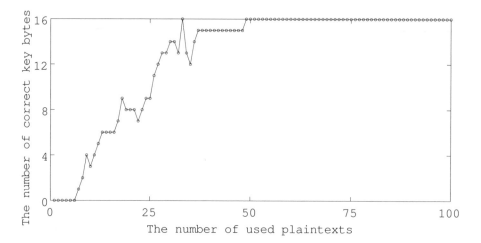

Fig. 5. Number of correct key bytes vs. number of used plaintexts

Compared to the DFA attacks, the FSA attacks do not require that the in-jected fault be restricted to a small subspace by assuming that only a few bits or bytes are disturbed. Even though the DFA attacks are likely to require fewer fault injections to retrieve the secret key [8, 10, 12, 16], it requires that the attacker have the ability to inject the expected fault in the first place. Since attackers have the device, there is no limitation on how many times that the non-invasive fault injections are performed; however, it requires much knowledge or investigation to inject the expected faults.

Furthermore, the FSA attacks do not require the values for faulty cipher-texts. So for the conventional fault-based attack countermeasures that provide a nonsense output or halt the calculation when a computational fault is detected, the FSA attack is still a potential threat. The fault sensitivity information is still available to attackers since they can distinguish whether or not a fault is injected. However, if the S-boxes are calculated in parallel, the fault detection is no longer byte-wise independent. The collected information only corresponds to the most sensitive part. Since the second requirement for the FSA attacks is not satisfied, it is difficult to retrieve the full key using the FSA attacks.

On the other hand, we note that the masking technique, which is shown not to be effective against the DFA attacks [3], is likely to be an effective countermeasure against the FSA attacks. Once all of the sensitive values are masked by uniformly distributed random numbers, the data-dependency between the intermediate val-ues and the fault sensitivity can no longer be used in the key retrieval.

4   FSA Attacks against WDDL-AES

Since the faulty output for WDDL circuit has no information regarding the key, it is concluded that the WDDL circuit is naturally protected from the setup-time violation attacks in FDTC 09 [11]. We note that WDDL-AES satisfies the attack requirements for FSA, so that theoretically it is potentially vulnerable to the FSA attacks.

4.1   WDDL "Protected" against Setup-Time Violation Attacks

WDDL was proposed by Tiri and Verbauwhede at DATE 2004 as a hiding countermeasure for power analysis. As a representative of the Dual-Rail Precharge Logic, each WDDL gate comprises two complementary operations. Every signal in WDDL has two complementary wires (true, false) as well, where the true wire has the actual value of the signal and the false wire has the complementary value. The logic values of two wires for a signal are either $(1,0)$ or $(0,1)$. Each clock cycle is divided into two phases, precharge and evaluation. In the precharge phase, all of the wires are set to be the precharge value, which is assumed to be 0 in this paper. In the evaluation phase, each pair of wires is set back to the logic values as either $(1,0)$ or $(0,1)$. As a result, exactly half of the wires will transit from 0 to 1, and the other half remain at 0 during each evaluation phase. Since the number of bit transitions is independent from the processed data, a WDDL circuit is likely to consume a constant amount of power for each clock cycle.

Under the fault injection caused by the setup-time violations, the two wires of a faulty signal in WDDL can only be $(0,0)$. Furthermore, an input faulty signal, $(0,0)$, is likely to spread to all of the output signals for a WDDL combinational logic tree. In the case of the fault injection at the beginning of the 8th round used in [10, 12, 16], attackers can only obtain a faulty output with all 0s. Since faulty ciphertext C' is not available for DFA attackers, it is concluded that WDDL is naturally immune to the setup-time violation attacks. However in this work, we show that in a practical implementation of WDDL-AES it is vulnerable to the FSA attacks based on setup-time violation.

4.2   Data-Dependency of Fault Sensitivity for WDDL-AES

For the WDDL-AES on SASEBO-R, we try to use the implementation itself to obtain the data-dependency of the fault sensitivity. With full knowledge of the secret key, we performed fault injections that shorten the evaluation period for the last round of AES. Then we found that the critical fault injection intensity of each byte is correlated with the value of one single bit of the ciphertext byte. Although the correlation between fault sensitivity information and the ciphertexts cannot be used for key retrieval, we understand the fault sensitivity for the WDDL circuits are dependent on its output.

4.3   Practical FSA Attack against WDDL-AES

As a practical attack, we performed another fault injection that shortens the evaluation period for the 9th round of AES, since the data-dependency based on H^9 can be used in the key retrieval. A modified key retrieval algorithm shown in Alg. 3 is applied. In Alg. 3, $f_{getbit}(A, b)$ represents a vector comprising the b-th bit of each element of A. The attack results after using 1200 plaintexts are shown in Fig. 6. The 6th key byte and the 11th key byte of K^{10} can be identified clearly. Also the correlation coefficient peak for the 4th key byte corresponds to the correct key as well.

Algorithm 3. Key Retrieval Procedure for WDDL-AES

Inputs: Bit length of (sub-)key: t, Ciphertexts: $CT[i]$, Critical fault injection intensity: $F^C[i]$

Outputs: Key

for $K_g = 0$ to $2^t - 1$ do
  for $i = 1$ to N do
    $F_g^C[i] \leftarrow InvSbox(CT[i] \oplus K_g)$
  end for
  for $bit = 0$ to 7 do
    $BitCor[bit] \leftarrow \rho(f_{getbit}(F_g^C, bit), F^C)$
  end for
  $Cor[K_g] \leftarrow Max(BitCor[bit])$
end for
$Key \leftarrow K_g$ where $Cor[K_g]$ is the maximum

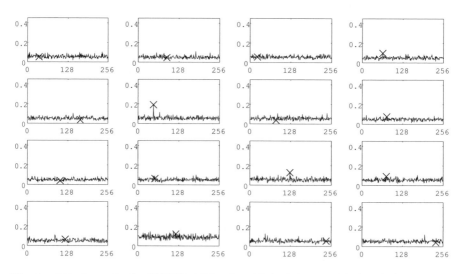

Fig. 6. Attack results for WDDL-AES using 1200 plaintexts. Each sub-figure corresponds to a key byte. The key guess is represented on the horizontal axis. The maximum 8-bit-based correlation coefficient is represented on the vertical axis. Each correct key byte is marked by an ×.

We believe that there are two reasons for this correlation. First, since the S-box of WDDL-AES is not based on S-boxes with clear gate arrays, even though the fault sensitivity for WDDL-AES is dependent on the input signals there is no clear correlation as in the case for PPRM1-AES. Second, assuming that the two complementary wires for the critical path have different timing delays, then the fault sensitivity will be correlated with this output signal since only $1 \rightarrow 0$ could occur for WDDL circuits under setup-time violations. For example, we assume that the timing delay of the true wire is longer than that for the false wire. As a result, the calculation with output signal $(1,0)$ is more sensitive to a fast clock than the one with output signal $(0,1)$, which leads to the correlation we observed in the experiments.

Compared to a single-rail circuit such as PPRM1, it is harder to apply the FSA attack to WDDL. Each path in the WDDL combinational logic tree has two wires that are supposed to have the same timing delay. However, practically the timing delays for two complementary wires cannot be exactly the same, so that the vulnerability to FSA attacks for practical WDDL implementations still exists. The unexploited key bytes may become exploitable by using a more precise experimental setup.

Another difficulty in attacking WDDL is that, since the fault injection is performed in the 9th round, there is influence from the key schedule and the MixColumns. In the proposed attack, we find that several bytes have the same fault sensitivity when we inject the fault in the 9th round, which indicates that the fault signal affects these bytes at the same time. This kind of fault is difficult to use in the FSA attacks. As future work, we plan to investigate in more detail the fault sensitivity of WDDL circuits.

5   Conclusions

This paper proposed a new fault-based attack called Fault Sensitivity Analysis, which has lower attack requirements than those for Differential Fault Analysis. The FSA attacks are based on the dependency between the sensitivity data and the critical conditions where faulty outputs begin to exhibit detectable characteristics. Two practical FSA attacks against ASIC implementations of AES were shown in the paper. For PPRM1-AES, less than 50 plaintexts were needed to retrieve the full key. For WDDL-AES, which was shown to be immune to DFA attacks based on setup-time violation, the proposed FSA attack successfully retrieved 3 out of 16 key bytes with 1200 plaintexts.

Acknowledgement

The authors would like to thank the anonymous referees for their valuable comments. This research was partially supported by the Strategic International Cooperative Program (Joint Research Type), Japan Science and Technology Agency.

References

1. Biham, E., Shamir, A.: Differential Fault Analysis of Secret Key Cryptosystems. In: Kaliski Jr., B.S. (ed.) CRYPTO 1997. LNCS, vol. 1294, pp. 513–525. Springer, Heidelberg (1997)
2. Blömer, J., Seifert, J.-P.: Fault Based Cryptanalysis of the Advanced Encryption Standard (AES). In: Wright, R.N. (ed.) FC 2003. LNCS, vol. 2742, pp. 162–181. Springer, Heidelberg (2003)
3. Boscher, A., Handschuh, H.: Masking Does Not Protect Against Differential Fault Attacks. In: Breveglieri, L., Gueron, S., Koren, I., Naccache, D., Seifert, J.-P. (eds.) FDTC, pp. 35–40. IEEE Computer Society, Los Alamitos (2008)
4. Dusart, P., Letourneux, G., Vivolo, O.: Differential Fault Analysis on A.E.S., Cryptology ePrint Archive, Report2003/010 (2003)
5. Research Center for Information Security (RCIS). Side-channel Attack Standard Evaluation Board (SASEBO), http://www.rcis.aist.go.jp/special/SASEBO/CryptoLSI-en.html
6. Giraud, C.: DFA on AES, Cryptology ePrint Archive, Report2003/008 (2003)
7. Li, Y., Gomisawa, S., Sakiyama, K., Ohta, K.: An Information Theoretic Perspective on the Differential Fault Analysis against AES, Cryptology ePrint Archive, Report2010/032 (2010)
8. Moradi, A., Shalmani, M.T.M., Salmasizadeh, M.: A Generalized Method of Differential Fault Attack Against AES Cryptosystem. In: Goubin, L., Matsui, M. (eds.) CHES 2006. LNCS, vol. 4249, pp. 91–100. Springer, Heidelberg (2006)
9. Morioka, S., Satoh, A.: An Optimized S-Box Circuit Architecture for Low Power AES Design. In: Kaliski Jr., B.S., Koç, Ç.K., Paar, C. (eds.) CHES 2002. LNCS, vol. 2523, pp. 172–186. Springer, Heidelberg (2003)
10. Mukhopadhyay, D.: An Improved Fault Based Attack of the Advanced Encryption Standard. In: Preneel, B. (ed.) AFRICACRYPT 2009. LNCS, vol. 5580, pp. 421–434. Springer, Heidelberg (2009)
11. Guilley, S., Graba, T., Selmane, N., Bhasin, S., Danger, J.-L.: WDDL is Protected Against Setup Time Violation Attacks. In: FDTC, pp. 73–83. IEEE Computer Society, Los Alamitos (2009)
12. Piret, G., Quisquater, J.-J.: A Differential Fault Attack Technique against SPN Structures, with Application to the AES and KHAZAD. In: Walter, C.D., Koç, Ç.K., Paar, C. (eds.) CHES 2003. LNCS, vol. 2779, pp. 77–88. Springer, Heidelberg (2003)
13. Saha, D., Mukhopadhyay, D., RoyChowdhury, D.: A Diagonal Fault Attack on the Advanced Encryption Standard, Cryptology ePrint Archive, Report2009/581 (2009)
14. Sakiyama, K., Yagi, T., Ohta, K.: Fault Analysis Attack against an AES Prototype Chip Using RSL. In: Fischlin, M. (ed.) CT-RSA 2009. LNCS, vol. 5473, pp. 429–443. Springer, Heidelberg (2009)
15. Tiri, K., Verbauwhede, I.: A Logic Level Design Methodology for a Secure DPA Resistant ASIC or FPGA Implementation. In: DATE, pp. 246–251. IEEE Computer Society, Los Alamitos (2004)
16. Tunstall, M., Mukhopadhyay, D.: Differential Fault Analysis of the Advanced Encryption Standard using a Single Fault, Cryptology ePrint Archive, Report2009/575 (2009)

A Experimental Setup for Fault Sensitivity Analysis Using Clock Glitch

The experimental setup for FSA is shown in Fig. 7. The fault injection technique used in the proposed attack is the clock glitch. We use two clock supplies in the experiment system. The first clock supply, $clk1$, is generated by a 24 MHz oscillator and provided to the control FPGA and the I/F of the LSI to ensure that they work appropriately. The second clock supply, $clk2$, is generated from a function generator that is controlled by a PC through GBIP. By multiplying $clk2$, higher frequency clock clk_{hf} is generated using the Digital Clock Manager (DCM) inside the control FPGA. Then, based on $clk2$, clk_{hf}, and the start signal from the LSI core, we use the control FPGA to generate a special clock, clk_{core}, which is provided to the LSI core. Most cycles of clk_{core} are the same as those for $clk2$, except one cycle is the same as the clk_{hf}, which triggers the computational fault at the time we want. Figures 8 and 9 show the power traces of PPRM1-AES and clocks clk_{core} without fault injection and with fault injection, respectively.

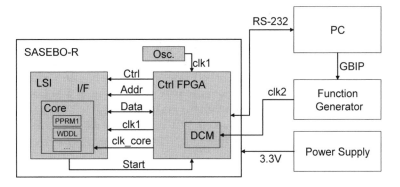

Fig. 7. Experimental setup for fault sensitivity analysis

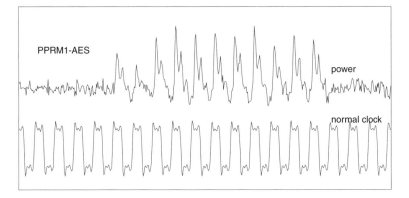

Fig. 8. Power trace of PPRM1-AES without fault injection (above) and clock supply without glitch (bottom)

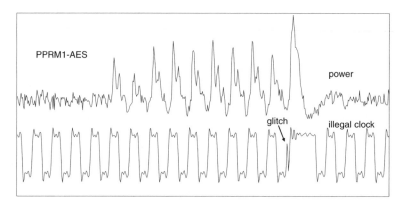

Fig. 9. Power trace of PPRM1-AES with fault injection (above) and clock supply with a glitch (bottom)

In order to reduce the total number of fault injections for a successful FSA attack, we first employ a binary search to determine a relatively high frequency for clk_{hf} that does not trigger any fault. Then we increase the frequency of clk_{hf} step-by-step and record the critical frequency for each byte of ciphertext. In the experiments, the period of clk_{hf} is decreased by approximately 35 picoseconds in each step. Furthermore, every plaintext is repeatedly used, until all of the bytes of ciphertext have been disturbed into a faulty value. In the worst case in the proposed attacks, a plaintext must be repeatedly used 120 times. In the experiment, we choose these parameters to make sure that 1) the recorded fault sensitivity has informative variations and 2) the level of efficiency in collecting the fault sensitivity information is tolerable. These parameters can be optimized to lead a more efficient FSA attack.

An Alternative to Error Correction for SRAM-Like PUFs

Maximilian Hofer and Christoph Boehm

Institute of Electronics, Graz University of Technology
maximilian.hofer@tugraz.at, christoph.boehm@tugraz.at
http://ife.tugraz.at

Abstract. We propose a new technique called stable-PUF-marking as an alternative to error correction to get reproducible (i.e. stable) outputs from physical unclonable functions (PUF). The concept is based on the influence of the mismatch on the stability of the PUF-cells' output. To use this fact, cells providing a high mismatch between their crucial transistors are selected to substantially lower the error rate. To verify the concept, a statistical view to this approach is given. Furthermore, an SRAM-like PUF implementation is suggested that puts the approach into practice.

Keywords: Physical Unclonable Functions, SRAM, Pre-Selection.

1 Introduction

Due to the widespread use of Smart Cards and radio frequency identification (RFID) devices, the demand for secure identification/authentication and other cryptographic applications is continuously increasing. For this purpose a "fingerprint" of a chip can be useful. Physical unclonable functions (PUFs) provide such an output. In 2001, Pappu et al. introduced the concept of PUFs [1]. In this approach a unique output is produced by evaluating the interference pattern of a transparent optical medium. Unfortunately, due to the way of pattern extraction, Pappu's approach turns out to be quite expensive. In [2,3], Gassend et al. introduce physical unclonable functions in silicon. The concept utilizes manufacturing process variation to distinguish between different implementations of the same integrated circuit (IC). This is done by measuring the frequency of self-oscillating loop circuits. These frequencies differ slightly between the realizations. However, the chip area is large and the current consumption is high. Another approach is to use the initial values of SRAM cells. [4,5] shows that there exist SRAM chips which deliver the same start-up value again and again which is the crucial property of a PUF. The best of them deliver an error rate of less than 3 %. So it seems that SRAM-like structures are feasible as dedicated PUF-cells [6].

One way to deal with errors in the PUFs' responses is to use error-correction codes (ECC) [7]. Here, redundace is added by storing parity bits during an initialization phase. These bits can be used afterwards to reconstruct the reference

S. Mangard and F.-X. Standaert (Eds.): CHES 2010, LNCS 6225, pp. 335–350, 2010.

value. Unfortunately, efficient decoding is difficult. If the error rate is high, the runtime increases strongly [8,9]. Other methods use statistical data of the fuzziness [10] (i.e. the degree of instability) of the PUF responses. In [10], Maes et. al. read out the response several times to collect data about the stability of the different PUF-cells. An advantage of this Soft Decision Data Helper Algorithm is that the number of PUF-cells can be reduced up to 58.4 %. A drawback is that the initialization phase needs a higher number of runs (e.g. 64 in [10]).

In this work we propose an alternative method to deal with unstable PUF-cells. The time needed for the read-out phase is reduced due to the fact that further post-processing of the PUF response becomes less complex or even needless depending on the application.

The remainder of the paper is organized as follows. Section 2 describes the idea behind the concept. In Section 3 a statistical analysis is given. Section 4 provides an approach to an implementation in silicon. Finally, section 5 concludes the paper. In the appendix some additional calculations and tables are given.

2   Idea

Figure 1 shows a CMOS SRAM-cell that can be used as a PUF-cell. A whole PUF consists of an application dependent number of such cells. We assume that the design and the layout of that PUF-cell are optimized in such a way that the PMOS transistors match and the NMOS transistors mismatch.[1] In an SRAM PUF, the output which is defined by the state of OUT after power-up, mainly depends on the threshold voltage (V_{th}) mismatch. Assuming identical initial potentials at OUT and $\overline{OUT}$, the mismatch of the NMOS transistors lead to a difference between i_1 and i_2 in the two branches of the SRAM cell. If i_2 is higher than i_1, the potential at OUT will move towards V_{SS}, the potential at $\overline{OUT}$ will move towards V_{DD}. If i_2 is lower than i_1, the cell behaves the other way round. This behavior at OUT and $\overline{OUT}$ should be an intrinsic property of the cell and should not change over time. If the mismatch is too small, the cell result will be unstable due to noise, temperature shifts, and other shifts in the working point, e.g. caused by changes in V_{DD}.

The idea is to select only the stable cells (i.e. those cells providing a high mismatch) to generate the PUF output. Before the PUF is used for the first time, during an initialization phase the stable PUF-cells are detected. These cells are marked. All the other PUF-cells are not used any longer. From now on, only stable PUF-cells generate a stable response.

This gives rise to the question of how to select the stable bits. An intuitive approach is to measure the results of a PUF-cell repeatedly and chose only those cells which always provide the same output. For various reasons this is not practicable. First of all, additional measurements must be done to get useful

---

[1] The mismatch's variance of the transistors can be controlled over the transistor area: Smaller area leads to higher mismatch. This means that the analog designer can influence the variance but not the individual value of the mismatch which defines the PUF-cell output.

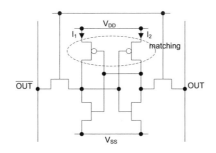

Fig. 1. Common SRAM-cell

statistical data and thus is not feasible for an initial production test flow where measurement time has proportional impact on the product costs. Another problem is that there are cells which show temperature depending behavior. So the initial measurements would have to be done over the whole temperature range. Furthermore, the influence of aging changes the mismatch behavior [5] and could cause additional errors after some time.

Another approach to find unstable PUF-cells is to use the fact that stable cells decide faster [6]. To detect the fast flipping cells, the decision time has to be measured. In figure 2 this concept is illustrated. After a certain time t_{useful} the

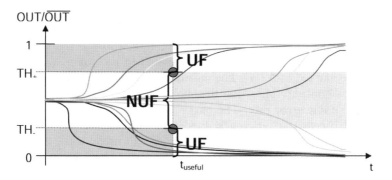

Fig. 2. Measurement of decision time (UF: useful, NUF: not useful)

cells above an upper threshold or under a lower threshold are marked as useful. All other cells which lie between the two thresholds are marked as not useful. Unfortunately, simulations show that the decision time strongly depends on the temperature. Therefore during the initialization phase a constant temperature is necessary to allow the use of an absolute time t_{useful}. Another solution to this problem could be to measure the time spans needed to reach a threshold value. The fastest cells are used. Since it may happen that the fastest cells of a chip are still not fast enough to meet the above requirements a stable behavior can not be expected in all cases.

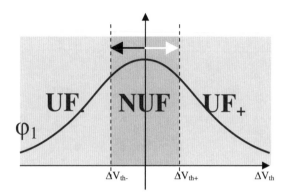

Fig. 3. The mismatch is divided in three classes: The useful PUF-cells with positive mismatch (UF_+), the useful PUF-cells with negative mismatch(UF_-) and the not useful PUFs (NUF)

The approach we propose in this paper is based on the selection of cells which provide a mismatch that exceeds a certain threshold. In the case of the shown SRAM-PUF, the mismatch of the NMOS transistors must be above such a threshold. In figure 3, the mismatch ΔV_{th} of two transistors is depicted schematically. Here, this distribution is assumed to be Gaussian. A positive and a negative threshold value (ΔV_{th+} and ΔV_{th-}, with $|\Delta V_{th+}| = |\Delta V_{th-}|$) are defined, which is necessary to divide the PUF-cells into three classes: the useful PUF-cells with positive mismatch (UF_+), the useful PUF-cells with negative mismatch (UF_-) and the not useful PUF-cells (NUF). In figure 3, the three sections are depicted. In the middle section, the mismatch is too small to provide a stable behavior. These bits are marked as NUF. The mismatch of the other bits is big enough to provide a stable output. Thus, the threshold value must be chosen correctly to reach an acceptable error rate. The larger the threshold value, the smaller the number of PUF-cells that are marked as stable and the smaller the error rate. Thus, to be able to provide the required number of useful cells, the number of initial PUF-cells has to be adapted to the chosen threshold value. For this reason, the threshold value is a trade-off between the ratio of the useful PUF-cells and all PUF-cells and the error rate.

One method to measure ΔV_{th} is to use a common analog to digital converter (ADC). In figure 4 a block diagram is shown. The disadvantage of this approach is the size of the ADC caused by the requirements on it. In order to get a balanced output, the ADC must have a small offset. Furthermore the ADC has to be fast and the result should not depend on the noise of the circuit.

The proposed concept to classify the cells into UF and NUF is to add a systematical V_{th} offset to the circuit (see figure 7): Two measurements per PUF-cell are needed. During the first measurement we add a negative offset. Thus the threshold is set to V_{th-}. During the second measurement we move the threshold to V_{th+}. This is illustrated in figure 3 denoted by the two arrows. The classification can be done as follows: If the mismatch of the transistors exceeds the

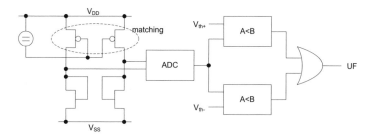

Fig. 4. Measurement of the mismatch using an ADC

threshold, the PUF-cell will provide the same output for both measurements and thus the cell is marked as useful. If the mismatch is too small, the output OUT of the cell will differ for the two measurements. The cell is marked as not useful. Problems will occur if the threshold value is chosen too big. In such a case, only a few or even no cells are marked as useful which can lead to severe problems. On the other hand, if the threshold is chosen too small, disturbances like noise will lead to output errors and make the whole pre-selection process useless.

3   Modeling and Statistical Aspects

To analyze the performance of this approach, Monte Carlo simulations are not feasible since the error rate after the pre-selection process (i.e. after the useful-PUF-marking) should be so small that the number of simulation runs to determine the error probability would exceed a tolerable number. So we prefer an analytic method to estimate the performance of the pre-selection process:

For all further analyses we assume that the distribution of the V_{th} mismatch as well as the distribution of the disturbances (noise, temperature-dependent errors, etc.) is Gaussian [11,12,13,14]. To determine the effect of the pre-selection process, we need the probability density function (PDF) $f(x)$ and its integral, the cumulative distribution function (CDF) $F(x)$ of a Gaussian:

$$f(x) = \phi_{\mu,\sigma}(x) = \frac{1}{\sigma\sqrt{2\pi}}e^{\frac{1}{2}\left(\frac{x-\mu}{\sigma}\right)^2} \tag{1}$$

$$F(x) = \Phi_{\mu,\sigma}(x) = \frac{1}{\sigma\sqrt{2\pi}}\int_{-\infty}^{x} e^{\frac{1}{2}\left(\frac{x-\mu}{\sigma}\right)^2} dx, \tag{2}$$

where μ is the mean and σ the variance of the Gaussian.

If there is no disturbance at the PUF-cell, the cell output will be the same whenever the PUF is read-out. In this case the output would be zero for all PUF-cells having a negative ΔV_{th} and one for all cells with positive ΔV_{th} (see figure 5a). If there are disturbances due to noise, temperature, etc., it may happen that if the mismatch is sufficiently small or the disturbance sufficiently large, the decision is defined by this disturbance. That effect can be seen in figure 5b where Φ_2 shows the mean output depending on ΔV_{th} taking the distribution of the

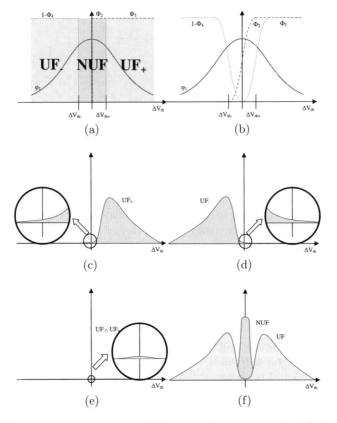

Fig. 5. (a) Ideal distributions $\Phi_1 - \Phi_4$.(b) Real distributions $\Phi_1 - \Phi_4$. (c) Useful positive PUF-cells(UF+). (d) Useful negative PUF-cells(UF-). (e) PUF-cells which occur in UF_+ and UF_-. (f) Useful PUF-cells(UF) and not-useful PUF-cells(NUF).

disturbance into account. At $\Delta V_{th} = 0$ the mean output equals 0.5. The same curve but biased with the threshold ΔV_{th-} and ΔV_{th+} depict Φ_4 and Φ_3. ϕ_1 is the distribution of the mismatch. After selecting the useful PUF-cells, the error rate can be decreased significantly. Figures 5c and 5d show the product of Φ_3 and ϕ_1, and the product of $(1 - \Phi_4)$ and ϕ_1 respectively. These curves depict the distribution of being selected as useful including disturbances, the distribution of the V_{th} mismatch and a certain V_{th} offset. Hence the figures represent the number of selected PUF-cells. Figure 5e shows those cells that are selected twice, i.e. that are declared to be useful for both offsets. To get correct results these double-selections have to be compensated for in the analysis. Figure 5f shows the distributions of selected and not selected cells.

Since $\sigma_2 = \sigma_3 = \sigma_4 = \sigma$, $\mu_1 = \mu_2 = 0$, and $\mu_3 = -\mu_4$ can be assumed, we get the following equation for the number of useful PUF cells α (see appendix):

$$\alpha = 1 - \frac{1}{\sigma_1 2\pi} \int_{\infty}^{-\infty} e^{-\frac{1}{2}\left(\frac{V_{th}}{\sigma_1}\right)^2} \left[\frac{1}{\sigma} \int_{-\infty}^{V_{th}} e^{-\frac{1}{2}\left(\frac{V'_{th}-\mu}{\sigma}\right)^2} dV'_{th} + \right.$$

$$-\frac{1}{\sigma} \int_{-\infty}^{V_{th}} e^{-\frac{1}{2}\left(\frac{V'_{th}-\mu}{\sigma}\right)^2} dV'_{th} + \frac{2}{\sigma^2\sqrt{2\pi}} \int_{-\infty}^{V_{th}} e^{-\frac{1}{2}\left(\frac{V'_{th}-\mu}{\sigma}\right)^2} dV'_{th} \cdot$$

$$\left. \cdot \int_{-\infty}^{V_{th}} e^{-\frac{1}{2}\left(\frac{V'_{th}-\mu}{\sigma}\right)^2} dV'_{th} \right] dV_{th} \tag{3}$$

The error rate e at ΔV_{th} can be derived using the following equation (see appendix):

$$e(\Delta V_{th}) = \phi_1 \Phi_2 - \phi_1 \Phi_2 \Phi_4 - \phi_1 \Phi_2 \Phi_3 \Phi_4 + \phi_1 \Phi_3 - \phi_1 \Phi_3 + \phi_1 \Phi_3 \Phi_4 - \Phi_2 \phi_1 \Phi_3 \Phi_4, \tag{4}$$

where all ϕ_i and Φ_i are evaluated at ΔV_{th}.

Example. The standard deviation of ΔV_{th} is $30\,\text{mV}$, the standard deviation of $\phi_{2,3,4}$ equals $6.16\,\text{mV}$. This coresponds to an error-rate of 5%.[2] In figure 6 the error rate and the ratio of useful PUF-cells α are shown in a diagram. It can be seen that selecting for example the best 50 % can decrease the error rate significantly. A table of different examples is shown in appendix B.

4   Implementation

Different circuits are possible to implement the approach described above. One of them is presented. To understand the circuit we consider an ordinary SRAM-cell depicted in figure 7. We assume that P_1 and P_2 match. Hence, the decision depends on the mismatch of the threshold voltage of N_1 and N_2 denoted ΔV_{th}. To mark the cells as introduced in section 2, we have to add an additional voltage source at the gate of one of the NMOS transistors to provide the bias we need for the threshold (see figure 7a). Since the implementation of such a circuit is difficult, the preferred way is to use its Norton equivalent - a current source - in parallel to one of the NMOS transistors (see figure 7b).

From figure 8, the equivalence of the two circuits can be seen. The characteristics of two different diode-loaded MOSFETS are shown. For the same V_{GS} and different threshold voltages, the amount of current through the transistors will be different. Thus, additional current at one of the branches of the SRAM-cell acts as a mismatch of V_{th}.

The circuit depicted in figure 9 is a practical implementation of the approach. During the first phase N_7 is switched-off. N_3, N_4 and P_1, P_2 are building a SRAM similar circuit. N_2 acts as a current limiter for this circuit. The circuit is designed, that the mismatch between P_1 and P_2 is small and should not affect the result. Due to the fact that the transistors N_3 and N_4 are diode loaded, the

---

[2] We meassured an error-rate of 4% in a dedicated PUF-cell in the temperature range from $0 - 80°C$. So this is a rather pessimistic value.

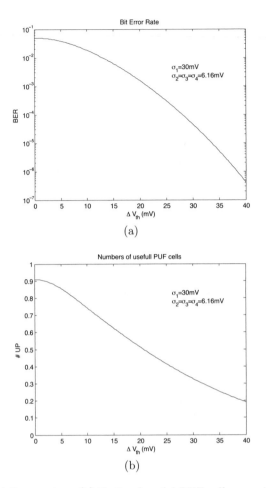

Fig. 6. a) Error rate e. (b) Ratio of useful PUF-cells α against $\mu_{3,4}$.

circuit does not flip as fast as the SRAM depicted in figure 1. During the second phase, N_7 is switched-on and the circuit flips completely to one direction. The bias transistors which are used for the PUF-cell selection (P_3 and P_4) are used during the first phase and switched-off during the second phase (P_7 and P_8 are switched-on; P_5 and P_6 are switched-off).

If we want to add a fictive negative offset voltage at the transistor N_4, P_8 is opened and P_6 is closed. Thus, the transistor P_4 is in parallel with transistor P_2. A higher current passes N_4. The same can be done on the right branch (P_7, P_5, P_3 and N_3). The truth table for the control of the transistors P_5, P_6, P_7, and P_8 is shown in table 1.

A further improvement of this circuit can be achieved by separating the mismatching transistors N_3 and N_4 from the evaluation circuit consisting of the transistors N_5 to N_7 and P_1 to P_8. Additionally, two transistors are required to connect each cell to the evaluation circuit. So, one PUF-cell consists of only five

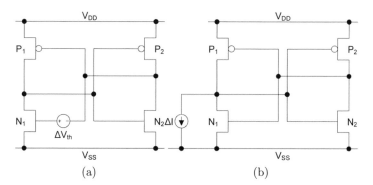

Fig. 7. (a) SRAM-cell with additional voltage source at the gate of N_1; (b) SRAM-cell with additional current source at the drain of N_1.

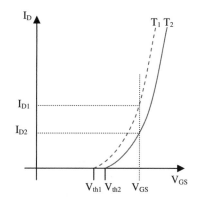

Fig. 8. Characteristics of two MOSFETS having different V_{th}

transistors as depicted in figure 10. The cells can be selected sequentially and evaluated using the same evaluation circuit (i.e. sense amplifier). Thus, the area of one PUF-cell is scaled-down to about the size of a common SRAM-cell. For the particular topology that is about $100F^2$ ('minimum featured size').

In such a circuit it could happen that the mismatch of the evaluation circuit influences the decision. Due to this fact, cells using the same evaluation circuit could tend to output the same value. To reduce the influence of an asymmetric sense amplifier, the number of PUF-cells which use the same evaluation logic should be chosen carefully.

The whole structure diagram of the system is depicted in figure 11. There are two modes: One for the initialization phase and another one for the nominal operation. The addresses of the useful cells are stored in a non-volatile memory (NVM). During the initialization phase this memory is filled with data: The outputs of the single PUF-cells after adding both bias currents are compared. If the output stays constant for both bias values, the address of the PUF-cell is written

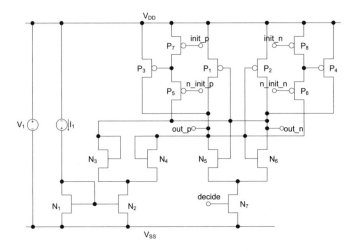

Fig. 9. Implementation example of an SRAM-like PUF with pre-selection transistors P_3 and P_4

Table 1. Control of the transistors P_5, P_6, P_7, P_8 for adding the current bias to the circuit

function	n_{th}	p_{th}	$init_p$	$ninit_p$	$init_n$	$ninit_n$
no threshold	0	0	0	1	0	1
p threshold	0	1	1	0	0	1
n threshold	1	0	0	1	1	0

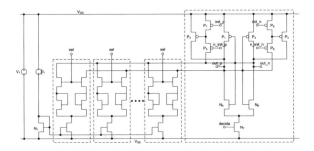

Fig. 10. PUF-cells with shared sense amplifier

into the NVM. The address of the NVM is incremented and the next PUF-cell is tested. This is done until the necessary number of outputs is reached. If not enough useful cells are provided an error occurs and the PUF must be considered to be defect. This indicates that the mismatch between the transistors is too small or that the ratio of required PUF-cells and available PUF-cells is too high. Possible solutions to this problem are to increase the number of PUF-cells or to reduce the upper and lower threshold values ΔV_{th-} and ΔV_{th+}. During the nominal mode, the PUF-cells stored in the NVM are read-out.

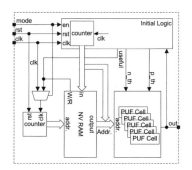

Fig. 11. Structure diagram with initialization logic

5   Conclusion

In this paper we introduced a pre-selection process for SRAM-like PUFs which can be implemented with little effort. We demonstrated that error-rates of 10E-6 are achievable. Due to the smaller error rate, using the marking procedure makes post-processing less complex or even unnecessary depending on the application. Hence, the area of the digital part of the circuit can be reduced. Furthermore, the smaller error rate leads to less power consumption and faster read-out. The additional effort caused by the initialization phase is small since the whole process can be done at one temperature and only two read-out cycles are necessary to separate the stable and the unstable PUF-cells.

References

1. Pappu, R., Recht, R., Taylor, J., Gershenfeld, N.: Physical one-way function. Science 297(5589), 2026–2030 (2002)
2. Gassend, B., Clarke, D., van Dijk, M., Devadas, S.: Silicon physical random functions. In: CCS '02: Proceedings of the 9th ACM conference on Computer and communications. security, pp. 148–160 (2002)
3. Blaise, G., Daihyun, L., Dwaine, C., van Dijk, M., Srinivas, D.: Identification and a2thentication of integrated circuits. Concurrency Computation: Pract. Exper. 16, 1077–1098 (2004)
4. Guajardo, J., Kumar, S.S., Schrijen, G.-J., Tuyls, P.: FPGA Intrinsic PUFs and Their Use for IP Protection. In: Paillier, P., Verbauwhede, I. (eds.) CHES 2007. LNCS, vol. 4727, pp. 63–80. Springer, Heidelberg (2007)
5. Boehm, C., Hofer, M.: Using SRAMs as Physical Unclonable Functions. In: Proceedings of the 17th Austrian Workshop on Microelectronics - Austrochip, pp. 117–122 (2009)
6. Ying, S., Holleman, J., Otis, B.P.: A Digital 1.6 pJ/bit Chip Identification Circuit Using Process Variations. IEEE Journal of Solid-State Circuits 43(1), 69–77 (2008)
7. Dodis, Y.: Fuzzy Extractors: How to Generate Strong Keys from Biometrics and Other Noisy Data. In: Naor, M. (ed.) EUROCRYPT 2007. LNCS, vol. 4515, pp. 523–540. Springer, Heidelberg (2007)
8. Hong, J., Vitterli, M.: Simple Algorithms for BCH Decoding. IEEE Transactions of Communications 43, 2324–2333 (1995)

9. Boesch, C., Guajardo, J., Sadeghi, A.R., Shokrollahi, J., Tuyls, P.: Efficient Helper Data Key Extractor on FPGAs. In: Oswald, E., Rohatgi, P. (eds.) CHES 2008. LNCS, vol. 5154, pp. 181–197. Springer, Heidelberg (2008)
10. Maes, R., Tuyls, P., Verbauwhede, I.: Low-Overhead Implementation of a Soft Decision Helper Data Algorithm for SRAM PUFs. In: Clavier, C., Gaj, K. (eds.) CHES 2009. LNCS, vol. 5747, pp. 332–347. Springer, Heidelberg (2009)
11. Pelgrom, M., Duinmaijer, A., Welbers, A.: Matching Properties of MOS-Transistors. IEEE Journal of Solid-State Circuits 24, 1433–1440 (1989)
12. Pelgrom, M.J.M., Tuinhout, H.P., Vertregt, M.: Transistor matching in analog CMOS applications. In: International Electron Devices Meeting, IEDM '98 Technical Digest., pp. 915–918 (1998)
13. Mizuno, T., Okumtura, J., Toriumi, A.: Experimental study of threshold voltage fluctuation due to statistical variation of channel dopant number in MOSFET's. IEEE Transactions on Electron Devices 41, 2216–2221 (1994)
14. Tsividis, Y.: The MOS Transistor. Oxford University Pres, New York (1999)

A   Calculations

Ratio of Useful PUF-Cells α: Partial probability of occurrence of selected PUF cells depending on ΔV_{th} (see figure 7(c) and 7(d)):[3]

$$UF_+ = \phi_1 \Phi_3 \tag{5}$$

$$UF_- = \phi_1(1 - \Phi_4) \tag{6}$$

Probability of occurrence PUF-cells being selected twice depending on ΔV_{th} (see figure 7(e)):

$$UF_+ \cap UF_- = \phi_1 \Phi_3 (1 - \Phi_4) \tag{7}$$

Total probability of occurrence depending on ΔV_{th}:

$$
\begin{aligned}
UF &= UF_+ + UF_- - 2(UF_+ \cap UF_-) = \\
&= \phi_1[\Phi_3 + (1 - \Phi_4) - 2\Phi_3(1 - \Phi_4)] = \\
&= \phi_1[1 - \Phi_4 - \Phi_3 + 2\Phi_3\Phi_4]
\end{aligned}
\tag{8}
$$

From UF the ratio of useful PUF-cells α can be determined:

$$
\begin{aligned}
\alpha &= \int_{\infty}^{-\infty} UF \, dV_{th} = \\
&= \int_{\infty}^{-\infty} \frac{1}{\sigma_1\sqrt{2\pi}} e^{-\frac{1}{2}\left(\frac{V_{th}-\mu_1}{\sigma_1}\right)^2} \left[1 - \frac{1}{\sigma_4\sqrt{2\pi}} \int_{-\infty}^{V_{th}} e^{-\frac{1}{2}\left(\frac{V'_{th}-\mu_4}{\sigma_4}\right)^2} dV'_{th} + \right. \\
&\left. - \frac{1}{\sigma_3\sqrt{2\pi}} \int_{-\infty}^{V_{th}} e^{-\frac{1}{2}\left(\frac{V'_{th}-\mu_3}{\sigma_3}\right)^2} dV'_{th} + \frac{2}{\sigma_3\sqrt{2\pi}} \int_{-\infty}^{V_{th}} e^{-\frac{1}{2}\left(\frac{V'_{th}-\mu_3}{\sigma_3}\right)^2} dV'_{th} \right. \cdot
\end{aligned}
$$

---

[3] The results of this section depend on the threshold values V_{th+} and V_{th-}.

$$\cdot \frac{1}{\sigma_4\sqrt{2\pi}} \int_{-\infty}^{V_{th}} e^{-\frac{1}{2}\left(\frac{V'_{th}-\mu_4}{\sigma_4}\right)^2} dV'_{th} \Bigg] dV_{th} =$$

$$= 1 - \frac{1}{\sigma_1 2\pi} \int_{\infty}^{-\infty} e^{-\frac{1}{2}\left(\frac{V_{th}-\mu_1}{\sigma_1}\right)^2} \Bigg[\frac{1}{\sigma_4} \int_{-\infty}^{V_{th}} e^{-\frac{1}{2}\left(\frac{V'_{th}-\mu_4}{\sigma_4}\right)^2} dV'_{th} +$$

$$- \frac{1}{\sigma_3} \int_{-\infty}^{V_{th}} e^{-\frac{1}{2}\left(\frac{V'_{th}-\mu_3}{\sigma_3}\right)^2} dV'_{th} + \frac{2}{\sigma_3\sigma_4\sqrt{2\pi}} \int_{-\infty}^{V_{th}} e^{-\frac{1}{2}\left(\frac{V'_{th}-\mu_3}{\sigma_3}\right)^2} dV'_{th} \cdot$$

$$\cdot \int_{-\infty}^{V_{th}} e^{-\frac{1}{2}\left(\frac{V'_{th}-\mu_4}{\sigma_4}\right)^2} dV'_{th} \Bigg] dV_{th} \tag{9}$$

In general we can assume that $\sigma_2 = \sigma_3 = \sigma_4 = \sigma$, $\mu_1 = \mu2 = 0$, and $\mu_3 = -\mu_4$. Then α becomes:

$$\alpha = 1 - \frac{1}{\sigma_1 2\pi} \int_{\infty}^{-\infty} e^{-\frac{1}{2}\left(\frac{V_{th}}{\sigma_1}\right)^2} \Bigg[\frac{1}{\sigma} \int_{-\infty}^{V_{th}} e^{-\frac{1}{2}\left(\frac{V'_{th}-\mu}{\sigma}\right)^2} dV'_{th} +$$

$$- \frac{1}{\sigma} \int_{-\infty}^{V_{th}} e^{-\frac{1}{2}\left(\frac{V'_{th}-\mu}{\sigma}\right)^2} dV'_{th} + \frac{2}{\sigma^2\sqrt{2\pi}} \int_{-\infty}^{V_{th}} e^{-\frac{1}{2}\left(\frac{V'_{th}-\mu}{\sigma}\right)^2} dV'_{th} \cdot$$

$$\cdot \int_{-\infty}^{V_{th}} e^{-\frac{1}{2}\left(\frac{V'_{th}-\mu}{\sigma}\right)^2} dV'_{th} \Bigg] dV_{th} \tag{10}$$

Ratio of Not-Useful PUF-Cells β: To verify the result of α, the ratio β of not selected PUF-cells is determined as well:

$$\beta = \int_{\infty}^{-\infty} NUF \, dV_{th} \tag{11}$$

$$\begin{aligned} NUF &= \phi_1[(1-\Phi_3)(1-(1-\Phi_4)+\Phi_3(1-\Phi_4))] \\ &= \phi_1[(1-\Phi_3)(\Phi_4)+\Phi_3(1-\Phi_4))] \\ &= \phi_1[\Phi_4 - \Phi_4\Phi_3 + \Phi_3 - \Phi_4\Phi_3))] \\ &= \phi_1[\Phi_4 + \Phi_3 - 2\Phi_4\Phi_3))] \end{aligned} \tag{12}$$

$$\beta = \int_{\infty}^{-\infty} \frac{1}{\sigma_1\sqrt{2\pi}} e^{-\frac{1}{2}\left(\frac{V_{th}-\mu_1}{\sigma_1}\right)^2} \Bigg[\frac{1}{\sigma_3\sqrt{2\pi}} \int_{-\infty}^{V_{th}} e^{-\frac{1}{2}\left(\frac{V'_{th}-\mu_3}{\sigma_3}\right)^2} dV'_{th} +$$

$$+ \frac{1}{\sigma_4\sqrt{2\pi}} \int_{-\infty}^{V_{th}} e^{-\frac{1}{2}\left(\frac{V'_{th}-\mu_4}{\sigma_4}\right)^2} dV'_{th} - \frac{2}{\sigma_3\sqrt{2\pi}} \int_{-\infty}^{V_{th}} e^{-\frac{1}{2}\left(\frac{V'_{th}-\mu_3}{\sigma_3}\right)^2} dV'_{th} \cdot$$

$$\cdot \frac{1}{\sigma_4\sqrt{2\pi}} \int_{-\infty}^{V_{th}} e^{-\frac{1}{2}\left(\frac{V'_{th}-\mu_4}{\sigma_4}\right)^2} dV'_{th} \Bigg] dV_{th} =$$

$$= \frac{1}{\sigma_1 2\pi} \int_{\infty}^{-\infty} e^{-\frac{1}{2}\left(\frac{V_{th}-\mu_1}{\sigma_1}\right)^2} \left[\frac{1}{\sigma_3} \int_{-\infty}^{V_{th}} e^{-\frac{1}{2}\left(\frac{V'_{th}-\mu_3}{\sigma_3}\right)^2} dV'_{th} + \right.$$

$$+ \frac{1}{\sigma_4} \int_{-\infty}^{V_{th}} e^{-\frac{1}{2}\left(\frac{V'_{th}-\mu_4}{\sigma_4}\right)^2} dV'_{th} - \frac{2}{\sigma_3\sigma_4\sqrt{2\pi}} \int_{-\infty}^{V_{th}} e^{-\frac{1}{2}\left(\frac{V'_{th}-\mu_3}{\sigma_3}\right)^2} dV'_{th} \cdot$$

$$\left. \cdot \int_{-\infty}^{V_{th}} e^{-\frac{1}{2}\left(\frac{V'_{th}-\mu_4}{\sigma_4}\right)^2} dV'_{th} \right] dV_{th} \tag{13}$$

In general we can assume that $\sigma_2 = \sigma_3 = \sigma_4 = \sigma$, $\mu_1 = \mu_2 = 0$, and $\mu_3 = -\mu_4$. Thus, β becomes:

$$\beta = \frac{1}{\sigma_1 \sigma 2\pi} \int_{\infty}^{-\infty} e^{-\frac{1}{2}\left(\frac{V_{th}}{\sigma_1}\right)^2} \left[\int_{-\infty}^{V_{th}} e^{-\frac{1}{2}\left(\frac{V'_{th}-\mu}{\sigma}\right)^2} dV'_{th} + \right.$$

$$+ \int_{-\infty}^{V_{th}} e^{-\frac{1}{2}\left(\frac{V'_{th}-\mu_4}{\sigma_4}\right)^2} dV'_{th} - \frac{2}{\sigma} \int_{-\infty}^{V_{th}} e^{-\frac{1}{2}\left(\frac{V'_{th}-\mu}{\sigma}\right)^2} dV'_{th} \cdot$$

$$\left. \cdot \int_{-\infty}^{V_{th}} e^{-\frac{1}{2}\left(\frac{V'_{th}-\mu}{\sigma}\right)^2} dV'_{th} \right] dV_{th} \tag{14}$$

Check: $1 = \alpha + \beta$

Estimation of the Error Rate e: An error occurs, if one of the PUF-cells which were marked useful provides the wrong output. Like the total ratio of selected PUFs, the total error $e(\Delta V_{th})$ is the sum of the two partial errors $e_-(\Delta V_{th})$ and $e_+(\Delta V_{th})$. The following errors are evaluated at a certain $e(\Delta V_{th})$:

$$e_-(\Delta V_{th}) = \frac{1}{\alpha} \Phi_2 [UF_- - (UF_+ \cap UF_-)] =$$

$$= \frac{1}{\alpha} \Phi_2 [\phi_1(1 - \Phi_4) - \phi_1 \Phi_3(1 - \Phi_4)] \tag{15}$$

$$e_+(\Delta V_{th}) = \frac{1}{\alpha}(1 - \Phi_2)[UF_+ - (UF_+ \cap UF_-)] =$$

$$= \frac{1}{\alpha}(1 - \Phi_2)[\phi_1 \Phi_3 - \phi_1 \Phi_3(1 - \Phi_4)], \tag{16}$$

where $\frac{1}{\alpha}$ is a normalization factor.

$$e(\Delta V_{th}) = e_+(\Delta V_{th}) + e_-(\Delta V_{th}) =$$

$$= \frac{1}{\alpha}(1 - \Phi_2)[\phi_1 \Phi_3 - \phi_1 \Phi_3(1 - \Phi_4)] +$$

$$+ \frac{1}{\alpha} \Phi_2 [\phi_1(1 - \Phi_4) - \phi_1 \Phi_3(1 - \Phi_4)] =$$

$$= \frac{1}{\alpha} \{[\phi_1 \Phi_2 - \phi_1 \Phi_2 \Phi_4 - \phi_1 \Phi_2 \Phi_3 + \phi_1 \Phi_2 \Phi_3 \Phi_4] +$$

$$+[\phi_1\Phi_3 - \phi_1\Phi_3(1-\Phi_4)] - \Phi_2\phi_1\Phi_3 + \Phi_2\phi_1\Phi_3(1-\Phi_4)\} =$$
$$= \frac{1}{\alpha}\{\phi_1\Phi_2 - \phi_1\Phi_2\Phi_4 - \phi_1\Phi_2\Phi_3\Phi_4 + \phi_1\Phi_3 - \phi_1\Phi_3 + \phi_1\Phi_3\Phi_4 -$$
$$-\Phi_2\phi_1\Phi_3 + \Phi_2\phi_1\Phi_3 - \Phi_2\phi_1\Phi_3\Phi_4\} \tag{17}$$

Table 2. Examples for the error rate e and the ratio of useful PUF-cells α. *The number in the brackets shows the BER without any pre-selection.

num	σ_1	$\sigma_2, \sigma_3, \sigma_4$	$\mu_3 = \mu_4$	α	e
1	30 mV	1 mV ($\approx 0.7\%$*)	5 mV	0.8677	2.19E-06
2	30 mV	1 mV ($\approx 0.7\%$*)	10 mV	0.7390	¡1e-12
3	30 mV	2 mV ($\approx 1.5\%$*)	10 mV	0.7394	5.09E-6
4	30 mV	2 mV ($\approx 1.5\%$*)	20 mV	0.5059	¡1e-12
5	30 mV	5 mV ($\approx 4\%$*)	10 mV	0.7422	0.0087
6	30 mV	5 mV ($\approx 4\%$*)	20 mV	0.5108	2.39E-4
7	30 mV	5 mV ($\approx 4\%$*)	40 mV	0.1884	1.03E-9

Table 3. Numeric examples of the error rate e and the ratio of useful PUF-cells α in dependence of $\mu_{3,4}(\sigma_1 = 30\,\text{mV}, \sigma_{2,3,4} = 6, 16$ mV). Without any pre-selection ($\mu_{3,4} = 0$mV) we get an error-rate of about 5%.

$\mu_{3,4}$(mV)	e	α	$\mu_{3,4}$(mV)	e	α
0	4.9965E-2	0.909	26	2.1295E-4	0.396
1	4.9435E-2	0.907	27	1.4701E-4	0.378
2	4.7882E-2	0.9	28	1.0037E-4	0.361
3	4.5412E-2	0.889	29	6.7765E-5	0.344
4	4.2189E-2	0.874	30	4.5242E-5	0.327
5	3.8415E-2	0.856	31	2.9867E-5	0.311
6	3.4307E-2	0.836	32	1.9495E-5	0.296
7	3.0078E-2	0.814	33	1.2581E-5	0.281
8	2.5917E-2	0.791	34	8.027E-6	0.267
9	2.1971E-2	0.767	35	5.0631E-6	0.253
10	1.8347E-2	0.743	36	3.1571E-6	0.24
11	1.5109E-2	0.719	37	1.9460E-6	0.227
12	1.2282E-2	0.695	38	1.1858E-6	0.215
13	9.8629E-3	0.671	39	7.1416E-7	0.203
14	7.8298E-3	0.648	40	4.2516E-7	0.192
15	6.1474E-3	0.624	41	2.5017E-7	0.181
16	4.7749E-3	0.601	42	1.4549E-7	0.17
17	3.6698E-3	0.579	43	8.363E-8	0.16
18	2.7909E-3	0.557	44	4.7509E-8	0.151
19	2.1004E-3	0.535	45	2.6674E-8	0.142
20	1.5641E-3	0.514	46	1.4800E-8	0.133
21	1.1525E-3	0.493	47	8.1156E-9	0.125
22	8.4015E-4	0.473	48	4.3978E-9	0.117
23	6.0592E-4	0.453	49	2.3550E-9	0.11
24	4.3229E-4	0.433	50	1.2462E-9	0.103
25	3.0507E-4	0.414			

To get the error e over all $e(\Delta V_{th})$, $e(\Delta V_{th})$ has to be integrated over all ΔV_{th}:

$$e = \int_{\infty}^{-\infty} e(\Delta V_{th}) \, d\Delta V_{th} \tag{18}$$

B   Numerical Examples

Table 2 shows some numeric examples for α and e. The number of useful PUF-cells depends mainly on the ratio $\frac{\sigma_1}{\mu_{3,4}}$. The main factors for the error are $\sigma_{2,3,4}$ and $\mu_{3,4}$. σ_1 influences the error rate only marginally. Table 3 shows e in dependence of $\mu_{3,4}$.

New High Entropy Element for FPGA Based True Random Number Generators

Michal Varchola and Milos Drutarovsky

Department of Electronics and Multimedia Communications,
Technical University of Kosice,
Park Komenskeho 13, 041 20 Kosice, Slovak Republic
michal@varchola.com, Milos.Drutarovsky@tuke.sk

Abstract. We demonstrate a new high-entropy digital element suitable for True Random Number Generators (TRNGs) embedded in Field Programmable Gate Arrays (FPGAs). The original idea behind this principle lies in the randomness extraction on oscillatory trajectory when a bi-stable circuit is resolving a metastable event. Although such phenomenon is well known in the field of synchronization flip-flops, this feature has not been applied for TRNG designs. We propose a new bi-stable structure – Transition Effect Ring Oscillator (TERO) where oscillatory phase can be forced on demand and be reliably synthesized in FPGA. Randomness is represented as a variance of the TERO oscillations number counted after each excitation. Variance is highly dependent on the internal noise of logic cells and can be used easily for reliable instant inner testing of each generated bit. Our proposed mathematical model, simulations and hardware experiments show that TERO is significantly more sensitive to intrinsic noise in FPGA logic cells and less sensitive to global perturbations than a ring oscillator composed from the same elements. The experimental TERO-based TRNG passes NIST 800-22 tests.

Keywords: TRNG, oscillatory metastability, randomness extraction, inner testability.

1 Introduction

Almost each cryptographic system contains a Random Number Generator (RNG) that produces random values for underlying algorithms. Random numbers are essential elements for secure transactions and therefore they should meet the highest strict requirements – they should be unpredictable, uniformly distributed on their range and independent [13].

RNGs can be divided into two main subgroups [9]: Pseudo RNG (PRNG) and True RNG (TRNG). The output of a PRNG is mathematically defined and all of its entropy is given by the random seed. On the other hand, entropy of a TRNG is increased by each generated bit. The TRNG operation is usually based on certain physical sources of entropy (e.g. thermal noise, timing jitter) that is present in modern electronic devices.

S. Mangard and F.-X. Standaert (Eds.): CHES 2010, LNCS 6225, pp. 351–365, 2010.

Field Programmable Gate Arrays (FPGAs) are a popular implementation platform for modern crypto-systems thanks to their reconfigurability [24]. Weak or obsolete cryptographic protocols or algorithms can be updated easily even in devices deployed in a hostile environment. Thus users and FPGA devices can better resist security treats. Moreover, an entire system should be implemented in the same chip due to security reasons. Due to mentioned security considerations, research on TRNGs for the FPGAs is still an area of active research [9].

Recent TRNGs for FPGAs employ two main randomness sources. First, timing jitter of Ring Oscillators (ROs) [20,7] or Phase Locked Loops (PLLs) [6] and, second metastability of logic cells [5,21,22]. A comprehensive survey of various TRNG principles was reported in [9]. However, serious disputes on reliability of RO-based TRNG [20] led to a chain of papers [17,3,4,19,23,1] reporting various merits of it. Designers should also consider other issues such as frequency injection attack [12] or TRNG evaluation methodology according [10].

Deep study on metastability was done in [8,15] where the main focus was synchronization issues rather than TRNGs. Short-time metastability of a bistable structure was forced by critical combination of input signals. It was noted that even a small perturbation can cause escape from this state. The resulting logic state and trajectory of approaching it was analyzed as well. The resulting state and/or resulting trajectory can possess random properties such as a jitter of temporarily oscillatory trajectory. Forcing metastability on demand is not trivial and represents a great challenge for synthesis in FPGA fabric. However, most published TRNG designs are targeted to ASIC technology and extract randomness from the "final logic state" after a metastable event.

The most recent design [22] uses a metastabe RO, where each inverter is in short circuit to first reach a metastable state. After a while, inverters are switched to the single chain in order to form a RO. Metastable issues provide random starting conditions for the RO. Oscillation of RO and signal sampling by a D-Flip-Flop (DFF) are used for a resolution of the foregoing metastable event. The proof that authors of [22] reached a metastable state in the FPGA is questionable. As evidence, they provide oscilloscope waveforms that in our opinion cannot be acquired reliably from internal FPGA gates.

Each approach for a metastable TRNG [22,21,5] is based mainly on forcing a system to a metastable state and then evaluating the state where the system converges when the metastability phase is over. However, the phase of convergence towards stable state by a temporally oscillatory trajectory was still neglected as a randomness extraction mechanism. As it will be shown in this paper, such phase is worthy to consider for random bit generation. We must point out that the main focus of the paper is to introduce features of the new high entropy element for FPGA-based TRNGs rather than present a complete TRNG design.

Paper is organized as follows: Section 2 brings design goals of the new entropy element. Section 3 introduces the new entropy element and its mathematical model accompanied by SPICE and VHDL macro-model simulations. A hardware implementation is analyzed in the Sect. 4. Experimental results are presented in the Sect. 5. Conclusion and future work is given in the Sect. 6.

2   New Entropy Element Design Goals

Each design of TRNG based on logic cells has pros and cons. Going by the recent
state of the art [9], there is still a gap for the TRNG based on better entropy
element, that is possible to synthesize in the FPGAs fabric. The design goals for
such an element are:

- sufficiently higher entropy rate than previous RO based designs,
- lower sensitivity on global interference and working conditions than previous
  RO based designs,
- ability to extract reliably intrinsic noise generated by logic elements,
- clear description of the mathematical model acceptable to the wide scientific
  community,
- inner testability feature in order to detect instantly when the entropy source
  is out of order and/or has weak statistical properties [18],
- ability to restart the element before each random bit generation period in
  order to utilize the stateless entropy concept [2],
- ability for several entropy elements operating independently and in parallel in
  order to place them into the same FPGA for enhancing statistical parameters
  and/or increasing the bit-rate,
- usage of least number of logic elements all implemented in the single block
  of logic to minimize signal paths, minimize interference, minimize resources
  utilization, and decrease the power consumption,
- element structure should have simple place and route strategies and clear
  recommendations on how to synthesize the structure,
- ability to utilize combinations of multiple known principles of randomness
  extraction, i.e. variation of time delay and the metastability phenomena.

3   Transition Effect Ring Oscillator

A novel structure capable of extracting noise from logic cells is depicted in Fig. 1.
Proposed structure was optimized for implementation on a single Spartan 3E
Complex Logic Block (CLB). An entire set of experiments has been carried out
using Xilinx Spartan 3E FPGA Starter Kit [25] for the purpose of this paper
research goals. Its physical behavior in FPGA, including the shape of control
waveforms, is shown in an oscilloscope screen-shot in Fig. 2. The XOR_1 – AND_1 –
XOR_2 – AND_2 loop begins to oscillate at each edge of the $ctrl$ signal. This effect
is a "transition" of the loop and therefore this structure will be referred the
Transition Effect Ring Oscillator (TERO) in this paper. When TERO circuit
operates in the topology shown in Fig. 1 its operation will be referred as a
"TERO mode" or just TERO. When $ctrl = {'1'}$ for XOR_1 and $ctrl = {'0'}$ for XOR_2
constantly the structure will behave as an RO of one inverting and three non-
inverting elements. This operation will be denoted as an "RO mode" in text for
the purpose of comparing TERO mode and RO mode features.

TERO operates as follows: The XOR gates act as inverters or buffers when the
$ctrl = {'0'}$ or $ctrl = {'1'}$ respectively. In other words, loop incorporates two buffers

and two inverters ($ctrl = {'0'}$) or just four buffers ($ctrl = {'1'}$) when assuming $rst = {'0'}$. Loop does not satisfy an oscillatory condition because it consists of even number of inverting elements in both cases. This is the reason why the loop does not oscillate by itself and settles to a steady state. However, when the edge of $ctrl$ is applied to XORs, they invert their actual output level. Such action disturbs the steady state of the loop because the newly reached XORs' output levels begin to circulate through the loop. The logic level that was previously stable in the entire loop is switched to the opposite level in the half of loop. A pulse is raised as a result of this process, and it begins to run along the loop. The pulse will disappear after several runs (from tens to hundreds) of oscillations. The number of oscillations generated by TERO varies during each $ctrl$ period. T-Flip-Flop (TFF) resolves if TERO made an odd or even number of oscillations during single $ctrl$ period, and that represents one random bit ($tffout1$ and $tffout2$ signal, or just $tffout$ in next text). The purpose of the ANDs is to force the same initial conditions at the end of each $ctrl$ period. The ANDs are controlled by the rst signal and their outputs are held in constant $'0'$ when $rst = {'1'}$ regardless the logic level on the other input. The $tffout$ is sampled at the falling edge of rst. After that $tffout$ is cleared by the clr signal. One can argue that there is no necessity of $tffout$ clearing because it does not affect volume of randomness, but despite this fact, it is cleared because of two reasons: first, it represents a Least Significant Bit (LSB) of an asynchronous counter used for measurements and the counter should start to increment from zero; and the second, more crucial reason is that slight bias will not be transformed to the correlation that enables the assumption of a stateless entropy concept[2].

TERO is a kind of bi-stable Flip-Flop (FF) with intentionally lengthened feedback paths. However, more common is to employ NAND gates or NOR gates instead of XORs for practical FFs. Employing NANDs or NORs in the proposed topology results in RS FF. Metastable behavior of RS FFs underwent rigorous analysis in [8] and [15]. Basically, when certain combination of signals appears at the inputs, RS FF can fall into the metastable state. After pico-to-nano-second time, this state is typically escaped by a trajectory, which may be temporarily oscillatory. Although TERO is not the RS FF regarding its functionality, it also escapes from the metastable state by the oscillatory trajectory due to lengthened feedback paths. This behavior is known as oscillatory metastable operation [8].

Practical evaluation of TERO by oscilloscope found that variation of the number of oscillations during each $ctrl$ period is extensive in comparison to RO mode. Also observed is that TERO oscillates on the double frequency in comparison to RO mode. TERO operation was confirmed by qualitative SPICE analysis of TERO structure, as presented in the next subsection.

3.1   Transistor Level SPICE Simulation

The purpose of the SPICE simulation was to confirm qualitatively that behavior of an approximately balanced TERO structure exhibits oscillatory transient character as was reported for quite general semiconductor bistable structures [8]. SPICE simulation was not performed to examine the precise behavior in

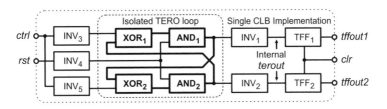

Fig. 1. Practical circuit of TERO used for the FPGA implementation. Entire TERO loop structure occupies just one CLB of Xilinx Spartan 3E. Usage of INV_1 – INV_5 enables such routing that signal directly connected to internal TERO loop will not be routed by off-CLB path. ANDs are used for forcing the same TERO initial conditions for each *ctrl* period by *rst* signal as well as for an additional delay needed for reliable oscillatory behavior of the circuit. TFFs are used for extraction of a random bits. TFFs are cleared at the end of each *ctrl* period by *clr* signal.

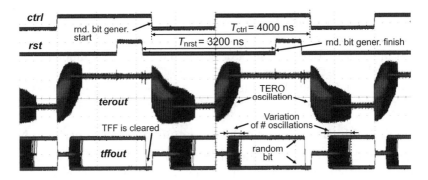

Fig. 2. The TERO operation oscilloscope screen-shot captured using infinite persistence mode and 20 MHz low-pass filter on *ctrl*, *rst*,and *tffout* channels. The image was acquired by the Tektronix MSO 4104 oscilloscope. Each edge of the *ctrl* signal causes oscillation of TERO loop (*terout* signal). The number of oscillations observed varies during each *ctrl* period. TFF resolves if TERO made odd or even number of oscillation periods during one *ctrl* period that represents one random bit (*tffout* signal). TERO is initialized to the same operating conditions at the end of each *ctrl* period by the *rst* signal. The *tffout* is sampled on the falling edge of *rst*. Then *tffout* is cleared.

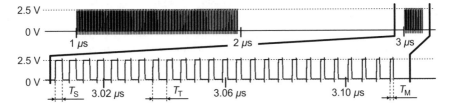

Fig. 3. LT Spice simulation of the TERO. TERO starts to oscillate at the rising ($1\,\mu s$) or falling ($3\,\mu s$) edge of the *ctrl* signal having a $4\,\mu s$ period. Zoomed region shows that excited pulse disappears due to its shortening each loop crossing. T_T stands for the mean value of the oscillation period, T_S and T_D denote time durations of logic $'1'$ level when oscillation behavior raises and disappears respectively.

particular FPGA or CMOS manufacturing process but to get a qualitative results. Therefore the SPICE simulation was based on publicly available Linear
Technology LT Spice IV tool [11] and 250 nm, 2.5 V CMOS transistors models
[14] that were used also in [4]. Only the isolated TERO loop that consists of
XOR$_1$ – AND$_1$ – XOR$_2$ – AND$_2$ (Fig. 1) was simulated. Tested XORs used standard
12 transistors layout and tested ANDs used standard 6 transistor layout. Wires
were implemented as buffers of certain time delay and time constant.

LT Spice simulation (Fig. 3) confirms oscillatory behavior of the TERO in
approximately balanced loop. Rising and falling edges of *ctrl* cause oscillations
excitation in the loop (at 1 μs and 3 μs time respectively) that is in a good
agreement with results of general bistable structures [8]. The zoomed region at
the bottom of Fig. 3 shows that the excited pulse disappears due to its shortening
in each loop passing. This phenomenon was also in a good agreement with [15].

Satisfactory results were obtained when the simulation was performed using
maximal time step of 0.5 ps, otherwise numerical computation errors caused non-
repeatability of that simulation. Simulation did confirm the double frequency of
TERO in comparison to RO. LT Spice transient simulation does not take into
account any contribution of transistor noises and therefore a simplified mathematical model is defined in the following section. This model allows for easier
analysis of oscillation number variation due to noise. Features observed due to
LT Spice simulation were used for the TERO basic model parameters derivation.

3.2   TERO Mathematical Model Based on Effects of Intrinsic Noise

LT Spice simulation provides a starting point for definition of a TERO mathematical model that takes intrinsic noise into account. Intrinsic noise, in which
the samples of amplitude are assumed to be iid (independent and identically
distributed) and follows a normal distribution $\mathcal{N}\left(0, \sigma^2\right)$, affects timing instability of signal edges passing through logic cells. We show that this noise has
substantial impact on TERO performance. Graphical representation of the proposed mathematical model is given in the Fig. 4. Model works as follows: when a
rising or falling edge of *ctrl* appears, TERO loop begins to oscillate (Fig. 2). The
mean value of TERO oscillation period is equal to total delay of TERO loop T_T.
An excited pulse of starting logic '1' level time length T_S is shortened in each
oscillation by T_D time due to slight intrinsic non-symmetry of the loop. Excited
pulse will disappear when instant logic '1' level time length reaches minimal
possible value T_M. Asymmetry T_D is assumed to be affected by a period jitter
$\Delta_{T_{ij}}$, where i and j stands for i-th T_T period and j-th T_{ctrl} period respectively.
The final number of oscillations executed for j-th T_{ctrl} is denoted as Y_{T_j}. The
basic mathematical model of TERO mode is expressed as:

$$T_S - T_M = \sum_{i=1}^{Y_{T_j}} \left(T_D + \Delta_{T_{ij}}\right) = T_D\, Y_{T_j} + \sum_{i=1}^{Y_{T_j}} \Delta_{T_{ij}} \ . \tag{1}$$

Both, T_S and T_M can be slightly affected by intrinsic noise and so considered
as a contribution to final randomness. Value of the former can be affected by

actual noise conditions when circuit is entering to oscillatory metastable state and value of the latter can be affected by actual noise conditions when the circuit does (or does not) allow to pass last pulse. However, according to our oscilloscope measurements the jitter accumulation over oscillatory trajectory exhibits the best entropy extraction and therefore this is the only focus in this paper. Investigation on T_S and T_M instability contribution to final randomness will be a subject of future research.

Similarly, it is possible to express the basic model of the same circuit in RO mode. In Fig. 2, T_{nrst} denotes a time period when RO is not reset and is oscillating at $T_R = 2\,T_T$ oscillation periods (Fig. 4). Each oscillation period is assumed to be affected by period jitter $\Delta_{R_{ij}}$. Final number of oscillations executed for j-th T_{ctrl} is denoted as Y_{R_j}. Thus, the basic mathematical model of the RO mode circuit is expressed as:

$$T_{nrst} = \sum_{i=1}^{Y_{R_j}} \left(T_R + \Delta_{R_{ij}} \right) = 2\,T_T\,Y_{R_j} + \sum_{i=1}^{Y_{R_j}} \Delta_{R_{ij}} \ . \tag{2}$$

Both, TERO mode (1) and RO mode (2) models were implemented in Matlab in order to evaluate them and to compare their sensitivity to intrinsic noise. The $\Delta_{T_{ij}} = \Delta_{R_{ij}} \approx \mathcal{N}\left(0, \sigma^2\right)$ simplification is assumed. The TERO mode model (1) was implemented using a pseudo-code Algorithm 1, where N stands for number of $ctrl$ periods. The RO mode model (2) was implemented similarly.

Accordingly, Fig. 5 shows that Y_{T_j} is affected in a greater manner than Y_{R_j} when exposing the circuit in TERO mode and in RO mode to the same noise conditions. The ratio between their standard deviations is derived in following section.

3.3 Analytical Comparison of the TERO and RO Modes

Analytical derivation of a ratio between standard deviations of Y_{T_j} and Y_{R_j} is the objective of this section. Denote the standard deviation and mean value of Y_{T_j} as σ_{Y_T} and $\overline{Y_T}$ respectively for the TERO mode. Similarly, denote the standard deviation and mean value of Y_{R_j} as σ_{Y_R} and $\overline{Y_R}$ respectively for the RO mode. It is possible to show, that according (1) and (2) and under assumption $\Delta_{T_{ij}} = \Delta_{R_{ij}} \approx \mathcal{N}\left(0, \sigma^2\right)$, good approximations of σ_{Y_T}, $\overline{Y_T}$, σ_{Y_R}, and $\overline{Y_R}$ for simplified comparison of TERO and RO sensitivities are:

$$\overline{Y_T} \approx \frac{T_S - T_M}{T_D} \ , \quad \sigma_{Y_T} \approx \frac{\sigma}{T_D} \sqrt{\frac{T_S - T_M}{T_D}} = \frac{\sigma}{T_D} \sqrt{\overline{Y_T}} \ , \tag{3}$$

$$\overline{Y_R} \approx \frac{T_{nrst}}{2\,T_T} \ , \quad \sigma_{Y_R} \approx \frac{\sigma}{2\,T_T} \sqrt{\frac{T_{nrst}}{2\,T_T}} = \frac{\sigma}{2\,T_T} \sqrt{\overline{Y_R}} \ . \tag{4}$$

The ratio $\frac{\sigma_{Y_T}}{\sigma_{Y_R}}$ is derived from combining (3) and (4):

$$\frac{\sigma_{Y_T}}{\sigma_{Y_R}} \approx \frac{2\,T_T}{T_D} \sqrt{\frac{2\,T_T\,(T_S - T_M)}{T_D\,T_{nrst}}} \ . \tag{5}$$

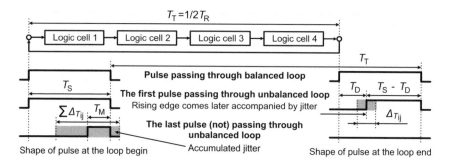

Fig. 4. Graphical representation of the mathematical model, where the instability of the pulse shortening during circulation around TERO is a key issue. Therefore the pulse will disappear after several oscillations.

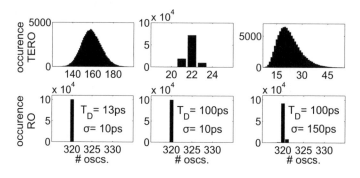

Fig. 5. Simulation of TERO (1) and RO (2) basic models performed in Matlab. Both models share the same parameters in order to compare their randomness extraction performance. Histograms show the occurrence of the recorded number of oscillations under three sets of different operating conditions, which vary in σ and T_D, where instability of T_D follows $\mathcal{N}\left(0, \sigma^2\right)$. Other parameters remain constant over all simulations: $T_T = 5\,\text{ns}$, $T_{\text{nrst}} = 3200\,\text{ns}$, $T_S - T_M = 0.4\,T_T$, $N = 10^5$.

Algorithm 1. TERO mathematical model simulation

Require: T_S, T_M, T_D, T_T, σ, N
Ensure: $Y_{T_1}, Y_{T_2} \ldots Y_{T_j} \ldots Y_{T_N}$
  for $j = 1$ to N **do**
    $Y_{T_j} \Leftarrow 0$
    $acc \Leftarrow 0$
    while $acc < T_S - T_M$ **do**
      $acc \Leftarrow acc + T_D + \mathcal{N}\left(0, \sigma^2\right)$
      $Y_{T_j} \Leftarrow Y_{T_j} + 1$
    end while
  end for

When applying practical values acquired from both hardware and LT Spice simulation: $T_{nrst} = 3200\,ns$, $T_T = 5\,ns$, $T_D = 0.013\,ns$, $T_S - T_M = 0.4\,T_T$ to (5), then $\frac{\sigma_{Y_T}}{\sigma_{Y_R}} \doteq 533$.

In other words, the proposed circuit is hundreds of times more sensitive in TERO mode to the period jitter than the same circuit in RO mode. At this point one can argue that this feature will increase vulnerability to external interferences or attacks. As it will be shown in more complex simulation that follows in next subsection, TERO thanks to its differential structure can decrease influence to the global (outside of CLB) perturbations while still maintaining high sensitivity to local (inside CLB) intrinsic noises.

3.4 TERO and RO Response under External Perturbations

The response of TERO to deterministic perturbations patterns was simulated using Macro-Model (MM) that in contrary to two previous subsections takes into account the same TERO loop structure as was implemented in real FPGA hardware (Fig. 1). MM was written in VHDL and simulated using ModelSim. The MM simulation setup including the simulated loop is shown in Fig. 6. The logic function of each component is computed instantly with an addition of a synthetic delay. The delay is implemented as a simple state machine which allows independent control of the delay of both the rising edge and the falling edge.

The noise pattern samples for each edge of each component are stored in separated file which was generated by Matlab. The simulated patterns are composed of deterministic perturbation and intrinsic noise. Deterministic perturbation represents global influence of power supply variations or electro-magnetic interference that can affect time delays of logic elements. It is assumed that deterministic perturbation affects same logic elements by the same manner. On the other hand intrinsic noise is assumed to be independent for each logic and follows the normal distribution $\mathcal{N}\left(0, \sigma^2\right)$.

Simulation results for various compositions of deterministic perturbation and intrinsic noise for both TERO and RO mode are shown in Fig. 7. Frequency of noise was assumed to be higher than frequency of TERO oscillations. It is possible to get qualitatively similar results when the noise is band limited as well. Results of the MM simulation confirm that TERO randomness extraction performance is superior to RO performance. Accordingly, results show in Fig. 8 that both the basic mathematical model simulation and the VHDL MM simulation are in good agreement with the results acquired from the FPGA hardware.

4 Hardware Implementation

Xilinx Spartan 3E Starter Board was used as an evaluation platform [25]. TERO shown in Fig. 1 was placed into one CLB of Xilinx Spartan 3E because of simpler proper routing, using only local, not global paths. Even though there are 9 logic functions and only 8 LUTs in single CLB, the TERO fits inside due to

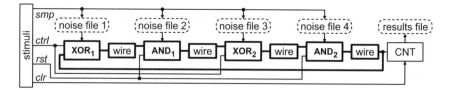

Fig. 6. The TERO macro-model implemented in VHDL. Constant delay of rising and falling edge can be set independently for each element, including wires. Non-symmetry was achieved by different rising edge delay and falling edge delay in the XOR₁ element. Signals *ctrl*, *rst* and *clr* have the same purpose as was described in Sect. 3 above. Signal *smp* controls noise sampling from the noise files 1–4. Noise files were generated by Matlab and contains data for both rising edge delay and falling edge delay instability. Final number of oscillations for each *ctrl* period is recorded in the results file.

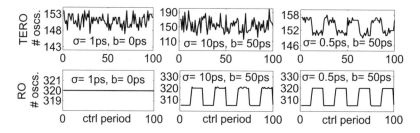

Fig. 7. VHDL macro-model simulation using ModelSim shows number of oscillations in 100 consecutive T_{ctrl} periods for both TERO (up) and RO (down) mode. Three different compositions of noise patterns (each column different) were used; where intrinsic noise follows $\mathcal{N}\left(0, \sigma^2\right)$ and b stands for amplitude of global deterministic perturbation of a square shape. It is obvious, that TERO is able to extract intrinsic noise of $\sigma = 0.5$ ps while RO is barely able to extract intrinsic noise of $\sigma = 10$ ps when both were exposed to the same operating conditions. Moreover, TERO is less sensitive to global deterministic perturbation than RO.

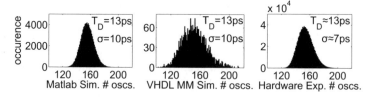

Fig. 8. Comparison of simulations and hardware experiment results. Histograms shows occurrences of TERO oscillations number for direct simulation of mathematical model using Matlab (left), VHDL macro-model simulation using ModelSim (midle) and hardware experiment (right). Intrinsic noise follows $\mathcal{N}\left(0, \sigma^2\right)$ in the simulations. Rest parameters were: $T_T = 5$ ns, $T_S - T_M = 0.4\,T_T$. $N = 10^5$, $N = 2600$, and $N = 10^6$ for the Matlab simulation, MM simulation, and the experiment respectively. Parameters T_D and σ of the hardware experiment are estimated according to VHDL MM simulation. The histogram of VHDL MM simulation is wider than the histogram of mathematical model simulation due to presence of the noise source $\mathcal{N}\left(0, \sigma^2\right)$ in each logic element.

using hardwired XORs in carry chain logic. There is indication, that this fast XOR causes more stable TERO performance. Consequently, a good constellation of placement and routing is locked by the user constrain file. Locking the circuit makes it portable through all CLBs with acceptable dispersion of the TERO parameters. The example of proper place and route of two TEROs in neighboring CLBs is shown in Fig. 9 b.

The entire system used for evaluating the TERO performance is shown in Fig. 9 a. There are two TEROs in order to investigate their cross-correlation. The number of oscillations are counted by asynchronous counters which are faster than synchronous counters. A typical frequency of TERO in four element loop structure is 200 MHz approximately. Asynchronous counters are implemented by the chain of TFFs. A control state machine ensures communication via USB that is used for transferring counter values to the computer for further analysis. The benefit of this structure is that expensive oscilloscopes are not necessary – just investigating of counter values is fairly enough for detailed analysis.

5   Experimental Results

Mean values of asynchronous counters, bias of extracted LSBs and autocorrelations of generated bit streams were used for fast evaluation of of TERO (RO) performance (dependency) in closely placed CLBs ("Next" configuration) as well as far-away placed CLBs ("Diag." configuration) in the target Xilinx FPGA. Mixing of two CLB outputs was performed by XOR operation that performs standard decimation by a factor of 2. Results shown in Fig. 10 and Table 1 were evaluated for 1 Mbit sequences acquired from the evaluation platform. More complex NIST [16] tests were performed for a 250 Mbit sequence in order to detetct potentially more complex deviations from the ideal one.

Autocorrelation test evaluates correlations between the sequence of extracted random bits and shifted (by number of *ctrl* periods) versions of it. Random bits are extracted as LSBs of number of oscillations of TERO (RO) at each period of *ctrl*. The statistic used in Fig. 10 for autocorrelations is normalized according to (5.5) on p.182 in [13] which approximately follows $\mathcal{N}(0,1)$ distribution for ideal random source. According to the 3σ rule, any value outside of the $< -3, 3 >$ interval indicates a probable deviation from ideal source of randomness.

From presented experimental results we can state that TERO can produce uncorrelated (or with high probability of independent) sequences which is not the case of RO composed of the same elements as TERO. Moreover, the XOR combination of two channels of TERO greatly improves statistical properties of generated sequences. This was confirmed by improving the mean value and especially passing of the NIST 800-22 or FIPS 140 statistical tests suites which indirectly indicate bit sequence independence. When evaluating XOR combination of two channels of RO the situation is worse as a consequence of the dependence of the examined bit sequences.

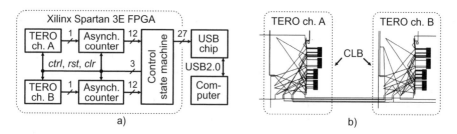

a)   b)

Fig. 9. (a) – Evaluation platform setup. (b) – Example of proper place and route of two TERO channels in two most closest CLBs; both TEROs are routed in the same way.

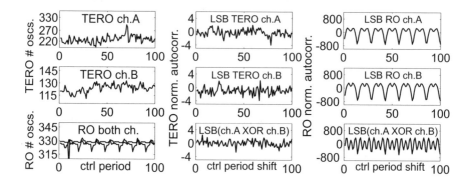

Fig. 10. TERO and RO mode comparison in hardware. Channels A and B are synthesized in the closest CLBs. Left column shows number of oscillations for both channels of TERO and RO in 100 consecutive *ctrl* periods. Normalized autocorrelation of TERO channels A and B and their XOR combination for 1–100 *ctrl* periods shift is given in the middle column. The same results, but for RO mode is depicted in right column.

Table 1. Results of the TERO evaluation in Xilinx Spartan 3E FPGA. Position "Next" means TERO (RO) A and B are placed in the closest CLBs, while "Diag." means A and B are placed in CLBs that are in opposite corners of the FPGA fabric.

TEST	Source	Next(TERO)	Diag.(TERO)	Next(RO)	Diag.(RO)
Mean	LSB A / LSB B	0.51/0.48	0.51/0.48	0.47/0.44	0.55/0.46
Value	LSB(A XOR B)	0.5002	0.4999	0.4539	0.7926
Normalized cross-correlation (for shift=0)	LSB (A,B)	0.4160	-0.0917	-94.3378	599.3945
NIST / FIPS	Only LSB A	F / P	F / F	- / F	- / F
Statistical	Only LSB B	F / F	F / F	- / F	- / F
tests result	LSB(A XOR B)	P / P	P / P	- / F	- / F

6   Conclusion and Future Work

A new high entropy element for FPGA-based TRNGs was introduced. This element was denoted as TERO and reasonably satisfies design goals formulated in Section 2. Its greatest advantage is high sensitivity to random processes inside FPGA logic cells, while rejecting global perturbation.

Moreover, TERO is straightforward inner testable. Instant evaluation of the number of oscillations and consequent estimation of noise parameters from them according to (3) allow instant detection of malfunctions when the random bit is generated. The implemented testing system can decide whether the bit will be used as a member of the resulting random sequence accordingly.

Furthermore, TERO has a clear basic mathematical model that was confirmed by LT Spice simulation, VHDL MM simulation and hardware evaluation. The TERO has hundreds times greater sensitivity to random processes inside logic gates then RO build up from the same components according to our proposed model. VHDL MM simulation shows that TERO can reject global perturbation better than RO due to its differential structure. In particular TERO can extract noise of $\sigma = 0.5\,\text{ps}$ standard deviation in the presence of $50\,\text{ps}$ global perturbation.

An experimental TRNG which uses an XOR combination of two TEROs was introduced. The source of randomness occupies just two CLBs and produce random sequence at $250\,\text{kbps}$ bit-rate of such quality that it can pass NIST 800-22 statistical tests without any need for further complex post-processor. This shows great potential of TERO for the FPGA based TRNGs designs.

Experiments showed that proper place and route strategies are essential for TERO and therefore further research will be focused on reliable lock of place and route synthesis constrains, analysis of TERO features in different FPGA internal positions, and evaluation of TERO operation in FPGAs of other vendors.

Although our entire investigation was carried out using one Spartan 3E board for the purpose of the paper, our latest experiments were processed using Actel Fusion FPGAs due to two reasons; first, to investigate TERO using different FPGA technology, and second, the availability of a greater number (ten) of equal Actel boards. Actel does not provide any tool for custom routing as Xilinx does so the routing is black-box in Actel. Nevertheless, preliminary results show that the variance of TERO oscillations in each tested Actel board at each tested position was satisfactory and also under nonstandard operation conditions through wide temperature and power supply range. The final random bit sequence composed of 16 TERO channels XOR combination passes the NIST 800-22 tests for every described setup. However, the quantity of results acquired from the Actel platform are excessive and will be a subject of an upcoming paper.

Future work will also include synthesis of a testing system that can estimate the statistical parameters of noise from the variance of oscillation number for the purpose of malfunction detection. This testing system will be incorporated in the final, ready-to-use TRNG.

Acknowledgement. This work has been done in the frame of the Slovak scientific project VEGA 1/0045/10 2010-2011 of Slovak Ministry of Education. Authors would like thank to Actel University Program for a donation of 10 Actel Fusion FPGA evaluation boards, which enabled us to confirm the TERO principle using this large number of FPGA boards.

References

1. Bochard, N., Bernard, F., Fischer, V.: Observing the randomness in RO-based TRNG. In: International Conference on Reconfigurable Computing and FPGAs, Cancun, Quintana Roo, Mexico, December 9-11, pp. 237–242. IEEE Computer Society, Los Alamitos (2009)
2. Bucci, M., Giancane, L., Luzzi, R., Varanonuovo, M., Trifilett, A.: A Novel Concept for Stateless Random Bit Generators in Cryptographic Applications. In: 2006 IEEE International Symposium of Circuits and Systems - ISCAS 2006, Island of Kos, Greece, May 21-24, pp. 317–320. IEEE Computer Society, Los Alamitos (2006)
3. Dichtl, M., Golić, J.: High-Speed True Number Generation with Logic Gates Only. In: Paillier, P., Verbauwhede, I. (eds.) CHES 2007. LNCS, vol. 4727, pp. 45–62. Springer, Heidelberg (2007)
4. Dichtl, M., Meyer, B., Seuschek, H.: SPICE Simulation of a "Provably Secure" True Random Number Generator (2008), http://eprint.iacr.org/2008/403.pdf
5. Epstein, M., Hars, L., Krasinski, R., Rosner, M., Zheng, H.: Design and Implementation of a True Random Number Generator Based on Digital Circuit Artifacts. In: Walter, C.D., Koç, Ç.K., Paar, C. (eds.) CHES 2003. LNCS, vol. 2779, pp. 152–165. Springer, Heidelberg (2003)
6. Fischer, V., Drutarovsky, M.: True Random Number Generator Embedded in Reconfigurable Hardware. In: Kaliski Jr., B.S., Koç, Ç.K., Paar, C. (eds.) CHES 2002. LNCS, vol. 2523, pp. 415–430. Springer, Heidelberg (2003)
7. Golic, J.: New Methods for Digital Generation and Postprocessing of Random Data. IEEE Transactions on Computers 55(10), 1217–1229 (2006)
8. Kacprzak, T.: Analysis of Oscillatory Metastable Operation of an R-S Flip-Flop. IEEE Journal of Solid-State Circuits 23(1), 260–266 (1988)
9. Koç, Ç.K. (ed.): Cryptographic Engineering. Springer, Heidelberg (2009)
10. Killmann, W., Schindler, W.: Functionality classes and evaluation methodology for true (physical) random number generators, Version 3.1 (September 2001), http://www.bsi.bund.de/zertifiz/zert/interpr/trngk31e.pdf
11. Linear Technology: LT Spice IV, http://www.linear.com/designtools/software/
12. Markettos, A.T., Moore, S.W.: The Frequency Injection Attack on Ring-Oscillator-Based True Random Number Generators. In: Clavier, C., Gaj, K. (eds.) CHES 2009. LNCS, vol. 5747, pp. 317–331. Springer, Heidelberg (2009)
13. Menezes, J., Oorschot, P., Vanstone, S.: Handbook of Applied Cryptography. CRC Press, New York (1997), http://www.cacr.math.uwaterloo.ca/hac/
14. Rabaey, J.M., Chandrakasan, A., Nilolic, B.: Digital Integrated Circuits, 2nd edn. Prentice-Hall, Englewood Cliffs (February 2010), http://bwrc.eecs.berkeley.edu/IcBook
15. Reyneri, L., Corso, D., Sacco, B.: Oscillatory Metastability in Homogenous and Inhomogeneous Flip-Flops. IEEE Journal of Solid-State Circuits 25(1), 254–264 (1990)

16. Rukhin, A., Soto, J., Nechvatal, J., Smid, M., Barker, E., Leigh, S., Levenson, M., Vangel, M., Banks, D., Heckert, A., Dray, J., Vo, S.: A Statistical Test Suite For Random And Pseudorandom Number Generators For Cryptographic Applications, NIST Special Publication 800-22 rev1a (April 2010), http://csrc.nist.gov/groups/ST/toolkit/rng/
17. Schellekens, D., Preneel, B., Verbauwhede, I.: FPGA Vendor Agnostic True Random Number Generator. In: Proceedings of the 16th International Conference on Field Programmable Logic and Applications (FPL), Madrid, Spain, August 28-30, pp. 1–6. IEEE Computer Society, Los Alamitos (2006)
18. Schindler, W., Killmann, W.: Evaluation Criteria for True (Physical) Random Number Generators Used in Cryptographic Applications. In: Kaliski Jr., B.S., Koç, Ç.K., Paar, C. (eds.) CHES 2002. LNCS, vol. 2523, pp. 431–449. Springer, Heidelberg (2003)
19. Sunar, B.: Response to Dichtl's Criticism (March 2008), http://ece.wpi.edu/~sunar/preprints/comment.pdf
20. Sunar, B., Martin, W.J., Stinson, D.R.: A Provably Secure True Random Number Generator with Built-in Tolerance to Active Attacks. IEEE Transactions on Computers 56(1), 109–119 (2007)
21. Tokunaga, C., Blaauw, D., Mudge, T.: True Random Number Generator With a Metastability-Based Quality Control. IEEE Journal of Solid-State Circuits, 78–85 (January 2008)
22. Vasyltsov, I., Hambardzumyan, E., Kim, Y.S., Karpinskyy, B.: Fast Digital TRNG Based on Metastable Ring Oscillator. In: Oswald, E., Rohatgi, P. (eds.) CHES 2008. LNCS, vol. 5154, pp. 164–180. Springer, Heidelberg (2008)
23. Wold, K., Tan, C.H.: Analysis and Enhancement of Random Number Generator in FPGA Based on Oscillator Rings. International Journal of Reconfigurable Computing 2009, 1–8 (June 2009), http://www.hindawi.com/journals/ijrc/2009/501672.html
24. Wollinger, T., Guajardo, J., Paar, C.: Security on FPGAs: State-of-the-art implementations and attacks. ACM Transactions on Embedded Computing Systems (TECS), 534–574 (2004)
25. Xilinx: Spartan-3E Starter Kit, http://www.xilinx.com/products/devkits/HW-SPAR3E-SK-US-G.htm

The Glitch PUF: A New Delay-PUF Architecture Exploiting Glitch Shapes

Daisuke Suzuki[1,2] and Koichi Shimizu[1]

[1] Information Technology R&D Center, Mitsubishi Electric Corporation
{Suzuki.Daisuke@bx,Shimizu.Koichi@ea}
.MitsubishiElectric.co.jp
[2] Graduate School of Environmental and Information Sciences,
Yokohama National University

Abstract. In this paper we propose a new Delay-PUF architecture that is expected to solve the current problem of Delay-PUF that it is easy to predict the relation between delay information and generated information. Our architecture exploits glitches that behave non-linearly from delay variation between gates and the characteristic of pulse propagation of each gate. We call this architecture Glitch PUF. In this paper, we present a concrete structure of Glitch PUF. We then show the evaluation results on the randomness and statistical properties of Glitch PUF. In addition, we present a simple scheme to evaluate Delay-PUFs by simulation at the design stage. We show the consistency of the evaluation results for real chips and those by simulation for Glitch PUF.

1 Introduction

1.1 Background

High-level security needs such as in financial transactions and Digital Rights Management (DRM) have widened the use of security chips as represented by smart cards and Trusted Platform Modules (TPMs). Security chips provide not only a variety of cryptographic functions but tampering countermeasures, which are mechanisms to protect sensitive information stored within the chips from physical attacks. Examples of tampering countermeasures include mounting sensors or mesh shielding within a chip.

Physical Unclonable Function (PUF) [1] is a technique in relation to tampering countermeasures which has been attracting wider attention in recent years. PUFs are designed to return responses to given challenges according to physical characteristics that are innately possessed by each artificial object such as an LSI. It is arguably difficult to clone an artificial object from the fact that its characteristics originate from manufacturing variation.

With the help of Fuzzy Extractor [2], which is a technique to extract stable secret information from noisy characteristics, it is even possible to generate device unique keys that are difficult to copy. The key information is resistant to analysis that directly reads data inside a chip by breaking it open, because the information does not need storing in nonvolatile memory to be reproducible.

S. Mangard and F.-X. Standaert (Eds.): CHES 2010, LNCS 6225, pp. 366–382, 2010.

PUFs are also advantageous in that they are feasible on general-purpose LSI such as FPGA and ASIC. There are many active research works on methods of PUF realization and generation of device unique keys [4–12].

SRAM-PUF is recognized as one of the most feasible and secure PUFs thus far because there have already been implementations of error correcting codes and universal hash functions optimized for it, which are needed for Fuzzy Extractor. It is, however, difficult to evaluate the information entropy and error rate of SRAM-PUF on ASIC chips before production because it, as well as Butterfly-PUF [8], exploits the metastable state of memory cells on power activation, where only a behavior model is available for the characteristics of the cells. The error rate is particularly changeable according to the process. In fact, Ref. [13] reports a much higher rate of error than that reported by the proposers [5], which implies the possibility that the error rate of SRAM-PUF changes on different target devices. On the other hand, the evaluation is possible on devices that are available before production such as FPGA.

As for Delay-PUF, security issues have been reported. It is shown that a machine learning attack can predict challenge-response pairs after a decent number of pairs are collected by self-evaluation [6]. Furthermore, although there have been proposed countermeasures such as Feed-Forward Arbiter PUF, which installs non-linear operations, and XOR-PUF, which is comprised of multiple Arbiter PUFs, it is shown that machine learning attacks are still applicable to those [14]. These issues originate from the simplicity of the circuit structure of Delay-PUF.

At the same time, however, Delay-PUF is advantageous in that delay information utilized by it has affinity with logic simulation, which is performed at the design stage. That enables to evaluate the amount of information of the PUF at an earlier stage of design process. At least, chip vendors must possess information about the delay variation since they need to embed the delay information in a cell library when they develop it.

On the other hand, Statistical Static Timing Analysis (SSTA), which is a design method for variation, has been intensely studied [15] now that increasing manufacturing variation prevents performance improvement as the process miniaturizes. SSTA is adopted by standard Electronic Design Automation (EDA) tools as PrimeTime. It is anticipated from these facts that logic circuit designers will be able to access information about delay variation in a near future.

1.2   Our Contributions

We propose a new Delay-PUF architecture, which is expected to solve the easiness of predicting the relation between delay information and generated information. The proposed architecture exploits glitch waveforms that behave non-linearly from delay variation between gates. We thus call this architecture *Glitch PUF*.

In this paper, we show a concrete structure of Glitch PUF. We also present the results of the evaluation on randomness and statistical properties of Glitch PUF performed on FPGA.

As the second contribution, we present a simple scheme to evaluate the characteristics of Glitch PUF with simulation at the design stage. We show the consistency between the results using the scheme, and those for real devices.

2   Simulating Behavior of Delay-PUFs

In this section, we discuss a concrete methodology to evaluate randomness and statistical properties of Delay-PUFs by simulation. The goal of this simulation is to evaluate the randomness and error rate of a Delay-PUF at its design stage. The reason that Delay-PUF circuits of the same design behave differently on each individual chip is that transistors hold characteristic variation (variation of threshold voltages V_{th}, to be concrete). The occurrence of errors even on the same chip results from the change of operating environment (static/dynamic IR-drop, change of temperature, etc.). By attributing these factors to the variation of gate delays, we attempt to realize the evaluation of the randomness and error rate.

The evaluation flow, shown in Fig. 1, is basically the same as an ordinary circuit design and timing evaluation. It is different in that delay variation is reflected before simulation.

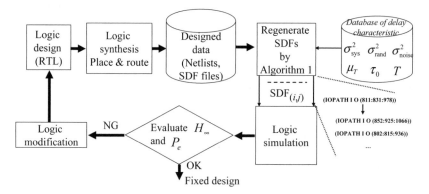

Fig. 1. PUF evaluation flow by simulation

A Standard Delay Format (SDF) [16] in Fig. 1 is a file that defines representative delays for a target device, and used for delay analysis for a circuit. It is thus possible to evaluate the operating delays of the device under corner conditions using the delay values that corresponds to several operating conditions of the circuit. However, it is not possible to evaluate PUFs with SDFs as they are because the delay values are fixed while PUFs assume delay variation.

In order to reflect individual difference and environmental change, we perform simulation with a large number of SDFs varied from the original SDF according to the previously-extracted characteristics and distributions of process, supply voltage and temperature (PVT) of a device.

Process variation is generally classified into systematic variation and random variation [17]. Systematic variation is correlated with location in a wafer or a chip. It is represented by the performance difference between chips such as the speed grade of FPGA. Random variation is not related with spatial location of transistors. It is known to result from the fluctuation of the concentration of impurities. Environmental change is parameterized representatively by voltage and temperature. These parameters are evaluated for TEG chips on a startup of LSI fabrication.

On the contrary, the information about PVT variation on FPGA is not disclosed. We hence try to extract the parameters by observing the delays in a chain of inverters under various layouts and environments as in [18] on 16 FPGAs. The parameters are as follows.

· Systematic delay variation between chips: σ_{sys}^2
· Random delay variation within chips: σ_{rand}^2
· Environmental random delay variation such as from dynamic IR-drop: σ_{noise}^2
· Average fraction of designed delays under 0 °C: τ_0
· Delay temperature coefficient: μ_T

The following assumptions are made to calculate each parameter from measured delays.

(1) Systematic delay variation within chips can be ignored.
(2) The distributions are normal (with variances σ_{sys}^2, σ_{rand}^2, and σ_{noise}^2).
(3) σ_{rand}^2, σ_{noise}^2, and μ_T are constant for all chips.

Note here that all the parameters are represented as fractions of designed delay values in a SDF. It is then possible to simulate individual difference and environmental change based on the delays that EDA tools output reflecting gate depths, numbers of fanouts and layout difference.

Designed delay values $d_1, \cdots, d_{\text{MAX}_{\text{NodeNum}}}$ for each node defined in a SDF are changed by the following formula (Algorithm 1), where sampling r from a distribution $N(0, \sigma^2)$ is denoted as $r \leftarrow N(0, \sigma^2)$.

3   Glitch PUFs

In this section, we explain the architecture of the proposed PUF, which exploits glitch waveforms. It is thus called Glitch PUF.

3.1   Basic Idea

We consider to simulate the behavior of a device at early design stages according to the characteristics of the device. The goal of this simulation is to estimate the amount of information of a PUF, especially its lower limit, without need to evaluate a large volume of real LSIs. As stated earlier, it is delay information that is most compatible with simulation at earlier stages of all characteristics.

Algorithm 1. Regeneration of SDFs with Individual and Environmental Difference

Setting: · $\text{MAX}_{\text{NodeNum}}$ nodes are included in the original SDF file.
　　　　· $\text{MAX}_{\text{ChipNum}}$ chips are simulated.
　　　　· The response data for each chip is regenerated $\text{MAX}_{\text{RepNum}}$ times to evaluate the error rate.
　　　　· T °C is the operating temperature.
Input: $T, (d_1, \cdots, d_{\text{MAX}_{\text{NodeNum}}})$
Output: $\text{SDF}_{(i,j)}, 0 \leq i \leq \text{MAX}_{\text{ChipNum}}, 0 \leq j \leq \text{MAX}_{\text{RepNum}}$.

```
 1: for i = 1 to MAX_ChipNum do
 2:     r_sys ← N(0, σ²_sys)
 3:     for j = 1 to MAX_RecNum do
 4:         for k = 1 to MAX_NodeNum do
 5:             r_rand ← N(0, σ²_rand)
 6:             r_noise ← N(0, σ²_noise)
 7:             d'_k := ((1+μ_T · T)(τ_0 + r_sys + r_rand) + r_noise) · d_k
 8:         end for
 9:         WriteSDF_(i,j)(d'_1, · · · , d'_MAX_NodeNum)
10:     end for
11: end for
```

Then it is probable to be able to evaluate the amount of information of a PUF within the current scheme of logic circuit design, if the delay variation among devices is closely connected to the change of reponse of the PUF.

We consider possible behavioral difference of the same logic circuits with different delays. Examples of such behaviors are shown in Fig. 2. Fig. 2-(a) shows a basic one that there is a time difference between output changes from an input change. The time from an input change to an output change is, however, difficult to be exploited as a device unique key because it depends not only on the variation of gate delays inherent from manufacturing but also largely on the operating temperature and voltage. On reflection, Arbiter PUF by Lin et al. exploits the time difference between two signals to ensure stability against such environmental changes as shown in Fig. 2-(b). But it is known about Arbiter PUF that it is possible to predict challenge-response pairs (CRPs) by machine learning if a sufficient number of CRPs have been collected. Feed-Forward Arbiter PUF, which introduces non-linear operations as a countermeasure, is also possible to be attacked by machine learning [14].

The examples thus far describes behaviors from a delay difference for very simple logic circuits. From here, we discuss more general circuits such as Fig. 2-(c) that perform AND and XOR to multiple inputs. In this kind of circuit there occurs a transient state of an output signal called a glitch from the delay difference between input signals, unless a particular condition holds. In the example in Fig. 2-(c), in case that input signals x_1, x_2, x_3 all change from 0 to 1, there is a convex glitch at the XOR output from the difference of transition time between x_1 and x_2. The glitch propagates to the AND output only if the transition of

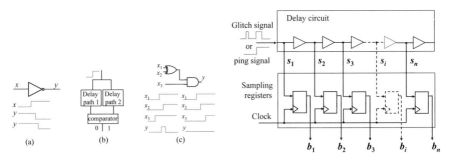

Fig. 2. Circuit behaviors for different delay values

Fig. 3. Sampling circuit

the input signal x_3 reaches the AND gate faster than the glitch. If not on the contrary, the glitch does not propagate to the output, in which case the output remains unchanged. Furthermore, even if x_3 is faster, the PATHPULSE [16] of the AND gate might prevent a narrow glitch from propagating. Notice here that for sufficiently wide glitches, their shapes are determined by the relation of delays, not by the absolute values of the delays. It is then anticipated that shapes of glitches are kept unchanged if the operating environment changes, like Arbiter PUF.

Now our attention is focused on glitches, which can take various shapes according to the order relation of delays between the inputs of each gate that consists in a logic circuit. We discuss a means of applying them to construct a PUF from here.

3.2   Overall Sequence

First of all, we describe a whole sequence of Glitch PUF. Glitch PUF consists largely of the three steps below.

STEP 1 Data input to a random logic
STEP 2 Acquisition of glitch waveforms at the output
STEP 3 Conversion of the waveforms into response bits

In the example of Fig. 2-(c), STEP 1 means causing changes to the inputs x_1, x_2, x_3. The accompanying glitch waveform at the output y is acquired as an n-bit data. The data is then transformed into a one-bit data r according to its shape. Changing the input in STEP 1 and iterating the steps, a bit sequence R is acquired as a response data R.

Each subsection below describes the details of necessary techniques to realize the operations from STEP 1 to STEP 3.

3.3   Acquisition of Glitch Waveforms

As described in Section 2.1, we attempt to construct a PUF using glitches, which can take various shapes according to delay variation. The issue here is how to

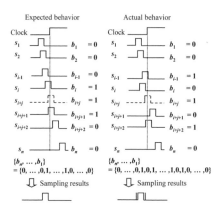

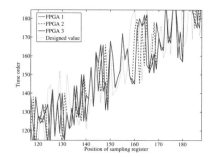

Fig. 4. Error in glitch acquisition   **Fig. 5.** Time order of sampling registers

accurately acquire the shape of a pulse signal that happens only for a tiny period of time. At the same time, the acquisition process must be realized as a digital circuit for the goal of this paper.

The phase-shift method is one of the general solutions to the issue, where multiple clocks with different phases are prepared to sample a tiny pulse. The sampling accuracy is heightend as the number of different phased clocks is increased. The method is, however, not practical since it needs too many clock lines. Particularly in FPGA, there is a limited number of global clock lines with little jitter, from several to several dozen. Although the number of clock lines can be reduced by introducing time-division, the speed of acquisition decreases then. We hence adopt a method where the target data is shifted by a tiny period of time, and sampled by the same clock. Fig. 3 shows a sampling circuit of this method. We call this operation *glitch acquisition* hereafter. Note here that since the sampling period must be short for acquisition accuracy, it is required to reduce the delay deference between signals loaded into flip-flops (FFs) as much as possible by decreasing the buffer depths between signals, or using other elements with shorter delays in Fig. 3. As the delay difference decreases, though, clock jitter between FFs and wire or gate delay cannot be ignored. In this case, the orders of sampling positions of FFs, and their actual delays do not match as in Fig. 4, thereby permuting the time order of the sampled data. It probably becomes impossible to recover the glitch shape accurately. This problem also occurs for the aforementioned phase-shifting of clock.

We thus introduce a preprocessing shown below before performing glitch acquisition in order to determine the time order of the sampling result. A signal wire is added to generate a simple rising edge, called a *ping signal* hereafter. First, a ping signal is input to the sampling circuit in Fig. 3 and sampled. Then, each FF latches 1 or 0 if the ping signal reaches it before or after the clock, respectively. On the other hand, there is a variable delay circuit in the clock line,

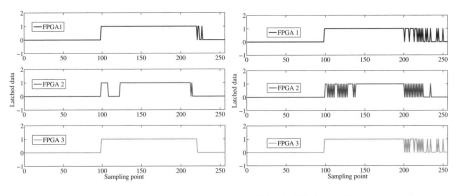

Fig. 6. With jitter correction   **Fig. 7.** Without jitter correction

with which the above process is performed multiple times with different clock delays. The number of latching 1 is thereby counted for each FF. Lastly, the time order of the samplings is determined according to the order of the numbers. An example time order is shown in Fig. 5. The glitch shape is recovered after sorting the sampled data according to this order.

From here, the above-mentioned preprocessing is called *jitter correction* and the sorting according to the result of the jitter correction is called *sorting*. The results of glitch acquisition with and without jitter correction are shown in Fig. 6 and Fig. 7 respectively.

3.4   Conversion to Response

One-bit value is converted from the glitch waveforms acquired as digital data by the above sampling method. We describe a means to convert the parity of the number of rising edges in a glitch waveform. The parity can be detected with differential and addition processing implemented in hardware or software. This detection is called *shape judgment*.

It is still difficult to acquire the time order of FFs completely even though the aforementioned preprocessing is performed. When the sorted time order is different from the actual order of the circuit, a glitch waveform like Fig. 4 is acquired, where there are narrow pulses near the edges. This kind of phenomenon is an unstable behavior occurring when the clock delays between FFs are close. As a result, the shape judgment can be different for each trial of glitch acquisition. In addition, an extremely narrow pulse can cause the same phenomenon due to the PATHPULSE mentioned previously.

We decide to perform a processing as shown in Fig. 8 to ignore pulses with widths less than a threshold w. This processing is called *filtering* hereafter.

3.5   Reliability Enhancement

In order to improve the error rate of the shape judgment, we utilize the feature that the same processing can be performed multiple times. That is, the final

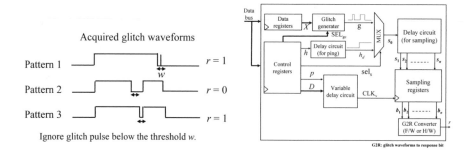

Fig. 8. Conversion from glitch waveform to response with filtering

Fig. 9. Whole structure of Glitch PUF

output is determined by majority after multiple trials of shape judgment are performed for the same input transition. In particular, when the initial key is generated such as by Generate (Gen) in Fuzzy Extractor [2], only such inputs as acquire the same outputs for M iterations are used. In this case, performing shape judgment M times for each of N input changes generates an N-bit mask as well as an N-bit response. Value 1 of a mask bit means that the bit position is used for key generation, and vice versa. The mask is output as part of Helper Data.

In methods such as suggested in Ref. [12], a probability distribution of errors is output as Helper Data when performing Gen and soft-decision is performed in the process of Reproduce (Rep) [2]. The amount of information is preserved although the error distribution is made public. On the contrary, the amount of information is reduced by the masking process. In Glitch PUF, bit positions with high error rates are determined for each chip and the number of them is small. We therefore choose to mask them and reduce the size of the necessary error correcting circuit.

3.6   The Architecture

Fig. 9 illustrates the circuit architecture of Glitch PUF. Glitch PUF consists mainly of control registers, data registers, a glitch generator, a sampling circuit, and two kinds of delay circuit. The control registers in Fig. 9 store the control parameters listed below.

- $\mathrm{SEL_{gc}}$: Selection signal to glitch generator ($\log u$ bits)
- $\mathrm{sel_s}$: Input selection signal to sampling circuit (1 bit)
- h: Ping signal (1 bit)
- D: Delay specifier signal to variable delay circuit ($q + q'$ bits)
- p: Trigger signal (1 bit)

The data registers store the data X (u-bit) input into the glitch generator. The glitch generator is comprised of a combinational circuit and a v-1 selector, where the circuit performs $Y = f(X)$ defined for X in Fig. 10 and the selector selects

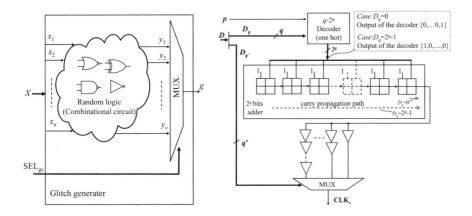

Fig. 10. Glitch generater   **Fig. 11.** Variable delay circuit

one bit out of v bits of Y according to a selection signal $\mathrm{SEL_{gc}}$. The delay circuit for the ping signal consists of a buffer chain, thereby outputting h_d, a delayed signal of h. The depth of the chain is determined at the design stage by simulation evaluating the occurrence timing of the glitch signals generated in the glitch generator. We describe the details in the next section. The sampling circuit, as discussed in Section 3.3, is constructed of a buffer chain and FFs shown in Fig. 3. It is noteworthy that when implementing a Glitch PUF on FPGA, the sampling resolution can be heightened by utilizing carry paths of addition circuits as a buffer chain rather than implementing the chain in Look-up tables (LUTs). The variable delay circuit is also implemented with carry paths to minimize the step size of delay by which the delay can be varied. At the same time, however, the range of the variable delay must be wider than that for sampling. Hence, the circuit requires large area if it is all constructed of carry paths. The issue can be avoided by combining delay circuits on carry paths and LUTs as in Fig. 11, thereby keeping a wide variable range and high resolution at the same time.

In this paper, the process until the sampling is implemented on hardware and the rest is on firmware in order to observe the behavior of the generated glitches.

3.7 Adjustment of the Design Parameter

In order to realize efficient glitch acquisition, parameters need to be adjusted for each circuit in Fig. 9 at the design stage. The parameters are as the following.

· n: The number of FFs in the sampling circuit
· $\mathrm{delay_s}$: The delay value of the buffers inserted between the signals of the sampling circuit
· $\mathrm{range_s}$: The sampling range of the sampling circuit
· $\mathrm{range_g}$: The range of glitch occurrence in the glitch generator

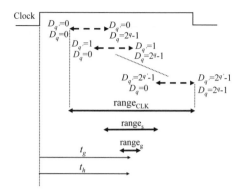

Fig. 12. Timing conditions for the designing parameters

· range$_{\text{CLK}}$: The variation range of the variable delay circuit
· t_g: The time of the central value in the range of glitch occurrence
· t_h: The time of rising of the delayed ping signal h_d

Fig. 12 illustrates the relationship between each parameter. n and delay$_s$ are related to the sampling range and resolution. range$_s$ is about $n \cdot$ delay$_s$. To acquire glitch shapes, it must hold that range$_g$ < range$_s$ < range$_{\text{CLK}}$. We discuss a design procedure to realize the relation in what follows.

At first, the time range range$_g$ where glitches can occur at the input of the sampling circuit, and the occurrence timing t_g are estimated by logic simulation with delay information. That is performed when the logic of the glitch generator is fixed. Second, n and delay$_s$ are determined such that the equation range$_g$ < range$_s$ holds. Here, the sampling resolution can be heightend by selecting a cell from the target platform as a buffer such that delay$_s$ is as small as possible. As a result, it is n that actually needs determining. The implementation in this paper sets range$_s$ to be more than twice range$_g$ as a design margin. Also, the buffer depth in the variable delay circuit is configured such that range$_{\text{CLK}}$ is around twice range$_s$.

Next, the buffer depth in the delay circuit for the ping signal is configured such that $t_g \simeq t_h$ for the previously determined t_g. The configuration is not only used to calculate the time order of the sampling results. It is also to determine the delay specifier signal Dg in the variable delay circuit, which is for acquiring glitch shapes. The details are as follows. Sampling the ping signal is performed with D being incremented from its minimum to maximum values. Dg is set to be the D when a certain FF around the center of the sampling results latches 1 for the first time. Glitch waveforms can thereby be sampled around the center of the sampling range since the glitch occurrence range is configured to be in the same range as the ping delay. However, range$_s$ and range$_{\text{CLK}}$ must have margins, like twice/half something as stated earlier, because it is generally difficult to accurately configure the delay between signals.

Table 1. Specification of experimental environment

Implementation environment	
Logic synthesis, P&R	Xilinx Platform Studio 10.1.03i
Simulator	NC-Verilog
Target FPGA	Xilinx XC3S400A-4FTG256C (16 boards)
$MAX_{ChipNum}$	1000
MAX_{RepNum}	1000
Number of bits of generated responses N	2048
Specification of Glitch PUF	
Glitch generator	AES SubByte (composite field)
Design parameter (n, u, v, q, q')	(256, 8, 8, 8, 2)
Filtering parameter w	2
Reliability Enhancement parameter M	10
SLICEs used	891/3584 (Whole SoC 3186/3584)
Operating frequency	50 MHz

Table 2. Delay characteristics

Systematic delay variation $\sigma_{sys}^2(\%^2)$	2.5037
Random delay variation $\sigma_{rand}^2(\%^2)$	5.3091
Environmental random delay variation $\sigma_{noise}^2(\%^2)$	0.0310
Average fraction of designed delays τ_0 (%)	56.9727
Delay temperature coefficient μ_T (%/°C)	0.1401

4 Experimental Results

This section presents the results of evaluating the randomness and statistical properties for an experimental implementation of Glitch PUF on FPGA. The experiment is performed for Spartan-3A evaluation boards by AVNET that mount one XC3S400A-4FT256, a Xilinx FPGA. 16 boards are used. We build a System-on-a-Chip (SoC), mounting on an FPGA a soft-macro CPU (MicroBlaze), UARTs, and memory controllers as well as a PUF circuit. Table 1 shows the specification of the implementation environment. The process after the shape acquisition mentioned above is performed by firmware on a MicroBlaze mounted on the same FPGA as the circuits are implemented on. AES SubBytes is used as the glitch generator since its logic is complex and circuit structure is well studied by designers of cryptographic hardware. The sampling circuit is implemented with 256 FFs. The variable delay circuit consists of a 256-bit addition circuit, and an LUT-based buffer chain whose depth of LUT can be 4, 8, 12, and 16 with a 4-1 selector.

We perform a basic experiment on delay characteristics described in Section 2 in order to extract the parameters for the same FPGA boards needed for the simulation of PUF. The parameters are shown in Table 2. The parameters are calculated as fractions of corresponding worst-case delays defined in SDF

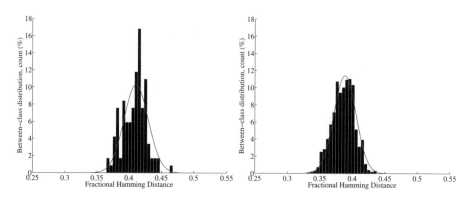

Fig. 13. Hamming distances of response data between FPGAs (actual chips)

Fig. 14. Hamming distances of response data between FPGAs (simulation)

Table 3. Change of information amount against change of variations

Simulation results $(24°\text{C})$		
Variation	$H_\infty(R)$	$H_\infty(R|\text{Mask})$
$(\sigma_{\text{sys}}^2, \sigma_{\text{rand}}^2)$	1,043	702
$((2 \cdot \sigma_{\text{sys}})^2, \sigma_{\text{rand}}^2)$	1,068	721
$((\frac{1}{2} \cdot \sigma_{\text{sys}})^2, \sigma_{\text{rand}}^2)$	1,046	703
$(\sigma_{\text{sys}}^2, (2 \cdot \sigma_{\text{rand}})^2)$	1,167	811
$(\sigma_{\text{sys}}^2, (\frac{1}{2} \cdot \sigma_{\text{rand}})^2)$	828	546
FPGAs	1,156	649

generated by an EDA tool after the layout. Using Table 2, we regenerate a number of SDFs according to Algorithm 1, and evaluate the randomness and statistical properties by simulation. The results are also shown in this section.

4.1 Inter-chip Variation

Fig. 13 is a histogram of Hamming distances between PUF outputs of two different FPGAs out of 16 (i.e. 120 combinations). This evaluation is a general way to show how different responses of chips are. The result shows that about 850 bits out of 2048 bits are different between chips. Fig. 14 shows the result of the same evaluation by simulation. Comparing Figs. 13 and 14, it is seen that the simulation is able to evaluate the randomness of responses generated by real chips.

Table 3 shows the min-entropy of the probability distribution of the response acquired through the experiment, and of the distribution of the masked response described in Section 3.5. Masking reduces the min-entropy of the original distribution since it discards the response bits that are judged to be unstable at Gen. However, the reduction rate is only around 30% for the experimental Glitch

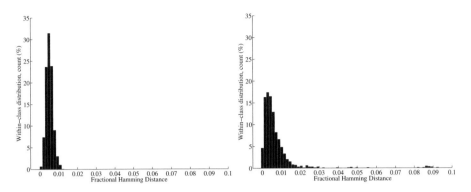

Fig. 15. Hamming distances between response data for the same FPGA (actual chips)

Fig. 16. Hamming distances between response data for the same FPGA (simulation)

PUF, indicating that the amount of information is still sufficient if the unstable bits for each chip are discarded. Table 3 also proves that the min-entropy loss from masking can be evaluated at the design stage by simulation.

It is also noteworthy in Table 3 that there is an interesting relationship between $H_\infty(R)$, σ_{sys}^2, and σ_{rand}^2. $H_\infty(R)$ changes sensitively to the change of σ_{rand}^2 while it does not to the change of σ_{sys}^2. In other words, Glitch PUF ensures the amount of information of the response data from the random delay variation within chips rather than from the systematic delay variation between chips. The result implies that Glitch PUF can guarantee the randomness for chips on the same wafer as well as for chips on different wafers, or from different lots. In addition, it is arguably possible that the randomness of Glitch PUF further improves for latest devices because random delay variation tends to enlarge as the process miniaturizes.

4.2 Intra-chip Variation

It is desirable for a PUF to stably generate the same response for the same FPGA. Fig. 15 plots the Hamming distances between the initial response and 100 responses measured thereafter, all of which are masked. The measurements are at normal temperature and voltage ($24°$C,$1.20V$), and averaged for 16 FPGAs. The mean error rate is around 1.3%. Next, as Fig. 17 shows, the maximum error rate in the rated temperature range is about 6.6 % at $80°$C, which is less than the half of 15% assumed in Ref. [11]. In addition it is shown by Figs. 16 and 17 that the change of the error rate with respect to the temperature can be evaluated by simulation with high accuracy.

Next, we discuss the effect of masking. Fig. 18 is a histogram of error rates for each bit of the 2048-bit response data at normal temperature and voltage. It is clear that there are many bits with error rates higher than 0.1 when masking is not performed. At the same time, most of these bits are removed by masking, which correctly treats the response data. Masking is effective for Glitch PUF

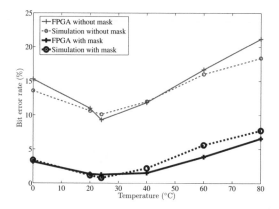

Fig. 17. Change of error rate accompanying temperature change

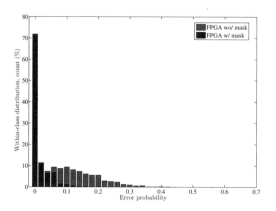

Fig. 18. Reduction of error rate by masking

since it greatly lowers the error rate, although the min-entropy decreases by about 30 % as stated earlier.

4.3   Secrecy Rate

In Ref. [19], the secrecy rate is defined to be $I(R, R')$, the mutual information of the response data at Gen R, and at Rep R'. The average secrecy rate of Glitch PUF is calculated to be 0.26 per bit from the aforementioned experimental results. This is about 1/3 that of SRAM-PUF. At the moment, Glitch PUF is thus inferior to SRAM-PUF in the efficiency to generate secret information. However, delay-PUFs including Glitch PUF have the advantage of being able to evaluate the secrecy rate by logic simulation, the same way as explained in the previous sections. On the contrary SRAM-PUF requires analog simulation like SPICE to evaluate the same thing.

5   Conclusions

In this paper, we propose Glitch PUF, which is a new Delay-PUF for the purpose of remedying a problem about the previous Delay-PUFs, that is, the easiness to predict the relationship between delay information and generated information. Glitch waveforms occurring at the output of a random logic behave non-linearly from delay variation between gates and the characteristic of pulse propagation of each gate. We present a method to accurately acquire the waveforms and to convert them into response bits. In addition, we prove the feasibility of Glitch PUF by evaluation of the randomness and statistical properties for an FPGA. Furthermore, we show a simple scheme to evaluate the characteristics of Glitch PUF. Using the scheme, we confirm the consistency of the evaluation results for real chips and those by simulation.

Lastly, we list open problems below.

· Construct a glitch generator that brings high amount of information and low error rate .
· Model machine learning attacks to Glitch PUF.
· Construct an error correcting code and universal hash function suitable for Glitch PUF.
· Model logic simulation for voltage change and aging degradation through acceleration test, and evaluate them on real chips

References

1. Pappu, R.S.: Physical One-way Functions. Ph.D. Thesis, M.I.T. (2001), http://pubs.media.mit.edu/pubs/papers/01.03.pappuphd.powf.pdf
2. Dodis, Y., Reyzin, M., Smith, A.: Fuzzy Extractors: How to Generate Strong Keys from Biometrics and Other Noisy Data. In: Cachin, C., Camenisch, J.L. (eds.) EUROCRYPT 2004. LNCS, vol. 3027, pp. 523–540. Springer, Heidelberg (2004)
3. Tuyls, P., Batina, L.: RFID-Tags for Anti-Counterfeiting. In: Pointcheval, D. (ed.) CT-RSA 2006. LNCS, vol. 3860, pp. 115–131. Springer, Heidelberg (2006)
4. Gassend, B., Clarke, D., van Dijk, M., Devadas, S.: Silicon Physical Random Functions. In: Proc., of the 9th ACM Conference on Computer and Communications Security (CCS 2002), pp. 148–160 (2002)
5. Guajardo, J., Kumar, S.S., Schrijen, G.J., Tuyls, P.: FPGA Intrinsic PUFs and Their Use for IP Protection. In: Paillier, P., Verbauwhede, I. (eds.) CHES 2007. LNCS, vol. 4727, pp. 63–80. Springer, Heidelberg (2007)
6. Lee, J.W., Lim, D., Gassend, B., Suh, G.E., van Dijk, M., Devadas, S.: A Technique to Build a Secret Key in Integrated Circuits for Identification and Authentication Applications. In: Proc. of the IEEE VLSI Circuits Symposium, pp. 176–179 (2004)
7. Suh, G.E., Devadas, S.: Physical Unclonable Functions for Device Authentication and Secret Key Generation. In: Proc. of the 44th annual Design Automation Conference (DAC 2007), pp. 9–14 (2007)

8. Kumar, S.S., Guajardo, J., Maes, R., Šchrijen, G.J., Tuyls, P.: Extended Abstract: The Butterfly PUF: Protecting IP on every FPGA. In: Proc. of the IEEE International Workshop on Hardware-Oriented Security and Trust 2008 (HOST 2008), pp. 67–70 (2008)
9. Majzoobi, M., Koushanfar, F., Potkonjak, M.: Lightweight secure PUFs. In: Proc. of the IEEE/ACM International Conference on Computer-Aided Design (ICCAD 2008), pp. 670–673 (2008)
10. Bosch, C., Guajardo, J., Sadeghi, A.R., Shokrollahi, J., Tuyls, P.: Efficient Helper Data Key Extractor on FPGAs. In: Oswald, E., Rohatgi, P. (eds.) CHES 2008. LNCS, vol. 5154, pp. 181–197. Springer, Heidelberg (2008)
11. Maes, R., Tuyls, P., Verbauwhede, I.: Low-Overhead Implementation of a Soft Decision Helper Data Algorithm for SRAM PUFs. In: Proc. of the 2009 IEEE International Symposium on Information Theory (ISIT 2009), pp. 2101–2105 (2009)
12. Maes, R., Tuyls, P., Verbauwhede, I.: A Soft Decision Helper Data Algorithm for SRAM PUFs. In: Clavier, C., Gaj, K. (eds.) CHES 2009. LNCS, vol. 5747, pp. 332–347. Springer, Heidelberg (2009)
13. Chopra, J., Colopy, R.L.: SRAM Characteristics as Physical Unclonable Functions. Worcester Polytechnic Institute Electric Project Collection (2009), http://www.wpi.edu/Pubs/E-project/Available/E-project-031709-141338/
14. Rührmair, U., Sölter, J., Sehnke, F.: On the Foundations of Physical Unclonable Functions. Cryptology ePrint Archive, 2009/277 (2009)
15. Najm, F.N., Menezes, N.: Statistical Timing Analysis Based on a Timing Yield Model. In: Proc. of the 41st annual Design Automation Conference (DAC 2004), pp. 460–465 (2004)
16. Standard Delay Format Specification version 3.0 (1995), http://www.eda.org/sdf/sdf_3.0.pdf
17. Hiramoto, T., Takeuchi, K., Nisida, A.: Variability of Characterisics in Scaled MOSFETs. J. IEICE 92(6), 416–426 (2009)
18. Berkelaar, M.: Statistical Delay Calculation, a Linear Time Method. In: Proc. of the International Workshop on Timing Analysis (TAU'97), pp. 15–24 (1997)
19. Ignatenko, T., Schrijen, G.J., Skoric, B., Tuyls, P., Willems, F.: Estimating the Secrecy-Rate of Physical Unclonable Functions with the Context-Tree Weighting Method. In: Proc. of the 2006 IEEE International Symposium on Information Theory (ISIT 2006), pp. 499–503 (2006)

Garbled Circuits for Leakage-Resilience: Hardware Implementation and Evaluation of One-Time Programs*

Kimmo Järvinen[1], Vladimir Kolesnikov[2],
Ahmad-Reza Sadeghi[3], and Thomas Schneider[3]

[1] Dep. of Information and Comp. Science, Aalto University, Finland
kimmo.jarvinen@tkk.fi**
[2] Alcatel-Lucent Bell Laboratories, Murray Hill, NJ 07974, USA
kolesnikov@research.bell-labs.com
[3] Horst Görtz Institute for IT-Security, Ruhr-University Bochum, Germany
{ahmad.sadeghi,thomas.schneider}@trust.rub.de***

Abstract. The power of side-channel leakage attacks on cryptographic implementations is evident. Today's practical defenses are typically attack-specific countermeasures against certain classes of side-channel attacks. The demand for a more general solution has given rise to the recent theoretical research that aims to build provably leakage-resilient cryptography. This direction is, however, very new and still largely lacks practitioners' evaluation with regard to both efficiency and practical security. A recent approach, One-Time Programs (OTPs), proposes using Yao's Garbled Circuit (GC) and very simple tamper-proof hardware to securely implement oblivious transfer, to guarantee leakage resilience.

Our main contributions are (i) a generic architecture for using GC/OTP modularly, and (ii) hardware implementation and efficiency analysis of GC/OTP evaluation. We implemented two FPGA-based prototypes: a system-on-a-programmable-chip with access to hardware crypto accelerator (suitable for smartcards and future smartphones), and a stand-alone hardware implementation (suitable for ASIC design). We chose AES as a representative complex function for implementation and measurements. As a result of this work, we are able to understand, evaluate and improve the practicality of employing GC/OTP as a leakage-resistance approach.

1 Introduction

Side-channels and protection. For over a decade, we saw the power and elegance of side-channel attacks on a variety of cryptographic implementations and devices. These attacks refute the assumption of "black-box" execution of cryptographic algorithms, allow the adversary to obtain (unintended) internal state

---

* This is a short version of the paper. The full version is available [7].
** Supported by EU FP7 project CACE.
*** Supported by EU FP7 projects CACE and UNIQUE, and ECRYPT II.

S. Mangard and F.-X. Standaert (Eds.): CHES 2010, LNCS 6225, pp. 383–397, 2010.

information, such as secret keys, and consequently cause catastrophic failures of the systems. Often the attacks are on the device in attacker's possession, and exploit physical side-channels such as power consumption or emitted radiation. Hence, from the hardware perspective, security has been viewed as more than the algorithmic soundness in the black-box execution model (see, e.g., [28]).

Today's practical countermeasures typically address known vulnerabilities, and thus target not *all*, but specific classes of side-channel attacks. The desire for a complete solution motivated the recent burst of theoretical research in *leakage-resilient cryptography*, the area that aims to define security models and frameworks that capture leakage aspects of computation or/and memory. Information leakage is typically modeled by allowing the adversary learn (partial) memory or execution states. The exact information given to the adversary is specified by the (adversarily chosen) leakage function. Then, the assumption on the function (today, usually the bound on the output length) directly translates into a physical assumption on the underlying device and the adversary. Proving security against such an adversary implies security in the real-world with the real device, subject to corresponding assumption (see [17] for a survey on this strand of research). We note that many of the results of this new line of research (i.e., leakage assumptions and leakage-resilient constructions), although clearly stated, have not yet been evaluated by practitioners and side-channel community.[1]

Secure Function Evaluation in hardware and leakage-resilience. Efficient Secure Function Evaluation (SFE) in an untrusted environment is a long-standing objective of modern cryptography. Informally, the goal of two-party SFE is to let two mutually mistrusting (polynomially-bounded) parties compute an *arbitrary* function on their private inputs without revealing any information about the inputs, beyond the output of the function. SFE has a variety of applications, particularly in settings with strong security and privacy demands. Deployment of SFE was very limited and believed expensive until recent improvements in algorithms, code generation, computing platforms and networks.

As advocated in numerous prior works [13,10,18,8], *Garbled Circuit* (GC) [29] is often the most efficient (and thus viable) SFE technique in the two-party setting. As we argue in the full version [7], the emerging fully homomorphic encryption schemes [3] are unlikely to approach the efficiency of GC.

Because of the execution flow of the GC solution (one party can non-interactively evaluate the function once the inputs have been fixed), the security guarantees of SFE are well-suited to prevent *all* side-channel leakage. Indeed, even GC evaluation in the open reveals no information other than the output. Clearly, it is safe to let the adversary see (as it turns out, even to modify) the entire GC evaluation process. The inputs-related stage of GC can also be made non-interactive with appropriate hardware such as Trusted Platform Modules (TPM) [6]. Goldwasser et al. [4] observed that very simple hardware is sufficient, one that, hopefully, can be manufactured tamper-resistant at low cost.

---

[1] Indeed, ongoing work of [21] investigates the practical applicability and usability of theoretical leakage models and the constructions proven secure therein.

They propose to use One-Time Programs (OTP), a combination of GC and above hardware, for leakage-resilient computation. Indeed, one of our goals is to evaluate today's performance of OTP in hardware.

Our objectives. Practical efficiency of SFE and leakage-resilient computing is critical. Indeed, in most settings, the technology can only be adopted if its cost impact is acceptably low. In this work, we pursue the following two objectives.

First, we aim to mark this (practical efficiency) boundary by considering *hardware-accelerated* state-of-the-art GC evaluation, and optimizing it for embedded systems. Implementing SFE (at least partially) in hardware promises to significantly improve computation speed and reduce power consumption. We evaluate costs, benefits and trade-offs of hardware support for GC evaluation.

Second, we use our GC hardware-accelerator to implement OTP and evaluate the efficiency of this provably leakage-resilient protection. The envisioned applications for OTPs mentioned in [4] are complex functionalities such as one-time proofs, E-cash, or extreme software protection. We make a first step towards estimating the costs of such complex OTP applications by implementing OTP evaluation of the AES function. We chose AES as it is relatively complex and allows comparison with existing (heuristic) leakage protection.

1.1 Our Contributions and Outline

In line with our objectives stated above, we implement a variant of OTP with state-of-the-art GC optimizations discussed in §2. As an algorithmic contribution, we propose an efficiency improvement for OTPs with multiple outputs in §3.1. Further, we describe a generic architecture for using GC/OTP in a modular way to protect against arbitrary side-channel attacks in §3.2.

In our implementation, we present a hardware architecture (§4.1) and optimizations (§4.2) for efficient evaluation of GC/OTP on memory-constrained embedded systems. We measure performance of GC/OTP evaluation of AES on our two FPGA-based implementations in §4.3: a system-on-a-programmable-chip with access to SHA-256 hardware accelerator (representative for smartcards and future smartphones) and a stand-alone hardware implementation. With optimization, GC evaluation of AES on our implementations requires about 1.3 s and 0.15 s, respectively. This shows that *provable leakage-resilience* via GC/OTP comes at a relatively high cost (but its use might still be justified in high-security applications): an *unprotected* implementation of AES in hardware runs in 0.15 μs, and requires 2.6 times smaller area than OTP-based solution. We note that the chip area for hardware-accelerated GC/OTP evaluation is independent of the evaluated function. As AES is a representative complex function, we believe that our results, in particular our performance measurements, will serve as reference point for estimating GC/OTP runtimes of a variety of other functions.

1.2 Related Work

Efficient implementations of Garbled Circuits (GC). We believe that our results are the first hardware implementation of garbled circuits (GC) and one-time programs (OTP) evaluation. While several implementations and measurements of GC exist in software, e.g., [13,18], the hardware setting presents different challenges. Our work allows to compare the approaches and estimate the resulting performance gain (our hardware implementation is faster than the software implementation of [18] by a factor of 10-17). Hardware implementation of GC *generation* in a cost-effective tamper-proof token with constant memory was shown in [8]. Our work is complementary, and our hardware accelerator for GC evaluation can be combined with the token of [8], or software frameworks.

One-Time Programs (OTP). The combination of GC with non-interactive oblivious transfer in the semi-honest setting was proposed in [6]. For malicious evaluator, OTP were introduced in [4] using minimal hardware assumptions. Subsequently, [5] showed how to build non-interactive secure computation unconditionally secure against malicious parties using tamper-proof hardware tokens. We extend and implement OTPs in hardware. Our extension is in the computational model with Random Oracles (RO), secure against malicious evaluator, and more efficient than the constructions of [4,5].

Protecting AES against side-channel attacks. We summarize current techniques for leakage-protecting AES. We stress that our implementation is provably leakage-free, but comes at a computational cost which we evaluate in this work.

A large amount of research has been done on countermeasures against side-channel attacks, e.g., to protect against power analysis attacks [9], the power consumption needs to be made independent of the underlying secrets by either randomizing the power consumption or making it constant [14]. Randomizing is done with masking, i.e., by adding random values. A variety of masking schemes for both algorithmic and circuit level have been proposed for AES, e.g., [1]. For constant power consumption one can use gates whose power consumption is independent of input values, e.g., with dynamic differential (dual-rail) logic (see, e.g., [25]). Countermeasures against power analysis have significant area overheads ranging from factor 1.5 to 5 [23]. Protecting implementations against other side-channel attacks or even fault attacks needs additional overhead. For instance, fault attack countermeasures require error detection techniques such as proposed in [20]. None of these countermeasures provides complete security. Indeed, countermeasures providing protection against simpler attacks have been shown to be useless against more powerful attacks, such as, template attacks [2] and higher-order differential power analysis [15].

2   Preliminaries

Garbled Circuits (GC). Yao's GC approach [29] allows sender S with private input y and receiver R with private input x, to securely evaluate a boolean circuit C on (x, y) without revealing any information other than the result $z = C(x, y)$ of the evaluation. We summarize the idea of Yao's GC protocol next.

The circuit *constructor* S creates a *garbled circuit* $\widetilde{C}$ from the circuit C: for each wire W_i of C, he randomly chooses two garblings $\widetilde{w}_i^0, \widetilde{w}_i^1$, where $\widetilde{w}_i^j$ is the *garbled value* of W_i's value j. (Note: $\widetilde{w}_i^j$ does not reveal j.) Further, for each gate G_i, $\mathcal{S}$ creates a *garbled table* $\widetilde{T}_i$ with the following property: given a set of garbled values of G_i's inputs, $\widetilde{T}_i$ allows to recover the garbled value of the corresponding G_i's output, but nothing else. $\mathcal{S}$ sends these garbled tables, called *garbled circuit* $\widetilde{C}$, to *evaluator* (receiver $\mathcal{R}$). Additionally, $\mathcal{R}$ (obliviously) obtains the *garbled inputs* $\widetilde{w}_i$ corresponding to the inputs of both parties: the garbled inputs $\widetilde{y}$ corresponding to the inputs y of $\mathcal{S}$ are sent directly: $\widetilde{y}_i = \widetilde{y}_i^{y_i}$. For each of $\mathcal{R}$'s inputs x_i, both parties run a 1-out-of-2 Oblivious Transfer (OT) protocol (e.g., [16]), where $\mathcal{S}$ inputs $\widetilde{x}_i^0, \widetilde{x}_i^1$ and $\mathcal{R}$ inputs x_i. The OT protocol ensures that $\mathcal{R}$ receives only the garbled value corresponding to his input bit, i.e., $\widetilde{x}_i = \widetilde{x}_i^{x_i}$, while $\mathcal{S}$ learns nothing about x_i. Now, $\mathcal{R}$ evaluates the garbled circuit $\widetilde{C}$ on the garbled inputs to obtain the *garbled output* $\widetilde{z}$ by evaluating $\widetilde{C}$ gate by gate, using the garbled tables $\widetilde{T}_i$. Finally, $\mathcal{R}$ determines the plain value z corresponding to the obtained garbled output value using an output translation table sent by $\mathcal{S}$.

Yao's garbled circuit protocol is provably secure ([12]) when both parties are semi-honest (i.e., follow the protocol but may try to infer information about the other party's inputs from the messages seen). We stress that each GC can be evaluated only once, i.e., a new GC $\widetilde{C}$ must be used for each invocation.

Improved Garbled Circuits. We use the improved GC construction of [18], summarized next. Each garbled value $\widetilde{w}_i = \langle k_i, \pi_i \rangle$ consists of a t-bit key k_i and a permutation bit π_i, where t denotes the symmetric security parameter. XOR gates are evaluated "for free", i.e., no garbled table and negligible computation, by computing the bitwise XOR of their garbled values [10]. For each non-XOR gate with d inputs the garbled table $\widetilde{T}_i$ consists of $2^d - 1$ entries of size $t + 1$ bits each; the evaluation of a garbled non-XOR gate requires one invocation of SHA-256 [18]. At the high level, the keys k_i of the non-XOR gate's garbled inputs are used to obtain the corresponding garbled output value by decrypting the garbled table entry which is indexed by the input permutation bits π_i. We present the detailed description of the construction in the full version [7].

Non-Interactive Garbled Circuits and One-Time Programs. The round complexity of Yao's GC protocol is exactly that of the underlying OT protocol. In [6] the authors suggested to extend the Trusted Platform Module (TPM) [26] for implementing non-interactive OT, resulting in a non-interactive version of Yao's protocol. Subsequently, One-Time Programs (OTP) were introduced in [4]. This approach considers malicious receivers and can be viewed simply as Yao's Garbled Circuit (GC), where the oblivious transfer (OT) function calls are implemented with One-Time Memory (OTM) tokens. An OTM token T_i is a simple tamper-proof hardware, which allows a single query of one of the two stored garbled values $\widetilde{x}_i^0, \widetilde{x}_i^1$ ([4] suggests using a tamper-proof one-time-settable bit b_i which is set as soon as the OTM is queried). Further, OTM-based GC execution can be non-interactive, in the sense that the sender can send the GC and corresponding OTMs to the receiver, who will be able to execute one instance of SFE on any input of his choice.[2] Finally, GC (and hence also OTP) is inherently a one-time execution object (generalizable to k-time execution by repetition).

A subtle issue in this context, noted and addressed in [4], is the following. Previous GC-based solutions were either in the semi-honest model, or used interaction during protocol execution, which precluded the receiver $\mathcal{R}$ from choosing his input adaptively, based on given (and even partially evaluated) garbled circuit. This possibility of adaptively chosen inputs results in possible real attacks by a malicious $\mathcal{R}$ in the non-interactive setting. The solution of [4] is to mask (each) output bit z_j of the function with a random bit m_j, equal to XOR of (additional) random bits $m_{i,j}$ contributed by *each* of the input OTMs T_i, i.e., $m_j = m_{1,j} \oplus m_{2,j} \oplus \ldots$ and $z'_j = z_j \oplus m_j$. The real-world adversary does not learn the output of the function before he had queried all OTMs with his inputs.

3   Extending and Using One-Time Programs

In §3.1 we present an extension of the OTP construction of [4], which results in improved performance for multiple outputs. Additionally we make several observations about uses, security guarantees and applicability of OTP, and present a generic architecture for using OTPs for leakage-resilient computation in §3.2.

3.1   Extending One-Time Programs

As mentioned in the previous section, the solution in [4] seems to require each OTM token to additionally store a string of the size of the output. We propose a practical performance improvement to the technique proposed in [4], which is beneficial for OTP evaluation of functions with many output bits. In our solution each OTM token (additionally) stores a random string r_i of length of the security parameter t. Consequently, our construction results in smaller OTMs when the function to be evaluated has more outputs than the size of the security parameter t. As a trade off, our security proof utilizes Random Oracles (RO), as we do not immediately see how to avoid their use and have OTM size independent of the number of outputs. We discuss RO, its uses and security guarantees in the full version [7].

Our main idea is to insert a "hold off" gate into each output wire W_j which can only be evaluated once *all* input OTMs had been queried, thus preventing malicious $\mathcal{R}$ from choosing his input adaptively. It can be implemented by requiring a call to a hash function H (modeled as a Random Oracle) with inputs which include data from all OTMs. To implement this, we secret-share a random value $r \in_R \{0,1\}^t$ over all OTMs for the inputs. That is, OTM T_i additionally stores a share r_i (released to $\mathcal{R}$ with $\tilde{x}_i$ upon the query), where $r = \bigoplus_i r_i$. Receiver $\mathcal{R}$ is able to recover r if and only if he queried all OTMs.

---

[2] If needed, the function can be fully hidden by evaluating a universal circuit [27,11,19] which simulates the function.

Fig. 1(b) depicts this contruction: Our version of OTM T_i, in addition to the two OT secrets $\widetilde{x}_i^0, \widetilde{x}_i^1$ and the tamper-proof bit b_i, contains a random share $r_i \in_R \{0,1\}^t$ which is released together with $\widetilde{x}_i^{x_i}$ once T_i is queried with input bit x_i. The GC is constructed as usual (e.g., as described in §2), with the following exception. On each output wire W_j with garbled outputs $\widetilde{z}_j^0, \widetilde{z}_j^1$, we append a one-input, one-output OT-commit gate G_j, with no garbled table. We set the output wire secrets of G_j to $\widehat{z}_j^0 = H(\widetilde{z}_j^0 \| r), \widehat{z}_j^1 = H(\widetilde{z}_j^1 \| r)$. To enable $\mathcal{R}$ to compute the wire output non-interactively, GC also specifies that $\widehat{z}_j^b$ corresponds to b.

We note that a full formal construction is readily obtained from the above description. Also note that a malicious $\mathcal{R}$ is unable to complete the evaluation of any wire of GC until all the OTMs have been queried, and his input has been specified in full. Further, he is not able to lie about the result of the computation, since he can only compute one of the two values $\widetilde{z}_j^0, \widetilde{z}_j^1$. Demonstration of knowledge of $\widetilde{z}_j^i$ serves as a proof for the corresponding output value.

Theorem 1. *The above protocol is secure against a semi-honest sender $\mathcal{S}$, who generates the OTM tokens and the garbled circuit, and malicious receiver $\mathcal{R}$, in the Random Oracle model.*

Proof. The proof of Theorem 1 is given in the full version [7]. □

3.2 Using One-Time Programs for Leakage Protection

Most of today's countermeasures to side-channel attacks are specific to *known* attacks. Protecting hardware implementations (e.g., of AES) usually proceeds as follows (e.g., see [1]). First, the inputs are hidden, typically by applying a random mask (this requires trusted operation, and often the corresponding assumption is introduced). Afterwards, the computation is performed on the masked data. To allow this, the functionality needs to be adapted (e.g., using amended AES S-boxes). Finally, the mask is taken off to reveal the output of the computation.

We use a similar approach with similar assumptions (cf. Fig. 1(a)) to provably protect *arbitrary* functionalities against *all* attacks, both known and unknown:

1. The private data x provided by $\mathcal{R}$ is masked in a trusted environment MASK. The masked data $\widetilde{x}$ does not reveal any information about the private data, but still allows to compute on it.
2. The computation on the masked data is performed in an untrusted environment where the adversary is able to arbitrarily interfere (passively and actively) with the computation. To compute on the masked data, the evaluation algorithm EVAL needs a specially masked version of the program $\widetilde{P}$. Additional masked inputs $\widetilde{y}$ of $\mathcal{S}$ that are independent of $\mathcal{R}$'s inputs can be provided as well. The result of EVAL is the masked output $\widetilde{z}$.
3. Finally, $\widetilde{z}$ is unmasked into the plain output z. The procedure UNMASK allows to verify that $\widetilde{z}$ was computed correctly, i.e., no tampering happened in the EVAL phase in which case UNMASK outputs the failure symbol $\perp$. For correctness of this verification, UNMASK is executed in a trusted environment where the adversary can observe but not modify the computation.

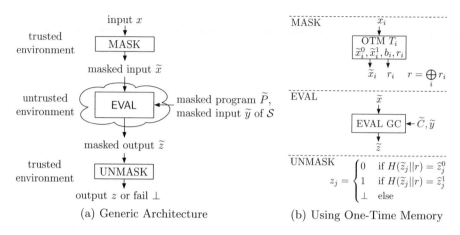

Fig. 1. Evaluating a Functionality Without Leakage

More specifically, the masked program $\widetilde{P}$ is a garbled circuit $\widetilde{C}$, masked values $\widetilde{x}, \widetilde{y}, \widetilde{z}$ are garbled values and the algorithms MASK, EVAL and UNMASK can be implemented as described next.

MASK: Masking the input data x of receiver $\mathcal{R}$ is performed by mapping each bit x_i to its corresponding garbled value $\widetilde{x}_i$, i.e., to one of two garblings $\widetilde{x}_i^0, \widetilde{x}_i^1$. This can be provided externally (e.g., by interaction with a party on the network). We concentrate on on-board *non-interactive* masking which requires certain hardware assumptions (see below). The following can be directly used as a (non-interactive) MASK procedure:

- OTMs [4]: For small functionalities, we favor the very cheap One-Time Memory (OTM), as this seems to carry the weakest assumptions (cf. §2). However, as OTMs can be used only once, a fresh OTM must be provided for each input bit of the evaluated functionality. For practical applications, OTMs (together with their garbled circuits) could be implemented for example on tamper-proof USB tokens for easy distribution.
- TPM [6]: Trusted Platform Modules (TPM) are low-cost tamper-proof cryptographic chips embedded in many of today's PCs [26]. TPM masking based on the non-interactive Oblivious Transfer (OT) protocol of [6] requires the (slightly extended) TPM to perform asymmetric cryptographic operations in form of a count-limited private key whose number of usages is restricted by the TPM chip. An interactive protocol allows re-initialization for future non-interactive OTs instead of shipping new hardware.
- Smartcard: In our preferred solution for larger functionalities, masking could be performed by a tamper-proof smartcard. The smartcard would keep a secure monotonic counter to ensure a single query per input bit. Another advantage of this approach is that the same smartcard can be used to generate GC as well, thus eliminating GC transfer over the network as done in [8]. Further, the smartcard can be naturally used for multiple OTP evaluations.

For non-interactive masking, the hardware that masks the inputs must be trusted and the entire input must be specified before anything about the output z is revealed to prevent adaptive input selection as discussed in §2 and §3.1.

EVAL: The main technical contribution of this paper, the implementation of EVAL (of the masked program $\widetilde{P}$ on masked inputs $\widetilde{x}$ and $\widetilde{y}$) in embedded systems is presented in detail in §4. Here we note that $\widetilde{P}$ and $\widetilde{y}$ (masked input of $\mathcal{S}$) can be generated offline by the semi-honest sender $\mathcal{S}$ and provided to EVAL by convenient means (e.g., via a data network or a storage medium). This is the scenario advocated in [4]; one of its advantages is that generation of $\widetilde{P}$ does not leak to EVAL. Alternatively, $\widetilde{P}$ and $\widetilde{y}$ could be generated "on-the-fly" using a cheap simple constant-memory stateless and tamper-proof token as shown in [8]. We reiterate that the masked program $\widetilde{P}$ can be evaluated exactly once.

UNMASK: Finally, the masked output $\widetilde{z}$ is checked for correctness and non-interactively decoded by $\mathcal{R}$ into the plain output z as follows (cf. §3.1 and Fig. 1(b)). For each output wire, the masked program $\widetilde{P}$ specifies the correspondence $\widehat{z}_j \rightarrow z_j$ in form of the two valid hash values $\widehat{z}_j^0$ and $\widehat{z}_j^1$. Even if EVAL is executed in a completely untrusted environment (e.g., processed on untrusted HW), its correct execution can be verified efficiently: when $H(\widetilde{z}_j || r)$ is neither $\widehat{z}_j^0$ nor $\widehat{z}_j^1$ the garbled output $\widetilde{z}_j$ is invalid and UNMASK outputs the failure symbol $\perp$. The reason for this verifiability property of GC is that a valid garbled output $\widetilde{z}_j$ can only be obtained by correctly evaluating the GC but cannot be guessed.

4   Efficient Evaluation of Garbled Circuits in Hardware

In this section we describe how GCs (and hence also OTPs) can be efficiently evaluated on embedded systems and memory-constrained devices. We first describe the HW architecture in §4.1. Then we present important compile-time optimizations and show their effectiveness in §4.2. Finally, we discuss technical details of our prototype implementation and timings in §4.3.

We stress that our designs and optimizations are generic. However, for concreteness and for meaningful comparison (e.g., with prior SW SFE of AES [18]), we take SFE of the AES function as our example for timings and other measurements. For AES evaluation, sender $\mathcal{S}$ provides AES key k as input y, and receiver $\mathcal{R}$ provides a plaintext block m as input x. $\mathcal{R}$ obtains the ciphertext c as output z, where $c = \text{AES}(k, m)$. Recall, during GC evaluation (EVAL), both key and message are masked (garbled) and hence cannot be leaked.

4.1   Architecture for Evaluating Garbled Circuits in Hardware

We describe our architecture for efficient evaluation of GC on memory-constrained devices, i.e., having a small amount of slow memory only.

To minimize overhead, we choose key length $t = 127$; with a permutation bit, garbled values are thus 128 bits long (cf. §2). In the following we assume that

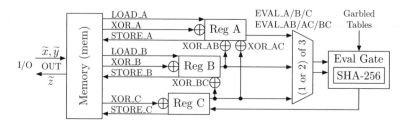

Fig. 2. Architecture for GC Evaluation (EVAL) on Memory-Constrained Devices

memory cells and registers store 128 bit garbled values. This can be mapped to standard hardware architectures by using multiple elements in parallel.

Fig. 2 shows a conceptual high-level overview of our architecture described next. At the high-level, EVAL, the process of evaluating GC, on our architecture consists of the following steps (cf. §3.2). First, the garbled input values $\widetilde{x}, \widetilde{y}$ are stored in memory using the I/O interface. Then, GC gates are evaluated, using registers A, B, and C to cache the garbled inputs and outputs of a single garbled gate. Finally, garbled output value $\widetilde{z}$ is output over the I/O interface.

As memory access is expensive (cf. §4.3) we optimize code to re-use values already in registers. Our instructions are one-address, i.e., each instruction consists of an operator and up to one memory address. Each of our instructions has length 32 bits: 5 bits to encode one of 18 instructions (described next) and 27 bits to encode an address in memory.

LOAD/STORE: Registers can be loaded from memory using instructions LOAD_A and LOAD_B. Register C cannot be loaded as it will hold the output of evaluated non-XOR gates (see below). Values in registers can be stored back into memory using STORE_A, STORE_B, and STORE_C respectively.

XOR: We evaluate XOR gates [10] as follows. XOR_A `addr` computes $A \leftarrow A \oplus$ mem[addr]. Similarly, the other one-operand XOR operations (XOR1) XOR_B and XOR_C xor the value from memory with the value in the respective register. To compute XOR gates where both inputs are already in registers (XOR2), the instruction XOR_AB computes $A \leftarrow A \oplus B$. Similarly, XOR_AC computes $A \leftarrow A \oplus C$ and XOR_BC computes $B \leftarrow B \oplus C$.

EVAL: Non-XOR gates [18] are evaluated with the Eval Gate block that contains a hardware accelerator for SHA-256 (cf. §2 for details). The garbled inputs are in one (EVAL1) or two registers (EVAL2), and the result is stored in register C. The respective instructions for 1-input gates are EVAL_A, EVAL_B, EVAL_C and for 2-input gates EVAL_AB, EVAL_AC, EVAL_BC. The required garbled table entry is read from memory.

I/O: The garbled inputs are always stored at the first $|x|+|y|$ memory addresses. The garbled outputs are obtained from memory with OUT instructions.

The full version [7] shows the sequence of instructions for an example circuit.

4.2   Compile-Time Optimizations for Memory-Constrained Devices

In this section, we summarize compile-time optimizations to improve performance of GC evaluation (EVAL) on our hardware architecture. We aim to reduce the size of GC (by minimizing the number of non-XOR gates), the size of the program (number of instructions), the number of memory accesses and memory size for storing intermediate garbled values. For concreteness, we use AES as representative functionality for the optimizations and performance measurements, but our techniques are generic.

Baseline [18]) Our baseline is the AES circuit/code of [18], already optimized for a small number of non-XOR gates. Their circuit consists of 11, 286 two-input non-XOR gates; thus, its GC has size ≈ 529 kByte. Without considering any caching strategies, this results in 113, 054 instructions, hence the program size is $113,054 \cdot 32$ bit ≈ 442 kByte, and the total amount of memory needed for EVAL is $34,136 \cdot 128$ bit ≈ 533 kByte.

We summarize our best optimization next and refer for a detailed description and intermediate optimization steps to the full version [7].

Optimized) First, we replace XNOR gates with an XOR gates and propagate the inverted output into the successor gates. For AES, this optimization results in the elimination of 4, 086 XNOR gates and reduces the size of AES GC to ≈ 338 kByte (improvement of 36%). Additionally, we re-use values already in registers to reduce the number of LOADs. Values in registers are saved to memory only if needed later. Finally, we randomly consider several orders of evaluation, and select the most efficient one for EVAL.

Result. Using our optimizations we were able to substantially decrease the memory footprint of EVAL. As shown in Table 1, our optimized circuit strongly improves over the circuit of [18] as follows. The size of the AES program P is only $73,583 \cdot 32$ bit ≈ 287 kByte (improvement of 34.9%). The amount of intermediate memory is $17,315 \cdot 128$ bit ≈ 271 kByte (improvement of 49.3%) and the number of memory accesses (read and write) is reduced by ≈ 35%.

Table 1. Optimized AES Circuits (Sizes in kB)

Circuit	Garbled Circuit $\tilde{C}$				Program P		Memory for GC Evaluation			
	non-XOR	1-input	XOR	Size	Instr.	Size	Read	Write	Entries	Size
Baseline [18]	11,286	0	22,594	529	113,054	442	67,760	33,880	34,136	533
Optimized	7,200	40	26,680	338	73,583	287	42,853	22,650	17,315	271

4.3   Implementation

We have designed two prototype implementations of the architecture of §4.1 – one for a System-on-a-Programmable-Chip with a hardware accelerator for SHA (reflecting smartcard and future smartphone architectures) and another for a stand-alone unit (reflecting a custom-made GC accelerator in hardware). Both prototype implementations are evaluated on an Altera/Terasic DE1 FPGA board

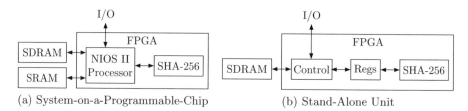

Fig. 3. Architectures for Hardware-Assisted GC Evaluation

comprising an Altera Cyclone II EP2C20F484C7 FPGA and 512kB SRAM and 8MB SDRAM running at 50 MHz (cf. full version [7] for details on our prototype environment) and are functionally equivalent: they take the same inputs (program P, garbled circuit $\widetilde{C}$, and garbled inputs $\widetilde{x}, \widetilde{y}$) and return the same garbled outputs $\widetilde{z}$; the only differences are the methods used in the implementation. The interfaces (I/Os in Fig. 3) allow the host to write to and read from the SDRAM. In the beginning, the host writes the inputs to the SDRAM and, in the end, the outputs are written into specific addresses from which the host retrieves them.

System-on-a-Programmable-Chip (SOPC). Our first implementation is a system-on-a-programmable-chip (SOPC) consisting of a processor with access to a hardware accelerator for SHA-256, which speeds up the most computational burden of the GC evaluation. This is a representative architecture for next generation smartphones or smartcards such as the STMicroelectronics ST33F1M smartcard which includes a 32-bit RISC processor, cryptographic peripherals, and memory comparable to our prototype system [22].

The architecture of our implementation is shown in Fig. 3(a). It consists of a NIOS II/e 32-bit softcore RISC processor (the smallest variation of NIOS II), a custom-made SHA-256 unit, the SRAM, and the SDRAM. The entire process is run in the NIOS II processor which uses the SHA-256 unit for accelerating gate evaluations. The SHA-256 unit is connected to the Avalon bus of the NIOS II as a peripheral component and it computes the hash for a 512-bit message in 66 clock cycles (excluding interfacing delays). The NIOS II program is stored in SRAM whereas OTP related data is stored in SDRAM.

Stand-Alone Unit. Our second implementation is a stand-alone unit consisting of a custom-made control state machine, registers (A, B, C), a custom-made SHA-256 unit, and SDRAM. This architecture could be used to design an Application Specific IC (ASIC) as high-speed hardware accelerator for GC evaluation. The architecture is depicted in Fig. 3(b).

When the host has written the inputs to the SDRAM, the stand-alone unit executes the program. The state machine parses the program and reads/writes data from/to SDRAM to/from the registers or evaluates the non-XOR gates using the SHA-256 unit according to the instructions (see §4.1 for details).

Area. The area requirements of our implementations are shown in Table 2. Both fit into the low-cost Cyclone II FPGA with 18,754 logic cells (LC), each containing a 4-to-1-bit look-up table (LUT) and a flip-flop (FF), and 52 4096-bit

Table 2. Areas of the Prototypes for GC Evaluation on an Altera Cyclone II FPGA

Design	LC	FF	M4K
SOPC	7501	4364	22
NIOS II	1104	493	4
SHA-256	2918	2300	8
Stand-Alone Unit	6252	3274	8
SHA-256	3161	2300	8
AES (unprotected)	2418	431	0

Table 3. Timings for Instructions on Prototypes (clock cycles, average)

Instruction	*SOPC*	*Stand-Alone Unit*
LOAD	291.43	87.63
XOR1	395.30	87.65
XOR2	252.00	1.00
STORE	242.00	27.15
EVAL1	1,282.30	109.95
EVAL2	1,491.68	135.05
OUT	581.48	135.09

embedded memory blocks (M4K). SHA-256 is the largest and most significant block in both prototypes. Table 2 also shows the area for an iterative implementation of AES-128 with no countermeasures against side-channel attacks on the same FPGA. Compared to an unprotected implementation, countermeasures against power analysis have area overheads ranging from factor of 1.5 to 5 [23] as discussed in §1.2; therefore, the area overheads of OTP evaluation are comparable with other side-channel countermeasures.

Timings. *Instructions.* The timings of instructions are summarized in Table 3. They show the average number of clock cycles required to execute an instruction excluding the latency of fetching the instruction. Gate evaluations are expensive in the SOPC implementation, although the SHA-256 computations are fast, because they involve a lot of data movement (to/from the SHA-256 unit and from the SDRAM) which is expensive. The dominating role of memory reads and writes is clear in the timings of the stand-alone implementation: the only instructions that do not require memory operations (XOR2) require only a single clock cycle and EVAL1 is faster than EVAL2 because it accesses the memory on average every other time (no access if the permutation bit is zero) compared to three times out of four (no access if both permutation bits are zeros).

AES. The timings to evaluate the optimized GCs for the AES functionality of §4.2 on our prototype implementations are given in Table 4. These timings are for GC evaluation only; i.e, they neglect the time for transferring data to/from the system because interface timings are highly technology dependent. The SHA-256 computations take an equal amount of time for both implementations as the SHA-256 unit is the same. The (major) difference in timings is caused by data movement, XORs, interface to the SHA-256 unit, etc. The runtimes for both implementations are dominated by writing and reading the SDRAM; e.g., 84.3% for the stand-alone unit and our optimized AES circuit. Hence, accelerating memory access, e.g., with burst reads and writes, is the key for further speedups.

Performance Comparison. A software implementation that evaluates the GC/ OTP of the unoptimized AES functionality (Baseline [18]) required 2 seconds on an Intel Core 2 Duo 3.0 GHz with 4GB of RAM [18]. Our optimized circuit evaluated on the stand-alone unit requires only 144 ms for the same operation

Table 4. Timings for the FPGA-based Prototypes for GC Evaluation

	System-on-a-Programmable-Chip				Stand-Alone Unit			
	Clock cycles		Timings (ms)		Clock cycles		Timings (ms)	
Circuit	SHA	Total	SHA	Total	SHA	Total	SHA	Total
Baseline [18]	744,876	94,675,402	14.898	1,893.508	744,876	11,235,118	14.898	224,702
Optimized	477,840	62,629,261	9.557	1,252.585	477,840	7,201,150	9.557	144.023

and, therefore, provides a speedup by a factor of 10.4–17.4 (taking the lack of precision into account). On the other hand, the unprotected AES implementation listed in Table 2 encrypts a message block in 10 clock cycles and runs on a maximum clock frequency of 66 MHz resulting in a timing of 0.1515 μs; hence, the GC/OTP evaluation suffers from a timing overhead factor of $\approx$ 950,000. For comparison, the timing overhead of one specific implementation with countermeasures against differential power analysis was factor of 3.88 [24].

Acknowledgements. We thank anonymous reviewers of CHES'10 for their helpful comments and co-authors of [18] for the initial AES circuit.

References

1. Akkar, M.-L., Giraud, C.: An implementation of DES and AES, secure against some attacks. In: Koç, Ç.K., Naccache, D., Paar, C. (eds.) CHES 2001. LNCS, vol. 2162, pp. 309–318. Springer, Heidelberg (2001)
2. Chari, S., Rao, J.R., Rohatgi, P.: Template attacks. In: Kaliski Jr., B.S., Koç, Ç.K., Paar, C. (eds.) CHES 2002. LNCS, vol. 2523, pp. 13–28. Springer, Heidelberg (2003)
3. Gentry, C.: Fully homomorphic encryption using ideal lattices. In: STOC'09, pp. 169–178. ACM, New York (2009)
4. Goldwasser, S., Kalai, Y.T., Rothblum, G.N.: One-time programs. In: Wagner, D. (ed.) CRYPTO 2008. LNCS, vol. 5157, pp. 39–56. Springer, Heidelberg (2008)
5. Goyal, V., Ishai, Y., Sahai, A., Venkatesan, R., Wadia, A.: Founding cryptography on tamper-proof hardware tokens. In: Micciancio, D. (ed.) TCC 2010. LNCS, vol. 5978, pp. 308–326. Springer, Heidelberg (2010)
6. Gunupudi, V., Tate, S.: Generalized non-interactive oblivious transfer using count-limited objects with applications to secure mobile agents. In: Tsudik, G. (ed.) FC 2008. LNCS, vol. 5143, pp. 98–112. Springer, Heidelberg (2008)
7. Järvinen, K., Kolesnikov, V., Sadeghi, A.-R., Schneider, T.: Garbled circuits for leakage-resilience: Hardware implementation and evaluation of one-time programs. Cryptology ePrint Archive, Report 2010/276(2010), http://eprint.iacr.org
8. Järvinen, K., Kolesnikov, V., Sadeghi, A.-R., Schneider, T.: Embedded SFE: Offloading server and network using hardware tokens. In: Sion, R. (ed.) FC 2010. LNCS, vol. 6052, pp. 207–221. Springer, Heidelberg (2010)
9. Kocher, P., Jaffe, J., Jun, B.: Differential power analysis. In: Wiener, M. (ed.) CRYPTO 1999. LNCS, vol. 1666, pp. 388–397. Springer, Heidelberg (1999)
10. Kolesnikov, V., Schneider, T.: Improved garbled circuit: Free XOR gates and applications. In: Aceto, L., Damgård, I., Goldberg, L.A., Halldórsson, M.M., Ingólfsdóttir, A., Walukiewicz, I. (eds.) ICALP 2008, Part II. LNCS, vol. 5126, pp. 486–498. Springer, Heidelberg (2008)

11. Kolesnikov, V., Schneider, T.: A practical universal circuit construction and secure evaluation of private functions. In: Tsudik, G. (ed.) FC 2008. LNCS, vol. 5143, pp. 83–97. Springer, Heidelberg (2008)
12. Lindell, Y., Pinkas, B.: A proof of Yao's protocol for secure two-party computation. Journal of Cryptology 22(2), 161–188 (2009)
13. Malkhi, D., Nisan, N., Pinkas, B., Sella, Y.: Fairplay — a secure two-party computation system. In: USENIX Security'04. USENIX Association (2004)
14. Mangard, S., Oswald, E., Popp, T.: Power Analysis Attacks: Revealing the Secrets of Smart Cards. Springer, Heidelberg (2007)
15. Messerges, T.S.: Using second-order power analysis to attack DPA resistant software. In: Paar, C., Koç, Ç.K. (eds.) CHES 2000. LNCS, vol. 1965, pp. 238–251. Springer, Heidelberg (2000)
16. Naor, M., Pinkas, B.: Efficient oblivious transfer protocols. In: SODA'01, pp. 448–457. Society for Industrial and Applied Mathematics (2001)
17. Pietrzak, K.: Provable security for physical cryptography. In: WEWORC'09 (2009), http://homepages.cwi.nl/~pietrzak/publications/Pie09b.pdf
18. Pinkas, B., Schneider, T., Smart, N.P., Williams, S.C.: Secure two-party computation is practical. In: Matsui, M. (ed.) ASIACRYPT 2009. LNCS, vol. 5912, pp. 250–267. Springer, Heidelberg (2009)
19. Sadeghi, A.-R., Schneider, T.: Generalized universal circuits for secure evaluation of private functions with application to data classification. In: ICISC 2008. LNCS, vol. 5461, pp. 336–353. Springer, Heidelberg (2008)
20. Satoh, A., Sugawara, T., Homma, N., Aoki, T.: High-performance concurrent error detection scheme for AES hardware. In: Oswald, E., Rohatgi, P. (eds.) CHES 2008. LNCS, vol. 5154, pp. 100–112. Springer, Heidelberg (2008)
21. Standaert, F.-X., Pereira, O., Yu, Y., Quisquater, J.-J., Yung, M., Oswald, E.: Leakage resilient cryptography in practice. Cryptology ePrint Archive, Report 2009/341 (2009), http://eprint.iacr.org/
22. STMicroelectronics. Smartcard MCU with 32-bit ARM SecurCore SC300 CPU and 1.25 Mbytes high-density Flash memory. Data brief (October 2008), http://www.st.com/stonline/products/literature/bd/15066/st33f1m.pdf
23. Tiri, K.: Side-channel attack pitfalls. In: DAC'07, pp. 15–20. ACM, New York (2007)
24. Tiri, K., Hwang, D., Hodjat, A., Lai, B.-C., Yang, S., Schaumont, P., Verbauwhede, I.: Prototype IC with WDDL and differential routing — DPA resistance assessment. In: Rao, J.R., Sunar, B. (eds.) CHES 2005. LNCS, vol. 3659, pp. 354–365. Springer, Heidelberg (2005)
25. Tiri, K., Verbauwhede, I.: A logic level design methodology for a secure DPA resistant ASIC or FPGA implementation. In: DATE'04, vol. 1, pp. 246–251. IEEE, Los Alamitos (2004)
26. Trusted Computing Group (TCG). TPM main specification. Technical report, TCG (May 2009), http://www.trustedcomputinggroup.org
27. Valiant, L.G.: Universal circuits (preliminary report). In: STOC'76, pp. 196–203. ACM, New York (1976)
28. Weingart, S.H.: Physical security devices for computer subsystems: A survey of attacks and defences. In: Paar, C., Koç, Ç.K. (eds.) CHES 2000. LNCS, vol. 1965, pp. 302–317. Springer, Heidelberg (2000)
29. Yao, A.C.: How to generate and exchange secrets. In: FOCS'86, pp. 162–167. IEEE, Los Alamitos (1986)

ARMADILLO: A Multi-purpose Cryptographic Primitive Dedicated to Hardware

Stéphane Badel[1], Nilay Dağtekin[1], Jorge Nakahara Jr[1,*], Khaled Ouafi[1,**],
Nicolas Reffé[2], Pouyan Sepehrdad[1], Petr Sušil[1], and Serge Vaudenay[1]

[1] EPFL, Lausanne, Switzerland
[2] Oridao, Montpellier, France
{stephane.badel,nilay.dagtekin,jorge.nakahara,pouyan.sepehrdad,petr.susil,
khaled.ouafi,serge.vaudenay}@epfl.ch, nicolas.reffe@oridao.com

Abstract. This paper describes and analyzes the security of a general-purpose cryptographic function design, with application in RFID tags and sensor networks. Based on these analyzes, we suggest minimum parameter values for the main components of this cryptographic function, called ARMADILLO. With fully serial architecture we obtain that 2923 GE could perform one compression function computation within 176 clock cycles, consuming 44 μW at 1 MHz clock frequency. This could either authenticate a peer or hash 48 bits, or encrypt 128 bits on RFID tags. A better tradeoff would use 4030 GE, 77 μW of power and 44 cycles for the same, to hash (resp. encrypt) at a rate of 1.1 Mbps (resp. 2.9 Mbps). As other tradeoffs are proposed, we show that ARMADILLO offers competitive performances for hashing relative to a fair Figure Of Merit (FOM).

1 Introduction

Cryptographic hash functions form a fundamental and pervasive cryptographic primitive, for instance, providing data integrity in digital signature schemes, and for message authentication in MACs. In particular, there are very few known hardware-dedicated hash function designs, for instance, Cellhash [6] and Subhash [5]. On the other hand, Bogdanov *et al.* [2] suggest block-cipher based hash functions for RFID tags using the PRESENT block cipher. Concerning block and stream ciphers, the most prominent developments include PRESENT [1], TEA [22], HIGHT [13], Grain [12], Trivium [4] and KATAN, KTANTAN family [3].

 We propose a cryptographic function dedicated to hardware which can be used for several cryptographic purposes.[1] Such functions rely on data-dependent bit transpositions [16]. Given a bitstring $x = x_{2k} \| \cdots \| x_1$, fixed permutations σ_0 and σ_1 over the set $\{1, 2, \ldots, 2k\}$, a bit string s, a bit $b \in \{0, 1\}$ and a permutation σ, define $x_{\sigma_s} = x$ when

---

* This work was supported by the National Competence Center in Research on Mobile Information and Communication Systems (NCCR-MICS), a center of the SNF under grant number 5005-67322.

** Supported by a grant of the Swiss National Science Foundation, 200021-119847/1.

[1] The content of this paper is subject to a pending patent by ORIDAO
http://www.oridao.com/

S. Mangard and F.-X. Standaert (Eds.): CHES 2010, LNCS 6225, pp. 398–412, 2010.

s has length zero, and, $x_{\sigma_{s\|b}} = x_{\sigma_s \circ \sigma_b}$, where x_σ is the bit string x transposed by σ, that is, $x_\sigma = x_{\sigma(2k)}\| \cdots \|x_{\sigma(1)}$. The function $(s,x) \mapsto x_{\sigma_s}$ is a data-dependent transposition of x. The function $s \mapsto \sigma_s$ can be seen as a particular case of the general semi-group homomorphism from $\{0,1\}^*$ to a group G. It was already used in the Zemor-Tillich construction [21] for $G = \mathsf{SL}_2$ and in braid group cryptography [10]. We observe that when σ_0 and σ_1 induce an expander graph on the vertex set $v = \{1,\ldots,2k\}$, then $(s,x) \mapsto x_{\sigma_s}$ has good cryptographic properties.

This paper is organized as follows: Sect. 2 describes a general-purpose cryptographic function called ARMADILLO. In Sect. 3 we analyze ARMADILLO. Sect. 4 contains design criteria for the bit permutation components of ARMADILLO. Sect. 5 suggests parameter vectors. Sect. 6 presents an updated design, called ARMADILLO2. Sect. 7 provides implementation results. Sect. 8 compares hardware implementations of ARMADILLO with other well-known hash functions.

Notations. Throughout this document, $\|$ denotes the concatenation of bitstrings, $\oplus$ denotes the bitwise XOR operation, $\bar{x}$ denotes the bitwise complement of a bitstring x; we assume the little-endian numbering of bits, such as $x = x_{2k}\| \cdots \|x_1$.

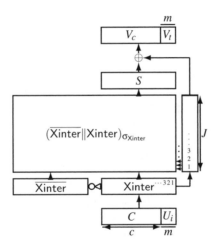

Fig. 1. The ARMADILLO function

2   The ARMADILLO Function

ARMADILLO maps an initial value C and a message block U_i to two values

$$(V_c, V_t) = \mathsf{ARMADILLO}(C, U_i)$$

By definition, C and V_c are of c bits, V_t as well as each block U_i are of m bits, a register Xinter is of $k = c + m$ bits. ARMADILLO is defined by integer parameters c, m, $J = c + m$, and two fixed permutations σ_0 and σ_1 over the set $\{1,2,\ldots,2k\}$. ARMADILLO(C,U) works as follows (see Fig. 1)

1: set $\mathsf{Xinter} = C\|U$;
2: set a $2k$-bit register $x = \overline{\mathsf{Xinter}}\|\mathsf{Xinter}$;
3: x undergoes a sequence of bit permutations, σ_0 and σ_1, which we denote by P. P maps a bitstring of k bits and a vector x of $2k$ bits into another vector of $2k$ bits. Assuming $J = k$, the output of this sequence of J bit permutations is truncated to the rightmost k bits, denoted S, by

$$S = P(\mathsf{Xinter}, x) = \mathsf{tail}_k((\overline{\mathsf{Xinter}}\|\mathsf{Xinter})_{\sigma_{\mathsf{Xinter}}})$$

4: set $V_c\|V_t$ to the value of $S \oplus \mathsf{Xinter}$.

The security is characterized by two parameters S_{offline} and S_{online}. Concretely, the best offline attack has complexity $2^{S_{\mathrm{offline}}}$, while the best online one, with practical complexity, has success probability $2^{-S_{\mathrm{online}}}$. Typically, we aim at $S_{\mathrm{offline}} \geq 80$ and $S_{\mathrm{online}} \geq 40$. However, we can only upper bound S_{offline} and S_{online}.

Application I: FIL-MAC. For challenge-response protocols (e.g. for RFID tags [17]), the objective is to have a fixed input-length MAC. Suppose that C is a secret and U is a challenge. The value V_t is the response or the authentication tag. We write

$$V_t = \mathsf{AMAC}_C(U)$$

Additionally, the V_c output could be used to renew the secret in a synchronized way or to derive an encryption key for a secure messaging session as specified in [17]. The security of challenge-response protocols requires that an adversary cannot extract from the RFID tag enough information that allows it to impersonate the tag with high probability. In this FIL-MAC context, the C parameter can be recovered by exhaustive search with complexity 2^c, where $c = |C|$; so, $S_{\mathrm{offline}} \leq c$. In addition to this, the adversary can try to guess V_t online with probability 2^{-m}, so $S_{\mathrm{online}} \leq m$.

Application II: Hashing and digital signatures. For variable-length input messages hashing, we assume a strengthened Merkle-Damgård [7,15] construction (with padding using length suffix) for ARMADILLO, with V_c as chaining variable, U as message block and V_c as hash digest. The initial value (IV) can use the fractional part of the square root of 3 truncated to c bits, similar to the values adopted in SHA-2 hash function family [20]. We write

$$V_c = \mathsf{AHASH}_{\mathsf{IV}}(\mathsf{message}\|\mathsf{padding}).$$

Generic birthday attacks are expected to find collisions in ARMADILLO with complexity $2^{\frac{c}{2}}$. So, $S_{\mathrm{offline}} \leq \frac{c}{2}$ when collisions are a concern. Preimages and second preimages are expected with probability 2^{-c}, so, $S_{\mathrm{offline}} \leq c$. Sometimes, free-start collisions or free-start second preimage attacks matter. In this case, we refer to Application II'.

Application III: PRNG and PRF. For pseudorandom generation, we take the first t bits of $V_c\|V_t$ after at least r iterations. We define

$$\mathsf{APRF}_{\mathsf{seed}}(x) = \mathsf{head}_t(\mathsf{AHASH}_{\mathsf{seed}}(x\|\mathsf{cste}))$$

with an input x with length multiple of m and cste a $(r-1)m$-bit constant. A relevant property for this application is indistinguishability. Assuming a secret seed, ARMADILLO could be used as a stream cipher. The keystream is composed of t-bit frames where the ith frame is $\mathsf{APRF}_{\mathsf{seed}}(i)$. The index i can be synchronized, or sent in clear in which case we have a self-synchronous stream cipher. In this setting, the output should be indistinguishable from a truly random string when the key is random.

3   Dedicated Attacks

Key recovery. Suppose $V_c \| V_t$ and U are known, and we look for C. Since U is known, the $\mathsf{tail}_m(\mathrm{Xinter})$ are known. Guessing the $\mathsf{tail}_{J-m}(C)$ gives access to the $\mathsf{tail}_J(S)$, since U is known. This fact motivates a meet-in-the-middle attack to recover $\mathsf{tail}_{J-m}(C)$. Let us split these $J-m$ bits of Xinter into two pieces of sizes $\lceil \frac{J-m}{2} \rceil$ and $\lfloor \frac{J-m}{2} \rfloor$. The heading $\lceil \frac{J-m}{2} \rceil$ bits are used to compute backwards the P permutation from S, with $m + \lceil \frac{J-m}{2} \rceil = \lceil \frac{J+m}{2} \rceil$ bits known. The tailing $\lfloor \frac{J-m}{2} \rfloor$ bits of C, together with the m bits of U form $m + \lfloor \frac{J-m}{2} \rfloor = \lfloor \frac{J+m}{2} \rfloor$ bits of Xinter. The meet-in-the-middle consists in checking consistency of known bits in the middle of the P permutation. We expect to find a solution with complexity $O(2^{\lceil \frac{J+m}{2} \rceil})$ and a single U. Thus, $S_{\mathsf{offline}} \leq \lceil \frac{J+m}{2} \rceil$ in Application I, II and III.

Free-start collision. We look for a triplet (C, U, U') that causes a collision, that is, $\mathsf{AHASH}_C(U) = \mathsf{AHASH}_C(U')$. For this, we look for (C, U, U') such that

$$P(U, \overline{C} \| \overline{U} \| C \| U) \approx P(U', \overline{C} \| \overline{U'} \| C \| U')$$

with $\approx$ meaning that the Hamming weight of the difference is some low value w. Then, we hope that the next P permutation will move all w different bits outside the window of the $c+m$ bits which are kept in S. Since the probability for a vector to have weight w is $\binom{2c+2m}{w} 2^{-2c-2m}$, the number of solutions we get is $\binom{2c+2m}{w} 2^{-c}$ on average. The probability that a solution leads to a collision is the probability that w difference bits are moved outside a window of c bits. Finally, the expected number of collisions we can get is $\binom{c+m}{w} 2^{-c}$. We can now fix $w = w_{\mathsf{opt}}$ such that $\binom{c+m}{w_{\mathsf{opt}}} \geq 2^c$ so that we can find one solution with complexity $2^{w_{\mathsf{opt}}}$. To implement the attack, for all U and U' we enumerate all C's such that $P(U, \overline{C} \| \overline{U} \| C \| U) \approx P(U', \overline{C} \| \overline{U'} \| C \| U')$. The complexity is $2^{2m} + \binom{2k}{w_{\mathsf{opt}}} 2^{-c}$ which is dominated by 2^{2m}. So, $S_{\mathsf{offline}} \leq 2m$ in Application II'.

A distinguisher. Assuming that the J iterations in the P permutation output a random $2k$-bit vector of Hamming weight k, we have $\binom{2k}{k}$ possible vectors. By extracting a window of t bits we do not have a uniformly distributed string. Indeed, any possible string of weight w has a probability of $p(w) = \binom{2k-t}{k-w} / \binom{2k}{k}$. There exists a distinguisher to tell whether a t-bit window comes from a random output from P or a truly random string, with advantage

$$\frac{1}{2} \sum_{w=0}^{t} \binom{t}{w} \left| \frac{\binom{2k-t}{k-w}}{\binom{2k}{k}} - \frac{1}{2^t} \right|$$

For $t = k = 160$, this is 0.1658. Here, the distinguisher recognizes P when the Hamming weight w is in the interval $[75, \ldots, 85]$, and a random string otherwise.

The final XOR hides this bias a bit but we can wonder by how much exactly. Assume that we hash a message of r blocks. The final output is the XOR of the initial value together with r outputs from P. Assuming that the initial value is known and that the P outputs are random and independent, we can compute the distribution of the final hash by convolution. Indeed, the probability that it is a given string x is $p_r(x)$ such that

$$p_r(x) = \sum_{x_1 \oplus \cdots \oplus x_r = x} p(\text{wt}(x_1)) \cdots p(\text{wt}(x_r))$$

Let us define the spectrum $\hat{p}_r(\mu)$ by $\hat{p}_r(\mu) = \sum_x (-1)^{\mu \cdot x} p_r(x)$. We have $\hat{p}_r(\mu) = (\hat{p}_1(\mu))^r$. We can now compute

$$\hat{p}_1(\mu) = \sum_{i=0}^{\text{wt}(\mu)} \sum_{j=0}^{t-\text{wt}(\mu)} \binom{\text{wt}(\mu)}{i} \binom{t-\text{wt}(\mu)}{j} (-1)^i p(i+j)$$

It only depends on $\text{wt}(\mu)$ so we write $\hat{p}_1(\text{wt}(\mu))$. Since $\sum_\mu \hat{p}_r(\mu)^2 = 2^t \sum_x p_r(x)^2$ we deduce that the Squared Euclidean Imbalance (SEI) of the difference of the hash of r blocks with the initial value is

$$\text{SEI}_r = 2^t \sum_x \left(p_r(x) - 2^{-t}\right)^2 = \sum_{\mu \neq 0} (\hat{p}_r(\mu))^2 = \sum_{w=1}^t \binom{t}{w} (\hat{p}_1(w))^{2r}$$

We have $S_{\text{offline}} \leq -\log_2 \text{SEI}_r$, where r is the minimal number of blocks which are processed in Application III. The SEI expresses as

$$\text{SEI}_r = \sum_{w=1}^t \binom{t}{w} \left(\sum_{i=0}^w \binom{w}{i} \sum_{j=0}^{t-w} \binom{t-w}{j} (-1)^i \frac{\binom{2k-t}{k-i-j}}{\binom{2k}{k}} \right)^{2r}$$

As an example, we computed SEI_r for four selections of $t = k$.

r	$t = k = 128$	$t = k = 160$	$t = k = 200$	$t = k = 275$
		SEI$_r$		
1	$2^{-2.70}$	$2^{-2.70}$	$2^{-2.70}$	$2^{-2.70}$
2	$2^{-18.99}$	$2^{-19.63}$	$2^{-20.28}$	$2^{-21.20}$
3	$2^{-34.98}$	$2^{-36.27}$	$2^{-37.56}$	$2^{-39.40}$
4	$2^{-50.96}$	$2^{-52.90}$	$2^{-54.81}$	$2^{-57.60}$
5	$2^{-66.95}$	$2^{-69.54}$	$2^{-72.12}$	$2^{-75.81}$
6	$2^{-82.94}$	$2^{-86.17}$	$2^{-89.40}$	$2^{-94.01}$
7	$2^{-98.93}$	$2^{-102.81}$	$2^{-106.68}$	$2^{-112.21}$

Given k and c, we look for r and t such that $\text{SEI}_r < 2^{-c}$ and r/t is minimal.

4   Permutation-Dependent Attacks

In this section we present security criteria for the σ_0 and σ_1 permutations.

Another distinguisher. Consider a set I of indices from $V = \{1, \ldots, 2k\}$. Let $\mathrm{swap}_I(\sigma) = \#\{i \in I; \sigma(i) \notin I\}$ and $\mathrm{wt}_I(x) = \sum_{i \in I} x_i$. We assume that $s_b = \mathrm{swap}_I(\sigma_b)$ is low for $b = 0$ and $b = 1$ to see how much the low diffusion between inside and outside I would lead to a distinguisher on $P(s, \cdot)$ with a random s of J bits. In the worst case we can assume that all indices in I are in the same half of x so that the distinguisher can choose the input on P with a very biased $\mathrm{wt}_I(x)$.

A permutation σ_b keeps $\#I - s_b$ of the bits inside I and introduce s_b bits from outside I. Assuming that all bits inside and outside I are randomly permuted, we have the approximation

$$E\left(\mathrm{wt}_I(x_{\sigma_b})\right) \approx (\#I - s_b) \frac{E\left(\mathrm{wt}_I(x)\right)}{\#I} + s_b \frac{k - E\left(\mathrm{wt}_I(x)\right)}{2k - \#I}.$$

Thus,

$$E\left(\mathrm{wt}_I(x_{\sigma_b})\right) - \frac{\#I}{2} \approx \left(1 - \frac{s_b}{\#I} - \frac{s_b}{2k - \#I}\right)\left(E\left(\mathrm{wt}_I(x)\right) - \frac{\#I}{2}\right).$$

On average over the control bits, we have

$$\frac{E\left(\mathrm{wt}_I(P(s,x))\right) - \frac{\#I}{2}}{E\left(\mathrm{wt}_I(x)\right) - \frac{\#I}{2}} \approx \left(1 - \frac{s_0 + s_1}{2} \times \frac{2k}{\#I(2k - \#I)}\right)^J.$$

The best strategy for the distinguisher consists of having either $\mathrm{wt}_I(x) = 0$ or $\mathrm{wt}_I(x) = \#I$. In both cases we have

$$\left| E\left(\mathrm{wt}_I(P(s,x))\right) - \frac{\#I}{2} \right| \approx \frac{\#I}{2}\left(1 - \frac{s_0 + s_1}{2} \times \frac{2k}{\#I(2k - \#I)}\right)^J.$$

The number of samples to significantly observe this bias is

$$T = \left(1 - \frac{s_0 + s_1}{2} \times \frac{2k}{\#I(2k - \#I)}\right)^{-2J}. \tag{1}$$

So, $S_{\mathrm{offline}} \leq \log_2 T$. This expression relates to the theory of expander graphs. We provide below a sufficient condition which can be easily checked.

To compute the minimal value of $\frac{s_0 + s_1}{2\#I}$ over all I we observe that if P_{σ_b} is the matrix of permutation σ_b and if x_I is the 0-1 vector whose coordinate of index in I are the ones set to 1, then

$$\frac{s_0 + s_1}{2\#I} = 1 - \frac{x_I \cdot \left(\frac{P_{\sigma_0} + P_{\sigma_1}}{2} x_I\right)}{x_I \cdot x_I}. \tag{2}$$

Let u be the vector with all coordinates set to 1. Clearly, the hyperplane $u^\perp$ orthogonal to u is stable by the matrix $M_0 = \frac{1}{2}(P_{\sigma_0} + P_{\sigma_1})$. Let $M = \frac{1}{2}(M_0 + M_0^t)$, where the superscript indicates the transpose matrix. We can easily see that $Mu = u$. Furthermore, we notice that $Mx = \lambda x$ with $x \neq 0$ implies $|\lambda| \leq 1$. Let λ be the second largest eigenvalue of M, or equivalently the largest eigenvalue of operator M restricted to $u^\perp$. Note that λ can be $\lambda = 1$ if the eigenvalue 1 has multiplicity higher than one. We can easily prove that $|\lambda| = 1$ and $Mx = \lambda x$ with $x \neq 0$ implies that x_i is constant for all $i \in I$, for all connected components I for the relation $i \sim j \iff \exists s \quad \sigma_{s_{|s|}} \circ \cdots \circ \sigma_{s_2} \circ \sigma_{s_1}(i) = j$. Hence, the only sets I which are stable by σ_0 and σ_1 at the same time are the empty one and the complete set if and only if eigenvalue 1 has multiplicity one. So, having $\lambda < 1$ is already a reasonable criterion but we can have a more precise one. We know that for any vector x orthogonal to u we have $\frac{x \cdot (M_0 x)}{x \cdot x} \leq \lambda$ with equality when x is an eigenvector for λ. Thus,

$$\frac{x \cdot (M_0 x)}{x \cdot x} \leq \frac{(1 - \lambda)\frac{(x \cdot u)^2}{u \cdot u} + \lambda (x \cdot x)}{x \cdot x}$$

for any $x \neq 0$. For $x = x_I$ we obtain

$$\frac{x_I \cdot (M_0 x_I)}{x_I \cdot x_I} \leq (1 - \lambda)\frac{\#I}{2k} + \lambda. \tag{3}$$

From (2), $\frac{s_0 + s_1}{2\#I} = 1 - \frac{x_I \cdot (M_0 \cdot x_I)}{x_I \cdot x_I} \geq 1 - (1 - \lambda)\frac{\#I}{2k} + \lambda = (1 - \lambda)(1 - \frac{\#I}{2k})$. Going back to the complexity (1) of our distinguisher we have $T \geq \lambda^{-2J}$. Hence, by having $\lambda \leq 2^{-\frac{S_{\text{offline}}}{2J}}$ for an offline complexity $2^{S_{\text{offline}}}$, we make sure that the distinguisher has complexity $T \geq 2^{S_{\text{offline}}}$. To conclude, if λ is the second largest eigenvalue of $M = \frac{1}{4}(P_{\sigma_0} + P'_{\sigma_0} + P_{\sigma_1} + P'_{\sigma_1})$ then we have an attack of complexity λ^{-2J}. So, $S_{\text{offline}} \leq -2J \log_2 \lambda$.

Yet another distinguisher. We define the vector x of dimension k such that the ith coordinate of x is the probability that x_i is set to 1. If x is fixed, we can consider that x is equal to x by abuse of notation. If $y = x_\sigma$, we have that y is obtained by multiplying a permutation matrix P_σ by x. We have $(P_\sigma)_{j,i} = 1$ if and only if $j = \sigma(i)$. Clearly, for $y = x_{\sigma_b}$ we can write

$$y = ((1 - b)P_{\sigma_0} + bP_{\sigma_1}) \times x$$
$$= \left(\frac{1}{2}(P_{\sigma_0} + P_{\sigma_1}) + \frac{(-1)^b}{2}(P_{\sigma_0} - P_{\sigma_1}) \right) \times x$$

We let $M_0 = \frac{1}{2}P_{\sigma_0} + \frac{1}{2}P_{\sigma_1}$ and $M_1 = \frac{1}{2}P_{\sigma_0} - \frac{1}{2}P_{\sigma_1}$. We have

$$\prod_{i=1}^{J}(M_0 + (-1)^{s_i}M_1) = \sum_{a_1=0}^{1} \cdots \sum_{a_k=0}^{1} (-1)^{a_1 s_1 + \cdots + a_k s_k} M_a = \hat{M}_s$$

where $M_a = M_{a_1} \times \cdots \times M_{a_k}$. So, the vector of $y = P(s,x)$ is $y = \hat{M}_s x$. The average $\langle \hat{M}_s \rangle$ of $\hat{M}_s$ over all $s_1, \ldots, s_k$ is M_0^k. We define a square matrix F in which all terms are equal to $\frac{1}{k}$. Clearly, if s is a uniformly distributed J-bit random string, the probability vector of $P(s,x)$ is $M_0^J \times x$. Since M_0 is a bi-stochastic matrix, we have $M_0 \times F = F \times M_0 = F$. Similarly, we have $M_1 \times F = F \times M_1 = 0$. We easily deduce that $(M_0 - F)^J = M_0^J - F$. Let θ be the second largest eigenvalue of $M_0^t M_0$, or equivalently, the largest eigenvalue of $M_0^t M_0 - F$. For any vector x such that $\sum x_i = w$ and $0 \leq x_i \leq 1$, we have $\|M_0^J x - wu\|_2^2 = \|(M_0 - F)^J x\|_2^2 \leq w\theta^J$, where $u = (1, \ldots, 1)$. So, the cumulated squared Euclidean imbalance of each component of $M_0^J x$ is bounded by $2k\theta^J$. Thus, the complexity is $\frac{1}{2k\theta^J}$, and $S_{\text{offline}} \le -J\log_2\theta - \log_2 2k$. The average $\left\langle \hat{M}_s\binom{\bar{s}}{s} \right\rangle$ of the image of Xinter $= s$ is

$$\sum_a M_a \left\langle (-1)^{a \cdot s} \binom{\bar{s}}{s} \right\rangle.$$

For $a = 0$ the average is $(\frac{1}{2} \cdots \frac{1}{2})$. For a of weight at least 2, the average is zero. For a of weight 1, e.g. $a = (1,0,\dots,0)$ the average is $(-\frac{1}{2},0,\dots,0,\frac{1}{2},0,\dots,0)$. We let e_i be the vector with coordinate 1 in its ith position and 0 elsewhere. We have

$$\left\langle \hat{M}_s\binom{\bar{s}}{s} \right\rangle = \begin{pmatrix} \frac{1}{2} \\ \vdots \\ \frac{1}{2} \end{pmatrix} + \frac{1}{2}\sum_{i=1}^{k} M_0^{i-1} M_1 M_0^{k-i}(-e_i + e_{k+i}).$$

Let

$$b = \frac{1}{2}\sum_{i=1}^{k} M_0^{i-1} M_1 M_0^{k-i}(-e_i + e_{k+i}). \tag{4}$$

The complexity is $\frac{1}{2\|b\|_2^2}$, so, we have $S_{\text{offline}} \le -2\log_2\|b\|_2 - 1$.

The parity of P. Let ε_i be the parity of σ_i. The $x \mapsto P(s,x)$ is a permutation whose parity is $\varepsilon_0^{|s|-\text{wt}(s)}\varepsilon_1^{\text{wt}(s)}$. If $\varepsilon_0 \ne \varepsilon_1$, an adversary with black-box access to $x \mapsto P(s,x)$ and knowing $|s|$ can thus easily deduce $\text{wt}(s)$. We thus recommend that $\varepsilon_0 = \varepsilon_1$.

5 Parameter Vectors

Here we suggest sets of parameters for four different applications, based on our analyzes. In all cases, we require $J = c + m$ and also that σ_0 and σ_1 have the same parity.

  I: in a challenge-response application: $S_{\text{offline}} \le \min(c, \frac{J+m}{2})$ and $S_{\text{online}} \le m$
 II: in a collision-resistance context: $S_{\text{offline}} \le \frac{c}{2}$
II': in a free-start collision context: $S_{\text{offline}} \le \min\left(\frac{c}{2}, 2m\right)$
III: $S_{\text{offline}} \le -\log_2 \text{SEI}_r$. If λ is the second largest eigenvalue of $M = \frac{1}{4}(P_{\sigma_0} + P_{\sigma_0}^t + P_{\sigma_1} + P_{\sigma_1}^t)$ then $S_{\text{offline}} \le -2J\log_2\lambda$. For σ_0 and σ_1, the second largest eigenvalue of $M_0^t M_0$, called θ, $S_{\text{offline}} \le -J\log_2\theta - \log_2 k - 1$. The bias b in (4) shall satisfy $S_{\text{offline}} \le -2\log_2\|b\|_2 - 1$. Moreover, $S_{\text{offline}} \le \frac{J+m}{2}$.

To match the ideal security, we need these bounds to yield $S_{\text{offline}} \le c$ and $S_{\text{online}} \le m$ for Application I, $S_{\text{offline}} \le \frac{c}{2}$ for Application II and II', and $S_{\text{offline}} \le c$ for Application III. So, we take $J = c + m$, $m \ge \frac{c}{2}$; r and t such that $\text{SEI}_r \le 2^{-c}$; σ_0 and σ_1 such that $-\log_2\lambda \ge \frac{c}{2(c+m)}$, $-\log_2\theta \ge \frac{c+\log 2k}{c+m}$, $-\log_2\|b\|_2 \ge \frac{c+1}{2}$, and $\varepsilon_0 = \varepsilon_1$. Our recommendations for the parameter values of ARMADILLO are given in Table 1. Note that c is the key length for Applications I and III and also the digest length for Application II.

Table 1. Parameter vectors

Vector	k	J	c	m	r	t
A	128	128	80	48	6	128
B	192	192	128	64	9	192
C	240	240	160	80	10	240
D	288	288	192	96	12	288
E	384	384	256	128	15	384

6   ARMADILLO2

Ever since the first version of ARMADILLO, we have developed an updated design, called ARMADILLO2, that is even more robust than the version presented in Fig. 1. In fact, ARMADILLO2 brings in a new compression function, called Q, which is not only more compact in hardware than P, but also addresses security concerns brought about during the continuous analyzes of ARMADILLO. For these reasons, ARMADILLO2 is our preferred design choice. Due to space limitations, further details about the security analysis of ARMADILLO2 are omitted. ARMADILLO2 is defined by

$$(V_c, V_t) = \mathsf{ARMADILLO2}(C, U) = Q(X, C\|U) \oplus X, \text{ where } X = Q(U, C\|U).$$

We call the new permutation Q, instead of P as in Fig. 1, to avoid confusion. The main novelties are:

- there is *no* complementation of the k-bit input $\mathsf{Xinter} = C\|U$ anymore; as a consequence, the σ_i permutations (and therefore Q) now operate on k-bit data $C\|U$, instead of $\overline{C}\|\overline{U}\|C\|U$, leading to a more compact design;
- a *new* permutation Q which interleaves σ_i's, $i \in \{0, 1\}$, with an xor using the k-bit constant bitstring $\gamma = 1010\cdots10$; Q is defined recursively as $Q(s\|b, X) = Q(s, X_{\sigma_b} \oplus \gamma)$ and $Q(\emptyset, X) = X$, for $b \in \{0, 1\}$ and bitstrings s and X;
- the outermost Q is controlled by a data-dependent value, $X = Q(U, C\|U)$, in contrast to simply $C\|U$ in Fig. 1;

In the new structure of Q, the output bias disappears and we can take $r = 1$ and $t = k$.

7   Hardware Implementation and Performance

There exist different demands on the implementation and the optimization meanings for various application scenarios. In this context, the scalability of ARMADILLO allows to deploy the implementation in a very wide realm of area and speed parameters, which constitutes the most essential trade-off in electronics circuits. The implementation of the P function, using the building block, is depicted in Fig. 2(b). It accepts an input vector of $2k$ bits and a key of J bits. It consists of a variable number N of permutation stages, all identical, and each stage essentially requires $2k$ multiplexers (Fig. 2(a)). One register of $2k$ bits is needed to hold the input and/or intermediate data, as well as one J-bit register to hold the permutation key. At each cycle, these registers are either loaded with new data/key or fed back the output data/key for a new permutation round, depending on the

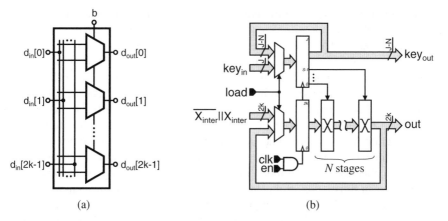

Fig. 2. Hardware implementation of the ARMADILLO function. (a) one permutation stage. (b) *P* function building block.

state of the load signal. The number N of permutations executed in each cycle can be adjusted, the only restriction being that J be an integer multiple of N. The output data is the $2k$ bits vector resulting from the permutation round, and the output key is the $J - N$ bits remaining to be processed. This building block can be flexibly assembled into a T-stage pipeline, where each stage performs a number $R = J/(N \cdot T)$ of permutation rounds (building blocks) before passing the results to the next stage and accepting new input from the previous stage. In that case, the throughput is $1/R$ items per cycle and the latency is J/N cycles, the parameters being linked by the equality $R \cdot N \cdot T = J$. The latency / throughput / cost trade-off can be adjusted, the two extreme cases being $R = 1$ (fully pipelined, resulting in a throughput of 1 item per cycle) and $T = 1$ (fully serial, resulting in a throughput of S/J items per cycle). Obviously, the more pipeline stages, the more hardware replication and therefore the higher the cost in area and power. To construct the complete hash function of Fig. 1, we essentially need to add a state machine (which is little more than a counter) around the permutation function block, and the final XOR operation.

Metrics for evaluating performance In order to compare different cryptographic functions, several metrics can be taken into account. The security is of course the primary concern. The silicon area, the throughput, the latency and the power dissipation are other metrics of interest, and can be traded-off for one another. For example, the power dissipation is nearly proportional to the clock frequency in any CMOS circuit, therefore, power can be reduced by decreasing the clock frequency and thus at the expense of throughput. Conversely, throughput can be increased by running at a faster clock frequency, up to a maximum clock frequency which is process- and implementation-dependent. Another example is serialization, where an operation is broken into several steps executed in series, allowing to reuse the same hardware, but again at the cost of a longer execution time. Through serialization, throughput and latency can be traded-off for area, down to a point where operations can not be broken into smaller operations anymore and we have reached a minimum area. Given this large design space, comparing the relative merit of different cryptographic functions is a challenging task.

The approach taken in [2] (and numerous other publications) includes comparing the area of synthesized circuits as reported in the literature or estimated by the authors in gate-equivalent (GE). It is notable though that the GE unit of measure, while being convenient because it is process-independent, is very coarse. For example, does the reported area after synthesis include the space needed for wiring? Typically, the utilization of a routed circuit can be in the range of 50%–80%, and is especially critical when using a limited number of metal layers for routing. A synthesis tool may report an estimated routing area, but in all cases it may vary to a large extent after physical implementation. Consider also that one design may have scan chains inserted while another may not, which may increase the register area by as much as 20–30% and require extra interconnections. Furthermore, different standard cells may be of varying area efficiency; as an illustration of this fact, a comparison of gate-equivalent figures from different standard-cell libraries can produce different results with a ratio up to $\frac{2}{3}$. For instance, a simple 2-input multiplexer can lead to 2.67 GE or 1.67 GE from one library to the other. Taking into account all these factors, it is clear that such a comparison can have a large margin of error, unless the circuits being compared have been implemented in the exact same conditions.

Besides comparing areas, the authors of [2] also use a metric called efficiency, which is defined as the ratio of the throughput (measured at a fixed clock frequency) over the area. It may seem at first sight that such a metric provides a more general measure of quality, since it may be fair to give up some area for a higher throughput, however it is flawed in that it does not consider the possibility of trading off throughput for power. Indeed, according to this metric, two designs A and B would be deemed of equal value if, for example, A's throughput and area were twice B's throughput and area, respectively. However, if B's power dissipation is half that of A at the same clock frequency, then by doubling B's operating frequency, its throughput can be made equal to that of A while consuming the same power and still occupying a smaller area. Clearly then, B should be recognized as superior to A, which can be captured by dividing the metric by the power dissipation, thus making it independent of the power/throughput trade-off. However, this does not come without its own problems, since the power dissipation is an extremely volatile quantity. Being subject to the same error factors as the area as described above, it also depends heavily on the process technology, the supply voltage, and the parasitic capacitances due to the interconnections. Furthermore, it can vary largely depending on the method used to measure it (i.e. gate-level statistical or vector-based simulation, or SPICE simulation). As if this were not enough, different standard-cell libraries also exhibit various power/area/speed trade-offs, for example, a circuit implemented with a high-density library is likely to result in a lower power figure than the same circuit implemented with a general-purpose library, for a similar gate count.

Nevertheless, a fairer figure of merit would need to include the influence of power dissipation. In order to keep process-independent metrics, we can assume that the power is proportional to the gate count.[2] This is reasonable since the dynamic power in CMOS

---

[2] In the same spirit as the GE unit of measure, a more interesting metric would be to divide the power by $C_{unit} \cdot V_{DD}^2$, where C_{unit} is the input capacitance of an inverter. However, this is not applicable to compare other published implementations since these quantities are usually unknown.

Table 2. Synthesis results at 1MHz

Algorithm	N=1				N=4			
	Area (GE)	Power (μW)	Throughput (kbps)	Latency (cycles)	Area (GE)	Power (μW)	Throughput (kbps)	Latency (cycles)
ARMADILLO-A	3972	69	375	128	5770	133	1500	32
ARMADILLO-B	6598	117	333	192	9709	237	1333	48
ARMADILLO-C	8231	146	333	240	12217	300	1333	60
ARMADILLO-D	8650	177	333	288	14641	368	1333	72
ARMADILLO-E	13344	228	333	384	19669	513	1333	96
ARMADILLO2-A	2923	44	272	176	4030	77	1090	44
ARMADILLO2-B	4353	65	250	256	6025	118	1000	64
ARMADILLO2-C	5406	83	250	320	7492	158	1000	80
ARMADILLO2-D	6554	102	250	384	8999	183	1000	96
ARMADILLO2-E	8653	137	250	512	11914	251	1000	128

circuits is proportional to the total switched capacitance, which correlates to the area. We propose therefore to use a figure of merit defined as FOM = throughput/GE^2. In practice, this is a coarse approximation, since it does not take into account switching activity or the influence of wire load; it is nevertheless fairer than not including power dissipation at all, since it tends to favor designs with smaller area (at equal throughput) which are very likely to dissipate less power.

Synthesis Results. Table 2 presents the results of synthesis for the hash function described above in a $0.18\mu m$ CMOS process using a commercial standard-cell library, with the parameters given in Sect. 5. Synthesis was performed with Synopsys Design Compiler in topographical mode, in order to obtain accurate wire loads. The power consumption was evaluated with Synopsys Primetime-PX using gate-level vector-based analysis.

In RFID applications, the latency is constrained by the communication protocols (though the constraint is relatively easily satisfiable) but a high throughput is not necessary, designating a fully serial implementation as the ideal candidate. Therefore T is set to $T = 1$. The number N of permutations per clock cycle in the permutation function is set to $N = 1$, which is favorable to smaller area and power consumption for the tight power budget associated with RFID applications. The clock frequency is set to 1MHz, which is a representative value for the target application.

In hash mode we hash m bits per compression. In encryption mode we encrypt t/r bits per compression. The throughput values given in Table 2 correspond to hash mode.

Our goal for selecting $T = 1$ and $N = 1$ was to minimize the hardware. The area in the proposed implementation is roughly proportional to

$$(k_{\text{reg}} * (2k + J) + k_{\log} * (2k(N+1) + J))T$$

for some constants k_{reg} and $k_{\log}$.

To maximize the FOM with T given, we can show that we should in theory pick

$$N = \left(\frac{k_{\text{reg}}}{k_{\log}} + 1\right)\left(1 + \frac{J}{2k}\right)$$

For $k_{\text{reg}} \approx 2k_{\log}$ and $J = k$, this is $N = 4.5$. In practice, the best choice is to take $T = 1$ and $N = 4$ for ARMADILLO2 in context A, for which we would get an area of 4030 GE, 77 μW, and a latency of 44 cycles (1.09 Mbps for hashing or 2.9 Mbps for encryption).

Table 3. Implementation comparison for hash functions with throughput at 100 kHz

Algorithm	Digest (bits)	Block (bits)	Area (GE)	Time (cycles/block)	Throughput (kb/s)	Logic (μm)	FOM (nanobit/cycle.GE^2)
ARMADILLO2-A	80	48	4030	44	109	0.18	67.17
ARMADILLO2-A	80	48	2923	176	27	0.18	31.92
H-PRESENT-128 [2]	128	128	4256	32	200	0.18	110.41
ARMADILLO2-B	128	64	6025	64	1000	0.18	27.55
MD4 [9]	128	512	7350	456	112.28	0.13	20.78
ARMADILLO2-B	128	64	4353	256	250	0.18	13.19
MD5 [9]	128	512	8400	612	83.66	0.13	11.86
ARMADILLO2-C	160	80	7492	80	100	0.18	17.81
ARMADILLO2-C	160	80	5406	320	250	0.18	8.55
SHA-1 [9]	160	512	8120	1274	40.18	0.35	6.10
ARMADILLO2-D	192	96	8999	96	100	0.18	12.35
C-PRESENT-192 [2]	192	192	8048	108	59.26	0.18	9.15
ARMADILLO2-D	192	96	6554	384	25	0.18	5.82
MAME [24]	256	256	8100	96	266.67	0.18	40.64
ARMADILLO2-E	256	128	11914	128	100	0.18	7.05
SHA-256 [9]	256	512	10868	1128	45.39	0.35	3.84
ARMADILLO2-E	256	128	8653	512	25	0.18	3.34

Table 4. Implementation comparison for encryption with throughput at 100 kHz

Algorithm	Key (bits)	Block (bits)	Area (GE)	Time (cycles/block)	Throughput (kb/s)	Logic (μm)	FOM (nanobit/cycle.GE^2)
DES [18]	56	64	2309	144	44	0.18	83.36
PRESENT-80 [1]	80	64	1570	32	200	0.18	811.39
Grain [11]	80	1	1294	1	100	0.13	597.22
KTANTAN64 [3]	80	64	927	128	50	0.13	581.85
KATAN64 [3]	80	64	1269	85	75	0.13	467.56
ARMADILLO2-A	80	128	4030	44	291	0.18	179.12
Trivium [11]	80	1	2599	1	100	0.13	148.04
PRESENT-80 [19]	80	64	1075	563	11	0.18	98.37
ARMADILLO2-A	80	128	2923	176	73	0.18	85.12
mCrypton [14]	96	64	2681	13	500	0.13	684.96
PRESENT-128 [1]	128	64	1886	32	200	0.18	562.27
HIGHT [13]	128	64	3048	34	189	0.25	202.61
TEA [23]	128	64	2355	64	100	0.18	180.31
ARMADILLO2-B	128	192	6025	64	300	0.18	82.64
ARMADILLO2-B	128	192	4353	256	75	0.18	39.58
AES-128 [8]	128	128	3400	1032	12	0.35	10.73
ARMADILLO2-C	160	240	7492	80	300	0.18	53.45
ARMADILLO2-C	160	240	5406	320	75	0.18	25.66
DESXL [18]	184	64	2168	144	44	0.18	94.56
ARMADILLO2-D	192	288	8999	96	300	0.18	37.04
ARMADILLO2-D	192	288	6554	384	75	0.18	17.46
ARMADILLO2-E	256	384	11914	128	300	0.18	21.13
ARMADILLO2-E	256	384	8653	512	75	0.18	10.02

8   Comparison

Table 3 shows a comparison of hardware implementations of ARMADILLO in the hash function setting, relative to other hash functions such as MD4, MD5, SHA-1, SHA-256, and MAME according to [2]. We computed the throughput in kbps at a clock rate of 100 kHz. We added the best FOM results for KATAN and KTANTAN with 64-bit blocks from [3]. Algorithms are categorized in terms of security by taking into account the digest size. In each category, we listed the algorithms by decreasing order of merit. To estimate the FOM we assumed that the power was proportional to the area. So, it is the speed divided by the square of the area. These figures show that different versions of ARMADILLO2 provide clear advantage for hashing, either in terms of area, or of throughput, or of overall merit.

9   Conclusions

This paper suggested a new hardware dedicated cryptographic function design called ARMADILLO. Applications for ARMADILLO include MACs, hashing for challenge-response protocols, PRNG and as a stream cipher.

References

1. Bogdanov, A., Knudsen, L.R., Leander, G., Paar, C., Poschmann, A., Robshaw, M.J.B., Seurin, Y., Vikkelsoe, C.: Present: a Ultra-Lightweight Block Cipher. In: Paillier, P., Verbauwhede, I. (eds.) CHES 2007. LNCS, vol. 4727, pp. 450–466. Springer, Heidelberg (2007)
2. Bogdanov, A., Leander, G., Paar, C., Poschmann, A., Robshaw, M.J.B., Seurin, Y.: Hash Functions and RFID Tags: Mind the Gap. In: Oswald, E., Rohatgi, P. (eds.) CHES 2008. LNCS, vol. 5154, pp. 283–299. Springer, Heidelberg (2008)
3. De Cannière, C., Dunkelman, O., Knežević, M.: KATAN & KTANTAN: a Family of Small and Efficient Hardware-Oriented Block Ciphers. In: Clavier, C., Gaj, K. (eds.) CHES 2009. LNCS, vol. 5747, pp. 272–288. Springer, Heidelberg (2009)
4. De Cannière, C., Preneel, B.: Trivium Specifications. eSTREAM technical report (2006), http://www.ecrypt.eu.org/stream/ciphers/trivium/trivium.pdf
5. Daemen, J., Govaerts, R., Vandewalle, J.: A Hardware Design Model for Cryptographic Algorithms. In: Deswarte, Y., Quisquater, J.-J., Eizenberg, G. (eds.) ESORICS 1992. LNCS, vol. 648, pp. 419–434. Springer, Heidelberg (1992)
6. Daemen, J., Govaerts, R., Vandewalle, J.: A Framework for the Design of One-Way Hash Functions Including Cryptanalysis of Damgård One-way Function based on a Cellular Automaton. In: Matsumoto, T., Imai, H., Rivest, R.L. (eds.) ASIACRYPT 1991. LNCS, vol. 739, pp. 82–96. Springer, Heidelberg (1993)
7. Damgård, I.B.: A Design Principle for Hash Functions. In: Brassard, G. (ed.) CRYPTO 1989. LNCS, vol. 435, pp. 416–427. Springer, Heidelberg (1990)
8. Feldhofer, M., Dominikus, S., Wolkerstorfer, J.: Strong Authentication for RFID Systems Using the AES Algorithm. In: Joye, M., Quisquater, J.-J. (eds.) CHES 2004. LNCS, vol. 3156, pp. 357–370. Springer, Heidelberg (2004)
9. Feldhofer, M., Rechberger, C.: A Case Against Currently Used Hash Functions in RFID Protocols. In: Meersman, R., Tari, Z., Herrero, P. (eds.) OTM 2006 Workshops. LNCS, vol. 4277, pp. 372–381. Springer, Heidelberg (2006)
10. Garber, D.: Braid Group Cryptography. CoRR, vol. abs/0711.3941, pp. 1–75 (2007)
11. Good, T., Chelton, W., Benaissa, M.: Hardware Results for Selected Stream Cipher Candidates. Presented at the State of the Art of Stream Ciphers SASC'07, Bochum, Germany (2007)
12. Hell, M., Johansson, T., Meier, W.: Grain: a Stream Cipher for Constrained Environments. International Journal of Wireless and Mobile Computing 2, 86–93 (2007)
13. Hong, D., Sung, J., Hong, S., Lim, J., Lee, S., Koo, B.S., Lee, C., Chang, D., Lee, J., Jeong, K., Kim, H., Kim, J., Chee, S.: HIGHT: a New Block Cipher suitable for Low-Resource Device. In: Goubin, L., Matsui, M. (eds.) CHES 2006. LNCS, vol. 4249, pp. 46–59. Springer, Heidelberg (2006)

14. Lim, C., Korkishko, T.: mCrypton: A Lightweight Block Cipher for Security of Lowcost RFID Tags and Sensors. In: Song, J.-S., Kwon, T., Yung, M. (eds.) WISA 2005. LNCS, vol. 3786, pp. 243–258. Springer, Heidelberg (2006)
15. Merkle, R.C.: One way Hash Functions and DES. In: Brassard, G. (ed.) CRYPTO 1989. LNCS, vol. 435, pp. 416–427. Springer, Heidelberg (1990)
16. Moldovyan, A.A., Moldovyan, N.A.: A cipher based on data-dependent permutations. Journal of Cryptology 1(15), 61–72 (2002)
17. Ouafi, K., Vaudenay, S.: Pathchecker: An RFID Application for Tracing Products in Supply-Chains. Presented at the International Conference on RFID Security 2009, Leuven, Belgium (2009)
18. Poschmann, A., Leander, G., Schramm, K., Paar, C.: New Lightweight DES Variants Suited for RFID Applications. In: Biryukov, A. (ed.) FSE 2007. LNCS, vol. 4593, pp. 196–210. Springer, Heidelberg (2007)
19. Rolfes, C., Poschmann, A., Leander, G., Paar, C.: Ultra-Lightweight Implementations for Smart Devices - Security for 1000 Gate Equivalents. In: Grimaud, G., Standaert, F.-X. (eds.) CARDIS 2008. LNCS, vol. 5189, pp. 89–103. Springer, Heidelberg (2008)
20. Secure Hash Standard. Federal Information Processing Standard publication #180-2. U.S. Department of Commerce, National Institute of Standards and Technology (2002)
21. Tillich, J.P., Zémor, G.: Hashing with SL_2. In: Desmedt, Y.G. (ed.) CRYPTO 1994. LNCS, vol. 839, pp. 40–49. Springer, Heidelberg (1994)
22. Wheeler, D.J., Needham, R.M.: TEA: a Tiny Encryption Algorithm. In: Preneel, B. (ed.) FSE 1994. LNCS, vol. 1008, pp. 363–366. Springer, Heidelberg (1995)
23. Yu, Y., Yang, Y., Fan, Y., Min, H.: Security Scheme for RFID Tag. Technical report WP-HARDWARE-022, Auto-ID Labs white paper (2006),
http://www.autoidlabs.org/single-view/dir/article/6/230/page.html
24. Yoshida, H., Watanabe, D., Okeya, K., Kitahara, J., Wu, J., Küçük, Ö., Preneel, B.: MAME: A Compression Function With Reduced Hardware Requirements. In: Paillier, P., Verbauwhede, I. (eds.) CHES 2007. LNCS, vol. 4727, pp. 148–165. Springer, Heidelberg (2007)

Provably Secure Higher-Order Masking of AES

Matthieu Rivain[1] and Emmanuel Prouff[2]

[1] CryptoExperts
matthieu.rivain@cryptoexperts.com
[2] Oberthur Technologies
e.prouff@oberthur.com

Abstract. Implementations of cryptographic algorithms are vulnerable to Side Channel Analysis (SCA). To counteract it, masking schemes are usually involved which randomize key-dependent data by the addition of one or several random value(s) (the *masks*). When dth-order masking is involved (*i.e.* when d masks are used per key-dependent variable), the complexity of performing an SCA grows exponentially with the order d. The design of generic dth-order masking schemes taking the order d as security parameter is therefore of great interest for the physical security of cryptographic implementations. This paper presents the first generic dth-order masking scheme for AES with a provable security and a reasonable software implementation overhead. Our scheme is based on the hardware-oriented masking scheme published by Ishai *et al.* at Crypto 2003. Compared to this scheme, our solution can be efficiently implemented in software on any general-purpose processor. This result is of importance considering the lack of solution for $d \geqslant 3$.

1 Introduction

Side Channel Analysis exploits information that leaks from physical implementations of cryptographic algorithms. This leakage (*e.g.* the power consumption or the electro-magnetic emanations) may indeed reveal information on the data manipulated by the implementation. Some of these data are *sensitive* in the sense that they are related to the secret key, and the leaking information about them enables efficient key-recovery attacks [7, 18].

Due to the very large variety of side channel attacks reported against cryptosystems and devices, important efforts have been done to design countermeasures with provable security. They all start from the assumption that a cryptographic device can keep at least some secrets and that only computation leaks [24]. Based on these assumptions, two main approaches have been followed. The first one consists in designing new cryptographic primitives inherently resistant to side channel attacks. In [24], a very powerful side channel adversary is considered who has access to the whole internal state of the ongoing computation. In such a model, the authors show that if a *physical* one-way permutation exists which does not leak any information, then it can be used in the pseudo-random number generator (PRNG) construction proposed in [4] to give a PRNG provably secure against the aforementioned side channel adversary. Unfortunately,

S. Mangard and F.-X. Standaert (Eds.): CHES 2010, LNCS 6225, pp. 413–427, 2010.

no such leakage-resilient one-way permutation is known at this day. Besides, the obtained construction is quite inefficient since each computation of the one-way permutation produces one single random bit. To get more practical constructions, further works focused on designing primitives secure against a *limited* side channel adversary [13]. The definition of such a limited adversary is inspired by the *bounded retrieval model* [10, 21] which assumes that the device leaks a limited amount of information about its internal state for each elementary computation. In such a setting, the block cipher based PRNG construction proposed in [29] is provably secure assuming that the underlying cipher is *ideal*. Other constructions were proposed in [13, 30] which do not require such a strong assumption but are less efficient [39]. The main limitations of these constructions is that they do not enable the choice of an initialization vector (otherwise the security proofs do not hold anymore) which prevents their use for encryption with synchronization constraints or for challenge-response protocols [39]. Moreover, as they consist in new constructions, these solutions do not allow for the protection of the implementation of standard algorithms such as DES or AES [14, 15].

The second approach to design countermeasures provably secure against side channel attacks consists in applying *secret sharing schemes* [2, 38]. In such schemes, the sensitive data is randomly split into several shares in such a way that a chosen number (called the *threshold*) of these shares is required to retrieve any information about the data. When the SCA threat appeared, secret sharing was quickly identified as a pertinent protection strategy [6, 16] and numerous schemes (often called *masking schemes*) were published that were based on this principle (see for instance [1, 3, 17, 22, 25, 28, 33, 37]). Actually, this approach is very close to the problem of defining Multi Party Communication (MPC) schemes (see for instance [9, 12]) but the resources and constraints differ in the two contexts (*e.g.* MPC schemes are often based on a *trusted dealer* who does not exist in the SCA context). A first advantage of this approach is that it can be used to secure standard algorithms such as DES and AES. A second advantage is that *dth-order masking schemes*, for which sensitive data are split into $d + 1$ shares (the threshold being $d + 1$), are sound countermeasures to SCA in *realistic leakage model*. This fact has been formally demonstrated by Chari et al. [6] who showed that the complexity of recovering information by SCA on a bit shared into several pieces grows exponentially with the number of shares. As a direct consequence of this work, the number of shares (or equivalently of masks) in which sensitive data are split is a sound security parameter of the resistance of a countermeasures against SCA.

The present paper deals with the problem of defining an efficient masking scheme to protect the implementation of the AES block cipher [11]. Until now, most of works published on this subject have focussed on first-order masking schemes where sensitive variables are masked with a single random value (see for instance [1, 3, 22, 25, 28]). However, this kind of masking have been shown to be efficiently breakable in practice by *second-order SCA* [23, 26, 41]. To counteract those attacks, *higher-order masking schemes* must be used but a very few have been proposed. A first method has been introduced by Ishai et al. [17] which enables to protect an implementation at any chosen order. Unfortunately, it is not suited for software implementations and it induces a prohibitive overhead for hardware implementations. A scheme devoted to secure the software implementation of AES at any chosen order has been proposed by Schramm and Paar [37] but it was subsequently shown to be secure only in the second-order case [8]. Alternative second-order masking schemes with provable security were further proposed in [33], but no straightforward extension of them exist to get efficient and secure masking scheme at any order. Actually, at this day, no method exists in the literature that enables to mask an AES implementation at any chosen order $d \geqslant 3$ with a practical overhead; the present paper fills this gap.

2   Preliminaries on Higher-Order Masking

2.1   Basic Principle

When higher-order masking is involved to secure the physical implementation of a cryptographic algorithm, every sensitive variable x occurring during the computation is randomly split into $d + 1$ shares $x_0, \ldots, x_d$ in such a way that the following relation is satisfied for a group operation $\perp$:

$$x_0 \perp x_1 \perp \cdots \perp x_d = x \ . \tag{1}$$

In the rest of the paper, we shall consider that $\perp$ is the exclusive-or (XOR) operation denoted by $\oplus$. Usually, the d shares $x_1, \ldots, x_d$ (called *the masks*) are randomly picked up and the last one x_0 (called *the masked variable*) is processed such that it satisfies (1). When d random masks are involved per sensitive variable the masking is said to be *of order d*.

Assuming that the masks are uniformly distributed, masking renders every intermediate variable of the computation statistically independent of any sensitive variable. As a result, classical side channel attacks exploiting the leakage related to a single intermediate variable are not possible anymore. However, a dth-order masking is always theoretically vulnerable to $(d+1)th\text{-}order \ SCA$ which exploits the leakages related to $d+1$ intermediate variables at the same time [23, 36, 37]. Indeed, the leakages resulting from the $d+1$ shares (i.e. the masked variable and the d masks) are jointly dependent on the sensitive variable. Nevertheless, such attacks become impractical as d increases, which makes higher-order masking a sound countermeasure.

2.2   Soundness of Higher-Order Masking

The soundness of higher-order masking was formally demonstrated by Chari *et al.* in [6]. They assume a simplified but still realistic leakage model where a bit b is masked using d random bits $x_1, \ldots, x_d$ such that the masked bit is defined as $x_0 = b \oplus x_1 \oplus \cdots \oplus x_d$. The adversary is assumed to be provided with observations of $d + 1$ leakage variables L_i, each one corresponding to a share x_i. For every i, the leakage is modelled as $L_i = x_i + N_i$ where the noises

N_i's are assumed to have Gaussian distributions $\mathcal{N}\left(\mu, \sigma^2\right)$ and to be mutually independent. Under this leakage model, they show that the number of samples q required by the adversary to distinguish the distribution $(L_0, \ldots, L_d | b = 0)$ from the distribution $(L_0, \ldots, L_d | b = 1)$ with a probability at least α satisfies:

$$q \geqslant \sigma^{d+\delta} \tag{2}$$

where $\delta = 4 \log \alpha / \log \sigma$. This result encompasses all the possible side-channel distinguishers and hence formally states the resistance against every kind of side channel attack. Although the model is simplified, it could probably be extended to more common leakage models such as the Hamming weight/distance model. The point is that if an attacker observes noisy side channel information about $d + 1$ shares corresponding to a variable masked with d random masks, the number of samples required to retrieve information about the unmasked variable is lower bounded by an exponential function of the masking order whose base is related to the noise standard deviation. This formally demonstrates that higher-order masking is a sound countermeasure especially when combined with noise. Many works also made this observation in practice for particular side channel distinguishers (see for instance [36, 37, 40]).

2.3   Higher-Order Masking Schemes

When dth-order masking is involved in protecting a block cipher implementation, a so-called *dth-order masking scheme* (or simply a *masking scheme* if there is no ambiguity on d) must be designed to enable computation on masked data. In order to be complete and secure, the scheme must satisfy the two following properties:

- *completeness*: at the end of the computation, the sum of the d shares must yield the expected ciphertext (and more generally each masked transformation must result in a set of shares whose sum equal the correct intermediate result),
- *dth-order SCA security*: every tuple of d or less intermediate variables must be independent of any sensitive variable.

If the dth-order security property is satisfied, then no attack of order lower than $d + 1$ is possible and we benefit from the security bound (2).

Most block cipher structures (*e.g.* AES or DES) alternate several rounds composed of a key addition, one or several linear transformation(s), and a non-linear transformation. The main difficulty in designing masking schemes for such block ciphers lies in masking the nonlinear transformations. Many solutions have been proposed to deal with this issue but the design of a dth-order secure scheme for $d > 1$ has quickly been recognized as a difficult problem by the community. As mentioned above, only three methods exist in the literature that have been respectively proposed by Ishai, Sahai and Wagner [17], by Schramm and Paar [37] (secure only for $d \leqslant 2$) and by Rivain, Dottax and Prouff [33] (dedicated to $d = 2$). Among them, only [17] can be applied to secure a non-linear transformation at any order d. This scheme is recalled in the next section.

2.4   The Ishai-Sahai-Wagner Scheme

In [17], Ishai *et al.* propose a higher-order masking scheme (referred to as ISW in this paper) enabling to secure the hardware implementation of any *circuit* at any chosen order d. They describe a way to transform the circuit to protect into a new circuit (dealing with masked values) such that no subset of d of its *wires* reveals information about the unmasked values[1]. For such a purpose, they assume without loss of generality that the circuit to protect is exclusively composed of NOT and AND gates. Securing a NOT for any order d is straightforward since $x = \bigoplus_i x_i$ implies $\text{NOT}(x) = \text{NOT}(x_0) \oplus x_1 \cdots \oplus x_d$. The main difficulty is therefore to secure the AND gates. To answer this issue, Ishai *et al.* suggest the following elegant solution.

Secure logical AND. Let a an b be two bits and let c denote $\text{AND}(a, b) = ab$. Let us assume that a and b have been respectively split into $d+1$ shares $(a_i)_{0 \leqslant i \leqslant d}$ and $(b_i)_{0 \leqslant i \leqslant d}$ such that $\bigoplus_i a_i = a$ and $\bigoplus_i b_i = b$. To securely compute a $(d+1)$-tuple $(c_i)_{0 \leqslant i \leqslant d}$ s.t. $\bigoplus_i c_i = c$, Ishai *et al.* perform the following steps:

1. For every $0 \leqslant i < j \leqslant d$, pick up a random bit $r_{i,j}$.
2. For every $0 \leqslant i < j \leqslant d$, compute $r_{j,i} = (r_{i,j} \oplus a_i b_j) \oplus a_j b_i$.
3. For every $0 \leqslant i \leqslant d$, compute $c_i = a_i b_i \oplus \bigoplus_{j \neq i} r_{i,j}$.

Remark 1. The use of brackets indicates the order in which the operations are performed, which is mandatory for security of the scheme.

The completeness of the solution follows from:

$$\bigoplus_i c_i = \bigoplus_i \left(a_i b_i \oplus \bigoplus_{j \neq i} r_{i,j} \right) = \bigoplus_i \left(a_i b_i \oplus \bigoplus_{j > i} r_{i,j} \oplus \bigoplus_{j < i} (r_{j,i} \oplus a_i b_j \oplus a_j b_i) \right)$$

$$= \bigoplus_i \left(a_i b_i \oplus \bigoplus_{j < i} (a_i b_j \oplus a_j b_i) \right) = \left(\bigoplus_i a_i \right) \left(\bigoplus_i b_i \right) .$$

In [17] it is shown that the AND computation above is secure against any attack of order lower than or equal to $d/2$. As stated in Section 4 (and proven the full version of the paper [35]) this scheme is actually dth-order secure.

Practical issues. Although the ISW scheme is an important theoretical result, its practical application suffers few issues. Firstly, it induces an important overhead in silicon area for the masked circuit. Indeed, every single AND gate is encoded using $(d + 1)^2$ AND gates plus $2d(d + 1)$ XOR gates, and it requires the generation of $d(d+1)/2$ random bits at every clock cycle. As an illustration, masking the compact circuit for the AES S-box described in [5] would multiply its size (in terms of number of gates) by 7 for $d = 2$, by 14 for $d = 3$ and by 22 for $d = 4$ (without taking the random bits generation into account). Secondly,

---

[1] Considering wires as intermediate variables, this is equivalent to the security property given in Section 2.3.

masking at the hardware level is sensitive to glitches, which induces first-order flaws although in theory every internal wire carries values that are independent of the sensitive variables [19, 20]. Preventing glitches in masked circuits imply the addition of synchronizing elements (*e.g.* registers or latches) which still significantly increases the circuit size (see for instance [31]).

Since software implementations of masking schemes do not suffer area overhead and are not impacted by the presence of glitches at the hardware level, a straightforward approach to deal with the practical issues discussed above could be to implement the ISW scheme in software. Namely, we could represent each non-linear transformation S to protect by a tuple of Boolean functions $(f_i)_i$ usually called *coordinate functions* of S, and evaluate the f_i's with the ISW scheme by processing the AND and XOR operations with CPU instructions. However, this approach is not practical since the timing overhead would clearly be prohibitive. The present paper follows a different approach: we generalize the ISW scheme to secure any finite field multiplication rather than a simple multiplication over $\mathbb{F}_2$ (*i.e.* a logical AND). We apply this idea to design a secure higher-order masking scheme for the AES and we show that its software implementation induces a reasonable overhead.

3   Higher-Order Masking of AES

The AES block cipher iterates a round transformation composed of a key addition, a linear layer and a nonlinear layer which applies the same substitution-box (S-box) to every byte of the internal state. As previously explained, the main difficulty while designing a masking scheme for such a cipher is the masking of the nonlinear transformation, which in that case lies in the masking of the S-box. Our method for masking the AES S-box is presented in the next section.

In what follows, we shall consider that a random generator is available which on an invocation rand(n) returns n unbiased random bits.

3.1   Higher-Order Masking of the AES S-Box

The AES S-box S is defined as the right-composition of an affine transformation Af over $\mathbb{F}_2^8$ with the power function $x \mapsto x^{254}$ over the field $\mathbb{F}_{2^8} \equiv \mathbb{F}_2[x]/(x^8 + x^4 + x^3 + x + 1)$. Since the affine transformation is straightforward to mask, our scheme mainly consists in a method for masking the power function at any order d. Our solution consists in a secure computation of the exponentiation to the power 254 over $\mathbb{F}_{2^8}$. Such an approach has already been described by Blömer *et al.* for $d = 1$ [3]. The core idea is to apply an exponentiation algorithm (*e.g.* the square-and-multiply algorithm) on the first-order masked input while ensuring the mask correction step by step. Compared to Blömer *et al.* 's solution, our exponentiation algorithm is able to operate on dth-order masked inputs and it achieves dth-order SCA security for any value of d. To perform such a secure exponentiation, we define hereafter some methods to securely compute a squaring and a multiplication over $\mathbb{F}_{2^8}$ at the dth order.

Masking the field squaring. Since we operate on a field of characteristic 2, the squaring is a linear operation and we have $x_0^2 \oplus x_1^2 \oplus \cdots \oplus x_d^2 = x^2$. Securely computing a squaring can hence be carried out by squaring every share separately. More generally, for every natural integer j, raising x to the power 2^j can be done securely by raising each x_i to the 2^j separately.

Masking the field multiplication. For the usual field multiplication we use the ISW scheme recalled in Section 2.4. Even if it has been described to securely compute a logical AND (that is a multiplication over $\mathbb{F}_2$), it can actually be transposed to secure a multiplication over any field of characteristic 2: variables over $\mathbb{F}_2$ are replaced by variables over $\mathbb{F}_{2^n}$, binary multiplications (*i.e.* ANDs) are replaced by multiplications over $\mathbb{F}_{2^n}$ and binary additions (*i.e.* XORs) are replaced by addition over $\mathbb{F}_{2^n}$ (that are n-bit XORs). This keep unchanged the completeness of the scheme recalled in Section 2.4. The whole secure multiplication over $\mathbb{F}_{2^n}$ is depicted in the following algorithm.

Algorithm 1. SecMult - dth-order secure multiplication over $\mathbb{F}_{2^n}$

INPUT: shares a_i satisfying $\bigoplus_i a_i = a$, shares b_i satisfying $\bigoplus_i b_i = b$
OUTPUT: shares c_i satisfying $\bigoplus_i c_i = ab$

---

1. **for** $i = 0$ **to** d **do**
2.   **for** $j = i + 1$ **to** d **do**
3. $r_{i,j} \leftarrow \mathsf{rand}(n)$
4. $r_{j,i} \leftarrow (r_{i,j} \oplus a_i b_j) \oplus a_j b_i$
5. **for** $i = 0$ **to** d **do**
6. $c_i \leftarrow a_i b_i$
7.   **for** $j = 0$ **to** $d,\, j \neq i$ **do** $c_i \leftarrow c_i \oplus r_{i,j}$

Masking the power function. Now we have a secure squaring and a secure multiplication over $\mathbb{F}_{2^8}$ it remains to specify an exponentiation algorithm. It is clear from Algorithm 1 that the running time of a secure multiplication is huge compared to that of a secure squaring. A secure squaring indeed requires $d + 1$ squarings while a secure multiplication requires $(d + 1)^2$ field multiplications, $2d(d+1)$ XORs and the generation of $d(d+1)/2$ random 8-bit values. Our goal is therefore to design an exponentiation algorithm using the least possible multiplications which are not squares. It can be checked that an exponentiation to the power 254 requires at least 4 such multiplications. The exponentiation algorithm presented hereafter achieves this lower bound and requires few additional squares. It involves three intermediate variables denoted y, z and w (note that x and y may be associated to the same memory address).

Algorithm 2. Exponentiation to the 254

INPUT: x
OUTPUT: $y = x^{254}$

1. $z \leftarrow x^2$ $\qquad\qquad\qquad\qquad\qquad\qquad\qquad\qquad\qquad\qquad\qquad$ $[z = x^2]$
2. $y \leftarrow zx$ $\qquad\qquad\qquad\qquad\qquad\qquad\qquad\qquad\qquad\qquad\qquad$ $[y = x^2 x = x^3]$
3. $w \leftarrow y^4$ $\qquad\qquad\qquad\qquad\qquad\qquad\qquad\qquad\qquad\qquad$ $[w = (x^3)^4 = x^{12}]$
4. $y \leftarrow yw$ $\qquad\qquad\qquad\qquad\qquad\qquad\qquad\qquad\qquad\qquad$ $[y = x^3 x^{12} = x^{15}]$
5. $y \leftarrow y^{16}$ $\qquad\qquad\qquad\qquad\qquad\qquad\qquad\qquad\qquad$ $[y = (x^{15})^{16} = x^{240}]$
6. $y \leftarrow yw$ $\qquad\qquad\qquad\qquad\qquad\qquad\qquad\qquad$ $[y = x^{240} x^{12} = x^{252}]$
7. $y \leftarrow yz$ $\qquad\qquad\qquad\qquad\qquad\qquad\qquad\qquad$ $[y = x^{252} x^2 = x^{254}]$

As argued in the full version of this paper [35], for the dth-order security to hold, it is important that the masks $(a_i)_{i \geqslant 1}$ and $(b_i)_{i \geqslant 1}$ in input of the SecMult algorithm are mutually independent. That is why we shall refresh the masks at some points during the secure exponentiation by calling a procedure RefreshMasks[2]. The whole exponentiation to the power 254 over $\mathbb{F}_{2^8}$ secure against dth-order SCA is depicted in the following algorithm.

Algorithm 3. SecExp254 - dth-order secure exponentiation to the 254 over $\mathbb{F}_{2^8}$

INPUT: shares x_i satisfying $\bigoplus_i x_i = x$
OUTPUT: shares y_i satisfying $\bigoplus_i y_i = x^{254}$

1. **for** $i = 0$ **to** d **do** $z_i \leftarrow x_i^2$ $\qquad\qquad\qquad\qquad\qquad\qquad\qquad$ $[\bigoplus_i z_i = x^2]$
2. RefreshMasks$(z_0, z_1, \ldots, z_d)$
3. $(y_0, y_1, \ldots, y_d) \leftarrow$ SecMult$\big((z_0, z_1, \ldots, z_d), (x_0, x_1, \ldots, x_d)\big)$ $\qquad$ $[\bigoplus_i y_i = x^3]$
4. **for** $i = 0$ **to** d **do** $w_i \leftarrow y_i^4$ $\qquad\qquad\qquad\qquad\qquad\qquad$ $[\bigoplus_i w_i = x^{12}]$
5. RefreshMasks$(w_0, w_1, \ldots, w_d)$
6. $(y_0, y_1, \ldots, y_d) \leftarrow$ SecMult$\big((y_0, y_1, \ldots, y_d), (w_0, w_1, \ldots, w_d)\big)$ $\quad$ $[\bigoplus_i y_i = x^{15}]$
7. **for** $i = 0$ **to** d **do** $y_i \leftarrow y_i^{16}$ $\qquad\qquad\qquad\qquad\qquad\qquad$ $[\bigoplus_i y_i = x^{240}]$
8. $(y_0, y_1, \ldots, y_d) \leftarrow$ SecMult$\big((y_0, y_1, \ldots, y_d), (w_0, w_1, \ldots, w_d)\big)$ $\quad$ $[\bigoplus_i y_i = x^{252}]$
9. $(y_0, y_1, \ldots, y_d) \leftarrow$ SecMult$\big((y_0, y_1, \ldots, y_d), (z_0, z_1, \ldots, z_d)\big)$ $\quad$ $[\bigoplus_i y_i = x^{254}]$

For completeness, we describe the RefreshMasks algorithm hereafter.

Algorithm 4. RefreshMasks

INPUT: shares x_i satisfying $\bigoplus_i x_i = x$
OUTPUT: shares x_i satisfying $\bigoplus_i x_i = x$

1. **for** $i = 1$ **to** d **do**
2. $\qquad tmp \leftarrow \mathsf{rand}(8)$
3. $\qquad x_0 \leftarrow x_0 \oplus tmp$
4. $\qquad x_i \leftarrow x_i \oplus tmp$

---

[2] Note that the masks resulting from the SecMult algorithm are independent of the input masks.

Table 1. Complexity of SecExp254

order	nb. XORs	nb. mult.	nb. $\hat{\ }2^j$	nb. rand. bytes	memory (bytes)
1	20	16	6	6	7
2	56	36	9	16	12
3	108	64	12	20	18
4	176	100	15	48	25
5	260	144	18	70	33
d	$8d^2 + 12d$	$4d^2 + 8d + 4$	$3d + 3$	$2d^2 + 4d$	$\frac{1}{2}d^2 + \frac{7}{2}d + 3$

Algorithm 3 involves of $8d(d+1) + 4d$ XORs, $4(d+1)^2$ multiplications (over $\mathbb{F}_{2^8}$), $d+1$ squares, $d+1$ raising to the 4 and $d+1$ raising to the 16. It uses $3(d+1) + d(d+1)/2$ bytes of memory[3] and it requires the generation of $2d(d+1) + 2d$ random bytes (see illustrative values in Table 1). In comparison, the 2nd-order countermeasures previously published [33, 37] require at least 512 look-ups and 512 XORs and have a memory consumption of at least 256 bytes (see [32, 34] for a detailed comparison).

Masking the full S-box. The affine transformation is straightforward to mask. After recalling that the additive part of Af equals 0x63, it can be checked that we have:

$$Af(x_0) \oplus Af(x_1) \oplus \cdots \oplus Af(x_d) = \begin{cases} Af(x) & \text{if } d \text{ is even,} \\ Af(x) \oplus \texttt{0x63} & \text{if } d \text{ is odd.} \end{cases}$$

Masking the affine transformation hence simply consists in applying it to every input share separately and, in case of an even d, in adding 0x63 to one of the share afterward. The full S-box computation secure against dth-order SCA is summarized in the following algorithm.

Algorithm 5. SecSbox

INPUT: shares x_i satisfying $\bigoplus_i x_i = x$
OUTPUT: shares y_i satisfying $\bigoplus_i y_i = \mathrm{S}(x)$

1. $(y_0, \ldots, y_d) \leftarrow \mathsf{SecExp254}(x_0, \ldots, x_d)$
2. **for** $i = 0$ **to** d **do** $y_i \leftarrow Af(y_i)$
3. **if** $(d \bmod 2 = 1)$ **then** $y_0 \leftarrow y_0 \oplus \texttt{0x63}$

Implementation aspects. Multiplications over $\mathbb{F}_{2^8}$ are typically implemented in software using log/alog tables (see for instance [11]). Note that for security reasons, such an implementation must avoid conditional branches in order to ensure a constant operation flow. The squaring and raisings to the 4 and 16 may be looked-up. Different time-memory tradeoffs are possible. If not much ROM is available, the squaring can be implemented using logical shifts and

---

[3] $3(d+1)$ bytes for the shares y_i's, z_i's and w_i's (Algorithm 3), and $d(d+1)/2$ for the intermediate variables $r_{i,j}$'s (Algorithm 1).

XORs (see for instance [11]), and the raising to the 2^j, $j \in \{2, 4\}$, can then be simply processed by j sequential squarings. Otherwise, depending on the amount of ROM available, one can either use one, two or three look-up table(s) to implement the raisings to 2^j, $j \in \{1, 2, 4\}$.

Remark 2. For the implementations presented in Section 5, we chose to implement the squaring by a look-up table, getting the raising to the 4 (resp. 16) by accessing this table sequentially 2 (resp. 4) times.

Our scheme may also be implemented in hardware. The sensitive part is the implementation of the SecMult algorithm (see Algorithm 1) which may be subject to glitches and which should incorporate synchronizing elements. In particular, the evaluation of the c_i shares should not start before the evaluation of all the $r_{i,j}$'s has been fully completed. Another approach would be to enhance the software implementation of the scheme with special purpose hardware instructions. For instance, the multiplication, squaring and raisings to powers 4 and 16 over $\mathbb{F}_{2^8}$ could be added to the instructions set of the processor.

3.2   Higher-Order Masking of the Whole Cipher

In the previous section, we have shown how the AES S-box can be masked at any chosen order d. Since the S-box is actually the most difficult part of AES to mask, and due to length constraints, we do not detail the masking of the whole AES cipher here. This description is given in the full version of this paper [35].

4   Security Analysis

In this section, we give a sketch of the security proof of our scheme. We first formally define the notion of *dth-order SCA security* and we introduce afterward our main security result (Theorem 1). The complete security proof is given in the full version of the paper [35].

We consider a *randomized encryption algorithm* $\mathcal{E}$ taking a plaintext p and a (randomly shared) secret key k as inputs[4] and performing a deterministic encryption of p under the secret key k while randomizing its internal computations by means of an external random number generator (RNG). The RNG outputs are assumed to be perfectly random (uniformly distributed, mutually independent and independent of the plaintext and of the secret key). Any variable that can be expressed as a deterministic function of the plaintext and the secret key, which is not constant with respect to the secret key, is called a *sensitive variable* with the exception of the ciphertext $\mathcal{E}_k(p)$ or any deterministic function of it. Note that every intermediate variable computed during an execution of $\mathcal{E}$ (except the plaintext and the ciphertext) can be expressed as a deterministic function of a sensitive variable and of the RNG outputs.

---

[4] The secret key k is assumed to be split into $d + 1$ shares k_0, k_1, ..., k_d such that $\bigoplus_i k_i = k$ and every d-tuple of k_i's is uniformly distributed and independent of k.

We shall consider the plaintext, the secret key and the intermediate variables of $\mathcal{E}$ as random variables. The distributions of the intermediate variables are induced by the algorithm inputs (p and k) distributions and by the uniformity of the RNG outputs. The joint distribution of all the intermediate variables of $\mathcal{E}$ thus depends on (p, k). On the other hand, some subsets of intermediate variables may be jointly independent of (p, k). This leads us to the following formal definition of dth-order SCA security.

Definition 1. *A randomized encryption algorithm is said to achieve dth-order SCA security if every d-tuple of its intermediate variables is independent of any sensitive variable.*

Equivalently, an encryption algorithm achieves dth-order SCA security if any d-tuple of its intermediate variables, except the plaintext and the ciphertext (or any function of one of them), is independent of the algorithm inputs (p, k).

The most sensitive part of our scheme is the masked multiplication algorithm based on the generalized ISW scheme (Algorithm 1). The theorem hereafter states that it achieves dth-order SCA security.

Theorem 1. *Let a and b be two sensitive variables. Let $(a_i)_{0 \leqslant i \leqslant d}$ and $(b_i)_{0 \leqslant i \leqslant d}$ be two families of intermediate variables in input of Algorithm 1 satisfying $a = \bigoplus_{0 \leqslant i \leqslant d} a_i$ and $b = \bigoplus_{0 \leqslant i \leqslant d} b_i$ with $(a_i)_{i \geqslant 1}$ and $(b_i)_{i \geqslant 1}$ being RNG outputs. Then, the distribution of every tuple of d or less intermediate variables in Algorithm 1 is independent of (a, b).*

Theorem 1 states that the generalized ISW scheme achieves dth-order SCA security whereas in [17] it is only proven that the ISW scheme achieves $(d/2)th$-order SCA security. This improvement is of practical interest since it enables to double the security order for any chosen complexity (in terms of timing and/or silicon area).

The proof of Theorem 1 as well as the security proof of our whole dth-order masking scheme for AES are given in the full version of the paper [35].

5   Implementation Results

To compare the efficiency of our proposal with that of other methods proposed in the literature, we applied them to protect an implementation of the AES-128 algorithm in encryption mode. We have implemented our new countermeasure for $d \in \{1, 2, 3\}$, namely to counteract either first-order SCA ($d = 1$) or second-order SCA ($d = 2$) or third-order SCA ($d = 3$). Among the numerous methods proposed in the literature to thwart first-order SCA we chose to implement only that having the best timing performance (the *table re-computation method* [22]) and that offering the best memory performance (the *tower field method* [27]). In the second-order case, we implemented the only two existing methods: the one proposed in [37][5] and the one proposed [33]. Eventually, since no countermeasure

---

[5] Initially, the method of [37] was devoted to thwart dth-order SCA for any chosen order d but it has been shown insecure for $d \geqslant 3$ [8].

Table 2. Comparison of secure AES implementations

Method	Reference	cycles	RAM (bytes)	ROM (bytes)
Unprotected Implementation				
No Masking	Na.	2×10^3	32	1150
First Order Masking				
Re-computation	[22]	10×10^3	$256 + 35$	1553
Tower Field in $\mathbb{F}_4$	[27, 28]	77×10^3	42	3195
Our scheme for $d = 1$	This paper	129×10^3	73	3153
Second Order Masking				
Double Re-computations	[37]	594×10^3	$512 + 90$	2336
Single Re-computation	[33]	672×10^3	$256 + 86$	2215
Our scheme for $d = 2$	This paper	271×10^3	79	3845
Third Order Masking				
Our scheme for $d = 3$	This paper	470×10^3	103	4648

against 3rd-order SCA was existing before that introduced in this paper, it is the single one in its category.

We wrote the codes in assembly language for an 8051-based 8-bit architecture. The implementations only differ in their approaches to protect the S-box computations. In Table 2, we list the timing/memory performances of the different implementations.

As expected, in the first-order case the countermeasures introduced in [22] and [27, 28] are much more efficient than ours. This is a consequence of the generic character of our method which is not optimized for one choice of d but aims to work for any d.

In the second-order case, our proposal becomes much more efficient than the existing solutions. It is 2.2 times faster than the countermeasure proposed in [37] with a RAM memory requirement divided by around 10. It is also 2.5 times faster than the countermeasure in [33] and requires 5.3 times less RAM. Memory allocation differences are merely due to the fact that the methods [37] and [33] generalize the table re-computation method and thus require the storage of one (for [33]) or two (for [37]) randomized representation(s) of the AES S-box. The differences in timing performances come from the fact that the methods in [37] and [34] process one loop over all the 256 elements of the S-box look-up table (each loop iteration processing itself a few elementary operations), which is more costly than the 36 field multiplications and 56 bitwise additions involved in our method (see Table 1).

Eventually, in the third-order case our method has acceptable timing/memory performances. For comparison, it stays faster than the second-order countermeasures proposed in [37] and [33] and it still requires much less RAM memory. For a chip running at 31MHz (which is today quite usual) an AES encryption of one block requiring 470×10^3 cycles, takes 91ms. For some use cases where the size of the message to encrypt/decrypt is not too long such a timing performance is acceptable (*e.g.* challenge-response protocols, Message Authentication Codes for one-block messages as in banking transactions).

6   Conclusion

In this paper, we have presented the first masking scheme dedicated to AES which is provably secure at any chosen order and which can be implemented in software at the cost of a reasonable overhead. We provided implementation results showing the practical interest of our scheme as well as its efficiency compared to the existing second-order masking schemes. In the full version of this paper [35], we further give a formal security proof of our scheme including an improved security proof for the scheme published by Ishai *et al.* at Crypto 2003.

References

1. Akkar, M.-L., Giraud, C.: An Implementation of DES and AES, Secure against Some Attacks. In: Koçc, çC.K., Naccache, D., Paar, C. (eds.) CHES 2001. LNCS, vol. 2162, pp. 309–318. Springer, Heidelberg (2001)
2. Blakely, G.: Safeguarding cryptographic keys. In: National Comp. Conf., New York, June 1979, vol. 48, pp. 313–317. AFIPS Press (1979)
3. Blömer, J., Merchan, J.G., Krummel, V.: Provably Secure Masking of AES. In: Matsui, M., Zuccherato, R. (eds.) SAC 2004. LNCS, vol. 3357, pp. 69–83. Springer, Heidelberg (2004)
4. Blum, M., Micali, S.: How to Generate Cryptographically Strong Sequences of Pseudo-Random Bits. SIAM J. Comput. 13(4), 850–864 (1984)
5. Canright, D.: A Very Compact S-Box for AES. In: Rao, J., Sunar, B. (eds.) CHES 2005. LNCS, vol. 3659, pp. 441–455. Springer, Heidelberg (2005)
6. Chari, S., Jutla, C., Rao, J., Rohatgi, P.: Towards Sound Approaches to Counteract Power-Analysis Attacks. In: Wiener, M. (ed.) CRYPTO 1999. LNCS, vol. 1666, pp. 398–412. Springer, Heidelberg (1999)
7. Chari, S., Rao, J., Rohatgi, P.: Template Attacks. In: Kaliski Jr., B., Koçc, çC.K., Paar, C. (eds.) CHES 2002. LNCS, vol. 2523, pp. 13–28. Springer, Heidelberg (2003)
8. Coron, J.-S., Prouff, E., Rivain, M.: Side Channel Cryptanalysis of a Higher Order Masking Scheme. In: Paillier, P., Verbauwhede, I. (eds.) CHES 2007. LNCS, vol. 4727, pp. 28–44. Springer, Heidelberg (2007)
9. Cramer, R., Damgård, I., Ishai, Y.: Share Conversion, Pseudorandom Secret-Sharing and Applications to Secure Computation. In: Kilian, J. (ed.) TCC 2005. LNCS, vol. 3378, pp. 342–362. Springer, Heidelberg (2005)
10. Crescenzo, G.D., Lipton, R.J., Walfish, S.: Perfectly Secure Password Protocols in the Bounded Retrieval Model. In: Halevi, S., Rabin, T. (eds.) TCC 2006. LNCS, vol. 3876, pp. 225–244. Springer, Heidelberg (2006)
11. Daemen, J., Rijmen, V.: The Design of Rijndael. Springer, Heidelberg (2002)
12. Damgård, I., Keller, M.: Secure Multiparty AES (full paper). Cryptology ePrint Archive, Report 20079/614 (2009), http://eprint.iacr.org/
13. Dziembowski, S., Pietrzak, K.: Leakage-resilient cryptography. In: FOCS, pp. 293–302. IEEE Computer Society, Los Alamitos (2008)
14. FIPS PUB 197. Advanced Encryption Standard. National Institute of Standards and Technology (November 2001)
15. FIPS PUB 46-3. Data Encryption Standard (DES). National Institute of Standards and Technology (October 1999)

16. Goubin, L., Patarin, J.: DES and Differential Power Analysis – The Duplication Method. In: Koçc, ç.K., Paar, C. (eds.) CHES 1999. LNCS, vol. 1717, pp. 158–172. Springer, Heidelberg (1999)

17. Ishai, Y., Sahai, A., Wagner, D.: Private Circuits: Securing Hardware against Probing Attacks. In: Boneh, D. (ed.) CRYPTO 2003. LNCS, vol. 2729, pp. 463–481. Springer, Heidelberg (2003)

18. Kocher, P., Jaffe, J., Jun, B.: Differential Power Analysis. In: Wiener, M. (ed.) CRYPTO 1999. LNCS, vol. 1666, pp. 388–397. Springer, Heidelberg (1999)

19. Mangard, S., Popp, T., Gammel, B.M.: Side-Channel Leakage of Masked CMOS Gates. In: Menezes, A. (ed.) CT-RSA 2005. LNCS, vol. 3376, pp. 351–365. Springer, Heidelberg (2005)

20. Mangard, S., Pramstaller, N., Oswald, E.: Successfully Attacking Masked AES Hardware Implementations. In: Rao, J., Sunar, B. (eds.) CHES 2005. LNCS, vol. 3659, pp. 157–171. Springer, Heidelberg (2005)

21. Maurer, U.: A provably-secure strongly-randomized cipher. In: Damgård, I. (ed.) EUROCRYPT 1990. LNCS, vol. 473, pp. 361–388. Springer, Heidelberg (1991)

22. Messerges, T.: Securing the AES Finalists against Power Analysis Attacks. In: Schneier, B. (ed.) FSE 2000. LNCS, vol. 1978, pp. 150–164. Springer, Heidelberg (2001)

23. Messerges, T.: Using Second-order Power Analysis to Attack DPA Resistant Software. In: Paar, C., Koçc, ç.K. (eds.) CHES 2000. LNCS, vol. 1965, pp. 238–251. Springer, Heidelberg (2000)

24. Micali, S., Reyzin, L.: Physically Observable Cryptography (Extended Abstract). In: Naor, M. (ed.) TCC 2004. LNCS, vol. 2951, pp. 278–296. Springer, Heidelberg (2004)

25. Nikova, S., Rijmen, V., Schläffer, M.: Secure Hardware Implementation of Non-linear Functions in the Presence of Glitches. In: Lee, P.J., Cheon, J.H. (eds.) ICISC 2008. LNCS, vol. 5461, pp. 218–234. Springer, Heidelberg (2009)

26. Oswald, E., Mangard, S., Herbst, C., Tillich, S.: Practical Second-order DPA Attacks for Masked Smart Card Implementations of Block Ciphers. In: Pointcheval, D. (ed.) CT-RSA 2006. LNCS, vol. 3860, pp. 192–207. Springer, Heidelberg (2006)

27. Oswald, E., Mangard, S., Pramstaller, N.: Secure and Efficient Masking of AES – A Mission Impossible? Cryptology ePrint Archive, Report 2004/134 (2004)

28. Oswald, E., Mangard, S., Pramstaller, N., Rijmen, V.: A Side-Channel Analysis Resistant Description of the AES S-box. In: Gilbert, H., Handschuh, H. (eds.) FSE 2005. LNCS, vol. 3557, pp. 413–423. Springer, Heidelberg (2005)

29. Petit, C., Standaert, F.-X., Pereira, O., Malkin, T., Yung, M.: A block cipher based pseudo random number generator secure against side-channel key recovery. In: Abe, M., Gligor, V.D. (eds.) Symposium on Information, Computer and Communications Security – ASIACCS 2008, pp. 56–65. ACM, New York (2008)

30. Pietrzak, K.: A Leakage-Resilient Mode of Operation. In: Joux, A. (ed.) EUROCRYPT 2009. LNCS, vol. 5479, pp. 462–482. Springer, Heidelberg (2010)

31. Popp, T., Kirschbaum, M., Zefferer, T., Mangard, S.: Evaluation of the Masked Logic Style MDPL on a Prototype Chip. In: Paillier, P., Verbauwhede, I. (eds.) CHES 2007. LNCS, vol. 4727, pp. 81–94. Springer, Heidelberg (2007)

32. Rivain, M.: On the Physical Security of Cryptographic Implementations. PhD thesis, University of Luxembourg (September 2009)

33. Rivain, M., Dottax, E., Prouff, E.: Block Ciphers Implementations Provably Secure Against Second Order Side Channel Analysis. In: Baignères, T., Vaudenay, S. (eds.) FSE 2008. LNCS, vol. 5086, pp. 127–143. Springer, Heidelberg (2008)

34. Rivain, M., Dottax, E., Prouff, E.: Block Ciphers Implementations Provably Secure Against Second Order Side Channel Analysis. Cryptology ePrint Archive, Report 2008/021 (2008), http://eprint.iacr.org/
35. Rivain, M., Prouff, E.: Provably Secure Higher-Order Masking of AES. Cryptology ePrint Archive (2010), http://eprint.iacr.org/
36. Rivain, M., Prouff, E., Doget, J.: Higher-Order Masking and Shuffling for Software Implementations of Block Ciphers. In: Clavier, C., Gaj, K. (eds.) CHES 2009. LNCS, vol. 5747, pp. 171–188. Springer, Heidelberg (2009)
37. Schramm, K., Paar, C.: Higher Order Masking of the AES. In: Pointcheval, D. (ed.) CT-RSA 2006. LNCS, vol. 3860, pp. 208–225. Springer, Heidelberg (2006)
38. Shamir, A.: How to Share a Secret. ACM Commun. 22(11), 612–613 (1979)
39. Standaert, F.-X., Pereira, O., Yu, Y., Quisquater, J.-J., Yung, M., Oswald, E.: Leakage resilient cryptography in practice. Cryptology ePrint Archive, Report 2009/341 (2009), http://eprint.iacr.org/
40. Standaert, F.-X., Veyrat-Charvillon, N., Oswald, E., Gierlichs, B., Medwed, M., Kasper, M., Mangard, S.: The World is Not Enough: Another Look on Second-Order DPA. Cryptology ePrint Archive, Report 2010/180 (2010), http://eprint.iacr.org/
41. Tillich, S., Herbst, C.: Attacking State-of-the-Art Software Countermeasures-A Case Study for AES. In: Oswald, E., Rohatgi, P. (eds.) CHES 2008. LNCS, vol. 5154, pp. 228–243. Springer, Heidelberg (2008)

Algebraic Side-Channel Analysis in the Presence of Errors

Yossef Oren[1], Mario Kirschbaum[2], Thomas Popp[2], and Avishai Wool[1]

[1] Computer and Network Security Lab, School of Electrical Engineering
Tel-Aviv University, Ramat Aviv 69978, Israel
{yos,yash}@eng.tau.ac.il

[2] Institute for Applied Information Processing and Communications
Graz University Of Technology, Inffeldgasse 16a, A-8010, Austria
{mario.kirschbaum,thomas.popp}@iaik.tugraz.at

Abstract. Measurement errors make power analysis attacks difficult to mount when only a single power trace is available: the statistical methods that make DPA attacks so successful are not applicable since they require many (typically thousands) of traces. Recently it was suggested by [18] to use algebraic methods for the single-trace scenario, converting the key recovery problem into a Boolean satisfiability (SAT) problem, then using a SAT solver. However, this approach is extremely sensitive to noise (allowing an error rate of well under 1% at most), and the question of its practicality remained open. In this work we show how a single-trace side-channel analysis problem can be transformed into a *pseudo-Boolean optimization* (PBOPT) problem, which takes errors into consideration. The PBOPT instance can then be solved using a suitable optimization problem solver. The PBOPT syntax provides for a more expressive input specification which allows a very natural representation of measurement errors. Most importantly, we show that using our approach we are able to mount successful and efficient single-trace attacks even in the presence of realistic error rates of 10%–20%. We call our new attack methodology *Tolerant Algebraic Side-Channel Analysis (TASCA)*. We show practical attacks on two real ciphers: Keeloq and AES.

Keywords: Algebraic attacks, power analysis, side-channel attacks, pseudo-Boolean optimization.

1 Introduction

1.1 Background

Side-channel cryptanalysis has been an active field of research for the last 15 years. For the simplest devices, that are susceptible to Simple Power Analysis attacks (SPA) [12], the secret key can be read directly from the shape of a side-channel trace (power consumption, EM radiation, etc.). More commonly, the cryptanalyst needs to use differential (DPA) analysis [12,14]. DPA techniques

S. Mangard and F.-X. Standaert (Eds.): CHES 2010, LNCS 6225, pp. 428–442, 2010.

typically require multiple traces, often hundreds or more, to overcome the measurement noise via signal processing and statistical estimation techniques. Obtaining all these traces places a significant burden on the attacker, and it is quite interesting to discover ways to extract the secret key data from a *single trace* from devices that are not susceptible to SPA.

Recently it was suggested by [18] to separate the problem into two separate phases: the first phase is the *estimation phase*, where information is extracted from the power traces using signal processing techniques, while the second is the *key recovery phase*, where this information is processed to return cryptanalytically significant results. In particular, [18] uses *algebraic methods* for the key recovery phase, converting the problem into a Boolean satisfiability (SAT) problem, then using a SAT solver. Algebraic cryptanalytic attacks using external solvers were first explored by Massacci and Marraro in [16] in the context of conventional cryptanalysis. However, these attacks are difficult to apply directly to side-channel attacks, since the SAT representation of a cryptosystem and its side-channel measurements is extremely sensitive to noise — indeed [18] were only able to solve problems with an error rate well under 1%, which is much lower than realistic noise on a single trace. Side-channel analysis using standard solvers was also suggested by [17]. Our goal in this paper is to demonstrate a more promising algebraic cryptanalysis approach, based on Pseudo-Boolean optimization, which is able to withstand much higher error rates.

Other non-algebraic methods have also been suggested for dealing with single-trace power analysis in the presence of noise. A side-channel attack using the Viterbi iterative algorithm [20] for dealing with errors, first presented in the context of elliptic-curve operations in [11], is one example.

1.2   Causes of Errors in Side-Channel Information

The side-channel information emitted by a cryptographic device is an analog high-frequency signal that is measured with a suitable instrument. In case of the power consumption side-channel, the logic cells in the digital circuit draw power from the supply according to their state and activity. This instantaneous power consumption signal is measured with an oscilloscope. The measurement process includes an analog-to-digital conversion of the sampled values. On their way from the logic cell to the oscilloscope's digital output, the power values are influenced by all kinds of physical effects and other signals. These influences are commonly denoted as *noise*. This noise can cause *decoding errors* when trying to estimate the original power consumption of the logic cell.

The overall noise that is present in measured power traces can be divided into electronic noise, quantization noise, and switching noise. Electronic noise is present in every measurement in practice. It includes the noise that occurs in conductors (e.g. thermal noise) and semiconductors (e.g. generation-recombination noise). Furthermore, sources of electronic noise are the conducted and radiated emissions from all components that are part of the control and measurement setup and from external components that operate in the vicinity of the measured cryptographic device. These components include the supply unit that powers the device and the oscilloscope. Another important source of electronic noise is the clock generator that supplies the digital circuit in the cryptographic device with the clock signal. Due to its typical rectangular shape, this signal contains high-frequency components that also influence the measured power values.

The digital oscilloscope contains another source of noise. The analog-to-digital conversion process that it performs introduces small errors in the measured values. The effect of these errors can be modeled as noise in the measured signal, commonly called *quantization noise*. The higher the resolution of the oscilloscope, the lower is the amount of quantization noise.

The third main type of noise is switching noise. Besides the power consumption of the logic cells we are interested in, typically also other cells contribute to the total power consumption value at a specific point in time. The power signals from these other cells are denoted as switching noise. The main parameters of the control and measurement setup that influence the amount of switching noise for a specific point in time are the bandwidth of the power measurement system and the clock frequency. The lower the bandwidth of the measurement path the more the distinct power consumption signals of individual logic cells get blurred together and the amount of switching noise increases. A higher clock frequency can also have such an effect[14].

1.3   Contributions

Our key observation is that a SAT representation does not offer a very convenient or efficient method to deal with errors in side-channel information. Instead, we suggest casting the problem in the more expressive language of non-linear *pseudo-Boolean optimization* (PBOPT).

A PBOPT representation offers several properties that are suitable for single-trace side-channel attacks in the presence of errors: (a) A side-channel measurement is typically the Hamming weight w of some hardware feature: this is naturally represented by equating a sum of state bits to the *integer w* – in contrast, representing integers in a SAT instance is quite awkward; (b) It is straight-forward to add variables representing error quantities to the side-channel equations; (c) Unlike a SAT, that is basically a decision ("yes/no") problem, a PBOPT instance includes an objective function, and the solver finds a solution that minimizes this objective.

Luckily, PBOPT offers more than a convenient representation formalism. Research on non-linear pseudo-Boolean equation solvers is a field which displays remarkable activity, and even has a highly-competitive yearly evaluation of solvers [15]. Thus, a PBOPT instance representing a single-trace side-channel attack with errors can actually be solved efficiently, leading to our new attack methodology: *Tolerant Algebraic Side-Channel Analysis (TASCA)*.

To demonstrate the viability of our TASCA approach, we mounted successful and efficient single-trace attacks, against real, fielded ciphers, even in the presence of realistic error rates of 10%–20%. We show a practical attack on the Keeloq system, and preliminary results on AES.

Organization: The next section describes the basics of algebraic side-channel attacks. Section 3 describes our new *Tolerant Algebraic Side-Channel Analysis (TASCA)* approach. Section 4 shows the effectiveness of TASCA against a power-simulated ASIC implementation of Keeloq, and Section 5 shows preliminary results against a power-simulated 8-bit microcontroller implementation of AES-128. Section 6 suggests some open problems. We conclude with Section 7.

2   Algebraic Side-Channel Attacks

2.1   General Structure of an Algebraic Attack

As stated in [18], the cryptanalytic problem needs to be transformed into a set of equations before being submitted to the equation solver. This equation set typically consists of a general description of the cryptographic algorithm, together with an assignment of any known inputs to the algorithm. If the equation set represents an algebraic side-channel attack, it will contain additional equations which describe the side-channel emanations of the system in addition to the standard known plaintexts and ciphertexts. Building on the results of [18], we can assume that an errorless description of the side-channel data will lead to successful key recovery. However, such an equation set is very sensitive to noise: a single errored side-channel measurement will create an equation set that is either unsatisfiable, or is satisfied by the wrong key.

The equations are presented to the solver using the solver's problem description language. The authors of [18] used a SAT solver which accepts its input in the form of conjugate normal form (CNF) SAT statements. As we shall see, we use the richer and more powerful pseudo-Boolean optimization representation.

2.2   Naïve Methods of Dealing with Errors

Assume that the vector z represents some side-channel information extracted from a certain cryptographic operation (for example Hamming weights or Hamming distances) under a certain key k_c, and that there exists some distance function $d(k, z)$ which indicates how likely a given vector z is to be the result of the operation under a certain key k. As noted in the introduction, the raw side-channel measurement (or *trace*) in itself does not typically have the form of a vector of Hamming weights and must pass some preprocessing before being used. We consider this process, called *estimation*, external to the attack itself.

A typical way of implementing the distance function $d(k, z)$ is to perform a power simulation of the cryptographic operation using a hardware model of the cryptographic device assuming the key k, obtain from this simulation a vector of simulated side-channel measurements z^k and return the mean-squared error (or L_2 distance) of the two vectors:

$$d(k, z) = \sum_i \left(z_i^k - z_i \right)^2 \tag{1}$$

We can assume that the measurement z was created from the "optimal" measurement z_c by the addition of some noise vector:

$$z = z_c + e \tag{2}$$

The magnitude of the vector e is defined by the noise model and the performance of the estimator. In this paper we assume a moderately effective estimator and limit our discussions to cases in which the maximum amplitude of e is ± 1 bits in each measurement. An estimator is a *hard estimator* if its outputs are discrete symbols without any confidence information. Under our assumptions the hard estimator will always have a measurement error of either -1, 0 or 1 bits. We can now quantify the errors by considering only P_{err}, the probability that e is nonzero in a given location.

We will now describe several well-known ways of attempting to identify and eliminate noise in decoding problems.

Random Subset Decoding. If P_{err} is very low, we can try to sample a random subset of measurement locations. If by chance none of the measurements are errored, we can attempt to recover k_c from the sample and verify its correctness using trial decryption. Assuming a vector with an i.i.d. probability of hard error of P_{err}, the probability that a set of m indices will contain no errors is $(1 - P_{err})^m$. If we assume, for example, that $P_{err} = 0.01$ and that 128 indices are required for an attack to succeed, the overall probability of success is only 27%. For higher error probabilities this method quickly becomes impractical.

Standard Algebraic Attack with Duplication. The algebraic attack presented in [18] requires that the measured side-channel information contains no errors. In such a model, a variant of the random subset method can be used: instead of selecting a subset of the data, we can enumerate over all possible locations of errors in the measured data, then create many duplicate instances, each "fixing" the anticipated measurement errors in a certain location and then attempting to carry out an algebraic attack. All duplicate instances are then combined, while we specify to the solver that a single one of the instances needs to be satisfied. Let us assume for example that we have 128 side-channel measurements and we assume that at most 2 locations out of the 128 contain single-bit errors. In this case we can create $\binom{128}{2}$ duplicate instances, each assuming the errors occurred at a certain pair of locations and "fixing" them. We then specify to the solver that only one out of all of the duplicate instances needs to be satisfied. While most of these duplicates will be unsatisfiable (or result in a wrong recovered key), in one of them the measurement error will indeed be cancelled out by our guess, leading to a successful key recovery. The duplication method is obviously only suitable for a very small amount of errors, since the number of additional instances grows exponentially with the amount of anticipated errors.

Iterative Methods. If the cipher uses the key bits sequentially (bit by bit) in the encryption or decryption process, an iterative Viterbi-like algorithm [20], which is described in detail in [11], can be used to recover errors. The Viterbi algorithm's main parameter is its data structure size, which controls the number of key candidates the algorithm maintains during its operation. Letting this size approach $2^{keysize}$, we can treat the iterative algorithm's output as an effective ordering of all key candidates with increasing distance from the measured side-channel information z. The index of the correct key candidate in this ordered list can be an indication of the effectiveness of the iterative approach for solving this specific problem.

The main disadvantage of the iterative method is that it operates in a greedy manner, and cannot return to a key candidate once it has been disqualified. Essentially, this limits the amount of usable side-channel data to a single use of each key bit. In addition, *diffusion* elements (such as the AES MixColumns operation) highly complicate the operation of iterative methods, since many state bits change almost simultaneously and affect every side-channel measurement.

3   Handling Errors by Pseudo-Boolean Representation

3.1   Side-Channel Analysis as a Pseudo-Boolean Problem

Before we present our approach, let us return to the fundamental problem of side-channel analysis, which can be described as follows:

> Given the algorithmic description of a cryptographic algorithm, the physical power model of the device under attack and the side-channel measurements, output a key assignment for which the expected side-channel information is as close as possible to the measured side-channel information.

When written in the above form, it is clear to see that side-channel analysis is naturally represented as an **optimization problem:**

> Find the **minimal** assignment to an **error vector** such that it is possible for the **cryptographic algorithm**, operating under a certain unknown **key** and in a certain **physical power model,** to produce the **measured side-channel information** affected by this error.

We call the class of attacks which performs cryptanalysis using an optimizer instead of a solver **Tolerant Algebraic Side-Channel Analysis (TASCA)**.

3.2   An Introduction to Pseudo-Boolean Optimizers

The field of pseudo-Boolean optimization (PBOPT) problems is a special case of integer programming problems [5]. Stated informally, a PBOPT instance consists of an *objective (goal) function* and a series of *inequality constraints*, both of which are defined over some set of Boolean variables. A solution to the PBOPT instance must satisfy all inequality constraints while minimizing the objective function. Unlike standard Boolean satisfiability (SAT) problems, a PBOPT problem instance admits multiple solutions, choosing the one solution that minimizes the objective function.

As stated formally in [4], a linear PB problem is an optimization problem over n binary (Boolean) variables $x_1 \cdots x_n$ having the following form:

$$\min c^{\mathrm{T}} x \tag{3}$$

$$Ax \geq b \tag{4}$$

$$x \in \{0,1\}^n \tag{5}$$

where all the coefficients are signed integers: $A \in \mathbb{Z}^{m \times n}, b \in \mathbb{Z}^m, c \in \mathbb{Z}^n$. The term $c^{\mathrm{T}} x$ is the *objective function* and the row inequalities in $Ax \geq b$ are the *linear constraints*. The solvers we are interested in can also accept *non-linear constraints* of the form $\sum_{i=1}^t d_i \prod_{j=1}^k \ell_{i,j} \geq r_i$, where $\ell_{i,j} \in \{x_{i,j}, \bar{x}_{i,j}\}$. Because all coefficients are signed values, equality constraints (of the form $\sum d_i \prod \ell_{i,j} = r_i$) and less-than constraints (of the form $\sum d_i \prod \ell_{i,j} \leq r_i$) can also be reduced to the above form.

Because of their relation to both SAT, linear programming, and integer programming, PB instances can be solved using a variety of approaches. Some solvers attempt to compile the PB instance into a SAT instance and apply a standard SAT solver, possibly multiple times; others map the problem into an integer programming instance; some solvers use a hybrid approach, combining the best features of the two.

The pseudo-Boolean description language is very expressive and allows relatively complex constraints to be described quite efficiently. Notably, each errored side-channel measurement can be efficiently written down as a single equation.

The solver we chose to use is SCIPspx version 1.2.0 [4,3,2]. SCIPspx won the first prize for non-linear optimizer in the Pseudo-Boolean Evaluation Contest of SAT 2009 [15]. SCIPspx solves the optimization problem by using integer programming and constraint programming methods. It performs a branch-and-bound algorithm to decompose the problem into sub-problems, solving a linear relaxation on each sub-problem and finally combining the results. The linear relaxation component of SCIPspx is the standalone LP solver SoPlex [21].

3.3   Elements of a TASCA Equation Set

To represent a side-channel attack as a PB optimization instance a TASCA equation set is written, consisting of the following four sections:

1. **A general description of the cryptographic algorithm as a set of equations:** The cryptosystem is described by writing down internal state transformations leading from plaintext to ciphertext. The specification is very hardware-minded, with each state bit/memory element (flip-flop) typically represented as a sequence of variables representing its evolution in time, and each combinational element (gate) finding its way into an equation connecting the variables. For example, the AES state has 16 bytes, each of which changes its value 4 times in each round (other than the first and the last). This means that the state of each subround is represented by 16×8 binary (0-1) variables, for a total of $16 \times 8 \times 41 = 5248$ variables for an entire AES

encryption. There will also be variables for every key bit and every subkey bit, and a set of equations representing the subkey expansion.

2. **An assignment of any known inputs to the algorithm:** These can be known plaintext or ciphertext, or even more subtle hints such as the relationship between two consecutive unknown plaintexts.

3. **A specification of the measurement setup:** The actual side-channel measurement is mapped to the internal state according to the structure of the physical hardware device. For example, an 8-bit microcontroller-based implementation will typically leak the Hamming weight of individual state bytes as they are accessed, while a parallelized ASIC will typically leak the Hamming distance between the former and present values of all bits in the device's internal state. Note that both in the case of Hamming distance and in the case of Hamming weight the measurement equation consists of an equality between a sum of Boolean variables on one side and an integer value on the other – $\sum_j state_{i,j} = m_i$, where $state_{i,j}$ is the value of state variable j at time i and m_i is the side-channel measurement at time i. This form of equation is natural to write down using the PB syntax. It should be noted that when attacking the same cipher running on different target architectures, the measurement setup is usually the only section of the equation set which needs to be modified.

4. **A set of potentially errored measurements:** This section matches the measurements described in the previous section to actual outputs of the estimation phase. As stated previously, the main point of the TASCA approach is to allow errors in the estimation. This is done in practice by adding additional *error variables* to the above-mentioned measurement equations. These error variables are used to cancel out errors in the measurements. In our implementation we included two error variables per measurement (one with a plus sign and one with a minus), which allow the true side-channel value to be within ±1 bits of the measured one. It should be noted that this section is the only part of the equation set which tolerates errors (all other sections are explicitly defined), and that this section only accounts for 1% to 5% of the entire set of equations for the cryptosystems we tested.

In addition to the equation set, the solver is provided with a **objective function** which it is required to minimize. In our case, our objective is to use as few error variables as possible.

4   An Attack on Keeloq

Keeloq is a block cipher which is most commonly used in remote keyless entry (RKE) systems, e.g. for cars. We chose to attack this cipher first since it has a very simple round structure which is relatively easy to represent as equations. Furthermore, a reduced version of Keeloq (using 140 rounds instead of the full 528) was already broken using standard algebraic techniques in [7] without requiring side-channel inputs. In [10] a physical ASIC implementation of the

Keeloq cipher was shown vulnerable to a standard DPA attack, an attack we were also able to reproduce.

Because it only operates on a single bit of the key in each round, Keeloq is very effectively attacked using the iterative approach described in 2.2.

4.1   The Keeloq Algorithm

The Keeloq algorithm [9] is a block cipher designed for efficient hardware implementation. Keeloq has a block size of 32 bits and a key size of 64 bits. As shown in Figure 1 (taken from [10]), its main components include an internal state register (32 bits) and a non-linear feedback function (NLF). In each round of the cipher, NLF operates on five bits from the cipher's current state. The output from NLF is mixed with some prior state bits and with one of the 64 key bits and finally shifted back into the state register. To perform encryption, the plaintext is loaded into the state register, the key is loaded into the key register, and the entire system is clocked for 528 rounds. After these 528 rounds the state register contains the ciphertext. To perform decryption the ciphertext is placed in the state register and the system is clocked 528 times in the opposite direction. The progression of the state register is typically modeled as a vector of bits $S_0 \cdots S_{559}$, with $S_0 \cdots S_{31}$ being the plaintext and $S_{528} \cdots S_{559}$ the ciphertext.

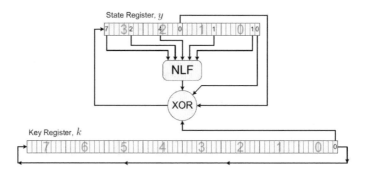

Fig. 1. Structure of the Keeloq cipher (taken from [10])

4.2   An Equation Set for Keeloq

As stated in Section 3.3, a TASCA equation set consists of four elements - the algorithm, the inputs, the measurement setup and finally the (potentially errored) measurements.

The algorithmic description of a single Keeloq round is a simple set of 2 PB equations:

$$NLF_i = NLF(S_{i+31}, S_{i+26}, S_{i+20}, S_{i+9}, S_{i+1}) \tag{6}$$

$$S_{i+32} = NLF_i \oplus S_i \oplus S_{i+16} \oplus K_{i \bmod 64} \tag{7}$$

The function NLF is a 5-to-1 non-linear function defined such that $NLF(a, b,$ $c, d, e)$ is bit number $abcde_b$ (binary) of the hexadecimal constant $3A5C742E$, where bit 0 is the least significant bit. It has no efficient linear or algebraic representation, and is represented as a single disjunctive normal form (DNF) equation based on the function's truth table (see the extended version of this paper for more details). The XOR function on 4 variables (effectively 5, since we realize the function $x_1 \oplus x_2 \oplus x_3 \oplus x_4 \oplus x_5 = 0$) is also represented by a single equation. Each of these equations is repeated 528 times to lead from the plaintext $(S_0 \cdots S_{31})$ to the ciphertext $(S_{528} \cdots S_{559})$.

In its most common mode of operation, Keeloq uses *rolling codes* which mandates that the ciphertext is known to the attacker but not the plaintext. Accordingly, the only known input to our solver was the ciphertext.

If we assume Keeloq is implemented on an ASIC, the power traces tend to be correlated with the *Hamming distance* of the entire 32 bits of the state register between the current round and the previous round (a similar attack can also be mounted if the device leaks the Hamming weight). To put this into equation form, we define the Hamming distance between each two consecutive bits of the state progression and group them in sets of 32. Finally, we add two additional Boolean variables to the measurement sums to allow for errors:

$$hd_i = S_i \oplus S_{i-1} \tag{8}$$

$$HD_i = \sum_{j=i}^{i+32} hd_j \tag{9}$$

$$\widehat{HD_i} = HD_i + e_i^+ - e_i^-$$

The number of rounds for which we produce side-channel measurement equations (m_{sc}) is a configuration parameter of the system: the following subsection shows how to select a proper value for m_{sc}. A Keeloq key recovery instance with side-channel measurements equations applied to the final $m_{sc} = 90$ rounds of encryption contains a total of 428 equations and has a file size of about 140K. A partial listing of a sample PB instance is provided for reference in the extended version of this paper.

4.3   Attack Results

We performed a power simulation of 300 ASIC-based Keeloq decryptions, corrupted the simulated power measurements with different probabilities of error and submitted them to the SCIP PB-solver. For each decryption and tested error probability, we selected $P_{err} \times 528$ rounds and corrupted the side-channel values measured in those rounds (specifically, the device-total Hamming distance) by ± 1. The attack used the 64-bit version of SCIPspx 1.2.0 on a quad-core Intel Core i7 950 at 3.06GHz with 8MB cache, running Windows 7 64-bit Edition. In each experiment the solver was asked to recover the 64-bit key from the errored side-channel outputs produced by the final 90 rounds of encryption.

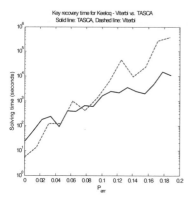

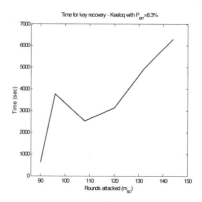

Fig. 2. TASCA key recovery from the final 90 rounds of Keeloq

Fig. 3. TASCA speed as a function of m_{sc}

Figure 2 shows our results. For reference we also show the performance of an iterative attack on the final 64 rounds of the encryption using comparable bit error rates. We emphasize that the attack is on the **full** 528-round cipher, even though it uses only a subset of the measured side-channel data.

It can be seen that the solver was able to find the key even with $P_{err} = 18.8\%$ with an average running time of 3.8 hours per instance, and that the time grows super-linearly with the error probability. The TASCA solver is 10 to 100 times faster than the iterative solver on instances with $P_{err} \geq 11\%$.

We also noted it was important to properly choose the number of side-channel measurements passed to the solver (m_{sc}). When too few rounds were passed, the optimal solution found was not necessarily the correct key. When too many rounds were passed, the computational burden involved slowed down the solver, as shown in Figure 3. In the case of Keeloq a good tradeoff was $m_{sc} = 90$ rounds which provided enough information to find the key in nearly all of the cases. For lower m_{sc} values the solver returned incorrect results for at least 25% of the instances.

As an aside note, the iterative decoder, which struggled with single-trace key recovery, had much better performance when attacking multiple traces. The algebraic solver did not perform as well with additional inputs, since each additional trace significantly increased the size of the equation set.

5   Preliminary Results on AES

5.1   The AES Algorithm

We chose to model our device under attack as naïve 8-bit microcontroller implemention of AES-128[8], is a block cipher with a 128-bit key and 16-byte (or 128-bit) input blocks. To perform encryption, the plaintext is first fed into a 16-byte state register. This state register is then manipulated 41 times during the sequence of the 10 rounds of AES-128 to produce the ciphertext. There are 4

Table 1. Instance size and performance of straight encryption

	Keeloq	AES (LUT)	AES (Canright)
Instance file size	553K	32873K	12569K
# of equations	1153	27344	93090
# of variables	13825	171208	229008
# of constraints	13825	173640	231506
Encryption time (sec.)	2.59	61.07	245.45

types of manipulations: SubBytes, AddRoundKey, ShiftRows and MixColumns, with SubBytes being the only non-linear operation in AES. AddRoundKey is performed 11 times during encryption, first with the supplied secret key and then 10 times with round keys derived from the secret key using a non-linear process which uses the SubBytes process as well.

5.2   An Equation Set for AES

The AES hardware realization can be modified and optimized in a variety of ways. Specifically when dealing with the S-box component of AES, which performs the SubBytes operation, there are a variety of hardware implementations offering various tradeoffs between better speed and more efficient hardware consumption.

Our first TASCA representation of AES implementation was based on a port of an OpenCores VHDL AES code [19]. This implementation models the S-box as a lookup table (LUT), leaving the compiler with the task of optimizing it to a minimal hardware footprint. A second TASCA representation was based on the efficient composite field representation of the S-box designed by Canright [6]. In this design, the S-box input is manipulated under a more efficient basis representation.

Table 1 summarizes the performance of the two cipher implementations, with the performance of the Keeloq encryption provided as reference. Since the encryption was described as an equation set, performance was similar whether the plaintext, the ciphertext or any intermediate state was supplied. Similarly, any round key can be substituted for the secret key with no effect on performance. Analyzing the results, it appears that the performance of the solver is dominated by the number of equations under consideration and not by the complexity of the equations themselves. This may be a property of the SCIP solver, and not of PB optimizers in general, since SCIP essentially performs a search over the tree of equations. Specifically, other PB solvers which compile their inputs into SAT instances may show better performance using the Canright S-box, since its reduced hardware complexity should make it easier to simulate.

Surprisingly, the solver had a very hard time inverting the SubBytes and MixColumns operations given their algebraic description. We found out that including equations for both SubBytes and for inverse SubBytes (that is, one equation stating that $S_{i+1} = SubBytes(S_i)$ and another stating that $S_i = SubBytes^{-1}(S_{i+1})$) sped up the solver dramatically.

Table 2. A TASCA attack on the AES key expansion phase

Rounds	AES (LUT)			AES (Canright)		
	instance size	# of equations	time (sec)	instance size	# of equations	time (sec)
1	765K	164	11	208K	1484	193
2	1529K	308	1341	414K	2948	10800
3	2293K	452	1690	620K	4412	345600

5.3   Initial Results

To date, the only attacks we have run on AES are reconstructions of the SPA attack of Mangard on the key expansion algorithm, as described in [13], without any errors. A secret key was recovered from 1, 2 and 3 rounds of expansion (16, 32 and 48 Hamming weights[1]). The key expansion was modeled using both S-box representations. The results are summarized in Table 2.

The time performance of the solver was worse than we estimated. We were also surprised to find that Canright representation yielded longer running times than the LUT representation, despite having instances that are 3 times smaller. Understanding these phenomena is a topic of future work.

6   Open Issues

6.1   Full Attack against AES and Other Ciphers

Our preliminary results thus far show that single-trace side-channel attacks against AES can be represented as PBOPT problems, and that the representations vary dramatically in size and complexity depending on the hardware implementation we start with. Furthermore, TASCA running time was not correlated with instance size - in fact the more compact Canright representation produced run times an order of magnitude slower than those produced by the LUT representation. We plan to try and better understand which instance types lead to faster TASCA attacks.

6.2   Better PB Solvers

The authors do not claim to be experts in the design and usage of PB solvers. In fact, the SCIP tool which we used has hundreds of configuration options which were left at their default values. It appears that the performance of the solvers can be increased by quite a large factor using careful design – the fact that a simple AES encryption took 60 seconds on our unoptimized platform is especially surprising. Since these solvers rely on heuristics to improve their performance, a set of heuristics for cryptanalysis needs to be developed. With a proper choice of heuristics we hope the performance of these attacks can be

---

[1] The published results in [13] show that 40 recovered Hamming weights are enough to uniquely determine the secret key.

increased by several orders of magnitude, either by using SCIP or by evaluating a different PB solver. To this end, we have shared our cryptanalytic instances with the PB design community [15].

6.3 TASCA as Part of the Design Tool Chain

The specification language used to define PB optimization problems is rich enough to allow description of arbitrary Boolean circuits. It seems possible to write a compiler that receives a hardware description in a high-level language such as VHDL [1] and outputs a PB-solver instance. Such a tool can be made part of a secure hardware design workflow, allowing designers to evaluate the susceptibility of their designs to side-channel attacks. By performing TASCA attacks with different subsets of the side-channel information, designers can assess the risk caused by exposure of various components of the internal state and so decide which components need a higher level of protection.

7 Conclusion

We showed a new attack methodology called *Tolerant Algebraic Side-Channel Analysis (TASCA)*. Our methodology transforms a single-trace side-channel analysis problem into a *pseudo-Boolean optimization problem* (PBOPT) form. The PBOPT syntax allows a very natural representation of measurement errors. We showed that using our approach we are able to mount successful single-trace attacks against real ciphers, even in the presence of realistic error rates.

Acknowledgements. Parts of the research described in this paper have been supported by the Austrian Science Fund (FWF) under grant number P22241-N23 ("Investigation of Implementation Attacks"). The authors wish to thank the anonymous reviewers for their encouraging and insightful comments.

References

1. IEEE standard VHDL language reference manual. IEEE Std 1076-2008 (Revision of IEEE Std 1076-2002), pp. c1–626 (26, 2009)
2. Achterberg, T.: Constraint Integer Programming. PhD thesis, Technische Universität Berlin (2007)
3. Berthold, T., Heinz, S., Pfetsch, M.E., Winkler, M.: SCIP – solving constraint integer programs. In: SAT 2009 competitive events booklet (2009)
4. Berthold, T., Heinz, S., Pfetsch, M.E.: Nonlinear pseudo-boolean optimization: Relaxation or propagation? In: Kullmann, O. (ed.) SAT 2009. LNCS, vol. 5584, pp. 441–446. Springer, Heidelberg (2009)
5. Bertsimas, D., Weismantel, R.: Optimization Over Integers. Dynamic Ideas (2005)
6. Canright, D.: A very compact S-Box for AES. In: Rao, J.R., Sunar, B. (eds.) CHES 2005. LNCS, vol. 3659, pp. 441–455. Springer, Heidelberg (2005)
7. Courtois, N., Bard, G.V., Wagner, D.: Algebraic and slide attacks on KeeLoq. In: Nyberg, K. (ed.) FSE 2008. LNCS, vol. 5086, pp. 97–115. Springer, Heidelberg (2008)

8. Daemen, J., Rijmen, V.: AES proposal: Rijndael (1998)
9. Dawson, S.: Code hopping decoder using a PIC16C56. Microchip confidential, leaked online in 2002 (1998)
10. Eisenbarth, T., Kasper, T., Moradi, A., Paar, C., Salmasizadeh, M., Manzuri Shalmani, M.T.: On the power of power analysis in the real world: A complete break of the Keeloq code hopping scheme. In: Wagner, D. (ed.) CRYPTO 2008. LNCS, vol. 5157, pp. 203–220. Springer, Heidelberg (2008)
11. Karlof, C., Wagner, D.: Hidden Markov model cryptoanalysis. In: Walter, C.D., Koç, Ç.K., Paar, C. (eds.) CHES 2003. LNCS, vol. 2779, pp. 17–34. Springer, Heidelberg (2003)
12. Kocher, P.C., Jaffe, J., Jun, B.: Differential power analysis. In: Wiener, M. (ed.) CRYPTO 1999. LNCS, vol. 1666, pp. 388–397. Springer, Heidelberg (1999)
13. Mangard, S.: A simple power-analysis (SPA) attack on implementations of the AES key expansion. In: Lee, P.J., Lim, C.H. (eds.) ICISC 2002. LNCS, vol. 2587, pp. 343–358. Springer, Heidelberg (2003)
14. Mangard, S., Oswald, E., Popp, T.: Power Analysis Attacks: Revealing the Secrets of Smart Cards (Advances in Information Security). Springer, New York (2007)
15. Manquinho, V., Roussel, O.: Pseudo-boolean competition 2009 (July 2009)
16. Massacci, F., Marraro, L.: Logical cryptanalysis as a SAT problem. J. Autom. Reason. 24(1-2), 165–203 (2000)
17. Potlapally, N.R., Raghunathan, A., Ravi, S., Jha, N.K., Lee, R.B.: Aiding side-channel attacks on cryptographic software with satisfiability-based analysis. IEEE Trans. on VLSI Systems 15(4), 465–470 (2007)
18. Renauld, M., Standaert, F.-X., Veyrat-Charvillon, N.: Algebraic side-channel attacks on the AES: Why time also matters in DPA. In: Clavier, C., Gaj, K. (eds.) CHES 2009. LNCS, vol. 5747, pp. 97–111. Springer, Heidelberg (2009)
19. Satyanarayana, H.: AES128 package (December 2004)
20. Viterbi, A.: Error bounds for convolutional codes and an asymptotically optimum decoding algorithm. IEEE Transactions on Information Theory 13(2), 260–269 (1967)
21. Wunderling, R.: Paralleler und objektorientierter Simplex-Algorithmus. PhD thesis, Technische Universität Berlin (1996)

Coordinate Blinding over Large Prime Fields

Michael Tunstall[1] and Marc Joye[2]

[1] Department of Computer Science, University of Bristol
Merchant Venturers Building, Woodland Road
Bristol BS8 1UB, United Kingdom
tunstall@cs.bris.ac.uk
[2] Technicolor, Security & Content Protection Labs
1 avenue de Belle Fontaine, 35576 Cesson-Sévigné Cedex, France
marc.joye@technicolor.com

Abstract. In this paper we propose a multiplicative blinding scheme for protecting implementations of a scalar multiplication over elliptic curves. Specifically, this blinding method applies to elliptic curves in the short Weierstraß form over large prime fields. The described countermeasure is shown to be a generalization of the use of random curve isomorphisms to prevent side-channel analysis, and our best configuration of this countermeasure is shown to be equivalent to the use of random curve isomorphisms. Furthermore, we describe how this countermeasure, and therefore random curve isomorphisms, can be efficiently implemented using Montgomery multiplication.

Keywords: Elliptic curve cryptography, side-channel analysis, countermeasures.

1 Introduction

Side-channel analysis can be used to try and derive unknown information used in cryptographic algorithms, such as cryptographic keys. The first side-channel described in the literature was based on the total time taken to compute a cryptographic algorithm [18]. Preventing this attack is well understood, as one just requires a regular algorithm to prevent any side-channel leakage.

Another side-channel that has been described in the literature is based on the observation that the power consumption of a microprocessor is dependent on the instruction being executed and on any data being manipulated [6,19]. An attacker can, therefore, observe where functions, and sequences of functions, occur in a power consumption trace. This allows information on cryptographic keys to be determined if the sequence of instructions is affected by the value of the key. An attacker can also determine if a value being manipulated by a microprocessor can be correctly predicted by computing the correlation between a set of predictions and the instantaneous power consumption. This allows information on cryptographic keys to be determined since one can verify a hypothetical set of values that occur after being combined with a key.

S. Mangard and F.-X. Standaert (Eds.): CHES 2010, LNCS 6225, pp. 443–455, 2010.

It was later observed that the electromagnetic field around a microprocessor also has this property [12,23]. Preventing an attacker from being able to use this information is more complex, as all the intermediate states of an algorithm need to be masked with some random value [11,19]. When implementing a block cipher this can be implemented by modifying the algorithm such that it operates in this manner by modifying each function [3].

When a public-key cryptographic algorithm, such as RSA [24], is implemented countermeasures are typically based on the structure of the entire function. For example, when generating a signature σ from a message m using RSA, one computes $\sigma = \mu(m)^d \bmod N$, where d is the private key and μ is an appropriate padding function. That is, a standard exponentiation algorithm in $(\mathbb{Z}/N\mathbb{Z})^*$. This can be changed such that the intermediate states of the calculation cannot be predicted by computing $\sigma = [(\mu(m) + r_1 N)^{r_2 \phi(N)+d} \bmod r_3 N] \bmod N$, where ϕ is Euler's totient function and r_i, for $i \in \{1, 2, 3\}$, are (small) random values. However, one would not want to directly apply this countermeasure to implementations of elliptic curve cryptosystems using prime fields. Increasing the size of the modulus used in RSA has a relatively small impact on the overall execution time. The impact on elliptic curves will be larger since the prime values used in the field arithmetic are much smaller.

Many different countermeasures for preventing side-channel analysis of elliptic curve cryptographic algorithms have been proposed in the literature. In this paper we describe a multiplicative blinding method for elliptic curve cryptographic algorithms over prime fields that is a generalization of previously proposed methods, and describe how it can be efficiently implemented.

The rest of this paper is organized as follows. In the next section, we introduce some background on elliptic curves and review some countermeasures against side-channel analysis. Section 3 is the core of our paper. We define a new addition using blinded coordinates. Detailed formulæ are provided for homogeneous and Jacobian representations. In Section 4 we describe how one could implement the proposed countermeasure. In Section 5 we discuss some further security considerations that one would need to take into account when implementing the proposed countermeasure. Finally, we conclude in Section 6.

2   Preliminaries

2.1   Elliptic Curves

Let $\mathbb{F}_q$ be a finite field. An elliptic curve $\mathcal{E}$ over $\mathbb{F}_q$ consists of points (x, y), with x, y in $\mathbb{F}_q$, that satisfy the full Weierstraß equation

$$\mathcal{E} : y^2 + a_1\, x\, y + a_3\, y = x^3 + a_2\, x^2 + a_4\, x + a_6$$

with $a_i \in \mathbb{F}_q$ ($1 \leq i \leq 6$), and the point at infinity denoted $\boldsymbol{O}$. The set $\mathcal{E}(\mathbb{F}_q)$ is defined as

$$\mathcal{E}(\mathbb{F}_q) = \{(x, y) \in \mathcal{E} \mid x, y \in \mathbb{F}_q\} \cup \{\boldsymbol{O}\},$$

where $\mathcal{E}(\mathbb{F}_q)$ forms an Abelian group under the chord-and-tangent rule and O is the identity element.

The addition of two points $P = (x_1, y_1)$ and $Q = (x_2, y_2)$ with $P \neq -Q$ is given by $P + Q = (x_3, y_3)$ where

$$x_3 = \lambda^2 + a_1 \lambda - a_2 - x_1 - x_2, \quad y_3 = (x_1 - x_3)\lambda - y_1 - a_1 x_3 - a_3 \quad (1)$$

$$\text{with } \lambda = \begin{cases} \dfrac{y_1 - y_2}{x_1 - x_2} & \text{if } P \neq Q \text{ [addition]} \\ \dfrac{3x_1{}^2 + 2a_2 x_1 + a_4 - a_1 y_1}{2y_1 + a_1 x_1 + a_3} & \text{if } P = Q \text{ [doubling operation]} \end{cases}.$$

Provided that the characteristic of field $\mathbb{F}_q$ is different from $2, 3$, we can take $a_1 = a_2 = a_3 = 0$. In the sequel we will also assume that $q = p$ is prime. We define the short Weierstraß form over prime field $\mathbb{F}_p$ by the equation

$$y^2 = x^3 + a x + b . \quad (2)$$

Note that the slope λ in the doubling then becomes $\lambda = (3x_1{}^2 + a)/(2y_1)$, which can be rewritten as $3(x_1 - 1)(x_1 + 1)/(2y_1)$ when $a = -3$.

The scalar multiplication of a given point is a fundamental operation in cryptographic algorithms that use elliptic curve arithmetic, i.e. $[k] P$ for some integer $k < |\mathcal{E}|$. This operation uses the above addition law in conjunction with algorithms analogous to standard exponentiation algorithms in $(\mathbb{Z}/N\mathbb{Z})^*$.

In this paper we concentrate on the short Weierstraß form since this is typically the form one will find in standards, and is, therefore, the most commonly used. For example, one can find standardized elliptic curves in the short Weierstraß form in FIPS 186-3 [22], WTLS [32] and ANSI X9.62 [33].

2.2 Side-Channel Resistant Scalar Multiplication

When implementing a scalar multiplication using elliptic curve arithmetic on a device that could potentially be attacked using side-channel analysis, there are a variety of considerations that need to be taken into account. The simplest type of side-channel analysis consists of timing or simple power analysis [18,19], where an attacker attempts to derive information from the time taken for an algorithm to execute or to identify operations from a few traces. Given that an attacker would expect to be able to distinguish an addition from a doubling operation, this requires an implementation to include one of the following countermeasures.

Regular Multiplication Algorithms. A variety of algorithms have been proposed that will always compute a regular sequence of additions and doubling operations (these methods are surveyed in [16]).

Unified Addition Formulae. The addition and doubling operations can be implemented such that the same operations are performed for both an addition and a doubling operation (e.g. [5,7]).

Dummy Operations. An alternative to unified addition formulae was proposed in [11,8], where the two operations are rendered indistinguishable using dummy operations. However, this approach can introduce the possibility
of a safe-error fault attack [34], although a discussion of fault analysis is
beyond the scope of this paper.

Furthermore, implementations need to be able to prevent an attacker from using
the observation that the power consumption (and electromagnetic field) is related
to the Hamming weight of the data being manipulated by a microprocessor at
any given point in time [12,19,23], referred to as differential side-channel analysis.
This requires an implementation to include further countermeasures to blind the
computation. The scalar itself can be protected using:

Multiplier Blinding. The scalar k can be modified by adding a random multiple of the order of the group $\mathcal{E}$ to the scalar k. This modifies the bits of k
without changing the output of a scalar multiplication [11].

There are numerous options for blinding the points being operated on. A summary of existing countermeasures is given below.

Point Blinding. If, for a given point R, where $S = [k]R$ is known, then $Q =
[k]P$ can be computed by calculating $Q = [k](P + R) - S$. Points R and S
can be stored in a device along with k and updated after each execution by
computing $R \leftarrow [r]R$ and $S \leftarrow [r]S$ for some small random value r [11,18].

Multiplier Splitting. A scalar can be divided into two values whose bitwise
representations are random. This allows a scalar multiplication to be conducted with two values whose combined effect is equivalent to that of the
desired scalar [9]. There are three methods of multiplier splitting:
  - **Additive Splitting.** If we define the scalar $k = r + (k - r)$ for some
    integer r that has a similar bit-length to k, then $Q = [k]P$ can be
    computed by calculating $Q = [r]P + [k - r]P$.
  - **Multiplicative Splitting.** For some elliptic curve $\mathcal{E}$ over $\mathbb{F}_q$ we define
    $k' = k r^{-1} \bmod |\mathcal{E}|$ for some integer r. Then $Q = [k]P$ can be computed
    by calculating $Q = [k']([r]P)$.
  - **Euclidean Splitting.** If we define the scalar $k' = \lfloor k/r \rfloor$ for some integer
    r, then $Q = [k]P$ can be computed by calculating $Q = [k']([r]P) +
    [k \bmod r]P$.

Randomized Projective Points. An affine point $P = (x, y)$ can, for example,
be represented as a homogeneous projective point $(\theta x, \theta y, \theta)$ for all $\theta \in
\mathbb{F}_p \setminus \{0\}$ (this is covered in more detail in Section 3). When computing a scalar
multiplication using projective coordinates a randomly generated $\theta \in \mathbb{F}_p \setminus \{0\}$
can be determined at the beginning of the computation so that an attacker
cannot guess what values are being manipulated [11].

Random Curve Isomorphisms. A given P on elliptic curve $\mathcal{E}$ can be randomized by computing $P^* \leftarrow \psi(P)$ on $\mathcal{E}^* \leftarrow \psi(\mathcal{E})$ for a random curve
isomorphism ψ. Then $Q = [k]P$ can be computed by calculating $Q =
\psi^{-1}([k]P^*)$ [17].

Of these countermeasures, the first two are not practical as they highly impact the execution time of a scalar multiplication or require dedicated operations not always readily available. Using randomized projective coordinates is much more efficient but does not allow θ to be set to one. It is for this reason that it is observed in [25] that using random curve isomorphisms is the most efficient of these countermeasures. However, when using random curve isomorphisms the parameters of $\mathcal{E}^*$ cannot be chosen and one cannot take advantage of algorithms that require curve parameters to be set to specific values.

3   Implementing Elliptic Curve Arithmetic

For elliptic curve arithmetic over $\mathbb{F}_p$ the use of projective coordinates is preferred as no inversion is required for an addition or a doubling operation [31]. A point on an elliptic curve can be represented with projective coordinates (X, Y, Z) that are not unique for a given affine point. For example, homogeneous projective coordinates $(\theta x, \theta y, \theta)$ represent the affine point (x, y) for all $\theta \in \mathbb{F}_p \setminus \{0\}$, and the point at infinity $\boldsymbol{O}$ is represented by $(0, \gamma, 0)$ for some $\gamma \in \mathbb{F}_p \setminus \{0\}$. The simplest countermeasure that can be applied to these coordinate systems is to choose some random $\theta \in \mathbb{F}_p \setminus \{0\}$ and use the point $(\theta x, \theta y, \theta)$ as a random representation of the affine point (x, y) [11] (referred to as randomized projective points in the previous section). However, when using this representation, the Z-coordinate cannot be chosen to be one. In the following sections we define addition rules for randomized projective coordinates where the Z-coordinate can be chosen to be one.

3.1   Homogeneous Projective Coordinates

As described above, homogeneous projective coordinates $(\theta x, \theta y, \theta)$ represent the affine point (x, y) for all $\theta \in \mathbb{F}_p \setminus \{0\}$, and the point at infinity $\boldsymbol{O}$ is represented by $(0, \gamma, 0)$ for some $\gamma \in \mathbb{F}_p \setminus \{0\}$. We define the map Φ as mapping a point $\boldsymbol{P} = (X, Y, Z) \in \mathcal{E}$ to the point $\boldsymbol{P}'$ where $\Phi(\boldsymbol{P}) = (X', Y', Z) = (f^\mu X, f^\nu Y, Z)$ for an arbitrary $f \in \mathbb{F}_p \setminus \{0\}$ and some small integers μ and ν. Note that $\boldsymbol{P}'$ is not necessarily on $\mathcal{E}$. The inverse of Φ can be computed without inverting f since $\boldsymbol{P} = \Phi^{-1}(\boldsymbol{P}') = (f^\nu X', f^\mu Y', f^{\mu+\nu} Z)$.

Consider the addition of two homogeneous projective points $\boldsymbol{R} = \boldsymbol{P} + \boldsymbol{Q}$. In order to blind the computation, we redefine the addition algorithm such that $\Phi(\boldsymbol{R}) = \boldsymbol{R}' = \boldsymbol{P}' + \boldsymbol{Q}' = \Phi(\boldsymbol{P}) + \Phi(\boldsymbol{Q})$. We define the point $\boldsymbol{R}' = (X_3', Y_3', Z_3)$, $\boldsymbol{P}' = (X_1', Y_1', Z_1)$ and $\boldsymbol{Q}' = (X_2', Y_2', Z_2)$. If $\boldsymbol{P} = \boldsymbol{Q}$, then $\boldsymbol{R}'$ can be computed from $\boldsymbol{P}'$ and $\boldsymbol{Q}'$ by calculating

$$X_3' = \lambda_{10}\lambda_4$$
$$Y_3' = f^{2\nu-3\mu}\lambda_3(\lambda_9 - \lambda_{10}) - 2\lambda_8 \, ,$$
$$Z_3 = \lambda_6$$

where $\lambda_1 = {X_1'}^2$, $\lambda_2 = {Z_1}^2$, $\lambda_3 = a\,f^{2\mu}\lambda_2 + 3\lambda_1$, $\lambda_4 = 2Y_1'Z_1$, $\lambda_5 = {\lambda_4}^2$, $\lambda_6 = \lambda_4\lambda_5$, $\lambda_7 = Y_1'\lambda_4$, $\lambda_8 = {\lambda_7}^2$, $\lambda_9 = (X_1' + \lambda_7)^2 - \lambda_1 - \lambda_8$ and $\lambda_{10} = f^{2\nu-3\mu}{\lambda_3}^2 - 2\lambda_9$ [5]. This requires an extra three multiplications with a power of f.

If $a = -3$ a faster doubling algorithm can be used and $\boldsymbol{R'}$ can be computed by calculating

$$X'_3 = \lambda_8 \lambda_2$$
$$Y'_3 = f^{2\nu-3\mu} \lambda_1 (\lambda_7 - \lambda_8) - 2\lambda_6 \,,$$
$$Z_3 = \lambda_4$$

where $\lambda_0 = f^\mu Z_1$, $\lambda_1 = 3(X'_1 - \lambda_0)(X'_1 + \lambda_0)$, $\lambda_2 = 2Y'_1 Z_1$, $\lambda_3 = \lambda_2{}^2$, $\lambda_4 = \lambda_2 \lambda_3$, $\lambda_5 = Y'_1 \lambda_2$, $\lambda_6 = \lambda_5{}^2$, $\lambda_7 = 2X'_1 \lambda_5$ and $\lambda_8 = f^{2\nu-3\mu} \lambda_1{}^2 - 2\lambda_7$ [5]. This also requires an extra three multiplications with a power of f.

If $\boldsymbol{P} \neq \boldsymbol{Q}$, then $\boldsymbol{R'}$ is computed by calculating

$$X'_3 = \lambda_6 \lambda_{10}$$
$$Y'_3 = \lambda_4 (\lambda_9 - \lambda_{10}) - \lambda_8 \lambda_1 \,,$$
$$Z_3 = \lambda_8 \lambda_3$$

where $\lambda_1 = Y'_1 Z_2$, $\lambda_2 = X'_1 Z_2$, $\lambda_3 = Z_1 Z_2$, $\lambda_4 = Y'_2 Z_1 - \lambda_1$, $\lambda_5 = \lambda_4{}^2$, $\lambda_6 = X'_2 Z_1 - \lambda_2$, $\lambda_7 = \lambda_6{}^2$, $\lambda_8 = \lambda_6 \lambda_7$, $\lambda_9 = \lambda_7 \lambda_2$ and $\lambda_{10} = f^{3\mu-2\nu} \lambda_5 \lambda_3 - \lambda_8 - 2\lambda_9$ [10]. This requires a single extra multiplications with a power of f.

3.2   Jacobian Projective Coordinates

Jacobian projective coordinates $(\theta^2 x, \theta^3 y, \theta)$ represent the affine point (x, y) for any $\theta \in \mathbb{F}_p \setminus \{0\}$, and the point at infinity $\boldsymbol{O}$ is represented by $(\gamma^2, \gamma^3, 0)$ for some $\gamma \in \mathbb{F}_p \setminus \{0\}$. We define the map Φ as mapping a point $\boldsymbol{P} = (X, Y, Z) \in \mathcal{E}$ to the point $\boldsymbol{P'}$ where $\Phi(\boldsymbol{P}) = (X', Y', Z) = (f^\mu X, f^\nu Y, Z)$ for an arbitrary $f \in \mathbb{F}_p \setminus \{0\}$ and some small integers μ and ν. Note that $\boldsymbol{P'}$ is not necessarily in $\mathcal{E}$. The inverse of Φ can be computed without inverting f since $\boldsymbol{P} = \Phi^{-1}(\boldsymbol{P'}) = (f^{\mu+2\nu} X', f^{3\mu+2\nu} Y', f^{\mu+\nu} Z)$.

Consider the addition of two Jacobian projective points $\boldsymbol{R} = \boldsymbol{P} + \boldsymbol{Q}$. In order to blind the computation, we redefine the addition algorithm such that $\Phi(\boldsymbol{R}) = \boldsymbol{R'} = \boldsymbol{P'} + \boldsymbol{Q'} = \Phi(\boldsymbol{P}) + \Phi(\boldsymbol{Q})$. We define the point $\boldsymbol{R'} = (X'_3, Y'_3, Z_3)$, $\boldsymbol{P'} = (X'_1, Y'_1, Z_1)$ and $\boldsymbol{Q'} = (X'_2, Y'_2, Z_2)$. If $\boldsymbol{P} = \boldsymbol{Q}$, then $\boldsymbol{R'}$ can be computed from $\boldsymbol{P'}$ and $\boldsymbol{Q'}$ by calculating

$$X'_3 = \lambda_7$$
$$Y'_3 = f^{2\nu-3\mu} \lambda_6 (\lambda_5 - \lambda_7) - 8\lambda_3 \,,$$
$$Z_3 = (Y'_1 + Z_1)^2 - \lambda_2 - \lambda_4$$

where $\lambda_1 = X'_1{}^2$, $\lambda_2 = Y'_1{}^2$, $\lambda_3 = \lambda_2{}^2$, $\lambda_4 = Z_1{}^2$, $\lambda_5 = 2((X'_1 + \lambda_2)^2 - \lambda_1 - \lambda_3)$, $\lambda_6 = 3\lambda_1 + af^{2\mu} \lambda_4{}^2$, $\lambda_7 = f^{2\nu-3\mu} \lambda_6{}^2 - 2\lambda_5$ [5]. This requires an extra three multiplications with a power of f.

If $a = -3$ a faster doubling algorithm can be used and $\boldsymbol{R'}$ can be computed by calculating

$$X'_3 = f^{2\nu-3\mu} \lambda_5{}^2 - 8\lambda_3$$
$$Y'_3 = f^{2\nu-3\mu} \lambda_5 (4\lambda_3 - X'_3) - 8\lambda_2{}^2 \,,$$
$$Z_3 = (Y'_1 + Z_1)^2 - \lambda_2 - \lambda_1$$

where $\lambda_1 = Z_1{}^2$, $\lambda_2 = Y_1'{}^2$, $\lambda_3 = X_1'\lambda_2$, $\lambda_4 = f^\mu\lambda_1$, $\lambda_5 = 3(X_1' - \lambda_4)(X_1' + \lambda_4)$ [4]. Again, this requires an extra three multiplications with a power of f.

If $P \neq Q$, then R' is computed by calculating

$$X_3' = f^{3\mu - 2\nu}\lambda_{10}{}^2 - \lambda_9 - 2\lambda_{11}$$
$$Y_3' = \lambda_{10}(\lambda_{11} - X_3') - 2\lambda_5\lambda_9 \quad ,$$
$$Z_3 = ((Z_1 + Z_2)^2 - \lambda_1 - \lambda_2)\lambda_7$$

where $\lambda_1 = Z_1{}^2$, $\lambda_2 = Z_2{}^2$, $\lambda_3 = X_1'\lambda_2$, $\lambda_4 = X_2'\lambda_1$, $\lambda_5 = Y_1'Z_2\lambda_2$, $\lambda_6 = Y_2'Z_1\lambda_1$, $\lambda_7 = \lambda_4 - \lambda_3$, $\lambda_8 = (2\lambda_7)^2$, $\lambda_9 = \lambda_7\lambda_8$, $\lambda_{10} = 2(\lambda_6 - \lambda_5)$, $\lambda_{11} = \lambda_3\lambda_8$ [5]. This requires a single extra multiplications with a power of f.

3.3 Choosing μ and ν

The above algorithms were defined to minimize the number of multiplications with f, or some power of f. However, one would wish to avoid a situation where the inverse of f is required. This means that, for the above algorithms, choices for μ and ν need to satisfy $2\nu \geq 3\mu$ and $3\mu \geq 2\nu$; that is, $2\nu = 3\mu$.

For both homogeneous and Jacobian projective coordinates the choice of $2\nu = 3\mu$ would allow for any multiplication with a power of f to be removed from the algorithm for computing the addition of two distinct points. That is, the countermeasure would have no impact on the performance of a point addition. Define $a' = a f^{2\mu}$. In the case of $a \neq -3$, if the cost of a multiplication by a' is the same to that of a multiplication by a, choosing $2\nu = 3\mu$ incurs no performance loss for the doubling operation in both homogeneous and Jacobian coordinates. If $a = -3$ the choice of $2\nu = 3\mu$ leads to only an extra multiplication by f^μ (in the evaluation of λ_0 for homogeneous coordinates and in the evaluation of λ_4 for Jacobian coordinates, respectively).

Case of $\mu = 2$ and $\nu = 3$. As a reminder, the elliptic curves $\mathcal{E} : y^2 = x^3 + a x + b$ and $\mathcal{E}^* : y^2 = x^3 + a^*x + b^*$ over $\mathbb{F}_p$ are isomorphic if and only if there exists some $f \in \mathbb{F}_p \setminus \{0\}$ such that $a^* = f^4 a$ and $b^* = f^6 b$. The isomorphism is given by $\psi : \mathcal{E} \xrightarrow{\sim} \mathcal{E}^* : P = (x, y) \mapsto P^* = (f^2 x, f^3 y)$ and $O \mapsto O$. It appears that the specific choice of $\mu = 2$ and $\nu = 3$ corresponds to the technique of using randomized curve isomorphisms [17].

4   Implementation Considerations

4.1 Using Montgomery Multiplication

When implementing an elliptic curve cryptographic algorithm over $\mathbb{F}_p$, it would be natural to use Montgomery multiplication [21], since the modular reduction is interleaved with the multiplication. As shown in Algorithm 1, the result of a Montgomery multiplication is not the product of x and y modulo p. The algorithm actually returns $x y R^{-1} \bmod p$, where $R^{-1} \bmod p$ is introduced by the algorithm ($R = b^n$, where the modulus consists of n words of size b). In order to use Montgomery multiplication x and y need to be converted to their Montgomery representation, i.e. $\tilde{x} \leftarrow x\,R \bmod p$ and $\tilde{y} \leftarrow y\,R \bmod p$. Then, when $\tilde{x}$ and $\tilde{y}$ are multiplied together using Montgomery multiplication, the result is $x\,y\,R \bmod p$.

Algorithm 1. Montgomery multiplication

Input: $p = (p_{n-1}, \ldots, p_1, p_0)_b$, $x = (x_{n-1}, \ldots, x_1, x_0)_b$, $y = (y_{n-1}, \ldots, y_1, y_0)_b$
  with $0 \leq x, y < p$, $R = b^n$, $\gcd(p, b) = 1$ and $p' = -p^{-1} \bmod b$.
Output: $A = x\,y\,R^{-1} \bmod p$.

$A \leftarrow 0$;
for $i = 0$ **to** $n - 1$ **do**
    $u_i = (a_0 + x_i\,y_0)p' \bmod b$;
    $A = (A + x_i\,y + u_i\,p)/b$;
end

if $A \geq p$ **then** $A \leftarrow A - p$;

return A

When implementing Montgomery multiplication for use in a group exponentiation one has to be aware that an attacker can use the final conditional subtraction to try and derive information on the exponent used. An attacker can potentially use the difference in time caused by the total number of subtractions [30] or by identifying individual subtractions in a power consumption trace (or other suitable side-channel) [28,29]. The final subtraction can be removed by increasing the number of iterations of the main loop [14,27]. However, these attacks and countermeasures are beyond the scope of this paper, since the arguments concerning the efficiency of the countermeasure described in Section 3 will remain unchanged.

Where a multiplication with a small value is required, such as the multiplication with the constant a in the short Weierstraß form, this value needs to be converted into its Montgomery representation. This means that the cost of such a multiplication will require the same number of single-precision multiplications as any other multiplication or squaring operations over $\mathbb{F}_p$, i.e. $n(2\,n+2)$ single-precision multiplications.

4.2   Generating f

A random value is typically generated for blinding purposes in a given instance of a side-channel resistant implementation of an algorithm. These values do are typically chosen to be relatively small, since the bit-length of the random value only needs to be large enough that an attacker cannot guess its value for multiple executions of the algorithm. That is, an attacker is required to guess this value for each acquisition in order to conduct a differential side-channel analysis [20]. For example, we can define f to be in $\{1, \ldots, b - 1\}$ but multiplying with f has the same problem as multiplying by a since f is not in its Montgomery representation. However, we can define the value that is used when multiplying with the mask to be $f' \equiv b f \pmod{p}$ with $f' \in \{1, \ldots, b-1\}$ and $f \in \mathbb{F}_p \setminus \{0\}$. This means that Algorithm 2, which is one iteration of the main loop of the Montgomery multiplication algorithm, can be used without having to correct the factor of $b^{-1} \bmod p$ that is introduced.

This allows a multiplication with the mask f using $2n + 2$ single-precision multiplications, and can be repeated for powers of f as required, i.e. multiplying with f^μ requires $\mu(2n+2)$ single-precision multiplications. In practice this means a random value in $\{1, \ldots, b-1\}$ can be generated and used in Algorithm 2.

We can note that a scalar multiplication using the algorithms described in Section 3 can be implemented using an arbitrarily chosen f' without knowing f. That is, all required multiplications with f can be conducted using f' and Algorithm 2. This includes converting a blinded point to a valid projective point by multiplying the coordinates by the required power of f. The further advantage of only using f' is that it is not necessary to store the montgomery representation of any powers of f in memory.

Algorithm 2. Montgomery multiplication with f

Input: $p = (p_{n-1}, \ldots, p_1, p_0)_b$, $f \in \{0, \ldots, b-1\}$, $y = (y_{n-1}, \ldots, y_1, y_0)_b$ with
$\qquad 0 \leq y < p$, $R = b^n$, $\gcd(p, b) = 1$ and $p' = -p^{-1} \bmod b$.
Output: $A = f y b^{-1} \bmod p$.

$u = f y_0 p' \bmod b$;
$A = (f y + u p)/b$;

if $A \geq p$ **then** $A \leftarrow A - p$;

return A

4.3 Performance

If the above optimization is applied to the algorithms described in Sections 3.1 and 3.2 the number of single-precision multiplications required can be reduced.

It is observed in Section 3.3 that the smallest penalty for using the proposed blinding method is incurred when $\mu = 2$ and $\nu = 3$, and no extra cost is observed for many of the operations. Where $a = -3$, the cost of using the proposed blinding method will incur an extra multiplication with f^μ can be computed with $(4n + 4)$ single-precision multiplications, rather than $n(2n + 2)$ for a full Montgomery multiplication.

If we consider randomize curve isomorphisms, the case where $a* = -3f^4$ (i.e. an isomorphic curve that would allow one to use the algorithm defined for $a = -3$) the necessary extra multiplication can be computed in the same way, and the same gain in performance would be observed. Using the above observation would also, therefore, allow the time required to compute operations on an isomorphic curve to be reduced.

5   Further Security Considerations

The algorithms defined in Section 3 can readily be used to implement a side-channel resistant scalar multiplication. These building blocks are, themselves, resistant to side-channel analysis (within certain bounds we will discuss in this section) and merely require a suitable multiplication algorithm to be chosen. We refer the reader to [16] for a discussion of this topic.

However, in [13] it is observed that elliptic curve arithmetic that uses multiplicative blinding will not necessarily prevent a scalar multiplication from being derived. If a point corresponding to the affine points $(0, y)$ or $(x, 0)$ exists, for some $x, y \in \mathbb{F}_p$, then an attacker could attempt to have this point produced as an intermediate state of a scalar multiplication, which could then be used to verify hypotheses concerning the scalar. This would be possible as multiplicative blinding will have no effect on a coordinate set to zero.

An extension to this attack was proposed in [1] that relied on the same observation, that the value zero cannot be blinded multiplicatively. They noted that the same attack could be conducted if any of the intermediate states could be equal to zero. That is, if there exists some combination of points where the point arithmetic will generate a zero as an intermediate state.

The simplest countermeasure to this attack would be simply use curves that do not have these points. However, it is noted in [13] that in many standardized curves a point exists where the x-coordinate is zero, but not where the y-coordinate is zero.

Countermeasures, therefore, need to be included when implementing a cryptosystem that uses an elliptic curve that can be attacked in this manner. Two such countermeasures are:

Linear Blinding. One countermeasure to this type of attack is described in [15], where the authors propose that the coordinates of a projective coordinate are modified by adding an extra coordinate. For example, to protect the x-coordinate one could define a projective point, for example, where an affine point $\boldsymbol{P} = (x, y)$ can be represented as a projective point $(\theta\,(x - \beta), \theta\,y, \theta, \beta)$ for all $\theta, \beta \in \mathbb{F}_p \setminus \{0\}$. This involves redefining the algorithms for addition and doubling operations and considerably increases the number of operations required.

Isogenies. It is pointed out in [26] that an isogenous curve can be selected. That is, an isogeny between elliptic curves $\mathcal{E}_1$ and $\mathcal{E}_2$ over the same field exists if a surjective morphism κ can be defined that preserves the identity element (i.e. the point at infinity $\boldsymbol{O}$). When implementing a scalar multiplication using an elliptic curve where zero-coordinates are possible, one can select an isogenous curve that does not have any points with zero-coordinates. Then $\boldsymbol{Q} = [k]\,\boldsymbol{P}$ can be computed by calculating $\boldsymbol{Q} = \kappa^{-1}([k]\,\kappa(\boldsymbol{P}))$. Further constraints on what isogenies can be used were defined in [2] to avoid intermediate states being attacked.

Of these two countermeasures, the use of isogenies is more efficient as the same algorithms can be used for point arithmetic with the addition of two transformations. Moreover, these transformations can be defined when a cryptosystem is implemented to minimize the impact on the time taken to compute a scalar multiplication. The principle problem with using linear blinding is that it has a large impact on the point addition algorithms.

6   Conclusion

In this paper we propose an multiplicative blinding method for protecting a scalar multiplication that is a generalization of the use of randomized curve isomorphisms. We also discuss how one could efficiently implement this countermeasure using Montgomery multiplication, and show that this would allow for a faster implementation than a naïve use of randomized curve isomorphisms.

The specific choice of $\mu = 2$ and $\nu = 3$ incurs only a small increase in the execution time of a scalar multiplication. However, as noted above, this corresponds to using an isomorphic curve, and that the optimizations presented in Section 4 also apply. That is, if we apply the same criteria in choosing f, i.e. such that $f' \equiv f\,b \bmod p$ is $\in \{1, \ldots, b-1\}$, the performance will be identical.

We note that Algorithm 2 could also be used to efficiently implement randomized projective coordinates [11]. The aim of this randomization is to make the intermediate values unpredictable by an attacker and it is not necessary to choose a random value with the same bit-length as the x and y-coordinates.

Acknowledgments

The work described in this paper has been supported in part by the European Commission IST Programme under Contract IST-2002-507932 ECRYPT and EPSRC grant EP/F039638/1 "Investigation of Power Analysis Attacks".

References

1. Akishita, T., Takagi, T.: Zero-value point attacks on elliptic curve cryptosystems. In: Boyd, C., Mao, W. (eds.) ISC 2003. LNCS, vol. 2851, pp. 218–233. Springer, Heidelberg (2003)
2. Akishita, T., Takagi, T.: On the optimal parameter choice for elliptic curve cryptosystems using isogeny. In: Bao, F., Deng, R.H., Zhou, J. (eds.) PKC 2004. LNCS, vol. 2947, pp. 346–359. Springer, Heidelberg (2004)
3. Akkar, M.-L., Giraud, C.: An implementation of DES and AES secure against some attacks. In: Koç, Ç.K., Naccache, D., Paar, C. (eds.) CHES 2001. LNCS, vol. 2162, pp. 309–318. Springer, Heidelberg (2001)
4. Bernstein, D.J.: A software implementation of NIST P-224 (2001), http://cr.yp.to/nistp224.html
5. Bernstein, D.J., Lange, T.: Faster addition and doubling on elliptic curves. In: Kurosawa, K. (ed.) ASIACRYPT 2007. LNCS, vol. 4833, pp. 29–50. Springer, Heidelberg (2007)

6. Brier, E., Clavier, C., Olivier, F.: Correlation power analysis with a leakage model. In: Joye, M., Quisquater, J.-J. (eds.) CHES 2004. LNCS, vol. 3156, pp. 16–29. Springer, Heidelberg (2004)
7. Brier, E., Joye, M.: Weierstraß elliptic curve and side-channel attacks. In: Naccache, D., Paillier, P. (eds.) PKC 2002. LNCS, vol. 2274, pp. 335–345. Springer, Heidelberg (2002)
8. Chevallier-Mames, B., Ciet, M., Joye, M.: Low-cost solutions for preventing simple side-channel analysis: Side-channel atomicity. IEEE Transactions on Computers 53(6), 760–768 (2004)
9. Clavier, C., Joye, M.: Universal exponentiation algorithm. In: Koç, Ç.K., Naccache, D., Paar, C. (eds.) CHES 2001. LNCS, vol. 2162, pp. 300–308. Springer, Heidelberg (2001)
10. Cohen, H., Miyaji, A., Ono, T.: Efficient elliptic curve exponentiation using mixed coordinates. In: Ohta, K., Pei, D. (eds.) ASIACRYPT 1998. LNCS, vol. 1514, pp. 51–65. Springer, Heidelberg (1998)
11. Coron, J.-S.: Resistance against differential power analysis for elliptic curve cryptosystems. In: Koç, Ç.K., Paar, C. (eds.) CHES 1999. LNCS, vol. 1717, pp. 292–302. Springer, Heidelberg (1999)
12. Gandolfi, K., Mourtel, C., Olivier, F.: Electromagnetic analysis: Concrete results. In: Koç, Ç.K., Naccache, D., Paar, C. (eds.) CHES 2001. LNCS, vol. 2162, pp. 251–261. Springer, Heidelberg (2001)
13. Goubin, L.: A refined power analysis attack on elliptic curve cryptosystems. In: Desmedt, Y.G. (ed.) PKC 2003. LNCS, vol. 2567, pp. 199–210. Springer, Heidelberg (2002)
14. Hachez, G., Quisquater, J.-J.: Montgomery exponentiation with no final subtractions: Improved results. In: Koç, C.K., Paar, C. (eds.) CHES 2000. LNCS, vol. 1965, pp. 293–301. Springer, Heidelberg (2000)
15. Itoh, K., Izu, T., Takenaka, M.: Efficient countermeasures against power analysis for elliptic curve cryptosystems. In: Quisquater, J.-J., et al. (eds.) Smart Card Research and Advanced Applications VI, pp. 99–113. Kluwer Academic Publishers, Dordrecht (2004)
16. Joye, M., Tunstall, M.: Exponent recoding and regular exponentiation algorithms. In: Preneel, B. (ed.) AFRICACRYPT 2009. LNCS, vol. 5580, pp. 334–349. Springer, Heidelberg (2009)
17. Joye, M., Tymen, C.: Protections against differential analysis for elliptic curve cryptography: An algebraic approach. In: Koç, Ç.K., Naccache, D., Paar, C. (eds.) CHES 2001. LNCS, vol. 2162, pp. 377–390. Springer, Heidelberg (2001)
18. Kocher, P.: Timing attacks on implementations of Diffie-Hellman, RSA, DSS, and other systems. In: Koblitz, N. (ed.) CRYPTO 1996. LNCS, vol. 1109, pp. 104–113. Springer, Heidelberg (1996)
19. Kocher, P., Jaffe, J., Jun, B.: Differential power analysis. In: Wiener, M.J. (ed.) CRYPTO 1999. LNCS, vol. 1666, pp. 388–397. Springer, Heidelberg (1999)
20. Mangard, S., Oswald, E., Popp, T.: Power Analysis Attacks — Revealing the Secrets of Smart Cards. Springer, Heidelberg (2007)
21. Montgomery, P.: Modular multiplication without trial division. Mathematics of Computation 44, 519–521 (1985)
22. National Institute of Standards and Technology (NIST). Recommended elliptic curves for federal government use. In: The appendix of FIPS 186-3 (June 2009), http://csrc.nist.gov/publications/fips/fips186-3/fips_186-3.pdf

23. Quisquater, J.-J., Samyde, D.: Electromagnetic analysis (EMA): Measures and counter-measures for smart cards. In: Attali, I., Jensen, T.P. (eds.) E-smart 2001. LNCS, vol. 2140, pp. 200–210. Springer, Heidelberg (2001)

24. Rivest, R., Shamir, A., Adleman, L.M.: Method for obtaining digital signatures and public-key cryptosystems. Communications of the ACM 21(2), 120–126 (1978)

25. Smart, N., Oswald, E., Page, D.: Randomised representations. IET Proceedings on Information Security 2(2), 19–27 (2008)

26. Smart, N.P.: An analysis of Goubin's refined power analysis attack. In: Walter, C.D., Koç, Ç.K., Paar, C. (eds.) CHES 2003. LNCS, vol. 2779, pp. 281–290. Springer, Heidelberg (2003)

27. Walter, C.D.: Montgomery exponentiation needs no final subtractions. Electronic Letters 35(21), 1831–1832 (1999)

28. Walter, C.D.: Longer keys may facilitate side channel attacks. In: Matsui, M., Zuccherato, R.J. (eds.) SAC 2003. LNCS, vol. 3006, pp. 42–57. Springer, Heidelberg (2004)

29. Walter, C.D.: Simple power analysis of unified code for ECC double and add. In: Joye, M., Quisquater, J.-J. (eds.) CHES 2004. LNCS, vol. 3156, pp. 191–204. Springer, Heidelberg (2004)

30. Walter, C.D., Thompson, S.: Distinguishing exponent digits by observing modular subtractions. In: Naccache, D. (ed.) CT-RSA 2001. LNCS, vol. 2020, pp. 192–207. Springer, Heidelberg (2001)

31. De Win, E., Mister, S., Preneel, B., Wiener, M.: On the performance of signature schemes based on elliptic curves. In: Buhler, J.P. (ed.) ANTS 1998. LNCS, vol. 1423, pp. 252–266. Springer, Heidelberg (1998)

32. Wireless Application Protocol (WAP) Forum. Wireless transport layer security (WTLS) specification, http://www.wapforum.org

33. ANSI X9.62. Public key cryptography for the financial services industry, the elliptic curve digital signature algorithm, ECDSA (1999)

34. Yen, S.-M., Joye, M.: Checking before output not be enough against fault based cryptanalysis. IEEE Transactions on Computers 49(9), 967–970 (2000)

Author Index

Printing: Mercedes-Druck, Berlin
Binding: Stein+Lehmann, Berlin